Second Edition

An Introduction to the
World's Oceans

Alyn C. Duxbury
University of Washington

Alison B. Duxbury
Seattle Community College

Illustrations by,
Tasa Graphic Arts, Inc.

D0068939

ɯɔb
William C. Brown Publishers
Dubuque, Iowa

Book Team

Editor *Jeffrey L. Hahn*
Developmental Editor *Lynne M. Meyers*
Production Editor *Harry Halloran*
Designer *David C. Lansdon*
Photo Research Editor *Carol M. Smith*
Art Editor *Barbara J. Grantham*
Visuals Processor *Vickie Werner*

ωcb group

Chairman of the Board *Wm. C. Brown*
President and Chief Executive Officer *Mark C. Falb*

ωcb

Wm. C. Brown Publishers, College Division

President *G. Franklin Lewis*
Vice President, Editor-in-Chief *George Wm. Bergquist*
Vice President, Director of Production *Beverly Kolz*
Vice President, National Sales Manager *Bob McLaughlin*
Director of Marketing *Thomas E. Doran*
Marketing Communications Manager *Edward Bartell*
Marketing Information Systems Manager *Craig S. Marty*
Marketing Manager *David F. Horwitz*
Executive Editor *Edward G. Jaffee*
Production Editorial Manager *Colleen A. Yonda*
Production Editorial Manager *Julie A. Kennedy*
Publishing Services Manager *Karen J. Slaght*
Manager of Visuals and Design *Faye M. Schilling*

An Introduction to the
World's Oceans

*T*o Andrew S. Duxbury, Alison Jean Duxbury, and Alec R. Duxbury.

Contents

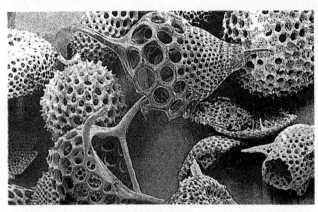

3

The Not-So-Rigid Earth 54

4

The Properties of Water 88

5

The Salt Water   *110*

6

Structure of the Oceans   *130*

7

Winds and Currents 156

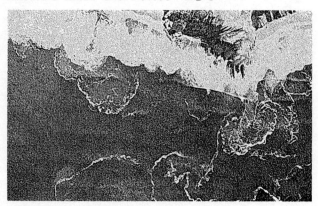

8

The Waves 180

9

The Tides 204

10

Coasts, Shores, and Beaches 232

11

Bays and Estuaries 250

12

Oceans: Environment for Life 270

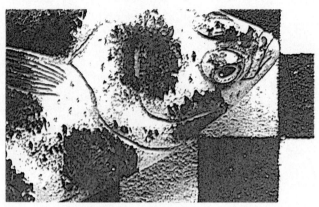

13
Production and Life 284

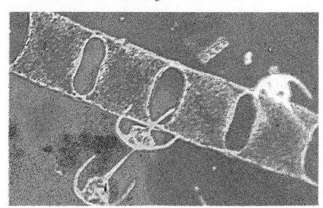

14
The Plankton: Drifters of the Open Ocean 304

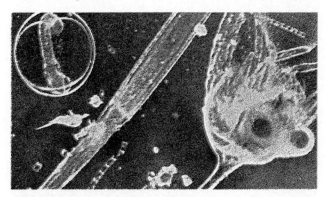

15

The Nekton: Free Swimmers of the Sea   322

16

The Benthos: Dwellers of the Sea Floor   348

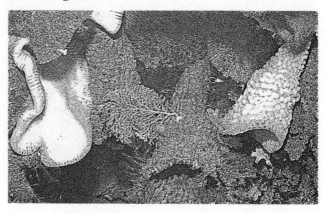

Preface

*T*his book is the result of many experiences. Lectures heard, books read, classes taught, and projects accomplished have all influenced our approach to the subject and our emphasis and choice of topics to be included. Our goals have been to write an introductory text on oceanography for the student without a background in mathematics, chemistry, physics, geology, or biology, and to emphasize the role of basic scientific principles in creating processes that govern the oceans and the earth.

Choices must be made in writing an introductory oceanography text: order of presentation, methods of explanation, types of information to be included. Because the breadth of the material is so great and the interdependency between subject areas is so strong, there is a continuous struggle over the order in which to present the material. We have chosen our own preferred order of presentation, but have tried to make each chapter stand as independently as possible. In this way we hope that individual instructors will be able to tailor their use of this book to their own particular course content and teaching style. Cross references from one chapter to another indicate more detailed discussion on topics elsewhere in the text. Keeping in mind nature's feedback mechanisms and spider-web relationships, we have tried to develop a text that does not lock the user into a fixed order, but allows flexibility and adaptation.

Methods of explanation have been keyed to the students who will read this book. For example, we have chosen to use centrifugal force to explain tidal principles since most students do not have the background to handle vectors. Also, more emphasis has been placed on the physical and geological aspects of the oceans than on the chemical and geochemical aspects that require more specific background knowledge. An ecological approach, as well as necessary descriptive material, is used to integrate the biological chapters with the other subjects. These choices were made to minimize the study of oceanography as a collection of subjects gathered under a marine umbrella and to emphasize the study of the oceans as a cohesive and united whole.

As in all introductory courses, there is an emphasis on the vocabulary of the discipline. Students must learn to speak the language if they are to be able to comprehend the explanation, interpret the result, and apply what is learned. Newspapers, magazines, television programs, and even radio bombard students with information concerning their planet at a time of rapid and increasing change. If the students are to be able to make intelligent decisions, they must be able to determine the worth of this information. To do so they need an understanding of language as well as process and principle. All terms are defined in the text; a student does not need a prior understanding of the terminology to use this text. Terms that are particularly important are printed in boldface at the point where they are first used and defined. A word list of important terms is at the end of each chapter, and a glossary of terms is included at the end of the book.

Summaries at the end of each chapter provide a quick review of key concepts for that chapter. Questions and problems are also included at the end of each chapter. These questions and problems are not intended merely for review, but to challenge the student to think further about the lessons of the chapter. The answers to the questions appear in the instructor's manual.

Information overload is a hazard in all beginning courses of this type. Students need to learn a set of principles that explain why oceanographic processes work the way they do; they also want to learn the oceanography that will help them to understand the things that they read and hear about in newspapers, magazines, and on television. For this reason we have tried to integrate topics of current interest and practical concern with the topic's basic information, instead of grouping all such items in one or two chapters at the end of the book. For example, seabed resources are discussed in the context of the geology of ocean basins and sediments, water quality is a topic in the chapter on estuaries and bays, and problems of overfishing and an overview of sea farming are presented in the appropriate biological chapters for the organisms discussed. Special interest items, such as the genetic manipulation of shellfish, changing ideas on open ocean production, and the effect of sunlight on seawater, are placed in boxes throughout the text.

This book is designed for a one-quarter or one-semester course. Since the experience and knowledge of those using this book will differ, it is expected that each instructor will emphasize and elaborate on some topics at the expense of others. The additional readings listed for each chapter can help extend and enrich subjects. Among the listed readings are papers on topics of current interest and reference works of lasting value, as well as some of oceanography's historical and classical books and papers. We recommend supplementing the text with films and videos from the earth sciences according to the resources available on campus.

As a book is the product of many experiences, it is also the product of people other than the authors. We owe many thanks to our friends and colleagues who have answered questions and provided information, and to Wm. C. Brown Publishers, without whose coordinated efforts this second edition could not have been completed. We sincerely thank you all.

We would also like to thank the following reviewers for their comments and suggestions: Stan Ulanski, James Madison University; John W. Winchester, Florida State University; Russ Flynn, Cypress College; Christopher Schmidt, Western Michigan University; James Reese, Orange Coast College; Ed Stroup, University of Hawaii at Manoa; Christina Emerick, University of Washington; James Ogg, Purdue University; and David Darby, University of Minnesota.

Society obtained the use of the naval corvette H.M.S. *Challenger.* The corvette's guns were removed, and the ship was refitted with laboratories, winches, and equipment. Wyville Thomson (1830–1882), a professor of natural history, was selected as the expedition's leader, and his assistant was a young geologist, John Murray (1841–1914). The *Challenger* sailed from Portsmouth, England, on December 21, 1872, for a voyage that was to last nearly three and a half years. The Challenger Expedition's purpose was scientific research, and during its voyage the vessel logged 113,000 km, took soundings at 361 ocean stations, collected deep-sea water samples, investigated deep-water motion, and made temperature measurements at all depths. Thousands of biological and sea-bottom samples were collected. On May 24, 1876, the *Challenger* returned to England; Queen Victoria conferred a knighthood on Thomson, and the Challenger Expedition was over.

The work of organizing and compiling information, however, was about to begin. It continued for twenty years, until the last of the fifty-volume *Challenger Reports* was issued. John Murray (later Sir John Murray) edited the reports after Thomson's death and wrote many of them himself. He is considered the first geological oceanographer. William Dittmar (1833–1892) prepared the information on seawater chemistry for the *Challenger Reports.* He identified the major elements present in the water and confirmed the findings of earlier chemists that in a seawater sample the proportion of the elements to each other is constant. Oceanography as a modern science is usually dated from the Challenger Expedition. The *Challenger Reports* laid the foundation for the science of oceanography.

The example of the Challenger Expedition stimulated other nations to mount ocean expeditions. Although their avowed purpose was the scientific exploration of the sea, national prestige was at stake in large measure. Norway explored the North Atlantic with the SS *Voringen* in the summers of 1876–1878; Germany studied the Baltic and North seas in the SS *Pomerania* in 1871 and 1872 and in the SS *Crache* in 1881, 1882, and 1884. The French government financed cruises by the *Travailleur* and the *Talisman* in the 1880s. The Austrian ship *Pola* worked in the Mediterranean and Red seas in the 1890s. The United States vessel *Enterprise* circumnavigated the earth between 1883 and 1886, as did Italian and Russian ships between 1886 and 1889.

During the latter years of the nineteenth century and the early years of the twentieth century, intellectual interest in the oceans increased. Theoretical models of ocean circulation and water movement were developed. Oceanography was changing from a descriptive science to a quantitative one. Oceanographic cruises now had the goal of testing hypotheses by gathering data. The Scandinavian oceanographers were particularly active in the study of water movement. One of them, Fridtjof Nansen (1861–1930), tested his ideas about the direction of ice drift in the Arctic by freezing his vessel, *Fram,* in the polar ice pack and drifting with it from 1893 to 1896.

Fluctuations in the abudance of commercial fish in the North Atlantic and adjacent seas, and the effect of these changes on national fishing programs, stimulated oceanographic research and international cooperation. As early as 1870 researchers began to realize their need for knowledge of ocean chemistry and physics in order to understand ocean biology. The study of the ocean and its fisheries required the crossing of national boundaries, and in 1902 Germany, Russia, Great Britain, Holland, and the Scandinavian countries formed the International Council for the Exploration of the Sea (ICES) to coordinate and sponsor research in the ocean and in fisheries.

Advances in theoretical oceanography sometimes could not be verified with practical knowledge until new instruments and equipment were developed. Lord Kelvin (1824–1907) invented a tide-predicting machine in 1872 that made it possible to combine tidal theory with astronomic predictions to produce predicted tide tables. Deep-sea circulation could not be systematically explored until approximately 1910, when Nansen's water sampling bottles were combined with thermometers designed for deep-sea temperature measurements, and when an accurate method for determining ocean water salinity was devised by the chemist Martin Knudsen (1871–1949). The reliable and accurate measurement of ocean depths had to wait until the development of the echo sounder, which was given its first scientific use on the 1925–1927 German cruise of the *Meteor* in the South Atlantic.

In the United States, government agencies related to the oceans had proliferated during the nineteenth century. These agencies were concerned with gathering information to further commerce, fisheries, and the Navy. After the Civil War, the replacement of sail by steam lessened government interest in studying winds and currents and in surveying the ocean floor. Private institutions and wealthy individuals took over the support of oceanography in the United States. Alexander Agassiz (1835–1910) financed a series of expeditions, and in the first twenty years of this century the Carnegie Institution funded a series of exploratory crusies. In 1927 the National Academy of Sciences established its first committee on oceanography, and in 1930 the Rockefeller Foundation allocated funds to stimulate programs and construct laboratories for marine research. Oceanography in the United States began to move onto university campuses. In order to teach oceanography, the subject material had to be consolidated, and in 1942 *The Oceans* by Harald U. Sverdrup, Martin W. Johnson, and

Richard H. Fleming was published. It captured between its covers nearly all the world's knowledge of oceanographic processes and was used to train a generation of ocean scientists.

Oceanography mushroomed during World War II, when practical problems of military significance had to be solved quickly. The United States and its allies needed to move men and materials by sea to remote locations, to predict ocean and shore conditions for amphibious landings, to know how explosives behaved in water, to chart beaches and harbors from aerial reconnaissance, and to find and destroy submarines. Academic studies ceased as oceanographers pooled their knowledge in the national effort.

After the war oceanographers returned to their classrooms and laboratories with an array of new, sophisticated instruments, including radar, sonar, automated wave detectors, and temperature depth recorders. They also returned with large-scale government funding for research and education. The earth sciences in general and oceanography in particular blossomed during the 1950s. The numbers of scientists, students, educational programs, research institutes, and professional journals all increased. The Office of Naval Research (ONR) funded applied research programs and research vessels; the National Science Foundation (NSF) underwrote basic research; the Atomic Energy Commission (AEC) financed ocean work at the South Pacific atoll sites of atomic tests. International cooperation brought about the 1957–1958 International Geophysical Year (IGY) program, in which sixty-seven nations cooperated to explore the seafloor and made discoveries that completely revolutionized geology and geophysics.

The decade of the 1960s brought giant strides in programs and equipment. As a direct result of the IGY exploration program, special research vessels and submersibles were built, and the Deep Sea Drilling Program, a cooperative venture between research institutions and universities, began to sample the earth's crust beneath the sea. Electronics developed by the space program were applied to ocean research. Computers went on board research vessels, and for the first time data could be sorted, analyzed, and interpreted at sea; experiments could be adjusted while in progress. Government funding allowed large-scale ocean experiments; ocean chemistry, water motion, and air-sea interaction were all studied by fleets of oceanographic vessels representing many institutions and nations. In these projects marine scientists shared their knowledge in the effort to understand the basic principles that control the oceans. The ability to predict ocean conditions depends on such knowledge.

During these years of expanding programs earth scientists began to recognize the need for policy and management decisions to keep the planet from further degradation. Students were attracted to the marine sciences in record numbers, and marine policy programs and ocean management courses were added to curricula. As more and more nations were turning to the sea for food and as technology was increasing our ability to harvest the sea, problems of dwindling fish stocks and the need for fishery management had to be faced as well.

Oceanography in the 1970s was faced with sharp inflation, which translated into a reduction in funding for ships and basic research; nonetheless, the discovery of deep-sea hot-water vents and their associated animal life and mineral deposits renewed the excitement over deep-sea biology, chemistry, geology, and ocean exploration in general. Instrumentation continued to become more sophisticated and expensive as deep-sea moorings, deep-diving submersibles, and the remote sensing of the ocean by satellite became possible, while cooperation among institutions for ship time was closely monitored by cost-conscious scientists and funding agencies. This cooperation led to the integration of research at sea between the disciplines of oceanography and has given rise to large-scale multifaceted research programs.

The targets for investigation in the 1980s and 1990s include the ocean's climate and circulation, the management of living and nonliving resources, the continental margins and the ocean seabed, the energy sources of the sea, methods of decreasing the cost of ocean transport, and food availability. At the same time, the use of refined sensors and techniques will allow scientists to continue to seek answers to the basic questions of how and why ocean processes occur and the relationship of these processes to sea resources and to ourselves.

The enormous increase in oceanographic information during the last thirty-five to forty years presents problems for student, instructor, and textbook author alike, especially at the introductory level. There is too much information over too broad a subject area in too great detail for any single textbook or course to cover, and therefore, some things must be emphasized at the expense of others. In particular, the plant and animal life of the oceans is so rich, so varied, and so profuse that extreme selectivity is required. Entire books are written concerning waves, beach processes, currents, seawater chemistry, and the many other aspects of ocean science. You are encouraged to explore the suggested readings given at the end of this book, or any other source that is available to you, so as to follow your interests in greater depth.

In this textbook, you will find that the emphasis has been placed on basic scientific principles and processes, illustrated when possible by the findings of recent research. Our intent is to help you build an understanding of ocean processes in time and space, integrating physical, chemical, geological, and biological principles. We hope that at the end of your introductory study you will have gained a perspective of the oceans as a dynamic system and an appreciation of the oceans' complexity and their relationship to your land environment.

An Introduction to the
World's Oceans

The Water Planet

1

*O*ne evening he asked the miller where the river went.

"It goes down the valley," answered he, "and turns a power of mills—six score mills, they say, from here to Unterdeck—and it none the wearier, after all. And then it goes out into the lowlands, and waters the great corn country, and runs through a sight of the fine cities (so they say) where kings live all alone in great palaces, with a sentry walking up and down before the door. And it goes under bridges with stone men upon them, looking down and smiling so curious at the water, and living folks leaning their elbows on the wall and looking over too. And then it goes on and on, and down through marshes and sands, until at last it falls into the sea, where the ships are that bring parrots and tobacco from the Indies. Ay, it has a long trot before it as it goes over our weir, bless its heart!"

"And what is the sea?" asked Will.

"The sea!" cried the miller. "Lord help us all, it is the greatest thing God made! That is where all the water in the world runs down into a great salt lake. There it lies, as flat as my hand and as innocent-like as a child, but they do say when the wind blows it gets up into water-mountains bigger than any of ours, and swallows down great ships bigger than our mill, and makes such a roaring that you can hear it miles away upon the land. There are great fish in it five times bigger than a bull, and one old serpent as long as our river and as old as all the world, with whiskers like a man, and a crown of silver on her head."

Robert Louis Stevenson, from The Merry Men

*B*illions of shimmering masses of stars, known as galaxies, move through the space we call the universe. About one-third of the way toward the center of one whirling mass of some 100 billion stars called the Milky Way galaxy there is a fairly ordinary star, the sun. Around the sun move nine specks, or planets, following predictable and nearly constant paths, or orbits. The sun and its nine planets are called the solar system, and the third planet from the sun is called earth (fig. 1.1). Although we call our planet earth, when we consider the qualities of this planet that set it apart from the other planets in the solar system we must consider it as the water planet. Water did not exist on the earth in the beginning, but its formation on a planet that was not too far from the sun nor too close, not too hot nor too cold, changed the earth and allowed the development of living organisms. Let us begin at the solar system's beginning to investigate the earth, the water planet, on which the largest bodies of water are known as the oceans.

1.1 The Beginnings

Origin of the Solar System

Modern theories attribute the beginning of our solar system to the collapse of a rotating interstellar cloud of gas and dust about 4.6 billion years ago. As the cloud collapsed its speed of rotation increased and, heated by its own gravitational energy, its temperature rose. The gas and dust, spinning faster and faster, contracted parallel to the axis of spin, forming a disk. At the center of the disk a star, our sun, was formed. Self-sustaining nuclear reactions kept the sun hot, but the outer regions began to cool, and in this cooler outer portion of the rotating disk the gases began to collide and chemically interact. These collisions and interactions formed grains or particles of matter and these particles, in turn, grew from other collisions and became large enough to have sufficient gravity to attract still other particles. The planets had begun to form. After a few million years the sun was orbited by nine planets (in order from the sun): Mercury, Venus, Earth, Mars, Jupiter, Saturn, Uranus, Neptune, and Pluto.

If Mercury, Venus, Earth, and Mars are compared with Jupiter, Saturn, Uranus, and Neptune, the four planets closer to the sun are seen to be much smaller in diameter and mass (see table 1.1). These four inner planets are rich in metals and rocky materials. The four outer planets are cold giants, dominated by ices of water, ammonia, and methane. Their atmospheres are made

Figure 1.1
The water planet earth as seen from space.

up of helium and hydrogen; the planets located nearer the sun lost these lighter gases because the higher temperature and intensity of solar radiation tends to push these gases out and away from the center of the solar system. If the mass of each planet in table 1.1 is divided by its volume, the results will show that the outer planets are composed of lighter or less dense materials than the inner planets.

Pluto, a little-known small planet at least five hundred times less massive than the earth, has an elliptical orbit that takes it inside the orbit of Neptune. It has a reflective surface that may be primarily composed of frozen methane. Because it is so different from the other four outer planets it has been suggested that Pluto might at one time have been a satellite of Neptune.

The Early Planet Earth

Before the beginning of the geologic record, during the first billion years of the earth's existence, the earth is thought to have been a conglomeration of silicon compounds, iron and magnesium oxides, and small amounts of other naturally occurring elements. According to this model, the earth formed originally from cold matter but events occurred that raised the earth's temperature and initiated processes that obliterated its earlier history and resulted in its present form. The early earth was bombarded by particles, each arriving with its own energy of motion, a portion of which was converted into heat on impact. Each new layer of material accumulated in

TABLE 1.1

Features of the Planets in the Solar System

Planet	Mean distance from sun (10^6 km)	Diameter (km)	Mass relative to earth mass	Rotation period[1] (hours, days)	Orbit period (years)	Mean temperature of surface (°C)	Principal atmospheric gases[2]
Mercury	57.9	4,878	0.055	58.6 d	0.24	−170 night 430 day	Na
Venus	108.2	12,102	0.815	−243 d	0.62	− 23 clouds 480 surface	CO_2, N
Earth	149.6	12,756	1.000	23.94 h	1.00	16	N, O
Mars	227.9	6,787	0.110	24.62 h	1.88	− 50(av)	Co_2, N, Ar
Jupiter	778.3	142,800	317.0	9.84 h	11.86	−150	H, He, CH_4, NH_3
Saturn	1427.0	120,660	95.2	10.32 h	29.46	−180	H, He, CH_4, NH_3
Uranus	2871	52,400	14.6	− 17.28 h	84.07	−210	H, He
Neptune	4497	50,000	17.2	16 h	164.82	−220	H, He
Pluto	5913	3,000	0.002	6.4 d	248.60	−230	?

[1]Negative rotation period indicates rotation opposite to orbit direction about the sun.
[2]H = hydrogen; He = Helium; CO_2 = carbon dioxide; N = nitrogen; O = oxygen; Ar = argon; Na = sodium, CH_4 = methane; NH_3 = ammonia.

this way buried the material below it, and the buried heat raised the temperature of the earth's interior. At the same time, the growing weight of the accumulating layers at the surface compressed the interior, and the energy of compression was converted to heat, raising the earth's internal temperature to approximately 1000° C. Atoms of radioactive elements, such as uranium and thorium, disintegrated by emitting atomic particles that were absorbed by the surrounding matter, and their energy of motion was converted into heat.

Some time during the first few hundred million years after the earth formed, it has been calculated that its interior could have reached the melting point of iron and nickel. When the iron and nickel in the planet melted, they migrated toward the center, generating heat and displacing lighter substances. In this way the temperature of the earth was raised to an average of 2000° C. Material from the partially molten interior moved upward and spread over the surface, cooling and solidifying. The melting and solidifying probably happened repeatedly, separating the lighter, less dense compounds from the heavier, more dense substances in the interior of the planet. The earth emerged from this period completely reorganized and differentiated into a layered system that will be explored in greater detail in chapter 3.

The earth's oceans and atmosphere are probably both by-products of this heating and differentiation. Originally any water would have been locked in the minerals as hydrogen and oxygen, but as the earth warmed and partially melted, the water was released and carried to the surface as water vapor mixed with other gases. As the earth's surface cooled, the water condensed to form the oceans.

An alternate theory, that the oceans are the product of the earth's collision with large showers of small, icy comets, has recently been proposed by University of Iowa physicist Louis Frank. Frank suggests that ten million small comets hit the earth's atmosphere every year. If these comets have bombarded the earth since its formation, they could have supplied enough water to fill the earth's oceans. Scientists are currently searching for evidence to prove or disprove this theory.

At first the earth must have been too small and had too little gravity to have accumulated an atmosphere. It is generally believed that during the process of differentiation, gases released from the earth's hot, chemically active interior formed the first atmosphere which was primarily made up of water vapor, hydrogen gas, hydrogen chloride, carbon monoxide, carbon dioxide, and nitrogen. Any free oxygen present would have combined with the metals of the crust to form compounds such as iron oxide. Oxygen gas could not accumulate in the atmosphere until its production exceeded its loss by chemical reaction. This did not occur until life evolved to a level of complexity in which green plants could convert carbon dioxide and water with the energy of sunlight into organic matter and free oxygen. This process and its significance to life is discussed in chapters 5 and 13.

1.2 Age and Time

The Age of the Earth

Over the centuries people have asked the question, "How old is the earth?" In the seventeenth century Archbishop Ussher of Ireland attempted to answer the question by counting the generations listed in the Bible; he arrived at 9:00 A.M., October 26, 4004 B.C., for the beginning of the earth. In the late 1800s the English physicist Lord Kelvin calculated the time necessary for molten rock to cool to present temperatures, and dated the earth as twenty million to forty million years old. In 1899 another physicist, John Joly, used the amount of salt in the oceans to determine the earth's age and calculated the time for the rivers to wash the salt from the land to be 100 million years.

It was not until scientists understood radioactive decay, and a method was developed to apply it to the dating of rock samples, that reliable age data became available. In this method, known as **radiometric dating,** use is made of radioactive **isotopes** of certain elements.

An atom of a radioactive isotope has an unstable nucleus. This unstable nucleus changes or decays and emits one or more particles plus energy. For example, the isotope carbon-14 decays or changes to nitrogen-14; uranium-235 decays to lead-207; and potassium-40 decays to argon-40. The time at which any single nucleus will decay is unpredictable, but if large numbers of atoms of the same isotope are present it is possible to predict that a certain fraction of the isotope will decay over a certain period of time. The time over which one-half of the atoms of an isotope will decay or change from one element to another element is known as its **half-life.** The half-life of each isotope is characteristic and constant. For example, the half-life of carbon-14 is 5730 years, uranium-235 is 704 million years, and potassium-40 is 1.3 billion years. Therefore, if a substance is found that was originally made up only of atoms of uranium-235, in 704 million years the substance is one-half uranium-235 and one-half lead-207. In another 704 million years three-quarters of the substance will be lead-207 and only one-quarter uranium-235. Because each isotopic system behaves differently in nature, data must be carefully tested, compared, and evaluated. The best data are those in which different isotopic systems give the same date.

The earth is an active planet and its original surface rocks no longer exist. The oldest rocks on the surface of the earth have been dated at 3.8 billion years old. Moon samples returned by the Apollo mission are dated at 4.2 billion years old. Meteorites that have survived the journey from space through the earth's atmosphere have been dated between 4.5 billion and 4.6 billion years old. The substances found in the meteorites all represent materials that condensed out of hot gases thought to be present at the beginning of the solar system. These ages agree with theoretical calculations made for the age of the sun. This information sets the accepted age of the earth at between 4.5 billion and 4.6 billion years.

Geologic Time

To refer to events in the history and formation of the earth, scientists use geologic time (see fig. 1.2 and table 1.2). The principal divisions are the eras: the Paleozoic era or era of ancient life, the Mesozoic era or era of intermediate life, and the Cenozoic era or era of recent life. Eras are subdivided into periods and epochs. Initially the appearance or disappearance of fossil types was used to set the boundaries of the time units; these have been refined by radiometric dating.

Very long periods of time are incomprehensible to most of us. We often have difficulty coping with time spans of more than ten years—what were you doing exactly ten years ago today? We have nothing with which to compare the 4.5 billion–4.6 billion-year age of the earth or the 500 million years since the first **vertebrates** appeared on this planet. In order to place geologic time in a framework which we can understand, let us divide the earth's age by one hundred million. If we do so, then we can consider that Mother Earth is forty-six years old, a middle-aged lady in whose past we have a strong and intimate interest. What has happened to her over that forty-six years?

The first seven years of her life leave no record; as she entered her eighth year a few fragments of her history can be read in some rocks of Africa and Greenland. By the time she was 12 years old the first living cells of bacteria-type organisms had appeared. Approximately ten or eleven years later, between ages 22 and 23, oxygen production by living cells began, but it took eight or more years, until about the earth's thirty-first year, to change the atmosphere sufficiently to support the first complex oxygen-requiring cells. The first fossils were laid down only six years ago, in her fortieth year, and by the time she was 41 the first vertebrates had developed. Eight and a half months later the first land plants appeared, quickly followed by an age in which the fish were the dominant life form. By age 43 the first reptiles, including the dinosaurs, had arrived, become abundant at the beginning of her forty-fourth year, and disap-

Figure 1.2
The history of the earth showing evolution of life forms.
After U.S. Geological Survey publication, "Geologic Time."

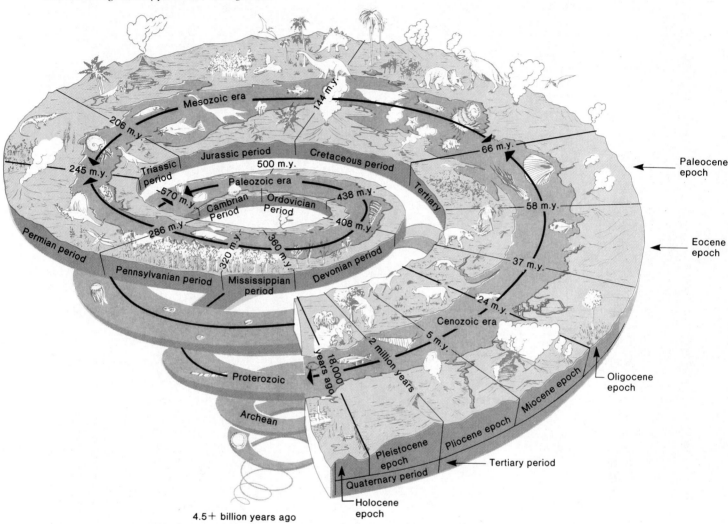

peared before her forty-fifth birthday. A little over a year ago plants with flowers began to develop; four months ago the mammals, birds, and insects became the dominant animal life forms. Our first human ancestors appeared twenty-five days ago, followed two weeks later by the first identifiable *Homo.* About half an hour ago modern humans began the long process we know as recorded civilization, and only one minute ago the Industrial Revolution began changing the earth and our relationship to her for all time.

Natural Time Periods

People first defined time by the natural motions of the earth, sun, and moon. Later, people grouped natural time periods for their convenience, and still later, artificial time periods were created for people's special uses.

Time is used to determine the starting point of an event, the event's duration, and the rate at which the event proceeds. An accurate measurement of time is required to determine location or position. This is discussed in the Location Systems section of this chapter.

The year is the time required by the earth to complete one orbit about the sun. The time required for this orbit is 365¼ days, adapted for convenience to 365 days with an extra day added every four years, except years ending in hundreds and not divisible by four. As the earth follows its orbit around the sun, those who live in temperate zones and polar zones are very conscious of the seasons and of the differences in the lengths of the periods of daylight and darkness. The reason for these

TABLE 1.2
The Geologic Time Scale

Era	Period	Epoch	Began millions of years ago	Life forms/Events
Cenozoic	Quaternary	Recent	0.01	Modern humans
		Pleistocene	2.5	Stone-age humans First humans
	Tertiary	Pliocene	7	
		Miocene	26	
		Oligocene	38	Flowering plants
		Eocene	54	Mammals, birds, and insects dominant
		Paleocene	65	
Mesozoic	Cretaceous		136	Last of dinosaurs; flowering plants begin
	Jurassic		190	Dinosaurs abundant; first birds
	Triassic		225	First mammals First dinosaurs
Paleozoic	Permian		280	Age of reptiles
	Carboniferous		345	Age of amphibians; first reptiles
	Devonian		395	First seed plants Age of fishes
	Silurian		430	First land plants
	Ordovician		500	Marine algae; vertebrate fish
	Cambrian		570	Primitive marine algae and invertebrates
Precambrian			3200–3500	Earliest bacteria and algae
			3800	Oldest surface rocks
			4500+	Oldest meteorites Formation of the earth (assumed)

seasonal changes is seen in figure 1.3. The earth moves along its orbit with its axis tilted 23½° from the vertical. Therefore, during the year the earth's North Pole is sometimes tilting toward the sun and sometimes tilting away from it. The earth's rotation on its axis as it moves along its orbit of the sun causes the northern hemi-sphere to receive the maximum hours of sunlight when the North Pole is tilted toward the sun; this is the northern hemisphere's summer. During the same period, the South Pole is tilted away from the sun, so that the southern hemisphere receives the least sun-light; this period is winter in the southern hemisphere.

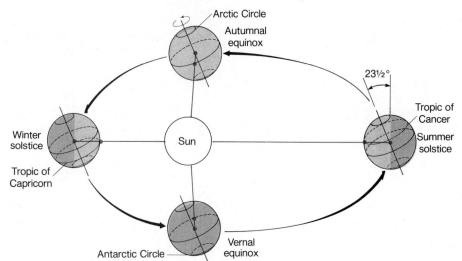

Figure 1.3
The earth's seasons. The fixed orientation of the earth's axis during the earth's orbit of the sun causes different portions of the earth to remain in shadow at different seasons. One year is the time increment between successive vernal equinoxes.

At the other side of the earth's orbit, the North Pole is tilted away from the sun, creating the northern hemisphere's winter, and the South Pole is inclined toward the sun, creating the southern hemisphere's summer. Note also that during summer in the northern hemisphere it is light around the North Pole and dark around the South Pole; the opposite is true during the northern hemisphere's winter.

As we are carried along by the spinning earth orbiting the sun, we are not conscious of any movement. What we sense is that the sun rises in the east and sets in the west daily and slowly moves up in the sky from south to north and back during one year. The periods of daylight in the northern hemisphere increase as the sun moves north in relation to the earth, to stand 23½° above the equator over the **Tropic of Cancer.** This position occurs at the **summer solstice** on or about June 22, the day with the longest period of daylight and the beginning of summer in the northern hemisphere; it is also the day on which the sun does not sink below the horizon above 66½° north of the equator, the **Arctic Circle,** nor does it rise above 66½° south of the equator, the **Antarctic Circle.** Then the sun appears to move southward until on about September 23, the **autumnal equinox,** it stands directly above the equator. On this day the periods of daylight and darkness are the same all over the world. The sun continues its southward movement until about December 21, when it stands 23½° below the equator over the **Tropic of Capricorn;** this position marks the **winter solstice** and the beginning of winter in the northern hemisphere. On this day the daylight period is the shortest in the northern hemisphere, and above the Arctic Circle the sun does not rise. In the southern hemisphere the reverse is true, and above the Antarctic Circle the sun does not set. The sun then begins to move northward, and about March 21, the **vernal equinox,** it stands again above the

equator; spring begins in the northern hemisphere, and the periods of daylight and darkness are once more equal around the world. Follow figure 1.3 around again, checking the position of the South Pole, and note how the seasons of the southern hemisphere are reversed from those of the northern hemisphere. Note also that the maximum effect of the annual variation in solar illumination that produces the four seasons occurs at the temperate zones. In the polar regions the seasons are dominated by the long periods of light and dark, but the annual change in solar heating of the oceans is small, as the sun is always low on the horizon. Between the Tropics of Cancer and Capricorn there is little seasonal change, as the sun never moves beyond these boundaries.

Weeks and months, as they presently exist, modify natural time periods. It requires 29½ days for the moon to orbit the earth. This period of 29½ days defines the **lunar month.** In the lunar month the moon passes through four phases: new moon, first quarter, full moon, and last quarter. The four phases match approximately the four weeks of the month. Days are grouped into twelve months of unequal length in order to form one calendar year. The present arrangement is known as the Gregorian calendar after Pope Gregory XIII, who in the sixteenth century made the changes necessary to correct the old Julian calendar, named after Julius Caesar, which had been adopted in 46 B.C. The Gregorian calendar was adopted in the United States in 1752, by which time the Julian calendar was eleven days in error. In that year, in both Great Britian and the United States, September 2 was followed by September 14 by parliamentary decree, accompanied by riots in which people demanded their eleven days back. Also in 1752, the beginning of the calendar year was changed from the original date of the vernal equinox, March 25, to January 1; 1751 had no months of January and February.

The day is derived from the earth's rotation on its axis as it orbits the sun. The average time for the earth to make one rotation relative to the sun is twenty-four hours; this is the mean **solar day,** our clock day. Another measure of a day is the time required for the earth to make a complete rotation with respect to a far-distant point in space. This is known as the **sidereal day** and is about four minutes shorter than the mean solar day; it gives the true rotational period of the earth. The sidereal day is useful in astronomy and navigation.

Living organisms respond to these natural cycles. In temperate zones flowers bloom and die back; forest trees lose their leaves, enter a period of dormancy, and then produce new leaves and buds as the length of the periods of daylight and darkness change and the temperatures increase or decrease; but tropic forests remain lush year-round. Some animals migrate and alternate periods of activity and hibernation with the seasons. Other animals set their internal clocks to the day-night pattern, hunting in the dark and sleeping in the light; still others do the reverse. Plants and animals of the sea also react to these rhythms, as do the physical processes that move the atmosphere and circulate the water in the oceans. As we investigate the oceans we shall return again and again to these natural cycles to see how they affect sea life and the processes of the sea.

1.3 The Shape of the Earth

As the earth cooled and turned in space, gravity and the forces of rotation produced its spherical shape. The earth sphere has a mean, or average, radius of 6371 km (3956 mi). Measurements show that the earth has a shorter polar radius (6356.9 km; 3947 mi) and a larger equatorial radius (6378.4 km; 3961 mi). This difference of 21.5 km, or about 15 mi, occurs because the earth is not a rigid sphere. Instead the earth is quite plastic, or deformable, beneath its crust. As it spins on its north-south axis it tends to bulge at the equator, much as a round ball of potter's clay bulges when spun on a stick (see fig. 1.4). Because the landmasses are presently concentrated in the middle region of the northern hemisphere and centered on the South Pole in the southern hemisphere, the earth's surface is depressed slightly at the South Pole and at the middle region of the northern hemisphere, while it is elevated at the North Pole and in the middle region of the southern hemisphere. This distribution causes the earth to have a very slight pear shape, but only very slightly so because these depressions and elevations average only about 15 meters (49.2 feet). The earth is in fact a nearly perfect sphere.

The earth is also quite smooth. The top of the earth's highest mountain, Mount Everest in the Himalayas, is about 8840 m (29,000 ft) above sea level, while

Figure 1.4

The rotation of the earth on its axis stretches the equator. Notice that the equatorial radius is larger than the polar radius.

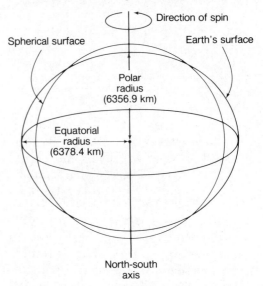

the deepest depth in the oceans, the bottom of the Challenger Deep in the Mariana Trench of the Pacific Ocean, is about 11,000 m (36,000 ft) deep. If these measurements are divided by the mean radius of the earth, which is 6,371,000 m (20,896,000 ft), the resulting ratios are 0.00139 for the mountain and 0.00173 for the trench. On a scale model of the earth with a radius of 50 centimeters (19.7 inches), the highest mountain would be about 0.07 cm (.027 in) high and the deepest hole would be about 0.086 cm (.034 in) in depth. The earth model's surface would feel rather like the skin of a grapefruit or the surface of a basketball.

1.4 Location Systems

Latitude and Longitude

In order to find our way around on the curved surface of our planet we need a reference (or location) system. Most of us use such a system daily consisting of the city name, street name or number, and building number. Armed with a city map, we confidently navigate to areas never visited before. Most of the earth's surface, however, is not provided with streets and building numbers, and we must use another system. To determine the location of a position on the earth we use a grid of reference lines that are superimposed on the earth's surface so that they cross each other at right angles. These lines are **latitude** and **longitude.** Lines of latitude begin at the **equator.** The equator is created by passing a plane through the earth halfway between the poles and at right

(a)

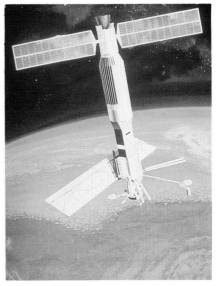

(b)

(c)

(d)

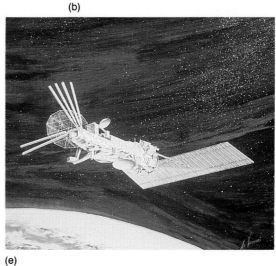

(e)

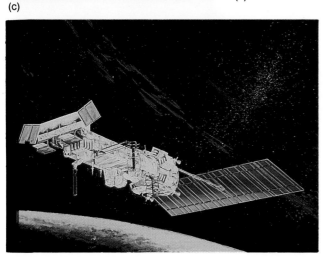

(f)

Colorplate 1
Past, present, and proposed U.S. satellites with ocean remote sensing systems. (a) NASA's TOPEX, to be launched in 1989 or 1990, will increase the resolution of radar altimetry, which will help determine ocean circulation patterns (see World Ocean Circulation Experiment box in chapter 9). (b) SEASAT, an ocean-observing satellite launched in 1978 by NASA, failed after 104 days of operation. During this period, it monitored sea surface topography, wave heights, and sea surface wind velocities and measured sea surface temperatures, atmospheric water vapor, and sea ice distribution. (c) NIMBUS-7, also launched in 1978, monitored sea surface temperatures, wind speed, and sea ice concentrations until 1987. Onboard devices also measured ocean color, water clarity, and concentration of small surface organisms (phytoplankton). (d) GEOSAT, a U.S. Navy satellite, continues to use radar altimetry to determine the topography of the ocean. (e) NROSS (Navy Remote Ocean Sensing System) is planned to observe sea surface winds, wave height and patterns, sea surface temperatures, sea ice distributions, and boundaries between oceanic water types for naval operations. (f) TIROS is one of a series of polar-orbiting meteorological satellites able to measure sea surface temperatures to 1° C and 1 km resolution; it carries the international Search and Rescue (SAR) emergency beacon system.

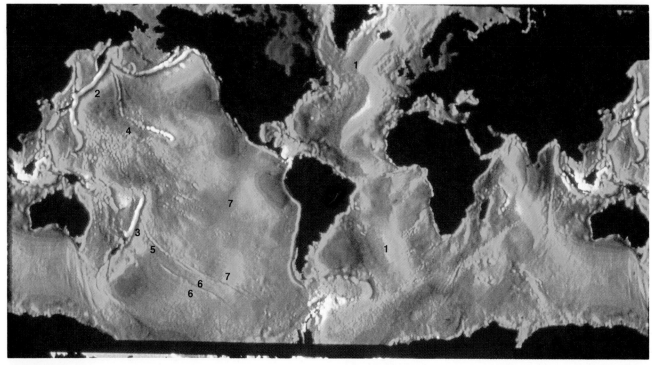

(a)

(b)

Colorplate 2
Sea surface topography. Seventy days' (a) worth of data collected by the short-lived SEASAT satellite were used to produce this computer-generated chart of mean sea surface topography. The satellite measured the elevation of the sea surface, which relates to changes in gravity caused by the topography of the sea floor. The sea surface may be depressed as much as 60 m (192 ft) over ocean trenches and may bulge as much as 5 m (16 ft) over seamounts. Marked are (1) the Mid-Atlantic Ridge, (2 and 3) trenches along the west and northwest margins of the Pacific,

(4) the Hawaiian Island-Emperor seamount system, (5) the Louisville Ridge of the South Pacific, (6) fracture zones produced by stresses acting on the earth's crustal plates, and (7) low-amplitude bumps which may mark upwellings of molten rock circulating in the earth's interior.

(b) Tropical skies. Only limited areas of the sea can be observed from a ship during any research cruise. Each sample and each piece of data is obtained with considerable time and effort.

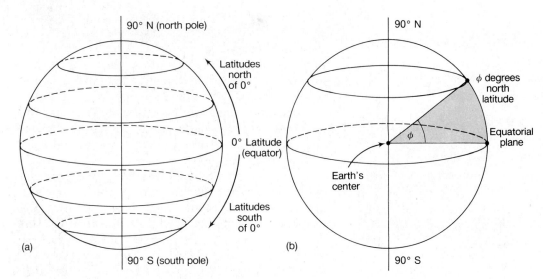

Figure 1.5
(a) Latitude lines are drawn parallel to the equatorial plane. (b) The value of a latitude line is expressed in angular degrees determined by the angle formed between the equatorial plane and the latitude line to the earth's center. This is the angle φ (phi). The degree value of φ must be noted as north or south of the equator.

angles to the earth's axis. This process is much like cutting an orange in two pieces halfway between the depressions marking the stem and the navel. The equator is marked at 0° latitude, and other latitude lines are drawn around the earth parallel to the equator, northward to 90° N, or the North Pole, and southward to 90° S, or the South Pole (see fig. 1.5a). Notice that the lines of latitude describe increasingly smaller circles as the poles are approached. Notice also that lines of latitude must be designated as either north or south of the equator. The latitude value is determined by the internal angle (φ or phi) between the latitude line, the earth's center, and the equatorial plane (see fig. 1.5b). Lines of latitude are also termed **parallels,** as they are parallel to the equator and to each other. The previously mentioned Tropics of Cancer and Capricorn correspond to latitudes 23½° N and S, respectively. Latitudes 66½° N and S respectively correspond to the Arctic and Antarctic Circles.

Lines of longitude, or **meridians,** are formed at right angles to the latitude grid (see fig. 1.6a). Longitude begins at an arbitrarily chosen point: 0° longitude is a line on the earth's surface extending from the North Pole to the South Pole and passing directly through the Royal Naval Observatory in Greenwich, England, just outside London. The 0° longitude line is shown outside the Greenwich Observatory in figure 1.7. On the other side of the earth, 180° longitude is directly opposite 0°. The 0° longitude line is known as the **prime meridian.** The 180° longitude line approximates the **international date line.** Longitude lines are identified by their angular displacement (θ or theta) to the east and west of 0° longitude, as shown in figure 1.6b. Note on figure 1.6a that all circles of longitude are the same size, much like the lines marking the segments of an orange. These circles are formed at the intersection of

Figure 1.6

(a) Longitude lines are drawn with reference to the prime meridian. (b) The value of a longitude line is expressed in angular degrees determined by the angle formed between the prime meridian and the longitude line to the earth's center. This is the angle θ (theta). The value of θ is given in degrees as east or west of the prime meridian.

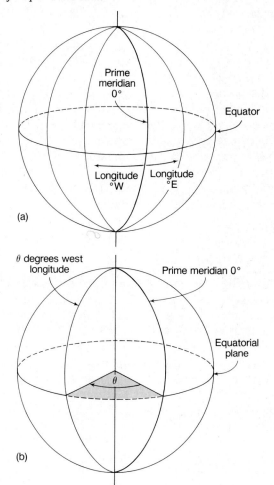

Figure 1.7

The Royal Naval Observatory at Greenwich, England. The brass strip set into the courtyard marks the prime meridian, the division between east and west longitudes.

a plane passing through the earth's center and the earth's surface. Such circles are called **great circles.** All longitude lines form great circles; in the case of latitude only the equator is a great circle.

To identify a particular spot on the earth's surface, we use the crossing of the latitude and longitude lines; for example, 158° W, 21° N is the location of the Hawaiian Islands, and 20° E, 33° S identifies the Cape of Good Hope at the southern tip of Africa. Since the distance expressed in whole degrees is large (1 degree of latitude equals 60 nautical miles), each degree of arc is divided into 60 minutes, each minute into 60 seconds, and each second into tenths of a second. One **nautical mile** is equal to 1.15 land miles (see Appendix). Using this system, positions on the earth can be specified with great accuracy.

Chart Projections

Charts and maps are used to show the earth's three-dimensional surface on a flat, or two-dimensional, surface. Maps usually portray the earth's land features or land and sea relationships worldwide, while similar displays of the sea and the sky are called charts. Any chart or map produces a distorted image of the curved surface of the earth. The task of the mapmaker, or cartographer, is to produce the most accurate, most convenient, and least distorted picture for the task for which the map or chart is to be used.

Maps are made by projecting the features of the earth, along with the latitude and longitude system, onto a surface; the resulting picture is called a map or chart **projection.** Consider a transparent globe with the continents and latitude and longitude lines painted on its surface. Place a light in the center of the globe and let the light rays shine out through the globe. The light will project the shadows of the continents and the latitude and longitude lines onto a piece of paper held up to the outside of the globe. Different projections are obtained by varying the position of the light and the type of surface on which the projection is made. Many chart and map projections have been constructed, but most of them are modifications of three basic types: cylindrical, conic, and tangent plane, all shown in figure 1.8. In these projections, the surface that is to be the map is rolled around the globe as a cylinder (fig. 1.8a), made into a cone (fig. 1.8b), or laid flat (tangent) against the sphere (fig. 1.8c). Although the cylindrical and conic surfaces may be placed around, over, or tangent to the earth at any location, they are usually placed so that the cylinder touches the earth at its equator and the cone is centered on the polar axis.

A comparison of the three parts of figure 1.8 shows that distortion increases on the resulting map or chart as the distance from the place of contact to the sphere increases. Consider Greenland. In the tangent plane projection (fig. 1.8c) its size and shape are very close to its true form on the earth's sphere. In the conic projection (fig. 1.8b) the island has grown larger, and in the cylindrical projection (fig. 1.8a) both its size and its shape are greatly distorted. The traditional and familiar world map used in many books and school classrooms is the **Mercator projection,** an adjusted form of the cylindrical type shown in figure 1.8a. Although distortion is great at high latitudes and the poles cannot be shown, the Mercator projection, unlike other projections, has the advantage that a straight line as drawn on the map is a line of true direction or constant compass heading, and therefore the Mercator projection is useful in navigation. Each type of chart or map has its own characteristics. It is up to the user to select the type of map with the least distortion that best suits his or her purpose.

Maps that show lines connecting points of similar elevation on land (known as **contours** of elevation) are **topographic** maps. Charts of the ocean showing contour lines connecting points of the same depth below the sea surface are **bathymetric** charts. The depth contours of these charts are called isobaths. A bathymetric chart is shown in figure 1.9. Color, shading, and perspective drawings may be used to indicate elevation changes, so as to produce a visual representation, or bird's-eye view, of the earth's elevation features. These

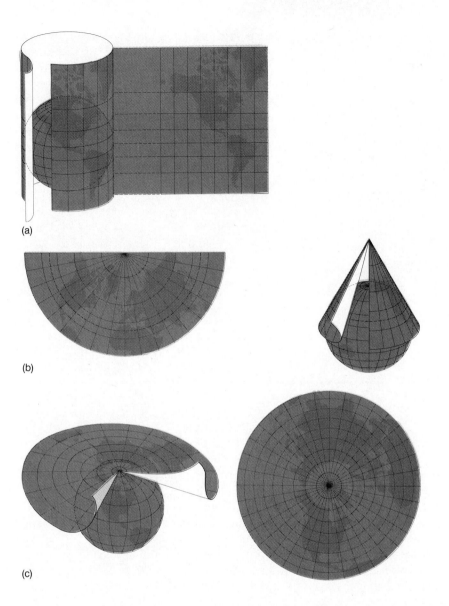

Figure 1.8
*(a) An equatorial cylindrical
projection. (b) A simple polar conic
projection. (c) A polar tangent
plane projection.*

(a)

(b)

(c)

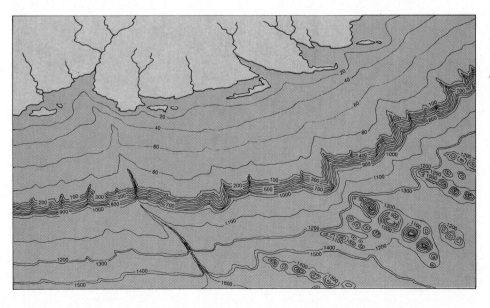

Figure 1.9
*A bathymetric, or contour, chart of
the sea floor along a section of
generalized coast. Changes in the
pattern and the spacing between
contour lines indicate changes in
depth.*

Figure 1.10

A physiographic chart of the same area shown in figure 1.9.

are **physiographic maps.** See figure 1.10 for an example of a physiographic map, and compare it with the bathymetric chart.

Measuring Latitude

Early maps show us that the first cartographers and navigators had considerable difficulty in precisely describing the then known earth features and their locations. When accurate measurement failed, artistic license appeared to fill the gaps (see fig. 1.11). The major problem was that early navigators were not able to determine their position accurately. As navigational techniques improved, so did the maps. It was known by early navigators that the North Star, **Polaris,** appeared to hang in the sky above the North Pole and did not appreciably move from this spot. In the northern hemisphere, therefore, measuring the angle of elevation of Polaris above the horizon gave a good estimate of one's latitude. Once a ship was out of sight of land, it sailed north or south until reaching the desired latitude then sailed east or west along this line of latitude until again reaching land. Adjustments north and south to reach the desired landfall of the voyage's end were made close to shore and according to visible landmarks.

Longitude and Time

Determining longitude was a much more difficult task. Since the longitude lines rotate with the turning earth it becomes necessary to know the position of the sun or the stars with time relative to one's longitude line. Because early clocks did not work well on rolling ships, precise longitude measurements were not possible although the theory for using time to determine longitude was put forth in the sixteenth century.

The relationship between time and longitude was proposed by the Flemish astronomer Gemma Frisius in 1530. If a clock is set to exactly noon when the sun is at its **zenith,** or highest elevation above a reference longitude, and if that clock is then carried to a new location and the zenith time of the sun is determined at this location, the clock's time difference between the sun's zenith at the reference longitude and at the new location is used to determine the longitude at the new location. Using this technique, a position that is 15° of longitude west of the reference longitude is directly under the sun one hour later, or at 1 P.M. by the clock, because the earth has turned eastward 15° during that hour. A position 15° east of the reference longitude is directly under the sun at 11 A.M., because it requires an hour to turn the 15° to bring the sun to its zenith over the reference longitude (see fig. 1.12).

The importance of longitude determination by this technique was quickly realized, in that commerce, exploration, and accurate chart making all needed this ability. As early as 1598 King Phillip III of Spain offered a reward of 100,000 crowns to any clock maker who could build a clock that would keep accurate time onboard ship. In 1714 the British Parliament offered 20,000 pounds sterling for a seagoing clock that could keep time with an error not greater than two minutes on a voyage to the West Indies from England. A Yorkshire clock maker, John Harrison, accepted the challenge. He built his first **chronometer** in 1735, but it was not until 1761 that his fourth model met the test, losing only 51 seconds on the 81-day voyage. In 1772, Captain James Cook took a copy of the fourth version of Harrison's chronometer (fig. 1.13) on his famous voyage of discovery to the south seas and with it was able to produce accurate charts of new areas and to correct previously charted positions. Harrison was awarded only a portion

Figure 1.11

A French map made by cartographer Nicolas Sanson in 1657. The distortion of continental shapes is due to inaccuracies in navigation as well as lack of knowledge.

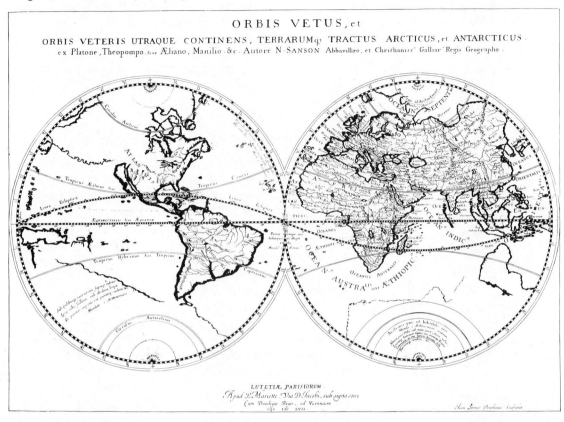

Figure 1.12

The time on a meridian relative to the sun changes by one hour for each 15° change in longitude.

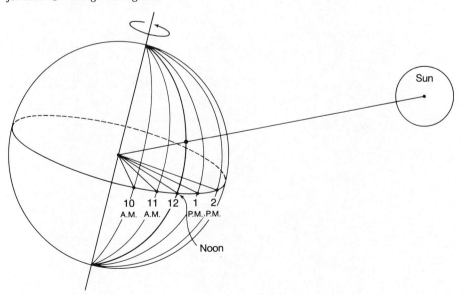

of the prize after his success in 1761, and it was not until 1775 at the age of eighty-three that he received the remainder from the reluctant British government. Accurate navigation based on time and celestial motions relative to the earth had become commonplace.

The reference longitude in use today is the prime meridian, or 0° longitude. The clock time is set to 12

noon when the sun is at its zenith above the prime meridian. This is **Greenwich mean time** (GMT), now called **coordinated universal time** or **zulu time** (for zero meridian time). Since sun time changes by one hour for each 15° of longitude, the earth has been divided into time zones that are 15° of longitude wide. The time zones do not follow lines of longitude exactly; they follow political boundaries when necessary for the convenience of the people living in those zones (see fig. 1.14).

Modern Navigational Techniques

Modern navigators still use chronometers and wait for clear skies to "shoot" the sun or stars with a sextant to determine positions at sea, but such measurements are primarily used to check their modern electronic navigational equipment. Chronometers are now calibrated, or reset, by broadcast time signals. When the vessels are near land, **radar** (radio detecting and ranging) gives an accurate picture of the shoreline and shows the position of vessels relative to the shore. The picture on the radar screen is formed by a pulse of radiation energy that is sent out by a transmitter, reflected from an object, returned to the antenna, and then displayed on a screen. **Loran** (long-range navigation) can be used farther out at sea. This electronic timing device is used to measure the difference in arrival time to the ship of the radio signals from two pairs of stations. The position of the ship may then be plotted on a chart that shows the time

Figure 1.13
John Harrison's fourth chronometer met the conditions required for time accuracy on board ship.

Figure 1.14
Distribution of the world's time zones. Degrees of longitude are marked along the bottom of the diagram. Time zones are positive (west zones) or negative (east zones). Time at Greenwich is determined by adding the zone number to the local time.

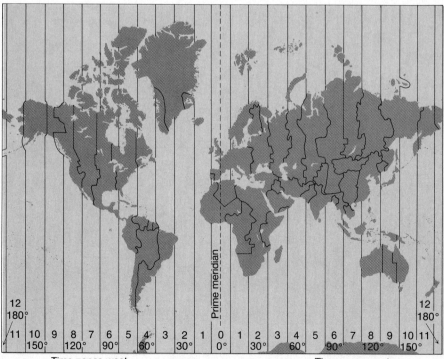

lines for these station pairs. Recent improvements in loran include receivers with built-in computers, which the navigator can program with the latitude and longitude of the desired destination. The loran receiver monitors the signals from the station pairs on land and reads out directly the course to be sailed and the distance to the desired destination. The computer can also monitor the signals and can continuously calculate the latitude and longitude, allowing the navigator to know the ship's position at all times.

The most accurate and sophisticated navigational aid available today is the **satellite navigation system.** A satellite orbiting the earth emits a coded signal of a precise frequency, which is picked up by a receiver on the ship. The shipboard receiver follows the frequency shift as the satellite passes and determines the exact instant in time at which the frequency is correct. At this instant the ship's path and the satellite's orbit are at right angles. Having this information and knowing the satellite's orbital properties, the computer can determine the ship's position to within 30 meters or less, using repeated measurements. All the techniques and devices used for navigation require the use of time. The more sophisticated the navigational instruments become, the more precise the requirement for time measurements will be.

The new electronic navigational aids and time devices are of tremendous help to the ocean scientist. It is now possible to determine a research vessel's position during all kinds of weather, and the vessel can return to the same exact station if further measurements are needed. Out on the open ocean, without landmarks and without a clear sky, it is possible to relocate precisely the same spot in order to time events as they occur at this position.

1.5 Earth: The Water Planet

Water on the Earth's Surface

As the earth and the other planets of the solar system cooled, the sun shone, sending its energy out into space to be intercepted by the planets. Gradually this solar radiant energy replaced the heat of planet formation to maintain the surface temperature of the planets. The earth developed a nearly circular orbital path at a mean distance of about 149 million kilometers (92.5 million miles) from the sun. Moving along this orbit the earth is 151 million kilometers from the sun in July and 146 million kilometers away in January, as shown in figure 1.15. At this distance from the sun, the nearly circular orbit keeps the earth's annual heating and cooling cycle within moderate limits. The earth's mean surface temperature is about 16° C, which allows water to exist as a gas, as a liquid, and as a solid.

The rotation of the earth on its axis is also important in moderating temperature extremes. The earth completes one rotation turning from west to east in twenty-four hours. If the earth rotated more slowly, the side of the earth toward the sun would be bathed in radiant energy for a longer period than it is at present and would become very hot, while the side in darkness would lose heat and become very cold. Temperature changes from day to night would be large. By contrast,

Figure 1.15
The earth's orbit around the sun is nearly circular. Note the change in distance between the earth and the sun during the year.

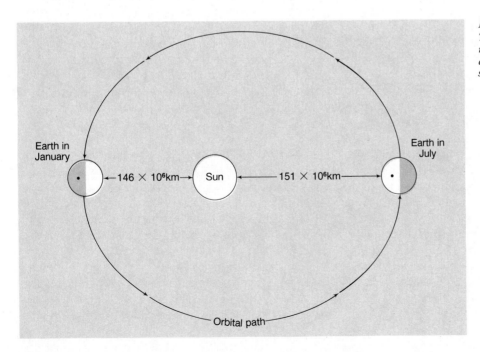

TABLE 1.3

The Earth's Water Supply

Reservoir	Volume (km³)	% of Total volume	Sphere depth (m)
atmospheric moisture expressed as water	15.3×10^3	0.001	0.03
rivers and lakes	510.0×10^3	0.036	1.0
ground waters	$5,100.0 \times 10^3$	0.365	10.0
glacial and other land ice	$22,950.0 \times 10^3$	1.641	45.0
oceanic water and sea ice	$1,370,323.0 \times 10^3$	97.957	2,686.0
totals	$1,398,898.3 \times 10^3$	100	2,742.0

a shorter period of rotation would decrease the present variation from day to night.

The earth's solar orbit, its rotation, and its blanket of atmospheric gases all act to produce surface temperatures that allow the existence of liquid water. The atmosphere covering the earth's surface acts as a protective shield between the earth and the sun. Without it, solar heating would evaporate water at a much increased rate. Compare the earth's distance to the sun, its period of rotation, and the time required for the earth to complete one orbit of the sun with those of other planets, as shown in table 1.1. Note the surface temperatures of the earth as compared to those of other planets.

There are a number of ways for expressing the amount of water on the earth's surface. For example, the oceans cover 139 million square miles (139×10^6 mi²) or 361 million square kilometers (361×10^6 km²) of the earth's surface. (If you are unfamiliar with scientific notation to express very large numbers, see the Appendix.) Because these numbers are so large, they do not convey a clear idea of size; therefore, a more meaningful concept is to remember that 71% of the earth's surface is covered by the world's oceans, and only 29% of the surface area is land above sea level.

The volume of water in the oceans is enormous: 1.37 billion cubic kilometers (1.37×10^9 km³). A cubic kilometer of seawater is very large indeed. Consider this: the largest building in the world in cubic capacity is the main assembly plant of the Boeing Company in Everett, Washington. This building is used for the manufacture of Boeing 747 airplanes and has a cubic capacity of 0.0847 km³. In other words, 11.8 of these buildings would fit into one cubic kilometer. Another way to express the oceanic volume is to think of a smooth sphere with exactly the same surface area as the earth (510 $\times 10^6$ km² or 316×10^6 mi²), uniformly covered with the water from the earth's oceans. The ocean water would be 2686 m (8800 ft) deep, a depth of about 1.7 miles. If the water from all other sources in the world

were added, including water from the land and from the atmosphere, the depth would rise 56 m to 2742 m (9000 ft). When water volumes are considered as depths over a smooth sphere, they are referred to as **sphere depths.** The ocean sphere depth is 2686 m and the total water sphere depth of all the earth's water is 2742 m. This information is summarized in the last column of table 1.3.

The Hydrologic Cycle

The earth's water is found as a liquid in the oceans, rivers, lakes, and below the ground surface; it occurs as a solid in glaciers, snow packs, and sea ice; it takes the form of water droplets and gaseous water vapor in the atmosphere. The places in which water resides are called **reservoirs,** and each type of reservoir, when averaged over the entire earth, contains a fixed amount of water at any one instant. But water is constantly moving from one reservoir to another, as liquid water evaporates from oceans and lakes into the air, icebergs melt in the oceans, rains fall on the land, and rivers flow back to the sea. This movement of water through the reservoirs is called the **hydrologic cycle.** Evaporation takes water from the surface of the ocean into the atmosphere; most of this water returns directly to the sea, but air currents carry some water vapor over the land. Precipitation returns this water to the earth's surface, where it percolates into the soil, fills rivers, streams, and lakes, or remains for longer time periods as snow and ice in some areas. Melting snow and ice, rivers, groundwater, and land runoff move the water back to the oceans. See figure 1.16 for a diagram of the hydrologic cycle. For a comparison of the water stored in the earth's reservoirs, see table 1.3 again.

In local areas excess water may evaporate from the land and excess precipitation may occur over the sea, but on a worldwide average, the cycle operates with a net removal of ocean water by evaporation, a net gain

Figure 1.16
The hydrologic cycle and annual transfer rates. Snow and rain together equal precipitation. Sublimation, the direct transfer of ice to water vapor, is included under evaporation. Surface flow and underground flow are considered as land runoff. Annual transfer rates are in thousands of cubic kilometers (10^3 km³).

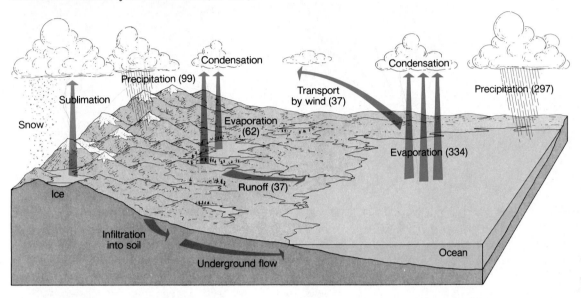

of water on land as a result of precipitation, and a return of the excess water on land to the sea by rivers and land drainage. The properties of climate zones are principally determined by the earth's surface temperature and its evaporation-precipitation patterns: the moist, hot equatorial regions, the dry, hot subtropic deserts, the cool, moist temperate areas, and the cold, dry high polar zones. Differences in the earth's surface temperature and in the evaporation and precipitation patterns in these climate zones, coupled with the movement of air between the zones, moves water through the hydrologic cycle at different rates from one reservoir to another. The transfer of water between the atmosphere and the oceans alters the salt content of the oceans' surface water. This, along with the seasonal and latitudinal changes in surface temperature, determines many of the characteristics of the world's oceans.

Reservoirs and Residence Time

Because the total amount of water in the earth's reservoirs is nearly constant, the hydrologic cycle must maintain a balance between the addition and removal of water from reservoirs if considerable changes are not to occur. The rate of removal of water from a reservoir must equal the rate of addition to it, for if the balance is disturbed one reservoir gains at the expense of another. The time that the water spends in any one reservoir is called its **residence time.** Large reservoirs have long residence times because it takes a long time to move all the water

in that reservoir through the hydrologic cycle, while the water in small reservoirs can be replaced comparatively quickly, so that they have short residence times. The size of the reservoir also determines how it reacts to changes in the rate at which the water is gained or lost. Large reservoirs show little effect from small rate changes, while small reservoirs may alter substantially when exposed to the same variation. For example, if the ocean volume decreased by 6½% and this volume of water were added to the land ice, it would produce a 400% increase in the present volume of land ice but only a 250 m (820 ft) drop in sea level. This example reflects the changes that have occurred in the earth's past history during the major ice ages.

About 396,000 km³ (94,644 mi³) of water move through the atmosphere each year. Since the atmosphere holds the equivalent of 15,300 km³ (3,580 mi³) of liquid water at any one time, a little arithmetic shows that the water in the atmosphere can be replaced twenty-six times each year. Atmospheric water has a very short residence time. The residence time for water in the other larger reservoirs is much longer. For example, it would take 37,027 years to evaporate and pass the water from all of the oceans through the atmosphere, to the land as precipitation, and return it to the oceans via rivers. Further study of water's movement shows us that annually 334,000 km³ (79,826 mi³) are evaporated from the oceans while the land loses 62,000 km³ (14,818 mi³). When the water returns as precipitation, 297,000 km³ (70,983 mi³) are returned directly to the sea surface, and

Figure 1.17
Continents and oceans are not distributed uniformly over the earth. The northern hemisphere (a) contains most of the land; the southern hemisphere (b) is mainly water.

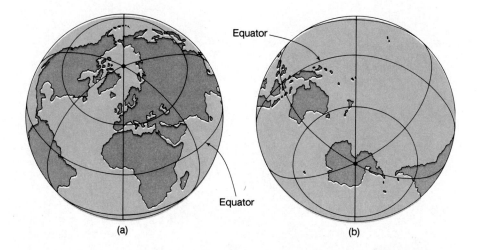

(a)

(b)

99,000 km³ (23,660 mi³) return to the land. However, the excess gained by the land (37,000 km³) (8,843 mi³) flows back to the oceans in the rivers, streams, and groundwater of the world (see fig. 1.16).

Distribution of Land and Water

The present distribution of land and water on the earth is also important. To understand this distribution consider the earth when it is viewed from the north (see fig. 1.17a) and from the south (fig. 1.17b). About 70% of the earth's landmasses are in the northern hemisphere, and most of this land lies in the middle latitudes. The southern hemisphere is the water hemisphere, with its land located mostly in the tropic latitudes and in the polar region. Recognizing this difference in the land distribution, we might give some thought to the history of the human race and wonder what if any significance can be attached to the land-latitude distribution and to the sequence of events occurring there. Would the history of the two hemispheres have been different if the temperate-latitude landmasses had been concentrated in the southern hemisphere and the water in the northern hemisphere? The details of this land-water-latitude distribution are presented in figure 1.18.

The Oceans

Oceanographers view the world's ocean as three fingers—the Atlantic, Pacific, and Indian oceans—stretching up from a common source around the Antarctic continent. The Arctic Ocean is considered an extension of the North Atlantic, and the Antarctic Ocean may be considered as the Southern Ocean below 50° S or it may be divided along chosen lines of longitude to become the southern portions of the Atlantic, Pacific, and Indian oceans (see fig. 1.19). Each of these three

oceans has its own characteristic surface area, volume, and mean (or average) depth. The Pacific Ocean has more surface area, a larger volume, and a greater mean depth than either the Atlantic or the Indian oceans, as shown in table 1.4. The Atlantic Ocean is the shallowest ocean and has the greatest number of shallow adjacent seas, such as the Arctic Ocean, Gulf of Mexico, Caribbean Sea, and Mediterranean Sea. The Indian Ocean is a southern-hemisphere ocean lacking a polar region in the northern hemisphere. It is the smallest ocean in terms of area, but it is quite deep. The volume of the Pacific Ocean is over twice that of either the Atlantic Ocean or the Indian Ocean.

A view of an ocean from above is a view of 100% of that ocean's area at the sea surface, or at a depth of zero meters. It is also a view of 100% of the volume of water contained in that ocean. If successive layers of water 1000 m (3280 ft) thick could be removed, then each view would show an ocean of reduced area and volume. This sequential removal would allow us to see how the ocean water is distributed with depth. A shallow ocean would reduce rapidly in both area coverage and volume content as the layers were removed, while a deep ocean would change more slowly.

This effect is shown in table 1.4 for each of the three principal oceans and for all the oceans combined. When the topmost 1000-m layer of water is removed, the Atlantic Ocean undergoes a large change in both volume and area compared to the Pacific and Indian oceans under the same conditions. Areas at given depths can be used to calculate percent changes in areas and volumes. As an example, for the Atlantic Ocean, the change in area from 0 to 1000 m ([106.4 − 84.7] × 10⁶ km²) divided by the total area at 0 m (106.4 × 10⁶ km²) and multiplied by 100 shows that 20.4% of the sea floor of the Atlantic falls within this depth range. This indicates that this ocean has extensive shallow regions. These shallow regions and the absence of any appreciable depths greater than 6000 m combine to produce

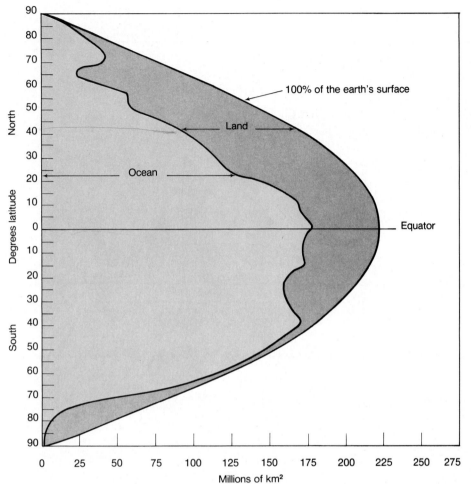

Figure 1.18
Distribution of land and ocean by latitude. In the northern hemisphere, middle latitude areas of land and ocean are nearly equal. Land is almost absent at the same latitudes in the southern hemisphere. The areas are calculated on the basis of 5° latitude intervals.

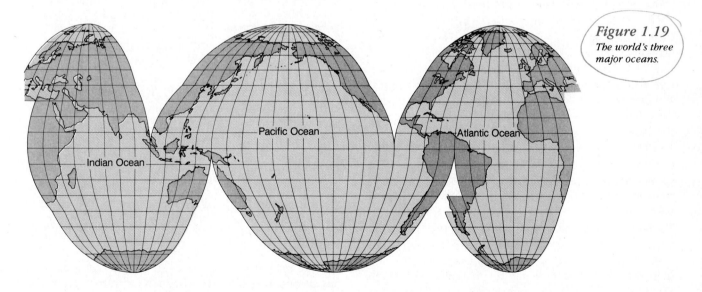

Figure 1.19
The world's three major oceans.

the relatively shallow mean depth of the Atlantic as compared to the Pacific and Indian oceans. The largest change in seabed area occurs when the layers between 4000 m and 6000 m are removed. This change indicates that the most commonly occurring depths in the oceans are found between these two values.

The Pacific and Indian oceans have much less area at the shallow depths, changing only 8.8% and 7.3%, respectively, between 0 m and 1000 m (3280 ft). This lack of shallow water area and the presence of depths greater than 6000 m (19,680 ft) combine to give both of these oceans mean depths greater than that of the Atlantic.

TABLE 1.4

Ocean Areas and Volumes versus Depth

Depth (m)	Atlantic (mean depth 3332 m)		Pacific (mean depth 4028 m)		Indian (mean depth 3897 m)		All oceans (mean depth 3795 m)	
	Area[1]	Volume[2]	Area	Volume	Area	Volume	Area	Volume
0	106.4	354.7	179.7	723.7	74.9	291.9	361.0	1370.3
1000	84.7	259.1	164.0	550.5	69.4	219.7	318.1	1029.3
2000	79.1	177.2	156.9	388.8	66.9	151.6	302.9	717.6
3000	69.7	102.8	147.5	236.6	61.3	87.5	278.5	426.9
4000	50.0	43.0	114.3	105.7	43.4	35.1	207.7	183.8
5000	22.6	6.7	51.0	23.1	14.8	6.0	88.4	35.8
6000	0.64	0	3.2	0.2	0.3	0.1	4.14	0.3
7000	0	0	0.36	0	0	0	0.36	0

[1]All areas are given in 10^6 km^2.
[2]All volumes are given in 10^6 km^3.

The Hypsographic Curve

Another method used by oceanographers to depict land-water relationships is shown in figure 1.20. This graph of depth or elevation versus area is called a **hypsographic curve.** Find the line indicating sea level and note that the elevation of land above sea level is given in meters along the left margin; the depth below sea level is given in meters along the right margin. The scale across the top of the figure indicates total earth area in 10^8 km^2. The scales along the bottom of figure 1.20 indicate percentages of the earth's surface area; note that the curve crosses sea level at the 29% mark, showing that 29% of the earth's surface is above sea level and the remaining 71% is below sea level. The lower of the two scales gives land and ocean areas as separate percentages. Referring to the land area percentage scale, note that only 20% of all land areas are at elevations above 2 km. The ocean area scale shows that approximately 85% of the ocean area is below 2 km in depth. Compare this value with the data in table 1.4. The hypsographic curve helps us to see not only that our earth is 71% covered with water but that the areas whose depths are well below the sea surface are much greater than the areas whose elevations are well above it; there are basins beneath the sea that are about four times greater in area than the area of land in mountains above sea level. Mount Everest, the highest land peak, reaches 8.84 km (5.49 mi) above sea level, while the ocean's deepest trench descends 11.02 km (6.84 mi) below sea level.

Since the hypsographic curve is constructed as a plot of area versus height, the diagram also shows volume, because volume is the product of area times height. The mean elevation of the land is 840 m (2750 ft), and the entire land volume above sea level fits within a box 840 m high, covering 29% of the earth's surface. The mean depth of the ocean is 3795 m (12,400 ft). In this case the ocean volume is the mean ocean depth times 71% of the earth's surface area.

Refer back to the discussion of sphere depths as a device for expressing water volumes and remember that, because humans are land dwellers, their reference line is sea level. But if we are considering the hypsographic curve for the total earth, a mean earth elevation becomes more suitable for describing the location of the earth's solid surface; this value is called the **mean earth sphere depth.** This level is 2440 m (8003 ft) below the present sea level, as shown in figure 1.20. This value represents the level at which the volume of crust above and water volume below the earth sphere depth are equal. To arrive at the mean earth sphere depth, think of moving all the elevated land and some of the sea bottom down into the ocean depressions until the earth is a perfectly smooth ball. When this is accomplished the mean earth sphere depth reference level would stand at the 2440-m (8003-ft) depth. Such an operation would also result in the displacement of seawater upward, until a new sea level 246 m (807 ft) above the present sea level was reached. This depth of water is the **mean ocean sphere depth,** 2686 m (8810 ft). It is the depth that the oceans would have if the volume of ocean water were spread over an area equal to the total earth area rather than over 71% of the earth's area. The oceans swallow the land, with only a comparatively small rise in water level, emphasizing again that earth is the water planet (or the ocean planet) and not the land planet.

10% of earth is higher than 3 (handwritten note)

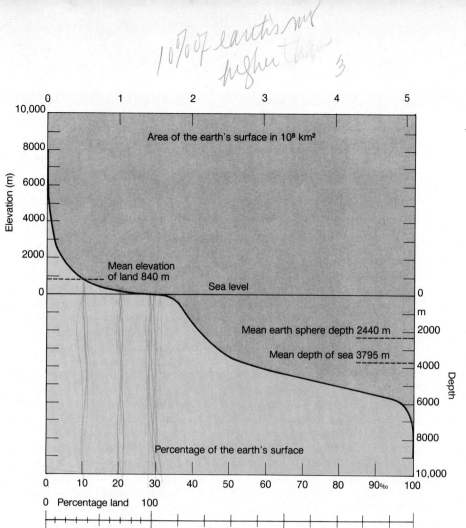

Figure 1.20

The hypsographic curve displays the area of the earth's surface at elevations above and below sea level.

Satellite Oceanography

Because the great area of the world's oceans makes it impossible to equip enough research vessels to study more than a small area of one ocean at one time, and because what happens in any one area of an ocean is dependent on processes at work in other parts of the world's oceans, oceanographers have needed the ability to study the oceans as a total system.

Oceanographers are using satellites with specialized sensors and measuring devices to provide total ocean surveillance and data on a global scale. These data are sent back to earth in the form of radio signals that are recorded, processed, and analyzed by computer to recreate the satellite's observations. Specific details may be enhanced to provide images emphasizing a researcher's interest.

If a satellite is set in high orbit about the earth's equatorial plane approximately 41,625 kilometers (22,500 nautical miles) above the earth, it will remain centered above one point on the earth's surface. This is a geosynchronous satellite that can be used to monitor changes at a single location. A satellite orbiting in a plane inclined to the earth's equatorial plane, at a speed allowing it to circle the earth every ninety minutes, and with an observational band of 22½° of longitude, will have a much lower orbit and cover a new band of earth surface each time the satellite crosses the equator. Complete earth surface coverage will be accomplished in twelve hours within the north-south limits of the satellite's orbit.

If the satellite is placed in a low polar orbit (925 km or 500 nm) above the earth with an orbital period of ninety minutes and if its sensors cover a band 22½° longitude-wide at the equator, an area of 100×10^6 km² (29.2×10^6 nm²) is observed during each ninety-minute period. Under these conditions each data point covers a relatively large area and a fuzzy average portrayal of the total earth's surface is achieved in twelve hours. If the sensors on the satellite are restricted to a narrower path of observation, a much more detailed picture can be obtained. These conditions require more orbits and a longer time to achieve full earth coverage.

The TIROS series of weather observation satellites, launched in 1960, gave us our first look at the earth's global cloud cover (colorplate 1f). By watching the movement of storm systems, weather prediction was greatly improved. The NIMBUS series of satellites, designed to observe the earth's surface features, followed. NIMBUS-7 (colorplate 1b), launched in 1978, monitors solar radiation reflected back from the earth's surface. This data allows detection of changes in the earth's surface temperature, detection of changes in the atmosphere's chemical composition, and the monitoring of chlorophyll abundance in land and ocean vegetation. This workhorse ceased to gather part of its data in June, 1986, and an attempt to reactivate its ocean chlorophyll scanners in December,

1986, was unsuccessful.

In June, 1978, SEASAT, a specialized oceanographic satellite, was launched and remained operational until October of that year (colorplate 1a). This disappointingly short-lived satellite carried a radar altimeter capable of measuring the distance between the satellite and the sea surface with an accuracy of about 5 cm (2 in) over a narrow observational width. The altimeter measured both wave heights and sea surface variation. Radar was also used to measure the reflection pattern of wind waves, from which wind speed and direction were calculated. Radiometers recorded radiation from the earth's surface, allowing the measurement of sea surface temperatures, atmospheric water vapor content, wind speed, and sea ice cover. SEASAT gave the first large-area coverage of oceanic winds and produced a picture of sea surface elevations that showed the effect of

local changes in the earth's gravity caused by variations in density of the earth's crustal material and the elevation of the sea floor. From this it was possible to produce a bathymetric map of the sea floor.

Currently, more satellites are being designed with more and different sensors, and data processing is being planned to obtain more accurate results with a greater variety of measurements. Remote sensing by satellite has the potential to help us forecast wave conditions, plan and design better coastal structures, identify hazards of sea ice and icebergs, find and forecast fish stocks, follow the distribution of pollutants and sediments, and much more. Such information will help us as we struggle to make the informed and wise decisions required to maintain a healthy marine environment. Satellite oceanography is costly, but it is the only way to achieve a global view of the earth, its oceans, and atmosphere.

Summary

The solar system began as a rotating cloud of gas. A series of events produced nine planets orbiting the sun, each planet having different characteristics. Over approximately 1½ billion years the earth heated, cooled, changed, and collected a gaseous atmosphere and an accumulation of liquid water.

Reliable age dates for earth rocks, meteorites, and moon samples are obtained by radiometric dating. The accepted age of the earth is between 4.5 billion and 4.6 billion years. Geologic time is used to express the time scale of the earth's history.

The distance between the earth and the sun, the earth's orbit, its period of rotation, and its atmosphere protect the earth from extreme temperature change and water loss. Because it rotates, the earth's shape is not perfectly symmetrical. Its exterior is relatively smooth. Natural time periods (the year, day, and month) are based on the motions of the sun, earth, and moon. Because of the tilt of the earth's axis as it orbits the sun, the sun moves annually between 23½° N and 23½° S, producing the seasons.

Latitude and longitude are used to form a grid system for the location of positions on the earth's surface. Different types of map and chart projections have been developed to show the earth's features on a flat surface. All of these projections distort the earth's features to some extent. Bathymetric and physiographic charts and maps use elevation and depth contours to depict the earth's topography.

In order to determine longitudinal position, we must measure time accurately. This need required the development of seagoing clocks for celestial navigation. Modern navigational techniques make use of radar, radio signals, computers, and satellites.

The earth is the water planet. Of the earth's surface, 71% is covered by its oceans. There is a fixed amount of water on earth. Evaporation and precipitation move the water through the reservoirs of the hydrologic cycle. Water's residence time varies in each reservoir and depends on the volume of the reservoir and the rate of addition and removal of the water.

The earth's northern hemisphere is the land hemisphere; the southern hemisphere is the water hemisphere. The earth has three large oceans extending north from the Southern Ocean. Each has a characteristic surface area, volume, and mean depth. The hypsographic curve is used to show land-water relationships of depth, elevation, area, and volume. It is also used to determine mean land elevation, mean ocean depth, earth sphere depth, and ocean sphere depth.

Key Terms

radiometric dating	lunar month	projection	zulu time
isotope	solar day	Mercator projection	radar
half-life	sidereal day	contour	loran
vertebrates	latitude	topography	satellite navigation system
Tropic of Cancer	longitude	bathymetry	sphere depth
summer solstice	equator	physiographic maps	reservoir
Arctic Circle	parallel	Polaris	hydrologic cycle
Antarctic Circle	meridian	zenith	residence time
autumnal equinox	prime meridian	chronometer	hypsographic curve
Tropic of Capricorn	international date line	Greenwich mean time	mean earth sphere depth
winter solstice	great circle	coordinated universal time	mean ocean sphere depth
vernal equinox	nautical mile		

Study Questions

1. How and why have estimates of the age of the earth changed over the past few hundred years? Do you think the present estimate of the earth's age will change in the future?

2. Describe the distribution of water and land on the earth.

3. Why does the earth's average surface temperature differ from the surface temperature of other planets in the solar system?

4. Why is the twilight period at sunset shorter at low latitudes than it is at high latitudes?

5. The route of a ship sailing a constant compass course on a Mercator projection is indicated by a straight line that cuts all longitude lines at the same angle. This is a rhumb line. Discuss how this line appears (a) on a polar conic projection, (b) on a globe, and (c) on a tangent plane projection centered on the polar axis.

6. Discuss how the hypsographic curve is used to determine the mean depth and sphere depth of the oceans.

7. Why are the Arctic and Antarctic circles displaced from the poles by 23½°, so that they are located at 66½° N and 66½° S?

Study Problems

1. Determine the distance between two locations: 110° W, 38.5° N and 110° W, 45.0° N. Express this distance in nautical miles and in kilometers.

2. The contour interval on a bathymetric chart is constant and equal to 100 m. Graph the slope of the sea floor across four evenly spaced contour lines if the distance between the first line and the fourth line is 2.5 km.

3. A plane leaves Tokyo, Japan, on June 6, at 0800 hours local Tokyo time and flies for nine hours, landing in San Francisco, California. What is the local time and date of arrival in San Francisco?

4. Show that the annual net evaporative loss of water from the world's oceans equals the annual net gain of water by precipitation on the land. Why does the ocean volume not decrease?

5. Show that the Atlantic Ocean has a larger percentage of its volume in the top layer (0–1000 m) than does the Pacific or Indian Ocean. Explain the significance of this difference.

6. Use the volume of the oceans and the earth's area to determine the sphere depth of the oceans.

The Sea Floor

2

The fog continued through the night, with a very light breeze, before which we ran to the eastward, literally feeling our way along. The lead was hove every two hours and the gradual change from black mud to sand, showed that we were approaching Nantucket South Shoals. On Monday morning, the increased depth and deep blue color of the water, and the mixture of shells and white sand which we brought up, upon sounding, showed that we were in the channel, and nearing George's; accordingly, the ship's head was put directly to the northward, and we stood on, with perfect confidence in the soundings, though we had not taken an observation for two days, nor seen land; and the difference of an eighth of a mile out of the way might put us ashore. Throughout the day a provokingly light wind prevailed, and at eight o'clock, a small fishing schooner, which we passed, told us we were nearly abreast of Chatham lights. Just before midnight, a light land-breeze sprang up, which carried us well along; and at four o'clock, thinking ourselves to the northward of Race Point, we hauled upon the wind and stood into the bay, north-north-west, for Boston light, and commenced firing guns for a pilot.

Richard Henry Dana, Jr.
from Two Years Before the Mast

*T*he sea floor, hidden by its covering of water, was an unknown environment to early mariners and the first curious scientists. They believed that the oceans were large basins or depressions in the earth's crust, but they did not conceive that these basins held features that were as magnificent as the major mountain chains, the deep valleys, and the great canyons of the land. As maps were created in greater detail and as ocean travel and commerce increased, it became essential to measure and record the bathymetry, or bottom features, in the shallower regions in order to maintain safe travel for sailing ships. The secrets of the deeper oceanic areas continued to lie hidden by water too deep to probe, and the bottom features of these areas were only gradually exposed.

It was not until the 1950s that improvements in technology made it relatively easy to sample the sea floor; the large numbers of vessels taking part in surveys accumulated sufficient data to give detail to this hidden terrain. What was found beneath the surface of the sea created more questions than answers and produced a demand for more and varied measurements to describe and explain the features of this portion of the earth's crust. Gradually, investigators began to relate the 71% of the earth's surface that lies beneath the oceans to the more familiar part above the sea and to study the crust of the earth as a single, continuous layer.

Not only are studies of the seabed valuable for their contribution to our knowledge of the earth's crustal geology and how it has changed with time, but they also give useful knowledge of mineral resources that can be used, if necessary, to supplement the dwindling supplies found on land. We need to know where deposits of minerals are located and how to remove them economically and with a minimal impact. We need information so that we can make wise and thoughtful decisions before irreversible steps are taken.

2.1 Bathymetry of the Sea Floor

The land below the sea surface is nearly as rugged as any land above it. The Grand Canyon, the Rocky Mountains, the desert mesas in the Southwest, and the Great Plains all have their undersea counterparts. In fact, the undersea mountain ranges are longer, the valley floors are wider and flatter, and the canyons are often deeper than those found on land. Topographic features on land (such as mountains and canyons) are continually and actively eroded by wind, rain, and ice, and are affected by changes in temperature and rock chemistry. Undersea mountains and canyons are called bathymetric features; here erosion is slow when compared to erosion on land. Only between the ocean surface and the land and on very steep underwater slopes are there active eroding forces, such as waves and scouring currents. Beneath the surface the water dissolves away certain materials, changing the rock chemistry, but the features, once formed, retain their shape and are only slowly modified as they are covered by a constant rain of particles, or **sediments,** falling from above. Movements of the earth's crust may displace features and fracture the sea floor, and the weight of some underwater volcanoes may cause them to sink, or **subside,** but the appearance of the bathymetric features of the ocean basins and sea floor has remained much the same through the last 100 million years. Computer-drawn profiles of crustal elevations across the United States and the Atlantic Ocean are shown in figure 2.1. Compare the topography of the Rocky Mountains and the undersea peaks shown in this figure.

The Continental Margin

The **continental margin** is made up of the continental shelf, shelf break, slope, and rise. The **continental shelf** lies at the edge of the continent. The continental shelves are nearly flatland borders of varying widths that slope very gently toward the ocean basins. Shelf widths average about 65 km (40 mi) and vary from only a few tens of meters across in some areas to more than 100 km (60 mi) in others. The average depth of the seaward edge of the continental shelf is considered to be 130 m (430 ft).

The distribution of the world's continental shelves is shown in figure 2.2. The width of the shelf is often related to the slope of the adjacent land, so that it is wide along low-lying land and narrow along mountainous coasts. Note the narrow shelf along the west coast of South America and the great wide expanses of continental shelf along the eastern and northern coasts of North America, Siberia, and Scandinavia. The continental shelves are geologically part of the continents and are underlaid by continental crust. Processes producing continental shelves include the storm waves that ceaselessly erode the edges of the continents and, in some cases, natural dams that trap sediments between the dam and the coast. Such dams include reefs, volcanic barriers, upthrust rock, and salt domes. In the Gulf of

Figure 2.1

Computer-drawn topographic profiles from the west coast of Europe and Africa to the Pacific Ocean. The elevations and depths above and below 0 meters are shown along a line of latitude by using the latitude line as zero elevation. For example, the depth at 40° N and 60° W is 5040 m (16,531 ft). The vertical scale has been extended about one hundred times the horizontal scale. If both the horizontal and vertical scales were kept the same, a vertical elevation change of 5000 m would measure only 0.05 mm (0.002 in).

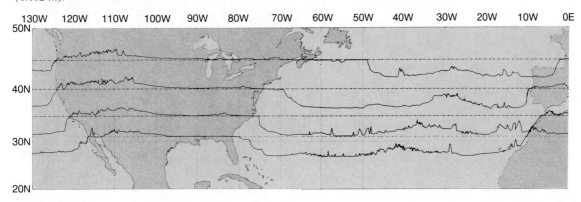

Figure 2.2

Distribution of the world's continental shelves. The seaward edges of these shelves are at an average depth of approximately 130 meters.

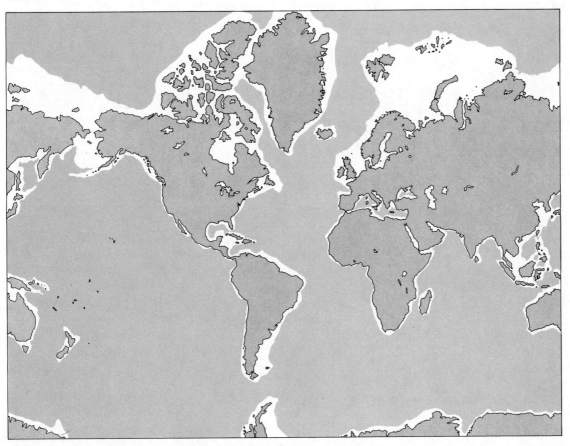

Figure 2.3
Examples of how continental shelves are formed by trapping land-derived sediments at the edge of the continental landmasses.

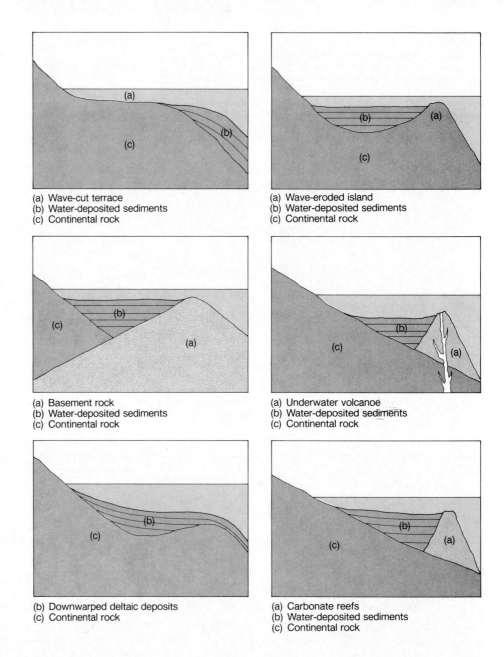

(a) Wave-cut terrace
(b) Water-deposited sediments
(c) Continental rock

(a) Wave-eroded island
(b) Water-deposited sediments
(c) Continental rock

(a) Basement rock
(b) Water-deposited sediments
(c) Continental rock

(a) Underwater volcanoe
(b) Water-deposited sediments
(c) Continental rock

(b) Downwarped deltaic deposits
(c) Continental rock

(a) Carbonate reefs
(b) Water-deposited sediments
(c) Continental rock

Mexico, sediments from the land are trapped behind folds and domes of the sea floor containing salt deposits. Sediments along the northeast coast of North America are trapped behind upturned rock near the outer edges of the shelves (see fig. 2.3).

During past ages, the shelves have been covered and uncovered by fluctuations in sea level. During the glacial epoch of the Pleistocene (2.5 million years ago) there were a number of short-term changes in sea level, some of which were greater than 120 m (400 ft). When the sea level was low erosion deepened valleys, waves eroded previously submerged land, and rivers brought sediments far out on the shelf. When the glacial ice melted, these areas were flooded and sediments built up in areas closer to shore. At present, although submerged, these areas still show the features they ac-

quired when exposed as part of the landmass, such as the scars of old riverbeds and glaciers. Today, some continental shelves are covered with thick deposits of silt, sand, and mud sediments derived from the land, as they are in the areas of the Mississippi and Amazon rivers, which deposit large amounts of such sediments annually.

Other shelves are bare of sediments, such as where the fast-moving Florida Current sweeps the tip of Florida, carrying the sediments northward to deeper water. The boundary of the continental shelf on the ocean side is determined by an abrupt change in slope and a rapid increase in depth. This change in slope is referred to as the **continental shelf break,** while the steep slope extending to the ocean basin floor is known as the **continental slope.** These features are shown in figure 2.4.

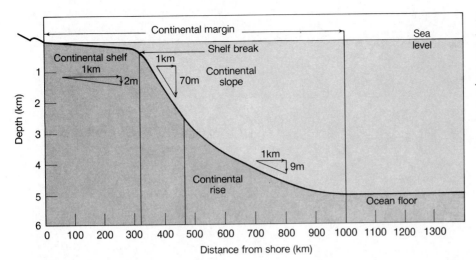

Figure 2.4
A typical profile of the continental margin. Notice both the vertical and horizontal extent of each subdivision. The slope is indicated for the continental shelf, slope, and rise. The vertical exaggeration is one hundred times greater than the horizontal scale.

TABLE 2.1
Major Submarine Canyons of the World

Pacific Ocean	Atlantic Ocean	Indian Ocean
Tokyo Canyon	Oceanographer Canyon	*Indus Canyon
Bering Canyon	*Hudson Canyon	*Ganges Canyon
*Columbia (or Astoria) Canyon	Wilmington Canyon	
Juan de Fuca Canyon	Norfolk Canyon	
Monterey Canyon	*Congo Canyon	
Arguello Canyon	*São Francisco Canyon	
Scripps Canyon	*Mississippi Canyon	
Coronados Canyon		

*Canyon has present-day relationship to a river. Other canyons are related to faults, inactive river beds, and drainage systems active at previous low sea levels.

The angle and extent of the slope vary from place to place. The slope may be short and steep (for example, the depth may increase rapidly from 200 m (656 ft) to 3000 m (9840 ft), as in figure 2.4), or it may drop as far as 8000 m (26,240 ft) into a great trench (for example, off the west coast of South America, where the narrow continental shelf is bordered by the Peru-Chile Trench). The continental slope may show rocky outcroppings, and it is relatively bare of sediments because of the steepness with which it drops to the sea floor.

The most outstanding features found on the continental slopes are **submarine canyons** (see table 2.1 for a list of major submarine canyons). These canyons sometimes extend up, into, and across the continental shelf as well. A submarine canyon is steep-sided and V-shaped, with tributaries similar to those of river-cut canyons on land. Figure 2.5a shows the Monterey and Carmel canyons off the coast of California. Figure 2.5a is a bathymetric chart; figure 2.5b compares the profile of the Monterey Canyon with the profile of the Grand

Canyon of the Colorado River. A submarine canyon is also shown in figures 1.9 and 1.10.

Many of these submarine canyons are associated with existing river systems on land and were apparently cut into the shelf during periods of low sea level, when the glaciers advanced and the rivers flowed across the continental shelves. Ripple marks presently observed on the floor of the submerged canyons and sediments fanning out at the ends of the canyons suggest that they have been formed by moving sediment and water. Scientists believe that they were cut by **turbidity currents.** Caused by earthquakes or the overloading of sediments on steep slopes, turbidity currents are fast-moving avalanches of mud, sand, and water that flow down the slope, eroding and picking up sediment as they gain speed. In this way the currents erode the slope and excavate the submarine canyon. As the flow reaches the bottom it slows and fans out, and the sediments settle. Because of their speed and turbulence, such currents can transport large quantities of mixed materials,

Figure 2.5
(a) Depth contours depict the Monterey and Carmel Canyons off the California coast as they cut across the continental slope and up into the continental shelf. (b) Cross-canyon profile of the Monterey Canyon shown in (a). Compare this profile with that of the Grand Canyon drawn to the same scale.

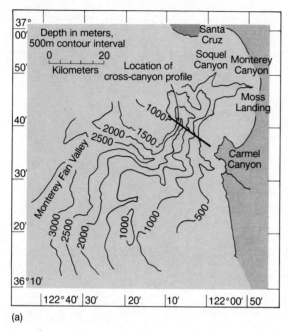

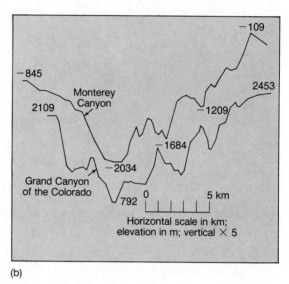

producing graded deposits of coarse materials overlaid by successive layers of decreasing particle size. These graded deposits are called **turbidites.**

Such large and occasional currents have never been directly observed, although similar but smaller and more continuous flows, such as sand falls, have been observed and photographed (see fig. 2.6). Research on turbidity currents began with laboratory experiments in the 1930s. These experiments were based on earlier observations of the silty Rhone River water moving along the bottom of Lake Geneva in Switzerland.

In 1929 an earthquake had broken the transatlantic telephone and telegraph cables on the continental slope and rise off the Grand Banks of Newfoundland. The pattern of the cable breaks was analyzed and it was shown that an initial series of rapid breaks occurred high on the continental slope during the earthquake, followed by a sequence of downslope breaks after the earthquake. These breaks were calculated to have been caused by a turbidity current that ran for 700 km (420 mi) at speeds of 40–55 km (25–35 mi) per hour. Later samples taken from the area showed a series of graded sediments at the end of the current's path. Searches of cable company records showed similar patterns of cable breaks in other parts of the world.

At present, researchers are developing computer and laboratory models to predict a turbidity current's speed, depth, sediment content, and the distance it will travel. One of the first models is being applied to the Scripps Submarine Canyon in California.

Canyons with U-shaped flat bottoms have been found off rivers supplying large quantities of sediments, such as the Ganges River. Continental shelves also show numerous gullies, probably formed by loose sediments moving down the slope.

At the base of the steep continental slope there may be a gentle slope formed by the accumulation of sediment. This portion of the sea floor is the **continental rise;** it is the product of deposit by turbidity currents, underwater landslides, and any other processes that carry sands, muds, and silt down the continental slope. The continental rise is a conspicuous feature in the Atlantic and Indian oceans and around the Antarctic continent. Few continental rises occur in the Pacific Ocean, where trenches are often located at the base of the continental slope. Refer back to figure 2.4 to see the relationship of the continental rise to the continental slope.

The Ocean Basin Floor

The true oceanic features of the sea floor occur seaward of the continental margin. The ocean basin floor covers more of the earth's surface (30%) than the continents (29%). In many places the ocean basin floor is a flat plain, flatter than any plain on land, extending seaward

Figure 2.6
Sand fall in San Lucas submarine canyon.

2 meters

Figure 2.7
An idealized portion of ocean basin floor with abyssal hills, a guyot, and a seamount on the abyssal plain. Seamounts and guyots are known to be volcanic in origin.

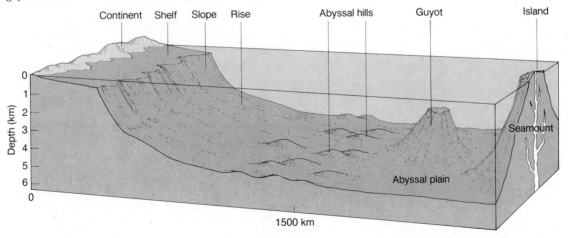

from the base of the continental slope. It is called the **abyssal plain.** The abyssal plains are caused by the sediment deposits of the turbidity currents and the continual fall of sediments from above, covering the irregular topography of the sea floor. Scattered across the sea floor in all the oceans are **abyssal hills,** less than 1000 m high, and steep-sided **seamounts,** rising abruptly and sometimes piercing the surface to be seen as islands.

These features are shown in figure 2.7. Abyssal hills are probably the earth's most common feature. They cover about 50% of the Atlantic sea floor and about 80% of the Pacific floor; they are also abundant in the Indian Ocean. Flat-topped seamounts, known as **guyots,** are found most often in the Pacific Ocean; a guyot is also shown in figure 2.7. These guyots, or table mounts, are 1000 to 1700 m (3280–5576 ft) below the surface, with many at the 1300 m (4264 ft) depth. Many guyots show the remains of shallow marine coral reefs and the evidence of wave erosion at their summits, indicating that at one time they were surface features. Their flat tops are the result of wind and wave erosion; they have since subsided owing to their weight, the accumulated sediment load bearing down on the oceanic crust, and crustal motion. Also as subsidence occurred, the sea level rose due to the melting of land ice. Seamounts are known to be underwater volcanoes; most abyssal hills are probably also volcanic, while others have been formed by other movements of the sea floor.

In the warm waters of the Atlantic, Pacific, and Indian oceans coral reefs and coral islands are formed in association with seamounts. Reef-building coral is a warm-water animal colony that requires a place of attachment and grows in intimate association with a single-celled plant. It is therefore confined to sunlit, shallow, tropic waters. When a seamount pierces the sea surface to form an island, it provides a base on which the coral can grow. The coral grows to form a **fringing reef** around the island. If the seamount sinks or subsides so slowly that the coral continues to grow upward at a rate that is not exceeded by the rising water, a **barrier reef** with a lagoon between the reef and the island is formed. If the process continues, eventually the seamount disappears below the surface and the coral reef is left as a ring, or **atoll.** This process is illustrated in figure 2.8.

These steps required to form an atoll were suggested by Charles Darwin, based on the observations he made during the voyage of the *Beagle* from 1831 to 1836. Darwin's ideas have been proved to be substantially correct by more recent expeditions that drilled through the debris on the lagoon floor and found the basalt peak of the seamount that once protruded above the sea surface.

The Ridges and Rises

The most remarkable features of the ocean basin floor are the midocean **ridge and rise systems.** This series of great continuous underwater volcanic mountain ranges stretches for 65,000 km (40,000 mi) around the world and runs through every ocean. See figure 2.9 for a bird's-eye view of the system. These ridge systems are about 1000 km (600 mi) wide and 1000–2000 m (3500–7000 ft) high. If the slopes of these mountain ranges are steep they are referred to as ridges (such as the Mid-Atlantic Ridge and the Mid-Indian Ridge); if the slopes are more gentle they are called rises (such as the East Pacific Rise). Lengthwise along some portions of the system's crest is a **rift valley,** 15–50 km (9–30 mi) wide and 500–1500 m (1500–5000 ft) deep. The rift valley is

Figure 2.8

Types of coral reefs and the steps in the formation of a coral atoll are shown in profile: (a) fringing reef; (b) barrier reef; and (c) atoll reef.

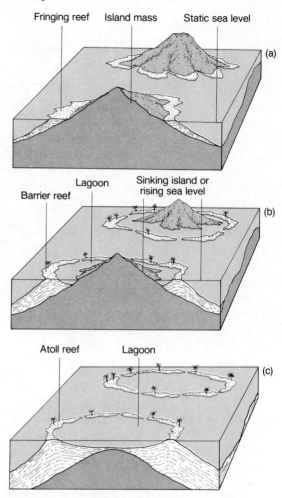

is volcanically very active and bordered by rugged rift mountains. **Fracture zones** with steep sides run perpendicular to the ridges and rises, displacing sections of the midocean ridges. These features are seen in the physiographic chart of the sea floor in figure 2.12.

The midocean ridges and rises separate the ocean basins into a series of sub-basins, which are shown in figure 2.10 and listed in table 2.2. The deep waters within these smaller basins are isolated from each other by the height of the ridges separating them. A low ridge allows good communication of deep water between basins, while a high, continuous ridge effectively isolates deep water in one basin from that in another. For example, deep water in the Angola Basin located in the bight of the west coast of Africa is cut off from the Brazil Basin to the west by the Mid-Atlantic Ridge and from the Cape Basin to the south by the Walvis Ridge.

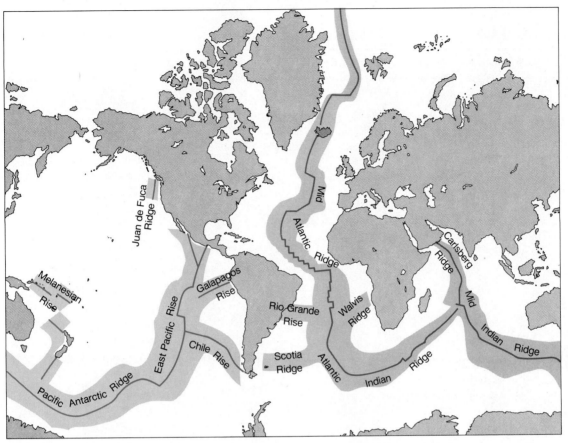

Figure 2.9
The midocean ridge and rise system with its central rift valley.

Midocean ridge   Mean position of rift valley

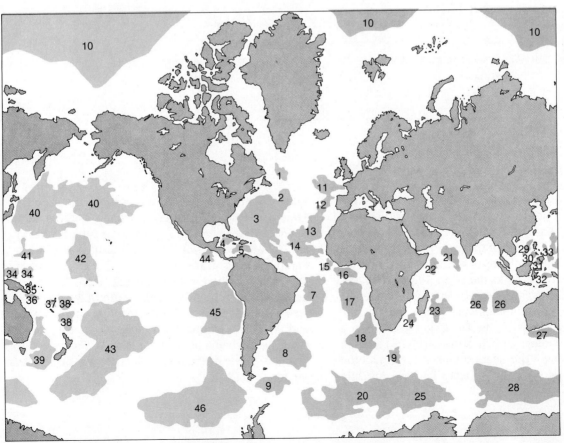

Figure 2.10
Major ocean basins of the world. Numbers are keyed to the basins listed in table 2.2.

TABLE 2.2
Major Ocean Basins of the World[1]

Ocean	Western side	Central	Eastern side
Atlantic	1. Labrador Basin		10. North Polar Basin
	2. Newfoundland Basin		11. West Europe Basin
	3. North American Basin		12. Iberia Basin
	4. Western Caribbean Basin		13. Canaries Basin
	5. Eastern Caribbean Basin		14. Cape Verde Basin
	6. Guiana Basin		15. Sierra Leone Basin
	7. Brazil Basin		16. Guinea Basin
	8. Argentina Basin		17. Angola Basin
	9. South Antilles Basin		18. Cape Basin
			19. Agulhas Basin
			20. Atlantic-Indian Antarctic Basin
Indian	21. Arabian Basin		26. India-Australia Basin
	22. Somali Basin		27. South Australia Basin
	23. Mascarenes Basin		28. Eastern Indian Antarctic Basin
	24. Madagascar Basin		29. South China Basin
	25. Atlantic-Indian Antarctic Basin		30. Sulu Basin
			31. Celebes Basin
			32. Banda Basin
Pacific	33. Philippines Basin	40. North Pacific Basin	44. Guatemala Basin
	34. Caroline Basin	41. Mariana Basin	45. Peru Basin
	35. Solomon Basin	42. Central Pacific Basin	46. Pacific Antarctic Basin
	36. Coral Basin	43. South Pacific Basin	
	37. New Hebrides Basin		
	38. Fiji Basin		
	39. East Australia Basin		

[1]The numbers are keyed to areas marked in figure 2.10.

The Trenches

Just as the Mid-Atlantic Ridge is the most overwhelming feature of the floor of the Atlantic Ocean, the narrow, steep-sided, deep-ocean **trenches** characterize the Pacific (see fig. 2.11). Some of these trenches are located on the seaward side of chains of volcanic islands known as **island arcs.** For example, the Japan-Kuril Trench, the Aleutian Trench, the Philippine Trench, and the deepest of all ocean trenches, the Mariana Trench, are all associated with island arc systems. The Challenger Deep, a portion of the Mariana Trench, has a depth of 11,020 m (6.84 mi), making it the deepest known spot in all the oceans. The longest of the trenches is the Peru-Chile Trench, stretching 5900 km (3664 mi) down the west side of South America. To the north of the Peru-Chile Trench, the Middle America Trench borders Central America. The Peru-Chile and Middle America trenches are not associated with volcanic islands, but are bordered by land volcanoes. In the Indian Ocean the great Java Trench runs for 4500 km (2794 mi) along Indonesia, while in the Atlantic there are only two comparatively short trenches: the Puerto Rico-Cayman Trench and the South Sandwich Trench, which are also associated with chains of volcanic islands. To view the bathymetry of the ocean floor as it is known to exist today, see figure 2.12.

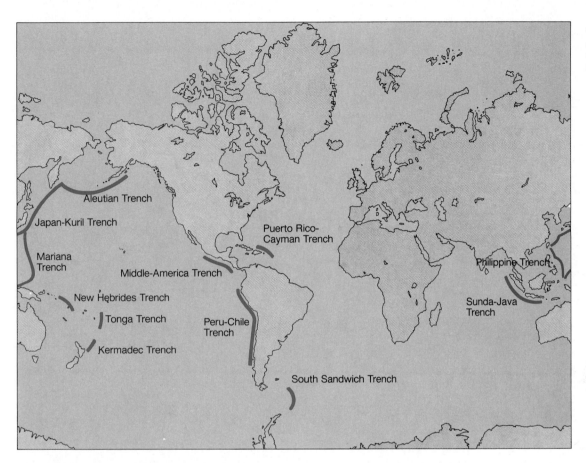

Figure 2.11
Major ocean trenches of the world.

Figure 2.12
A physiographic chart of the world's oceans.

2.2 *Measuring the Depths*

The discovery of the deep ocean's bathymetric features and the accumulation of accurate depth measurements in the deep sea are both accomplishments of recent technology. Early ocean explorers were not concerned with knowing how deep the ocean was; they were interested in how shallow the water was in the path of their ships. There were no practical reasons to measure the great depths; only the curiosity of the scientist required satisfaction.

Early mariners made their **soundings,** or depth measurements, using a hemp line or rope marked in equal distances (usually **fathoms,** which was the length between a person's fully outstretched hands, standardized at six feet) with a greased lead weight at one end. The depth was measured by the amount of line required for the weight to touch bottom. The change in tension when the weight touched allowed the sailor to sense bottom, and the particles from the bottom adhering to the grease confirmed the contact and brought a bottom sample to the surface. This method was quite satisfactory in shallow water, and the experienced captain used the properties of the bottom sample to aid in navigation, particularly at night or in heavy fog. In deep water, however, the weight of the hemp line was so great that it was difficult to sense when the lead weight touched the bottom. The sediments adhering to the grease could still confirm a touch but there was no way of knowing how much slack line lay on the bottom. For this reason the deeper areas measured by this technique were often thought to be greater than their real depth.

Later, piano wire with a cannonball attached to one end was used. The heavy weight of the ball compared to the weight of the wire made it easier to sense the bottom, but the time (eight to ten hours) to winch the wire out and in and the effort consumed for each measurement were so great that by 1895 only about 7000 depth measurements had been made in water greater than 2000 m and only 550 measurements had been made of depths greater than 9000 m over all the world's oceans.

It was not until the 1920s, when acoustic sounding equipment was invented, that deep-sea depth measurements became routine. The **echo sounder,** or **depth recorder,** which measures the time required for a sound pulse to leave the surface vessel, reflect off the bottom, and return, allows continuous measurements to be made easily and quickly when the ship is under way. The behavior of sound in seawater and its uses as an oceanographic tool are discussed in chapter 4. A trace from a depth recorder is shown in figure 2.13.

In 1925, the German *Meteor* expedition made the first large-scale use of an echo sounder on a deep-sea

Figure 2.13
Depth recorder trace. A sound pulse reflected from the ocean floor traces a depth profile as the ship sails a steady course.

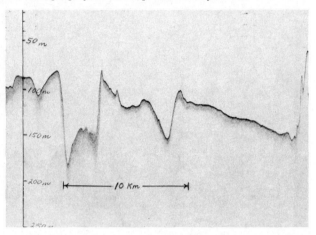

Figure 2.14
Underwater camera system. The camera and lights in waterproof pressure cases are mounted inside the steel frame. The system is towed slowly behind the ship. A sound source also mounted in the frame sends out a sound pulse. Sound pulses and the echoes from the sea floor are received by the ship and used to maintain the camera's position 5 meters above the sea floor.

oceanographic research cruise. After this expedition, depth measurements gradually accumulated at an ever-increasing rate. As the acoustic equipment improved and was used more frequently, knowledge of the bathymetry of the ocean floor expanded and improved, culminating in the 1950s with the detailed mapping of the midocean ridge system. Today, data from precision depth recorders, underwater photography, and television, and direct observation by submersibles are used together to produce even more detailed knowledge of the bathymetry of the sea floor (see fig. 2.14).

2.3 Sediments

The margins of the continents and the ocean basin floors receive a continuous supply of particles from many sources. Whether these particles have their origin in living organisms, the land, the atmosphere, or the sea itself, they are called sediment. Heavy deposits of sediment are found along the continental margins, while the deep sea floor receives a constant but slow accumulation of sediment.

Oceanographers study the rate at which sediments accumulate, the distribution of sediments over the sea bottom, their sources, their chemistry, and the history they record in layer after layer as they slowly but continuously accumulate on top of the oceanic crust. In order to describe and catalogue the sediments found on the ocean floor, geological oceanographers have classified the sediments by their source, place of deposit, chemistry, particle size, thickness of deposit, and color.

Sediment Sources

Classification by source relates the sediments to their origins from rocks, water, air, and earth's living things. Sediments derived from rock are **lithogenous sediments.** Rock at or above sea level is weathered by wind, water, and continual freezing and thawing. The particles produced are transported from their place of origin by water, wind, gravity, and ice. Most of the lithogenous sediments are deposited on the continental margins; smaller amounts are carried to the deep sea. Dust from arid areas of the continents and ash from active volcanoes are carried to sea by the winds and are additional sources of lithogenous particles.

Sediments that were derived from living organisms are called **biogenous sediments** and may include shell and coral fragments as well as the hard skeletal parts of single-celled plants and animals that live in the surface waters. Biogenous sediments are discussed in more detail later in this chapter.

In certain areas sediments are derived from the seawater itself by chemical processes. These sediments are called **hydrogenous sediments** and include **carbonates** (limestone-type deposits), **phosphorites** (phosphorus in phosphate form in crusts and nodules), and **manganese nodules.** The manganese nodules are deposits of manganese, iron, copper, cobalt, and nickel in the form of variously shaped nodules scattered across the ocean floor and lying on top of the other sediments, as shown in figure 2.15. Nodules develop very slowly (1–10 mm per million years), accumulating layer upon layer, often around a hard skeletal piece such as a shark's tooth or the ear bone of a whale. If there is a rapid accumulation of other sediments they either do not form or are buried quickly by the constant rain of particles

Figure 2.15

Manganese nodules on the floor of the Pacific off Hawaii. The suspended object is a compass with a vane to show the direction of the current.

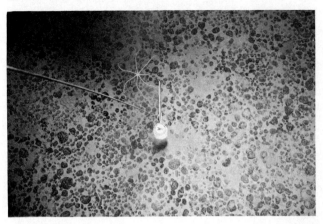

from above. Therefore they form in deep water or in areas of rapid bottom currents.

Salt deposits occur when a high rate of evaporation removes most of the water and leaves a very salty brine in shallow areas. Chemical reactions occur in the brine and salts are deposited on the bottom. In such a process carbonate salts are formed first, followed by sulfate salts, and then rock salt.

Outer space is an intriguing source of ocean sediment. Particles from space constantly bombard the earth's surface; the particles are very small, stay in suspension for a very long time, and most are thought to dissolve as they drift to the sea floor. These iron-rich **cosmogenous sediments** are found in small amounts in all oceans scattered on the surface of the other sediments. The pattern of related cosmic materials indicates the direction of the particle shower that supplied them. The particles become very hot as they pass through the earth's atmosphere and partially melt; this melting gives the particles a characteristic rounded or teardrop shape. Cosmic bodies can disintegrate and melt surface materials as they strike the earth. Their impact can cause a splash of melted particles that spray outward and produce splash-form **tektites** (see fig. 2.16). Microtektites are found on the ocean floor and on land.

Biogenous Oozes

Deep-sea biogenous sediments are composed almost entirely of the hard remains of single-celled organisms (fig. 2.17). These remains are chemically either calcareous (calcium carbonate or shell material) or siliceous (clear, hard, glasslike silicate). If the sediment of an area is more than 30% fine biogenous sediment by weight, then it is known as an **ooze**—a **calcareous ooze** or a **siliceous ooze.**

Figure 2.16
Tektites collected from Southeast Asia.
Courtesy Virgil E. Barnes.

Figure 2.17
Photomicrographs of biogenous sediments: (a) radiolarians; (b) diatoms and coccolithophores. The small disks are detached coccoliths.

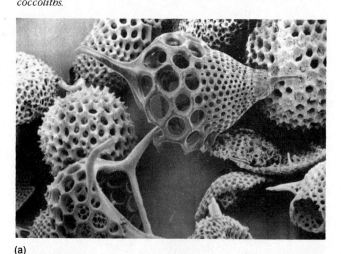

(a)

(b)

The distribution of calcareous and siliceous remains of organisms in the sediments is related to the supply of organisms, the dissolving capabilities of the water through which the organisms fall, the dissolution rates of material making up the organisms' mineral remains, and dilution with other minerals. In the Atlantic Ocean, the waters from the surface down to a depth of about 4000 m (12,400 ft) are saturated with respect to calcite, a form of calcium carbonate, and dissolution of calcareous materials is retarded. Waters below this depth are undersaturated and dissolution occurs. In the North Pacific Ocean, water saturation with calcite only occurs from the surface to a few hundred meters depth. This means that calcite-type calcareous remains dissolve very little as they fall through the upper 4000 m (13,100 ft) of the Atlantic and the upper 500 m (1600 ft) of the northern Pacific. The South Pacific has water saturated with calcite down to about 2500 m. Below these depths the higher carbon dioxide content of aged water makes the water more acidic and capable of dissolving the easily broken down calcite particles.

Calcite solubility has an unusual temperature dependence: calcite is more soluble in cold water than in warm water. This factor causes the surface water at higher latitudes, where surface water temperatures are low, to have an increased ability to dissolve calcite carbonates, further limiting the accumulation of carbonates in the underlying sediments. This cold surface water effect coupled with the water's changing carbonate-dissolving ability with depth and age produces the observed distribution of carbonate sediments. Calcareous deposits are found at temperate and tropical latitudes in such shallower areas of the sea floor as the Carribean Sea, on elevated ridge systems, and in coastal regions.

The waters of both oceans are undersaturated for biogenic silica over their entire depth. However, the siliceous remains of marine organisms are resistant to dis-

solution and break down slowly, even though the entire water column through which they fall is undersaturated.

The siliceous remains of organisms are found at all depths because of their slow dissolution rates at either warm or cold temperatures. In regions where there is a high supply rate of siliceous biogenic remains, and where calcareous remains are more readily dissolved, the siliceous content of the accumulating sediments is high. The diatomaceous sediments surrounding Antarctica are an example.

Siliceous oozes have one principal source in the temperate latitudes: single-celled plants called **diatoms** that produce the **diatomaceous ooze** found around the Antarctic and in a band across the North Pacific. (See fig. 2.17b for a picture of diatom skeletons.) Because the diatoms are plants, they require sunlight and fertilizers, or **nutrients,** for growth. The sunlight is available at the ocean's surface; the nutrients are pro-

Figure 2.18

Distribution of the principal sediment types on the deep sea floor. Sediments are usually a mixture but are named for their major component.

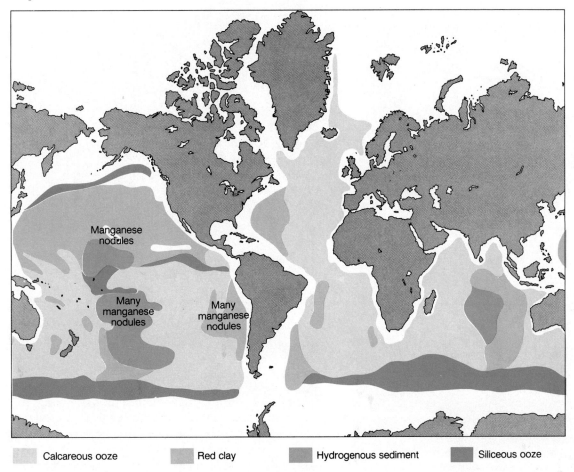

Manganese nodules

Many manganese nodules

Many manganese nodules

| | Calcareous ooze | | Red clay | | Hydrogenous sediment | | Siliceous ooze |

duced by the decomposition of all plant and animal life in the ocean, and these nutrients are liberated in the deeper water as the decomposition takes place. Only at certain locations are these nutrients returned to the surface by a large-scale upward flow of deeper water. Where this upward flow occurs, the sunlight combines with the nutrients to provide the conditions needed for plant production. Production is discussed in chapter 13. Because diatoms reproduce rapidly in cold water, large populations are found in the areas that combine light, nutrients, and suitable temperatures: the water of the North Pacific and Antarctic oceans. See the chart in figure 2.18 to trace the distribution of diatomaceous oozes.

Siliceous sediments are also found in equatorial latitudes (again, see fig. 2.18). These sediments are not produced by diatoms because the water at these latitudes is too warm. Here, the principal source is a single-celled, amoeba-like animal, the **radiolarian,** which produces a siliceous outer shell that is often covered with long spines. These delicate, glasslike external cov-

erings may accumulate in sufficient quantity to produce **radiolarian ooze** in tropic areas. Figure 2.17a shows radiolarian sediments at high magnification.

Calcareous oozes are formed at the shallower oceanic depths from the remains of other small animals and plants, with worldwide distribution. Look once more at figure 2.18 to locate the deposits of calcareous oozes that are often found on the major ridge systems. Both the microscopic **foraminifera,** another single-celled relative of the amoeba, and the **pteropods,** small, graceful, swimming snails, contribute their shells to these sediments. Minute, single-celled plants, **coccolithophores,** covered with calcareous plates called coccoliths, also play a role in the formation of these bottom deposits. Each of these is named for its principal constituent: **foraminifera ooze, pteropod ooze, and coccolithophore ooze.**

Figure 2.19
Classification of sediments by location of deposit. The distribution pattern is in part controlled by the sediments' proximity to source and rate of supply.

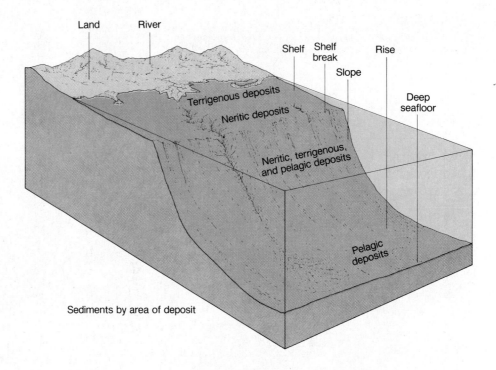

Red Clay

On the deep ocean floor, far away from the land and under water that holds little marine life, a very fine lithogenous material accumulates very slowly. This fine rock powder may remain suspended in the water for many years; it is believed to have been blown out to sea by wind or to have been swept out of the atmosphere by rain. It is known variously as **red clay, brown clay,** or **brown mud.** The red-brown color is due to the oxidizing (or rusting) of the iron present in the sediment. Although traces of red clay are found throughout the oceans, it only becomes a substantial component of the sediment in areas where sediments from other sources do not mask its presence.

Patterns of Deposit

Ocean scientists classify sediments by the area in which they are deposited as well as by their source. **Terrigenous sediments** are found close to their land source (for example, silt, sand, gravel, wood chips, and sewage sludge). **Neritic sediments,** or coastal sediments, are found deposited under the shallow waters of the continental shelf, and **pelagic sediments** are deep-sea sediments that have been deposited remote from the direct influence of land. Areas of sediment deposits are shown in figure 2.19.

The patterns formed by the sediments on the sea floor are related to the distance from their source and to the processes that control the rates at which they are produced, transported, and deposited. Of the mass of

marine sediments, 75% are terrigenous; the majority of these terrigenous sediments are deposited on the continental margins. The terrigenous sediments of coastal regions are primarily lithogenous, supplied by rivers and wave erosion along the coasts. Coarse sediments are concentrated close to their sources in high energy environments. Beaches with swift currents and breaking waves are an example. The waves and currents can move quite large rock particles in the shore zone, but these larger particles settle out quickly. Finer particles are held in suspension and are carried farther away from their source. This results in a gradation by particle size: coarse particles close to shore and to their source, with finer and finer particles predominating as the distance from the source increases.

Finer sediments are deposited in low energy environments, offshore away from the currents and waves or in quiet bays and estuaries. In higher latitudes, deposits of rock and gravel carried along by glaciers are found in coastal environments, while in equatorial latitudes fine sediments predominate and are considered to be products of large rivers, heavy rainfall, and loose surface soils.

At the present time, most of the land-derived sediments are accumulating off the world's river mouths and in estuaries. Estuaries and river deltas serve as sediment traps, preventing these terrigenous sediments from reaching the deep-sea floor in such places as the Chesapeake and Delaware bay systems along the North Atlantic coast, in the Georgia Strait of British Columbia, and in California's San Francisco Bay along the West Coast. If sediments are supplied to a delta faster than

(a)

(b)

(c)

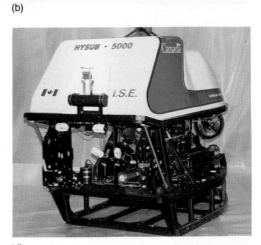

(d)

Colorplate 3

Submersibles and Remotely Operated Vehicles (ROVs). These vessels are used in scientific research as well as for industrial purposes. (a) The BRUKER Sea Horse II, used for research and inspection of underwater facilities, is the largest German-built submersible. It is 15.5 m long and operates to a 500 m depth. (b) The BRUKER diving bell, Flying Bell, is tethered to a mother ship at the surface but can be maneuvered by the operator within the bell. The crew can transfer from the bell to the submersible while underwater. (c) The AMETEK ASD/620 and (d) the ISE HYSUB-5000 are ROVs. They are maneuvered from the surface in response to the view from their video cameras. Both have robotic capabilities with tools and samplers.

Colorplate 4
Physiographic map of the world's ocean floor. (Courtesy of Marie Tharp.)

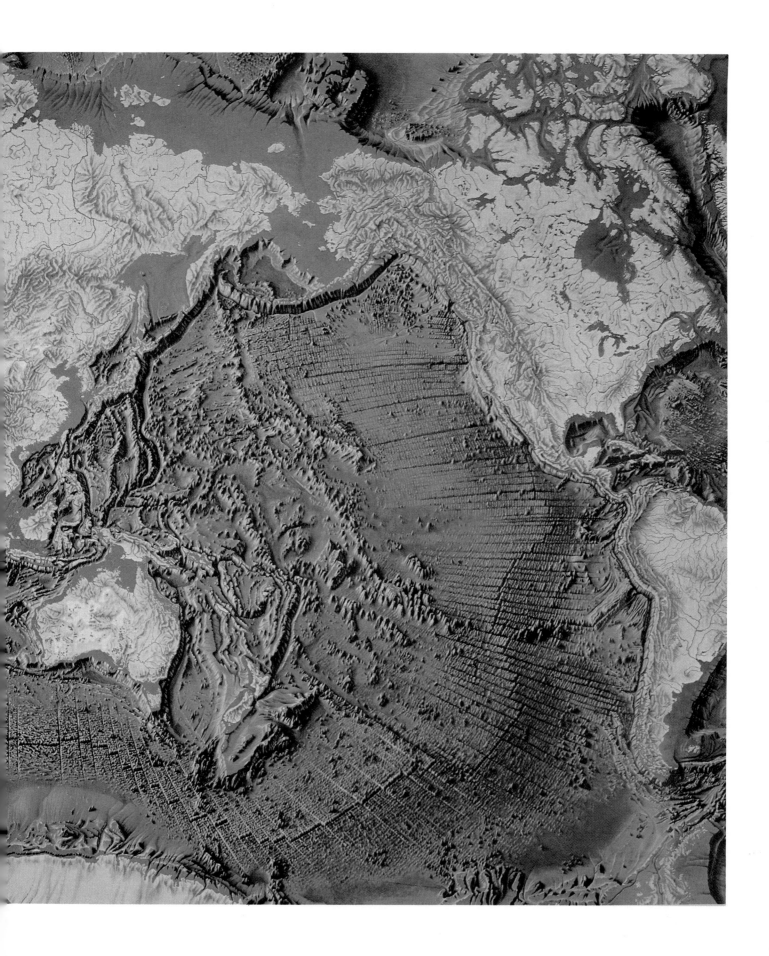

Colorplate 5
(a) *The submersible* Alvin *prior to being lowered to the sea surface for a 1984 dive along the Juan de Fuca Ridge in the northeastern Pacific.* (b) *Cameras on board* Alvin *photographed a black smoker with its cloud of metallic sulfide. Identification data is added to the photo as the picture is taken.*
(c) *Preparing a piston corer for use by loading it with weights.*
(d) *A box corer is used to obtain large, undisturbed surface samples from the sea floor.*

they can be retained, the sediments will move across the shelf into the deeper water environments. This is currently the case with the sediments of the Mississippi River (see colorplate 9). Much of the total thickness of sediments on the outer continental shelf is sediment laid down when the sea level was lower during the ice ages; for example, Georges Bank southeast of Cape Cod. Little is currently being added to these outer regions of the continental shelf.

The rapid accumulation of sediments on the continental shelf results in large, unstable, steep-sided deposits that may slump, sending a flow of terrigenous sediment moving rapidly down the continental slope in a turbidity current (see the previous discussion of turbidity currents). Turbidity currents move coarse materials of land origin a long way out to sea; in doing so they distort the general deep-sea sediment pattern. Near shore, the spring flooding of the rivers alternates with periods of low discharge from the rivers in summer and fall. The floods bring large quantities of sediment to the coastal waters, and a record of the contribution of this flooding is seen in the layering of the sediments. Sudden contributions of sediment material, originating from the collapse of a cliff or the eruption of a volcano, are seen in the sediment pattern as a specific addition of large quantities of sand or ash.

Oceanic sediments are also laid down in layers. The layers are recognized by changes in color, particle size, and kinds of particles (see fig. 2.20). Layering occurs if the sediment type varies over time or if the rate at which a sediment type is supplied varies. Seasonal variations and the pattern of long and short growing seasons in the surface waters can also be determined from the properties and thicknesses of the layers of biogenous material. Over long periods of geologic time, climatic changes such as the ice ages have altered the biological populations that produce sediment and have left their record in the sediment layers.

Along the continental margins, biogenous sediments are diluted by the large amounts of lithogenous sediment washing from the land. When marine life is abundant in coastal areas, biogenous sediments may be formed from shells and corals. In the more homogeneous environment of the deep sea, biogenous sediments make up the majority of the pelagic deposits. There is little dilution with terrigenous materials and few environmental changes disturb the bottom deposits. Seafloor conditions may stay relatively unchanged for long periods of time in contrast to the cycles of climate change, including ice ages, which cause processes to fluctuate along the continental margins. Calcareous oozes are found where the production of organisms is high, dilution by other sediments small, and depths are less than 4000 m (13,000 ft). See the areas including the midocean ridges and the warmer shal-

Figure 2.20
Layered marine sediments.

lower areas of the South Pacific (refer back to fig. 2.18). Siliceous oozes cover the deep-sea floor beneath the colder surface waters of 50–60°N and S latitude, and in equatorial regions where cold deeper water is brought to the surface by vertical circulation processes. Deep basin areas of the Pacific have extensive deposits of red clay (again, refer back to fig. 2.18).

Large rock particles of land origin are also moved out to sea by a process known as **rafting.** Glaciers carry sand, gravel, and rocks with them as they cut through the earth's crust. If the glacier reaches the sea, parts of it break off and fall into the water as icebergs. The icebergs are carried away from land by the currents and winds, and as the ice melts, the rocks and gravel that were frozen in the ice sink to the seafloor, far from their original source. Sometimes large, brown seaweeds known as kelp, which grow attached to rocks in coastal areas, are dislodged by storm waves. The kelp may have enough buoyancy to float away, carrying with it the rock to which it is attached. When the plant dies or sinks, the rock is deposited on the ocean floor at some distance from its origin. The deposit of larger rocks by this process of rafting is an irregular and isolated process.

In some parts of the world the wind is an effective agent for the movement of lithogenous materials out to sea. Figure 2.21 indicates the frequency with which winds carry airborne dust, or haze, out to sea. Winds blowing offshore from the Sahara Desert or other arid regions transfer sand particles directly from land to sea, sometimes 1000 km (600 mi) or more offshore. A similar process can occur between sand dunes and coastal waters.

Particle Size

Scientists also classify the sediments by the size of their individual particles (see table 2.3). Note that such familiar terms as gravel, pebble, and clay have specific size ranges. The geological oceanographer determines the percentage by weight of sediment in each size category in a sediment sample. This process effectively identifies the sediment and allows comparisons to be made with deposits at other locations; it also shows any changes in

Figure 2.21
Frequency of haze as a result of airborne dust during the northern hemisphere's (a) winter and (b) summer. Values are given in percentages of total observations.

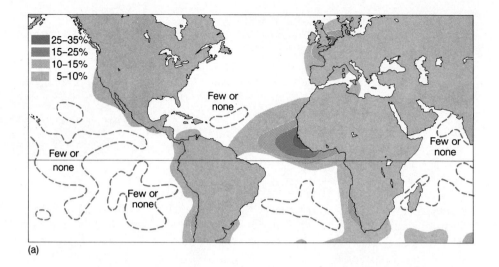

25–35%
15–25%
10–15%
5–10%

(a)

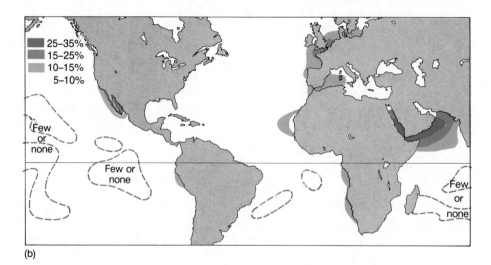

25–35%
15–25%
10–15%
5–10%

(b)

Classification of Sediments by Particle Size

Descriptive name		Diameter ranges (mm)
gravel	boulder	> 256
	cobble	64–256
	pebble	4–64
	granule	2–4
sand	very coarse sand	1–2
	coarse sand	0.5–1
	medium sand	0.25–0.5
	fine sand	0.125–0.25
	very fine sand	0.0625–0.125
mud	silt	0.0039–0.0625
	clay	< 0.0039

the sediments at a location sampled over a period of time. Figure 2.22 shows how sediments appear when they are mixtures of different particle sizes.

If the source material of a sediment contains particles of mixed sizes, the resulting deposit is said to be poorly sorted (fig. 2.22b). But if the smaller and finer particles are removed from the source material by turbulence and current flow before it is laid down, then the sediment becomes a well-sorted, coarse deposit. Figure 2.22a shows the fewer size categories in the graded series that are typical of a well-sorted sample. The washed-out finer material may then be deposited as a well-sorted fine deposit in another location. Size frequency graphs of the type shown in figure 2.22 are made from deposit samples and are compared with graphs from sediment sources. Such comparisons indicate to geologists the net result of the processes that acted on the sediment before they were deposited on the sea floor.

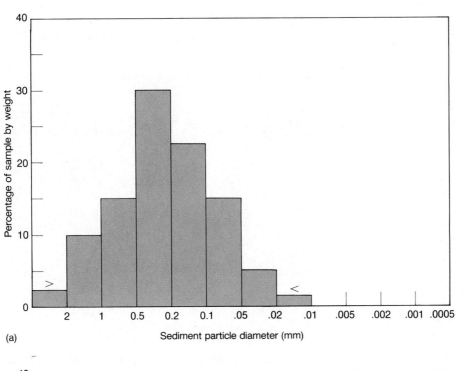

(a)

Figure 2.22
(a) A well-sorted sample. One size range predominates in a limited distribution of sizes. (b) A poorly sorted sample. Particles fall into a wide variety of size ranges in approximately equal amounts.

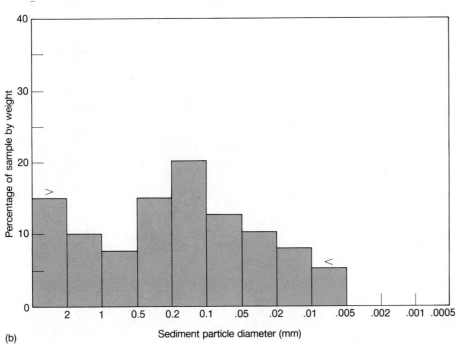

(b)

Rates of Deposit

The rates at which sediments accumulate vary enormously, owing to the variability in the processes already discussed. An average accumulation value for the deep oceans can be given as 0.5–1.0 cm (.2–.4 in)/1000 years, while at the ocean margins the rates are considerably greater. In river estuaries the rate may be more than 800,000 cm (315,000 in)/1000 years (800 cm (315 in)/

year); the rivers of Asia, such as the Ganges, the Yangtse, the Yellow, and the Brahmaputra, contribute more than one-quarter of the world's land-derived marine sediments each year. In quiet bays, the rate may be 500 cm (197 in)/1000 years, while on the continental shelves and slopes values of 10–40 cm (3.9–15.7 in)/1000 years are appropriate, with the flat continental shelves receiving the larger amounts. It should be noted that many of the sediments found covering the continental shelves away from river mouths are sediments laid down by processes that no longer exist at that location. Such sediments are called **relict sediments** and represent conditions that existed 3000 to 7000 years ago or longer, when the sea level was lower than it is today owing to the accumulation of water in ice caps and glaciers.

Although deep-sea sedimentation rates are excessively slow, there has been plenty of time during geologic history to accumulate the average deep-sea sediment thickness of approximately 500 to 600 m (1640–1968 ft). At a rate of 0.5 cm(.2 in)/1000 years, it takes only 100 million years to accumulate 500 m (1640 ft) of sediment, and the oldest seafloor crustal rocks are known to have existed for twice as long, or 200 million years. Because very small particles sink extremely slowly (for example, a particle 0.001 mm (.0004 in) in diameter will take about 100 years to sink 3000 m (9840 ft)), scientists have puzzled over the close correlation in distribution between the particles found in the surface waters and those found on the sea floor. If the particles truly sank at such a slow rate, they would be carried very long distances from their source before reaching the bottom. Therefore, some mechanisms must be working to aggregate the tiny particles into larger particles, thus increasing the sinking rate, in order to explain how the particle distribution on the ocean floor could be so similar to that found in the overlying surface waters. Scientists have observed that small particles often attract each other owing to their electrical charges. This attraction forms larger particles that sink more rapidly. This process is important in the formation of the abundant sediment deposits in river deltas. Also, the small organisms at the surface ingest these tiny particles, which are then expelled with the organism's wastes as fecal pellets. The fecal pellets themselves may be microscopic, but they are large enough to reach the seafloor in ten to fifteen days. Once the pellets are deposited, the breaking down of the organic portion of the pellets liberates the small particles.

Sampling Methods

In order to analyze sediments the geological oceanographer must have an actual bottom sample to examine. A variety of devices have been developed to take a sample from the seafloor and return it to the laboratory for analysis. **Dredges** are net or wire baskets that are dragged across a bottom to collect loose bulk material, surface rocks, and shells in a somewhat haphazard manner (see fig. 2.23). **Grab samplers** are hinged devices that are spring- or weight-loaded to snap shut when the sampler strikes the bottom. See figure 2.24 for an example of this type of device. Grab samplers sample surface sediments from a fixed area of the sea floor from a single known location.

Figure 2.23

A geological dredge with heavy-duty chain bag. The dredge picks up loose rock and breaks off fragments from rock outcroppings on the sea floor.

Figure 2.24

Grab samplers: vanveen (left) and orange peel (right), both in open positions. Grabs take sediment samples.

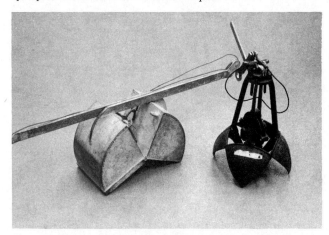

A **corer** is essentially a hollow pipe with a sharp cutting end. The free-falling pipe is forced down into the sediments by its weight or, for longer cores, by a piston device that enables water pressure to help drive the core barrel into the sediment; coring devices are shown in figure 2.25. The result is a cylinder of mud, usually 1 to 20 m (3.3–65.6 ft) long, that contains undisturbed sediment layers (refer back to fig. 2.20). Box corers are used when a large and nearly undisturbed sample of surface sediment is needed (see colorplate 5). These corers drive a rectangular metal box into the sediment, with doors that close over the bottom before the sample is retrieved. Longer cores may be obtained by drilling. For a discussion of the highly sophisticated drilling techniques used by the research vessel, *Joides Resolution,* see chapter 3.

Figure 2.25

(a) A simple gravity corer in use. (b) The Phleger corer is a free-fall gravity corer. The weights help to drive the core barrel into the soft sediments. Inside the core is a plastic liner. The sediment core is removed from the corer by removing the plastic tube, which is capped to form a storage container for the core. (c) A sketch of a piston corer in operation. The corer is allowed to fall freely to the sea bottom. The action of the piston moving up the core barrel owing to the tension on the cable then allows water pressure to force the core barrel into the sediments.

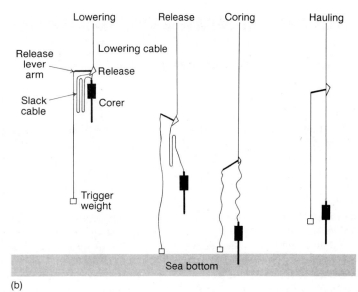

(b)

(a)

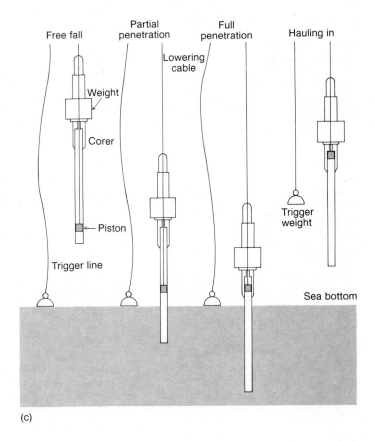

(c)

Submersibles and Remote Operating Vehicles

*I*t is said that Alexander the Great (356–323 B.C.) had himself lowered into the sea in a container equipped with an air hose to satisfy his curiosity about the world below the sea surface. Since then, the human desire for adventure and knowledge has continued to drive people to build environments in which they can explore the ocean deeps.

William Beebe was a pioneer of underwater exploration, making a record descent in 1934 of 918 m (3028 ft) in a steel sphere supplied with air and power by its mother ship and equipped with portholes and lights. The first free, self-contained deep-diving craft was the bathyscaphe, invented by Auguste Piccard (1884–1962). This vehicle had a spherical compartment for two persons suspended beneath a huge float filled with gasoline. In 1960 Auguste's son Jacques and U.S. Navy Lieutenant Don Walsh, in the bathyscaphe *Trieste,* descended 10,800 m (35,800 ft), to the bottom of the Challenger Deep.

The manned deep-sea submersibles of today are sophisticated, mobile vessels equipped with the latest in photographic equipment and robotic sampling arms. The best-known U.S. submersible is the three-person *Alvin,* which completes more than 150 dives per year, with an average dive-depth in 1985 of 2413 m (8000 ft) in the Pacific Ocean and 2559 m (8500 ft) in the Atlantic Ocean. It was in the *Alvin* that Robert Ballard first directly viewed the hydrothermal vents, with their exciting systems of animals in 1977 (see chapters 3 and 16). It was the *Alvin,* again used by Ballard, in

which he first viewed the *Titanic* in 1985.

The U.S. Navy's *Sea Cliff* is a research, recovery, and rescue vessel certified to operate to 6100 m (20,000 ft). In 1986 *Sea Cliff* was used by the U.S. Geological Survey to conduct surveys of mineral deposits off the coasts of California and Oregon (see chapter 3).

The *Johnson Sea-Link 1* (JS-L1) and the *Johnson Sea-Link 2* (JS-L2) are four-person submersibles. Both Sea-Links participated in the space shuttle *Challenger* recovery, bringing back the first photos and pieces of *Challenger.*

Manned submersibles have also been used to photograph sharks off Bermuda, survey coral reefs at Biscayne National Park in Florida, and produce TV documentaries about ocean life. They are used commercially in offshore petroleum and engineering projects (see colorplate 3).

Easy to maneuver, carrying the latest photographic and sampling technology, the one-person microsubmersible *Deep Rover* has a depth capability of 1000 m (3300 ft) and has been used in biological research off California and in commercial work in its Canadian home waters. A sister vehicle, *Deep Flight,* will have a depth ability of 1500 m (5000 ft), and will "fly" through the water at four knots.

Manned submersibles are expensive to use and not without risk. If what is visible to the vehicle can be transmitted to the sea surface, an onboard human observer is not necessarily required. A Remotely

Operated Vehicle (ROV), controlled from the surface, can be used to carry both still and television cameras, and remotely operated robotic arms, samplers, and other devices. The need for power and information transfer through an umbilical cord to the mother ship is a disadvantage, but that disadvantage may be offset by the decreased risk factor and lower costs of construction and operation when human lives are not involved. During the exploration of the *Titanic* by *Alvin* in 1986, the *Alvin* carried with it a small ROV, *Jason Jr.,* or JJ, which was tethered to and operated from the *Alvin.* Its small size and maneuverability made it possible to explore the interior of the *Titanic* through JJ's camera eyes, something the larger, more cumbersome *Alvin* could not do.

The cost of ROVs is dropping and they are being designed with a variety of capabilities. ROVs are being used more and more in offshore petroleum work and on engineering projects, such as bridges, dams, and pipelines. Military and law enforcement agencies are interested in ROVs, and there is a growing demand for their use in routine inspection and maintenance in marine transportation. Although nothing can completely replace the on-the-spot, intelligent, and trained human mind, ROVs do offer scientists the ability to survey much larger areas of the sea floor for much longer periods of time and at much less cost than can be accomplished by manned submersibles.

2.4 *Practical Considerations: Seabed Resources*

Long ago, people began to exploit the materials of the seabed. The ancient Greeks extended their lead and zinc mines under the sea, medieval Scottish miners followed veins of coal under the Firth of Forth, and more recently, coal has been mined from undersea veins off Japan, Turkey, and Canada. As technology has developed, and as people have become concerned about the depletion of onshore mineral reserves, a new interest in seabed minerals and mining has developed. During the past twenty years a reasonable but still incomplete inventory of these minerals and materials has been taken. The current situation for a number of these materials will be summarized and reviewed in this section. Keep in mind that each potential deep-sea source is in competition with an onshore supply. Whether the seabed source will be developed depends largely on international markets, needs for strategic materials, and whether offshore production costs can compete with onshore costs.

Sand and Gravel

The largest of all superficial seafloor mining operations is for sand and gravel, used in cement and concrete for buildings, for landfills, and to construct artificial beaches. The technology and cost required to mine sand and gravel in shallow water differ very little from land operations. This is a high-bulk, low-cost material tied to the economics of transport and the distance to the market. Annual world production is approximately 112 billion metric tons; the reported potential reserve is more than 600 billion metric tons. The United Kingdom and Japan each take 20% of their total annual sand and gravel requirements from the sea floor.

Sand and gravel mining is the only significant seabed mining done by the United States at this time. It is estimated that the United States has a reserve of 450 billion tons of sand off its northeast coast, and large deposits of gravel along Georges Bank off New England and also in the area off New York City. Along the coasts of Louisiana, Texas, and Florida, calcareous shell deposits are mined for use in the lime and cement industries, as a source of calcium oxide used to remove magnesium from seawater, and, when crushed, as a gravel substitute along roads and highways.

Sands are mined as a source of calcium carbonate throughout the Bahama Islands, which have an estimated reserve of 100 billion metric tons. Coral sands are mined in Fiji, in Hawaii, and along the U.S. Gulf Coast. Other coastal sands contain iron, tin, uranium, platinum, gold, and diamonds. The "tin belt" stretches for 3000 km (1800 mi) from northern Thailand and western Malaysia to Indonesia. Here, sediments rich in tin have been dredged for hundreds of years and supply more than 1% of the world's market. Iron-rich sediments are dredged in Japan, where the reserve of iron in shallow coastal waters is estimated at 36 million tons. The Soviet Union has been working sediments containing uranium in the Black Sea since 1972. The United States, Australia, and South Africa recover platinum from some sands, and gold is found in river delta sediments along Alaska, Oregon, Chile, South Africa, and Australia. Diamonds, like gold, are found in sediments washed down the rivers in some areas of Africa and Australia. Muds bearing copper, zinc, lead, and silver also occur on the continental slopes, but they lie too deep for exploitation, considering the present demand and their market value.

Phosphorite

Phosphorite, which can be mined to produce the phosphates needed for fertilizers, is found in shallow waters as phosphorite muds and sands containing 12% to 18% phosphate, and as nodules on the continental shelf and slope. The nodules contain about 30% phosphate, and large deposits are known to exist off Florida, California, Mexico, Peru, Australia, Japan, and northwestern and southern Africa. Recently a substantial source of phosphorite appears to have been located in Onslow Bay, North Carolina. Eight beds have been located and five are thought to be economically valuable; they have been estimated to contain 3 billion metric tons of phosphate concentrates.

The world's ocean reserve of phosphorite is estimated at about 50 billion tons. Readily available land reserves are not in short supply, but most of the world's land reserves are controlled by relatively few nations. Therefore, political considerations may make these marine deposits attractive as mining ventures for some countries. No commercial phosphorite mining occurs at present in the oceans.

Sulfur

In the Gulf of Mexico, sulfur is mined by injecting high-pressure steam to melt the sulfur, which is then pumped ashore to processing plants. Today, the cheaper and easier recovery of sulfur waste from pollution-control equipment has replaced all but one of the mining operations, and it is thought that the remaining mining operation may be closed down completely by the year 2000. Sulfur is necessary for the production of sulfuric acid, which is needed in nearly all industrial processes. If sulfur must be mined again, millions of tons of sulfur reserves are known to exist in the Gulf of Mexico and the Mediterranean Sea.

Coal

Coal was produced by land vegetation in tropic climates, but changes in sea level and land geography over geologic time have caused some coal deposits to become submerged. Coal deposits under the sea floor are mined when the coal is present in sufficient quantity and quality to make the operation worthwhile. In Japan, the undersea coal deposits are reached by shafts that stretch under the sea from the land or descend from artificial islands.

Oil and Gas

Oil and gas represent 90% of mineral value presently taken from the sea. Major offshore oil fields are found in the Gulf of Mexico, the Persian Gulf, and the North Sea, and off the northern coast of Australia, the southern coast of California, and the coasts of the Arctic Ocean. Bringing the North Sea oil fields into production has required the development of massive drilling platforms and specialized equipment to withstand its heavy seas and fierce storms, and to allow drilling and development of wellheads at great depths (see fig. 2.26). The Beaufort Sea of Alaska and the Canadian Arctic are both promising areas; test wells are now being drilled by Canada. There are still many areas of the world that are relatively unexplored in the search for oil and gas, for example, the continental shelves of east Asia, south Asia, east Africa, northwest Africa, parts of South America, and Antarctica. New discoveries have been made near the Philippines and off the mouth of the Amazon River.

At present, the annual revenues from offshore gas and oil production are $80 billion, nearly equal to the total world market revenue for all other potential seabed materials combined (about $90 billion). Continental margin deposits account for nearly one-third of the world's estimated gas and oil resources. Although the cost of drilling and equipping an offshore well is three to four times greater than a similar venture on land, the large size of the deposits allows offshore ventures to compete in the marketplace. However, offshore wells in the Arctic will be several times more costly than those in the Gulf of Mexico. The gas and oil potential of the deeper areas of the sea floor is still relatively unknown, but the deeper the water in which the drilling must be done, the higher the cost. The new methods and equipment used and developed for deep-sea oceanographic drilling and research have provided the prototypes for new generations of deep-sea commercial drilling systems. Even though legal restraints, environmental concerns, and worldwide political uncertainties will continue to contribute to the slow development of offshore deposits, petroleum exploration and development will undoubtedly continue to be the main focus of ocean mining in the near future.

Figure 2.26
The Belier oil production platform off the Ivory Coast.

Manganese Nodules

Manganese nodules are hydrogenous, pelagic deposits found scattered across the world's deep-ocean floors, with particular concentrations in the red clay regions of the northeast Pacific. The nodules were discovered by the *Challenger* expedition (1873–1876), and in the last twenty-five years have been the focus of intense research and development of mining and extraction techniques by large mining corporations and multinational consortia (see fig. 2.27). The valuable mineral content varies from place to place, but the nodules in some areas contain 30% manganese, 1% copper, 1.25% nickel, and 0.25% cobalt, much higher concentrations than are usually found in land ores. Although the nodules grow at an estimated rate of 0.1 mm/1000 years, they are already present in vast quantities and an estimated 16 million additional tons accumulate each year.

Since the 1960s, large multinational consortia have spent over $600 million to locate the highest nodule concentrations and develop technologies for their col-

Figure 2.27
The Deep Sea Miner II, *a converted ore carrier, is used to make deep-ocean mining tests.*

lection. However, their expectations of rapid development have been disappointed, and in the 1980s some of these consortia have withdrawn completely while others are dormant. The primary reason for the lack of development of this industry is the presently depressed international market in metals. Another reason involves the history of ownership of the pelagic nodules (see the section in this chapter on Laws and Treaties). In some areas manganese crusts rich in cobalt are formed. These crusts have four to five times the cobalt content of the nodules. Japan and West Germany have shown an interest in exploring these deposits, but at present there are no existing techniques to break, sort, or lift these hard-rock deposits.

Sulfide Mineral Deposits

In the 1970s and 1980s deposits of minerals combined with sulfur to form sulfides of zinc, iron, copper, and possibly silver, molybdenum, lead, chromium, gold, and platinum have been found along the East Pacific Rise near the Gulf of California, the Galapagos Ridge off the west coast of South America, the Juan de Fuca Ridge off the coast of Washington state, and the Gorda Ridge off Oregon and northern California. Along the rift valleys in these areas hot molten material from beneath the earth's crust rises, fracturing and heating the rock. Seawater percolates into and through the fractured rock to form mineral-rich hot solutions. When these solutions cool the metallic sulfides precipitate to the sea floor.

Deposits may be tens of meters high and hundreds of meters long. Too little is known about these deposits at this time to know whether they might be of economic importance at some future date. No technology exists as yet for practical, selective sampling; little is known about the thickness of the deposits, and there is no technology to retrieve them.

In the 1960s metallic sulfide muds were discovered in the Red Sea. Deposits of mud 100 m (330 ft) thick were found in small basins at depths of 1900 to 2200 m (6200–7200 ft). High amounts of iron, zinc, and copper, and smaller amounts of silver and gold were found. The salty brines over these muds were hundreds of times richer in some metals than normal seawater. In 1985, a pilot mining and processing operation was set up under a Saudi-Sudanese joint commission.

Laws and Treaties

Because of the potential value of these minerals, specifically manganese nodules, and because the nodules are found in international waters, outside the usual 200-mile economic zones of coastal nations, the developing nations of the world feel they have as much claim to this wealth as those countries that are presently technically able to retrieve the nodules. The developing nations want access to the technology and a share in the profits. For nearly ten years, the United Nations' Law of the Sea Conferences worked to produce a treaty to regulate deep-ocean exploitation, including mining. The Law of the Sea Treaty was completed in April, 1982. The treaty recognizes the nodules as the heritage of all humankind, to be regulated by a U.N. seabed authority, which would license private companies to mine in tandem with a U.N. company. The quantities removed would be limited, and the profits would be shared.

The United States chose not to sign the treaty. In 1984 the United States, Belgium, France, West Germany, Italy, Japan, and the Netherlands signed a separate Provisional Understanding Regarding Deep Seabed Matters, and, under this provisional understanding, four international consortia have been awarded exploration licenses by the United States, West Germany, and the United Kingdom. Other countries have applied for licenses under the Law of the Sea Treaty. In any case, it is unlikely that any deep-sea mining will occur until the twenty-first century. The high costs of sea mining, low metal prices, and still undeveloped land sources combine to make rapid commercialization unlikely under any legal system.

In the United States, there is no offshore mining except for the sand and gravel operations in the state-owned waters of a few coastal states. Mining for hard minerals, which include manganese nodules, cobalt crusts, metallic sulfides, and phosphorites, would be carried out within U.S. waters under provisions of U.S. domestic laws. Under these laws the Department of the Interior is authorized to issue leases for mineral exploration and development on the continental shelf. No leasing or regulatory program has yet been developed, but federal and state task forces are currently working to develop programs for cobalt-rich crusts in the Hawaiian Islands, metallic sulfides on the Gorda Ridge off the coast of Oregon and northern California, and phosphorites off North Carolina.

Summary

The bathymetric features of the ocean floor are as rugged as the topographic features of the land but erode more slowly. The continental margin includes the continental shelf, slope, and rise. The continental shelf break is located at the change in steepness between the continental shelf and the continental slope. Submarine canyons are major features of the continental slope and, in some cases, the continental shelf. Some canyons are associated with rivers; others are believed to have been cut by turbidity currents. Turbidity currents deposit graded sediments known as turbidites.

The ocean basin floor is a flat abyssal plain, but it is interrupted by scattered abyssal hills, volcanic seamounts, and flat-topped guyots. In warm shallow water, corals have grown up around the seamounts to form fringing reefs. A barrier reef is formed when a seamount subsides while the coral grows. An atoll results when the seamount's peak is fully submerged. Great, continuous volcanic mountain ranges, called midocean ridges and rises, extend through all the oceans and separate the ocean floor into basins. Rift valleys run along the ridge and rise crest; fracture zones run perpendicular to the ridge system. Trenches are long depressions in the ocean floor that are associated with island arcs; they are found mainly in the Pacific Ocean.

Depth soundings were made first with a hand line, then with wire, and, since the 1920s, with echo sounders.

Sediment classifications are based on their sources, areas of deposit, and particle size. Sediments formed from rock are lithogenous, those from living organisms are biogenous, those from seawater are hydrogenous, and cosmogenous sediments are particles from outside the atmosphere. Sediments arising from and deposited close to land are terrigenous; those arising from processes over the continental shelf are neritic; and pelagic sediments are found on the deep-sea floor, away from the influence of land. Red clay is a pelagic, lithogenous sediment that accumulates slowly. Sediments made up of 30% or more biogenous material are oozes. Siliceous oozes are insoluble; calcareous particles are soluble at depths below 4000 m.

Patterns of sediment deposit result from the distance from their source, the abundance of living forms contributing their remains, the seasonal variations in river flow, the waves and currents including turbidity currents, the variability in land sources, the prevailing winds, and sometimes rafting.

Coarse sediments are concentrated close to shore; finer sediments are found in quiet offshore or nearshore environments. Terrigenous sediments are found mainly along coastal margins; most deep-sea sediments come from biogenous sources. Mixtures of particle sizes reveal the processes that formed the deposit. Sediment layers provide clues to ancient climate patterns.

In general, sedimentation rates are slowest in the deep sea and greatest near the continents. Relict sediments were deposited under conditions that are no longer in existence. Mechanisms that increase the sinking rates of particles include clumping and incorporation of sediment particles into the larger fecal pellets of small marine organisms.

Sediments are sampled with dredges, grabs, corers, and by drilling.

Seabed resources include sand and gravel used in construction and landfills. Sands and muds that are rich in mineral ores are mined. Phosphorite nodules have potential as fertilizer. Oil and gas are the most valuable of all seabed resources. Manganese nodules are rich in copper, nickel, and cobalt; they are present on the ocean floor in huge numbers. The status of manganese-nodule mining is clouded by disputes over international law with reference to mining claims and shared technology. Sulfide mineral deposits have been discovered along rift valleys; their economic importance is unknown.

Key Terms

sediment	barrier reef	phosphorite	pteropod ooze
subsidence	atoll	manganese nodule	coccolithophore
continental margin	ridge and rise system	cosmogenous sediment	coccolithophore ooze
continental shelf	rift valley	tektite	red clay
continental shelf break	fracture zone	ooze	brown clay
continental slope	trench	calcareous ooze	brown mud
submarine canyon	island arc	siliceous ooze	terrigenous sediment
turbidity current	fathom	diatom	neritic sediment
turbidite	soundings	diatomaceous ooze	pelagic sediment
continental rise	echo sounder	nutrient	relict sediment
abyssal plain	depth recorder	radiolarian	rafting
abyssal hill	lithogenous sediment	radiolarian ooze	dredge
seamount	biogenous sediment	foraminifera	corer
guyot	hydrogenous sediment	foraminifera ooze	grab sampler
fringing reef	carbonate	pteropod	

Study Questions

1. Are calcareous oozes more common in the southern Pacific Ocean or the northern Pacific Ocean? Why is this so?

2. What is a turbidity current? Where would you expect a turbidity current to occur? How does the structure of a sediment deposit left by a turbidity current differ from other sediment deposits?

3. List the four basic sediment types classified by source. Where is each sediment type most likely to be found?

4. Discuss the probability of commercial development and exploitation of deep-sea mineral resources during the next twenty-five years.

5. Imagine that you are in a submersible on the ocean bottom. You leave New York and travel across the North Atlantic to Spain. Draw a simple ocean bottom profile showing each major bathymetric feature you see as you move across the ocean. Name each feature. Do the same for the South Pacific between the coast of Chile and the west coast of Australia. Compare the two profiles.

6. Which of the sampling devices discussed in this chapter takes a relatively undisturbed bottom sample? Why is this important to a marine geologist?

7. What is the continental margin? What pattern of sediment deposit would you expect to find associated with it? What processes produce these patterns of deposit?

8. What combination of factors is required to form a coral atoll?

Study Problems

1. If underwater cables are spaced 14 km apart on the sea floor, and if monitoring equipment shows that they break in sequence from shallow to deeper water at fifteen-minute time intervals, what can you determine about the event causing the breaks?

2. If the average concentration of suspended sediment in the water is 1 g/m³ and the volume of water in a harbor is 158 km³, what is the average residence time of sediment in the water of this harbor? The daily sediment supply rate averages 1×10^7 kg.

3. In how many years will each of the particles listed in the following table reach the sea floor if the particles fall through 4000 m of seawater? All the particles are derived from land rock of the same density (2.8 g/cm³). Settling rate is calculated from Stokes's law:

$$V \text{ cm/sec} = 2.62 \times 10^4 r^2$$

where r is the radius expressed in centimeters. How does the settling rate change if the diameter remains constant but the density of a particle changes?

Particle Type	Particle Diameter (mm)	Setting Rate (V) (cm/sec)
granule	4	1048
sand	0.25	4.09
clay	0.004	10.48×10^{-4}

4. Assuming a constant sedimentation rate of 0.4 cm/1000 years, how thick will the sediments be in a portion of an ocean basin where the underlying crust is 130 million years old?

The Not-So-Rigid Earth

3

*I*n its entire length, the basin of this sea [the Atlantic] is a long trough, separating the Old World from the New, and extending probably from pole to pole.

This ocean-furrow was scored into the solid crust of our planet by the Almighty hand, that there the waters which "he called seas" might be gathered together, so as to "let the dry land appear," and fit the earth for the habitation of man. . . .

Could the waters of the Atlantic be drawn off, so as to expose to view this great sea-gash, which separates continents and extends from the Arctic to the Antarctic, it would present a scene the most rugged, grand, and imposing. The very ribs of the solid earth, with the foundations of the sea, would be brought to light, and we should have presented to us at one view, in the empty cradle of the ocean, "a thousand fearful wrecks," with that dreadful array of dead men's skulls, great anchors, heaps of pearl and inestimable stones which, in the poet's eye, lie scattered in the bottom of the sea, making it hideous with sights of ugly death.

Matthew Fontaine Maury
from The Physical Geography of the Sea, *1855*

*M*any features of our planet have presented scientists with contradictions and puzzles. For example, the remains of warm-water coral reefs are found off the coast of the British Isles; marine fossils occur high in the Alps and the Himalayas; and coal deposits that were formed in warm, tropic climates are found in northern Europe, Siberia, and northeast North America. At the same time, similar patterns were observed in widely scattered places but for no known reason. The volcanoes known as the ring of fire border the east and west coasts of the Pacific Ocean, and deep ocean trenches are found adjacent to long island arcs, such as the Aleutian Islands and Trench, the Mariana Islands and Trench, and the Japanese Islands and Japan-Kuril Trench. No single coherent theory explained all these features, until new technology and the exploration of a great underwater mountain range in the 1950s combined to trigger a complete reexamination of the earth's history.

In this chapter we explore the history of an idea and the evidence that allowed this idea to become, first, theory and, at last, generally accepted fact. During the last thirty-five years our understanding of our planet has changed completely; new tools and new knowledge have allowed us to gain insights into the past, appreciate the present, and even look forward into the future.

3.1 *The Interior of the Earth*

Internal Layers

Living as we do on the surface of the outer crust of the earth, we cannot directly observe the interior of the planet. However, scientists have been able to learn a great deal about the structure and composition of the earth's interior using indirect methods.

We understand the interior of the earth to be a series of layers. If the earth were cut and a piece were removed, the layers would appear as in figure 3.1. At the planet's center is a core; its radius is 1070 km (665 mi). This **inner core** is solid, very dense, magnetized, rich in iron and nickel, and very hot (5000° C). The solid inner core is surrounded by a transition zone about 700 km (435 mi) thick, which is in turn surrounded by a 1700-km (1056-mi) thick layer of liquid material; together they form the **outer core.** The liquid material of the outer core is similar in composition to the inner core but cooler (4000° C). The next layer contains the largest mass of material of any of the layers (about 70%);

this is the **mantle.** This layer is 2835 km (1761 mi) thick; it is less dense than the core and still cooler (1500° to 3000° C). The outer part of the mantle is thought to be rigid, while the inner region is plastic, or deformable, and can flow slowly. The earth's outermost layer is the cold, rigid, thin (10 to 65 km) (6–40 mi), familiar surface, or **crust.**

Evidence for the Internal Structure

The evidence that has allowed scientists to build this model of the earth's interior has come from many sources. Using the spherical shape, the mean (or average) radius, and the mass of the earth, it is possible to determine the average density of the earth: 5.52 g/cm³. **Density** is defined as mass per unit volume and is usually given in grams per cubic centimeter, written g/cm³. This calculated density is considerably greater than the average density of the earth's crust found by direct measurement, which is approximately 2.8 g/cm³. In order to bring the average of the earth's density to such a high value, the material below the crust must have a density much greater than that of the crust. Since the earth wobbles only very slightly as it rotates and since the acceleration due to gravity over the earth's surface is quite uniform, the earth's mass must be distributed uniformly about the earth's center, as in a series of concentric layers. Based on gravity, density, and the earth's dimensions, it is possible to calculate the pressures within the earth and also the temperatures that can be reached under these pressures. Also, because there is a magnetic field present around the earth, the materials in the earth's core must be magnetic.

Another clue is furnished by meteorites that occasionally hit the earth and are considered to be the remains of unknown planets. More than half of the meteorites that have been found are "stony" silicate or rocky lumps, another large group is made mainly of iron, nickel, and other metals, and a few are "stony-iron" with metal inclusions. Radiometric dating of meteorites gives a maximum age of 4.6 billion years, the same as the age of the solar system. They allow us to directly analyze the density, chemistry, and minerology of the nickel-iron cores of bodies that we believe to have a similar composition to that of the earth.

A major tool in the understanding of the earth's interior is the **seismic wave.** A seismic wave is an underground earth shock wave, or vibration, that is produced by an earthquake or underground explosion. All over the surface of the earth geologists and geophysicists monitor seismic recording stations, which pick up the vibrations caused by earthquakes, volcanic eruptions, landslides, and deliberately caused detonations. As the seismic waves move through one earth layer and into another, the waves bend, or **refract.**

Figure 3.1

The layered structure of the earth.

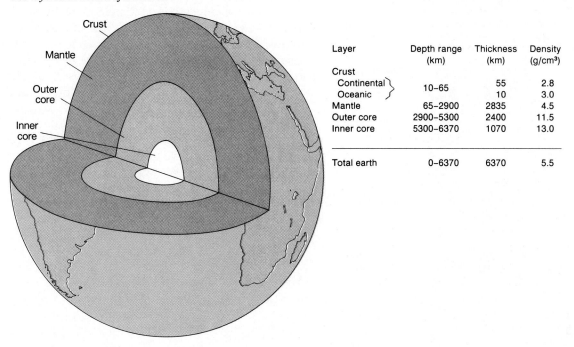

Layer	Depth range (km)	Thickness (km)	Density (g/cm³)
Crust			
Continual	10–65	55	2.8
Oceanic		10	3.0
Mantle	65–2900	2835	4.5
Outer core	2900–5300	2400	11.5
Inner core	5300–6370	1070	13.0
Total earth	0–6370	6370	5.5

There are two basic kinds of seismic waves that are associated with the earth's movements and that travel through the earth's interior. These kinds are pressure waves, or **P-waves,** which oscillate in the same direction as they move (similar to sound waves), and shear waves, or **S-waves,** which oscillate at right angles to their direction of movement (similar to water waves). These wave types are shown in figure 3.2. P-waves are able to move through solids and liquids. S-waves cannot pass through a liquid, and so they do not pass through the outer core of the earth.

The speeds of the waves change as the waves pass from layer to layer, because each layer has different properties. These changes in speed cause the waves to refract as shown in figures 3.3a and b. The paths taken by the seismic waves as they pass through the earth provide information about the dimensions, structure, and physical properties of each of the internal layers. If the interior of the earth had uniform properties, the waves would follow straight lines and their speed would not change, as shown in figure 3.3c. Measurements of travel time and wave direction confirm that there are abrupt changes of speed and direction at certain depths corresponding to the interior layers.

Because of the paths followed by the S-waves (shown in fig. 3.3b), we know that the outer core is liquid. The P-waves (fig. 3.3a) are refracted, but not stopped, by changes in density as they pass through the layers. In both diagrams (fig. 3.3a and 3.3b), shadow

Figure 3.2

Waves passing through a solid are of two types: pressure waves, or P-waves (shown above), and shear waves, or S-waves (shown below).

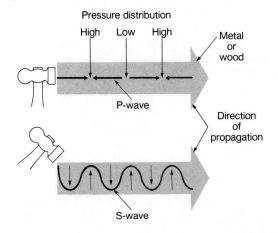

zones are formed where waves are absent. From this evidence, a picture of the earth's layered structure has been formed (refer back to fig. 3.1).

The availability of data from an increasingly densely spaced and sophisticated array of seismic-recording stations, and the computer capacity to analyze the travel time of thousands of seismic waves from the world's earthquakes, have been used to produce a three-dimensional map of the boundary between the

Figure 3.3

Movement of seismic waves through the earth. (a) Refraction of P-waves and shadow zones produced by the earth's interior structure. (b) Refraction of S-waves and shadow zones produced by the earth's interior structure. (c) No refraction and no shadow zone occurs in an earth with a uniform structure.

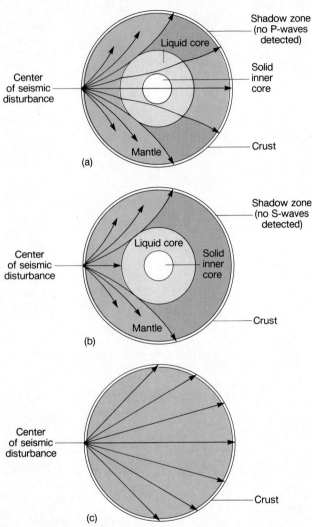

core and the mantle. This process, called **seismic tomography,** is beginning to give us a more detailed description of the interior of the earth.

The new data indicates that the core is not smooth, but has major peaks and valleys over its surface. These features extend as much as 11 km (7 mi) above and below the mean surface of the outer core. At this time, it is thought that the regions above a peak are areas where the mantle has excess heat and a plume of mantle material rises toward the crust, drawing the core with it. Cooler, more viscous mantle material sinks to cause the depressions in the core's surface. It is likely that these peaks and valleys last only as long as it takes a

rising plume to lose its excess heat and sink back toward the core, perhaps a hundred million years.

Seismic tomography is combined with model studies to try to understand how plumes of mantle material form. Using layered fluids of different densities and viscosities, researchers build rising plume models with a lower density fluid underlying a more dense fluid. The structure of these model plumes, their distribution, and the speeds at which they ascend are controlled by the differences in density and viscosity of both liquids and the thickness of the overlying layer. Continuing research will eventually give us a more complete understanding of the processes at work in the earth's interior and a better window into the heart of our planet.

3.2 The Lithosphere

The Layers

The crust has been studied in great detail. The large continental landmasses are formed primarily from **granite**-type rock, which has a high content of aluminum and magnesium silicate; quartz and feldspar are the principal mineral components. The crustal rock layer lying under the ocean is **basalt**-type rock, which is lower in silica and higher in iron and magnesium. Continental crust has a density of 2.8 g/cm³ and is approximately 55 km (34 mi) thick, while oceanic crust is only about 10 km (6 mi) thick and has a density of 3.0 g/cm³.

The upper edge of the mantle is just below the crust; the boundary between the crust and the mantle is the Mohorovičić discontinuity, named for its discoverer, and usually called the **Moho.** The Moho is a boundary at which there is a sudden change in the speed of seismic waves. At one time it was thought that the Moho was the structure at which the earth's rigid crust moves relative to the mantle. Current research emphasizes the function or behavior of the crust and upper mantle rather than the structures that might be detectable, and places the zone of movement between the rigid crust and upper mantle from 70 km (43 mi) below the ocean crust to 150 km (93 mi) below the continental crust. The mantle just below the crust is considered to be rigid, solidified, basalt-type rock, fused to the crust but at the same time separated from it by the Moho. This rigid layer of crust and upper mantle is known as the **lithosphere.** Seismic waves pass through it at high speeds, implying that the lithosphere is both strong and rigid. The region of the mantle extending about 250 km (155 mi) below the lithosphere is the **asthenosphere**; this region of the mantle passes seismic waves more slowly and is considered to be partly melted, or plastic, allowing it to flow when stressed. The lithosphere is less

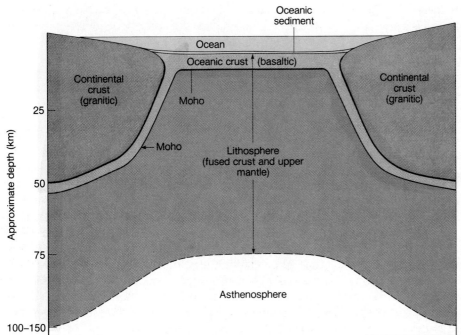

Figure 3.4
The lithosphere is formed from the
fusion of crust and upper mantle.
Notice that the Moho is relatively
close to the earth's surface under the
ocean's basaltic crust but is
depressed with the mantle under the
granite continents.

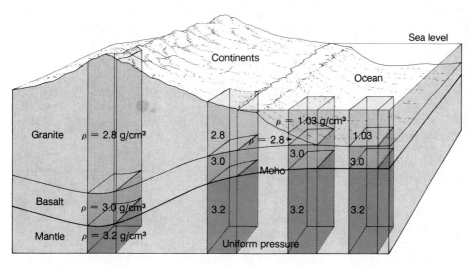

Figure 3.5
Isostasy. Columns of crustal material
are unequal in height and density
but generate the same pressure at
the same depth within the mantle.

dense than the asthenosphere; both continental and oceanic lithosphere float on the asthenosphere. The lithosphere and asthenosphere are shown in figure 3.4.

Isostasy

The distribution of elevated continents and depressed ocean basins over the crust requires that a balance be kept between the internal pressures under the land blocks and those under the ocean basins. This is the principle of **isostasy.** The balance is possible because the greater thickness of low-density granitic crust in the continental regions is compensated for by the elevated higher-density mantle material under the oceans (see fig. 3.5). The situation is often compared to the floating of an iceberg. The top of the iceberg is above the sea surface, supported by the buoyancy of the displaced water below the surface. The deeper the ice extends below the surface, the higher the iceberg reaches above the water. The less-dense continental land blocks are often said to float on the more-dense mantle in the same way, with most of the continental volume below sea level.

In other words, the asthenosphere offers buoyant resistance to a section of lithosphere sagging under the weight of a mountain range. Because the lithosphere is cooler, it is more rigid and stronger than the asthenosphere. If a thick section of lithosphere has a large area and it is mechanically strong, it depresses the asthenosphere only slightly and is able to support a mountain range such as the Himalayas or the Alps. In other places where the crust is fractured and mechanically weak, growing continents and mountains must penetrate deeper into the mantle in order to provide buoyancy and keep the continental mass from sinking; the Andes are a mountain range with roots deep in the mantle.

If material is removed from or added to the continents, isostatic adjustment will occur. For example, parts of North America and Scandinavia continue to rise as the continents readjust to the lost weight of the ice sheets that receded at the close of the last ice age ten thousand years ago. Newly formed volcanoes protruding above the sea surface as islands often subside as their weight depresses the oceanic crust, and they sink back under the sea. The plastic portion of the mantle gradually changes its shape in response to weight changes in the overlying, more-rigid crust.

3.3 The Movement of the Continents

History of a Theory

As world maps became complete and more accurate, observant individuals were intrigued by the shapes of the continents on either side of the Atlantic Ocean. The possible ''fit'' of the bulge of South America into the bight of Africa was noted by the English scholar and philosopher Francis Bacon (1561–1626), the French naturalist George Buffon (1707–1788), the German scientist and explorer Alexander von Humboldt (1769–1859), and others. In the 1850s the idea was expressed that the Atlantic Ocean had been created by a separation of the two continents during some unexplained cataclysmic event early in the history of the earth. Thirty years later another suggestion was made, that a portion of the earth's continental crust had been torn away to form the moon, creating the Pacific Ocean and triggering the opening of the Atlantic. As scientific studies of the earth's crust continued, patterns of rock formation, fossil distribution, and mountain range placement began to show even greater similarities between the now separated continents. Based on this evidence, an Austrian geologist, Edward Suess, in a series of volumes published between 1885 and 1909, proposed that the southern continents had been joined into a single continent he called **Gondwanaland.** He assumed that isostatic changes had allowed portions of the continents to sink and create the oceans between the continents. This idea was known as the subsidence theory of sep-

aration. At the beginning of this century, Alfred L. Wegener and Frank B. Taylor independently proposed that the continents were slowly drifting about the earth's surface. Taylor soon lost interest, but Wegener, a German meteorologist, astronomer, and arctic explorer, continued to pursue this concept until his death in 1930.

Wegener's **theory of drifting continents** proposed the existence of a single supercontinent he called **Pangaea** (see fig. 3.6). He thought that forces arising from the rotation of the earth began Pangaea's breakup. First, the northern portion composed of North America and Eurasia, which he called **Laurasia,** separated from the southern portion formed from Africa, South America, India, Australia, and Antarctica, for which he retained the earlier name Gondwanaland. Laurasia and Gondwanaland are shown in figure 3.6. The continents as we know them today then gradually separated and moved to their present positions. Wegener based his ideas on the geographic fit of the continents and the way in which some of their older mountain ranges and rock formations appeared to relate to each other when the landmasses were assembled to form Pangaea. He also noted that fossils more than 150 million years old collected on different continents were remarkably similar, implying the ability of land organisms to move freely from one landmass to another. Fossils dated after this period from different places showed quite different forms, suggesting that the continents and their populations had separated from one another.

Wegener's theory provoked considerable debate in the decade of the 1920s, but most geologists agreed that it was not possible to move the continental rock masses through the rigid basaltic crust of the ocean basins. There was no mechanism to cause the drift, and the theory was not regarded very seriously in the scientific community; it became a footnote in geology textbooks.

Evidence for a New Theory

Armed with new sophisticated instruments and the technology developed during World War II, earth scientists returned to a study of the earth's crust in the 1950s. For the first time, scientists were able to examine the floor of the deep ocean in detail, and explore the world's largest mountain range running through the oceans. Refer back to figures 2.9 and 2.12 to trace this remarkable and continuous underwater feature. The theories current at the time of its discovery had not predicted the existence of such an extensive midocean mountain range and could not explain it, but a new theory was advanced that made the old idea of continental drift seem plausible.

Based on the earlier work of Arthur Holmes, H. H. Hess of Princeton University proposed in the early 1960s

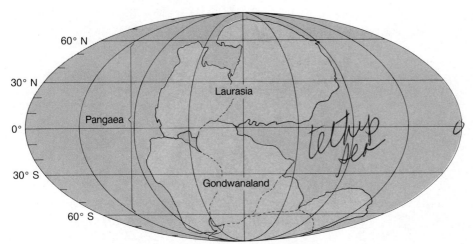

Figure 3.6

Pangaea 200 million years before the present. Wegener's supercontinent is composed of the two subcontinents, Laurasia and Gondwanaland.

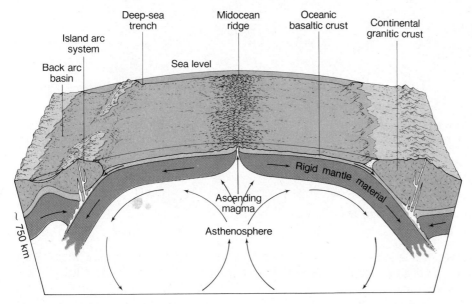

Figure 3.7

Seafloor spreading creates new crust at midocean ridges and loses old crust in deep-sea trenches. This process is shown being driven by convection cells in the asthenosphere.

that deep within the earth's mantle there are currents of low density, molten material heated by the earth's natural radioactivity. When these upward-moving currents of mantle material reach the lithosphere they move along under it, cooling as they do so until they become cool enough and dense enough to sink down toward the core again. These patterns of moving mantle material are called **convection cells** (see fig. 3.7). The convection cells are generally believed to occur in the asthenosphere, although there is no definite evidence that the convection cells are confined to this layer. Indeed, some scientists believe that they extend throughout the mantle.

If the upward-moving mantle material, or **magma,** breaks through the crust of the seafloor instead of continuing to flow underneath it, underwater volcanoes are produced and a mountain range forms along the crack in the crust. As the cooling magma, or lava, oozes out along the underwater mountain range, or ridge system, it hardens and is added to the earth's surface as new oceanic-type basaltic crust. If new crust is being produced in this manner, a mechanism is needed to remove old crust since there is no appreciable change in the size of the earth. The great, deep trenches of the Pacific were proposed as areas where the older crust dips down and disappears back into the earth's interior, eventually to be recycled into the mantle and the convection cell system. Figure 3.7 shows this process of producing new crust at the ridges and losing old crust at the trenches in a system driven by the motion of convection cells.

Although some of the ascending molten material breaks through the crust and solidifies, most of the rising material is turned aside under the rigid lithosphere and moves away toward the descending sides of the convection cells, dragging pieces of the lithosphere with it. This lateral movement of the crust is called **seafloor**

Figure 3.8
The world's major earthquake belts coincide with midocean ridge
systems and the deep-ocean trenches.

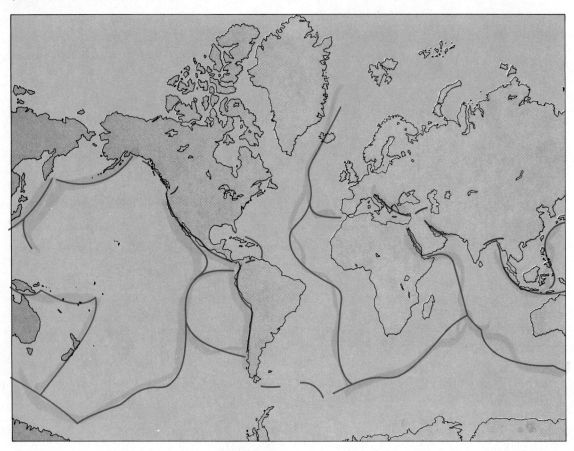

spreading (see fig. 3.7). Areas in which new crust is formed above rising magma are **spreading centers;** areas of descending crustal material are **subduction zones.** The seafloor spreading mechanism provides the forces to cause continental drift. The continents are not moving through the basalt of the seafloor; instead, the continents are being carried as passengers on the basaltic crust, similar to boxes on a conveyor belt.

Evidence for Crustal Motion

Additional evidence was needed to support the idea of seafloor spreading, and as oceanographers, geologists, and geophysicists explored the earth's crust on land and under the oceans, the facts began to accumulate.

Earthquakes were known to be distributed around the earth in narrow and distinct zones. These zones were found to correspond to the areas along the ridges (or spreading centers) and the trenches (or subduction zones). Compare figure 3.8 with figures 2.9, 2.11, and 2.12.

Researchers sank probes into the sea floor to measure the heat from the interior of the earth moving through the crust. These measurements produced data with a high degree of variability, even over closely spaced intervals. In part, this variation in the data is blamed on seawater seeping down through porous or fractured portions of the crust at one location and rising as heated water at another. This circulation does not occur over regions of the sea floor that are sealed by a thick layer of sediment. In such areas, the measured heat flow shows a regular pattern. In these areas, the heat flow is found to be highest in the vicinity of the midocean ridges and is found to decrease away from the ridge system (see fig. 3.9).

Radiometric dating of the age of rocks from the land and from the sea floor shows that the oldest rocks from the oceanic crust are only about 200 million years old, while the rocks from the land are much older. The sea floor formed by convection cell processes at the spreading centers is new, young, and short-lived, for it is lost at the subduction zones, where it plunges back down into the asthenosphere.

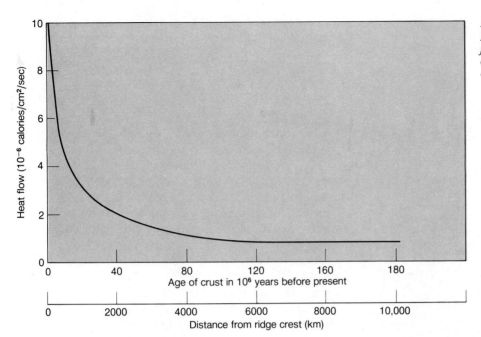

Figure 3.9
Heat flow through the Pacific Ocean floor. Values are shown against age of crust and distance from the ridge crest.

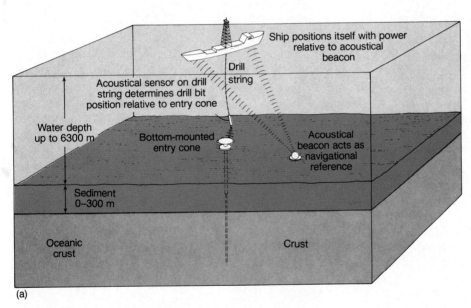

Figure 3.10
Deep-ocean drilling technique. Acoustical guidance systems are used to maneuver the drilling ship over the bore hole and to guide the drill string back into the bore hole.

Cores were drilled down through the sediments that cover the ocean bottom and into the ocean floor. This process required the development of a new technology, and in the late summer of 1968 the specially constructed drilling ship, *Glomar Challenger,* was used for this study. This ship is 122 m (400 ft) long with a beam of 20 m (65 ft) and a draft of 8 m (27 ft); it displaces 10,500 tons when loaded. Its specialized bow and stern thrusters and its propulsion system respond automatically to computer-controlled navigation, using acoustic beacons on the sea floor. This allows the ship to remain for long periods of time in a nearly fixed position over a drill site in water too deep to anchor. A

sonar guidance system enables it to replace drill bits and reenter the same bore holes in water about 6000 m (20,000 ft) deep. This method is illustrated in figure 3.10.

In 1983 the *Glomar Challenger* was retired, after logging 600,000 km (375,000 mi), drilling 1092 holes at 624 drill sites, and recovering a total of 96 km (57.6 miles) of deep-sea cores for study. A new deep-sea drilling program with a new drill vessel, the *JOIDES Resolution,* began in 1985. The *JOIDES Resolution* is about the same size as the *Glomar Challenger,* complete with the world's most sophisticated, state-of-the-art scientific and drilling equipment (see fig. 3.11). Information on the current research being conducted by

Figure 3.11
The deep-sea drilling vessel, Joides Resolution.

the *JOIDES Resolution* will be found in the Current Research section later in this chapter.

The cores taken by the *Glomar Challenger* provided much of the factual data needed to establish the existence of seafloor spreading. No rocks older than 180 million years old were found, and sediment age and thickness were shown to increase with distance from the ocean ridge system (see fig. 3.12). Note that the sediments closest to the ridge system are thin; this area comprises new crust and has not had long to accumulate its sediment load. The crust farther away from the ridge system is older and is more heavily loaded with sediments.

Although each of these pieces of evidence fits the theory that the earth's crust along the ridge system is new and young, the most elegant proof for seafloor spreading came from a study of the magnetic evidence locked into the oceans' floors.

The earth has the familiar north and south geographic poles at 90° N and 90° S latitude, marking the axis about which it rotates. It also has a magnetic North Pole located in the Hudson Bay area and a magnetic South Pole directly opposite, in the South Pacific Ocean.

The earth behaves as if it had a giant bar magnet embedded in its interior. Associated with any magnet is its magnetic field, and we can visualize the earth's magnetic field as lines of force surrounding the planet. The lines of force converge and dip toward the earth at the magnetic poles. These lines of magnetic force are parallel to the earth at the magnetic equator. At other positions on the earth's surface the lines of force have both horizontal and vertical components.

When materials that can be magnetized are heated in a magnetic field they become magnetized as they cool, with their north and south magnetic poles lined up with the poles of the magnetic field. In the same way, as molten volcanic material cools and solidifies, its iron particles become magnetized and permanently aligned with both the horizontal and vertical components of the earth's magnetic field (see fig. 3.13).

Research done on age-dated layers of volcanic rock found on land shows that the polarity, or north-south orientation, of the earth's magnetic field reverses for varying periods of time. This means that at different times in the earth's history the present north magnetic pole has been the south magnetic pole and the present south magnetic pole has been the north magnetic pole. Each time a layer of volcanic material solidified, it trapped the magnetic orientation and polarity of the time period in which it occurred. The dating and testing of samples taken through a series of volcanic layers have allowed scientists to build up a calendar of these events (see fig. 3.14). No one knows what triggers these **polar reversals,** but during a reversal the earth's magnetic field gradually collapses, removing this barrier to cosmic radiation, and recent work has suggested a correlation between such periods and a decrease in populations of delicate, single-celled organisms that live in the surface layers of the ocean. Nearly 170 reversals have been identified during the last 76 million years, and our present magnetic orientation has existed for 600,000 years. How long our current polarity will last we do not know.

When magnetometers were towed over the sea floor by scientists from the Scripps Institution of Oceanography in 1960, a pattern of stripes appeared, caused by changes in the polarity of the vertical magnetic component locked into the crust of the sea floor. These stripes ran parallel with the midocean ridge system (see fig. 3.15). Their significance was not understood until 1963, when F. J. Vine and D. H. Matthews of Cambridge University proposed that these stripes represented a recording of the polar reversals of the vertical component of the earth's magnetic field, frozen into the sea floor. As the molten basalt rose along the crack of the ridge system and solidified, it locked in the direction of the prevailing magnetic field. Seafloor spreading moved this material off on either side of the ridge, to be replaced

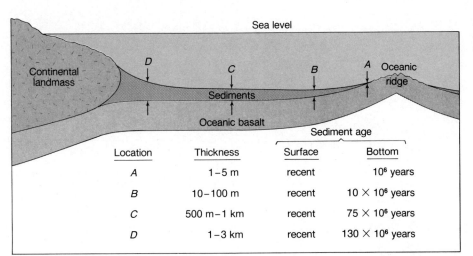

Figure 3.12
Age and thickness of seafloor sediments.

Location	Thickness	Sediment age	
		Surface	Bottom
A	1–5 m	recent	10^6 years
B	10–100 m	recent	10×10^6 years
C	500 m–1 km	recent	75×10^6 years
D	1–3 km	recent	130×10^6 years

Figure 3.13

Lines of magnetic force surround the earth and converge at the magnetic poles. The lines of force have components that are parallel and perpendicular to the earth's surface. The vertical components are not present at the earth's magnetic equator and are at their maximum at the magnetic poles, where the horizontal components disappear. A compass orients its needle with the horizontal component to indicate the direction to the magnetic pole. NP and SP indicate the north and south geographic poles, respectively.

Figure 3.14

Polarity reversal time scale during the Cenozoic era. Time is given in millions of years.

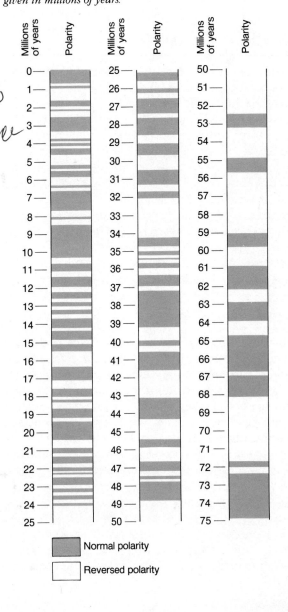

by more molten materials. Each time the earth's magnetic field reversed, the direction of the magnetic field was recorded in the new crust. Vine and Matthews proposed that if such were the case, there should be a symmetric pattern of magnetic polarity stripes centered at the ridges and becoming older away from the ridges. The polarity and age of these stripes should correspond to the same magnetic field changes found in the dated layers on land. Experimental work confirmed their ideas and thus dramatically confirmed seafloor spreading. The present ocean basins are not old but new, created by seafloor spreading during the past 200 million years, or the last 5% of the earth's history.

Figure 3.15
Reversals in the earth's magnetic polarity cause the symmetrically striped pattern centered on the Mid-Atlantic Ridge. The age of the sea floor increases with the distance from the ridge. The spreading rate along the Mid-Atlantic Ridge is about one centimeter per year.

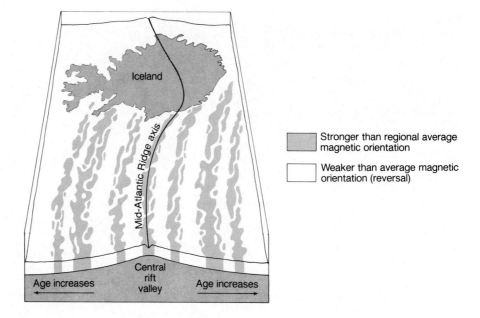

Stronger than regional average magnetic orientation

Weaker than average magnetic orientation (reversal)

3.4 Plate Tectonics

When the theory of continental drift was joined with the idea of seafloor spreading, bolstered by the previously discussed evidence, it led to the formation of a single concept of crustal movement: **plate tectonics.** The earth's lithosphere is seen as a series of rigid plates outlined by the major earthquake belts of the world (refer back to fig. 3.8). These earthquake belts are the plate boundaries and coincide with the trenches, ridges, and faults. At present, seven major lithospheric plates are recognized—the Pacific, Eurasian, African, Australian, North American, South American, and Antarctic—as well as numerous smaller plates off South and Central America, in the Mediterranean area, and along the northwest United States (see fig. 3.16).

Plate Boundaries

Sections of lithosphere are known as plates; each plate is made up of continental and/or oceanic crust. The direction of plate motion at each boundary depends on whether lithospheric plates move apart, move together, or slide past each other. Plate boundaries move apart at the midocean ridges; these are known as **divergent plate boundaries.** At the trenches plates move toward each other, forming **convergent plate boundaries.** Plates move past each other along **faults,** where there is a break in the rocky crust with displacement of one side relative to the other. There are many types of faults; in some the movement is vertical, in some horizontal, and in some oblique. Close inspection of the midocean ridge and rise system shows sections of the ridge which are offset or displaced laterally from each other along a

special kind of fault called a **transform fault.** The opposite sides of a transform fault are two different plates that are moving in opposite directions (see area *A* in fig. 3.17). This movement creates a fault zone that is active and may be the seat of frequent and severe but shallow earthquakes. Where this same fault line extends outside both ridge axes, the crustal plates are moving in the same direction, and differential motion along the fault is much reduced (see area *B* in fig. 3.17). These regions of the transform faults are quiet, but the faults can still be detected.

The lateral motion of the ridge along these transverse faults produces sharp vertical displacements called **escarpments** across the width of the ridge. These are regions where there are sudden changes in the depth of the ocean. The escarpments and transform faults are boundaries where one piece of crust moves relative to another.

An excellent example of one of the larger transform faults is the San Andreas Fault extending up through the Gulf of California, to the San Francisco Bay region, and out to sea to the north (see fig. 3.18). In this particular case, where the East Pacific Rise intersects the west coast of Mexico a transform fault extends from this location northward through California and out to sea off the northwest United States. Here it terminates at another ridge system pointed to the northeast, the Gorda and Juan de Fuca Ridges. The San Andreas Fault system allows crustal sections on either side to move relative to each other. In this case, the land on the Pacific side, including the coastal area from San Francisco to the tip of Baja California, is actively moving northward past the

Figure 3.16
Major lithospheric plates of the world.

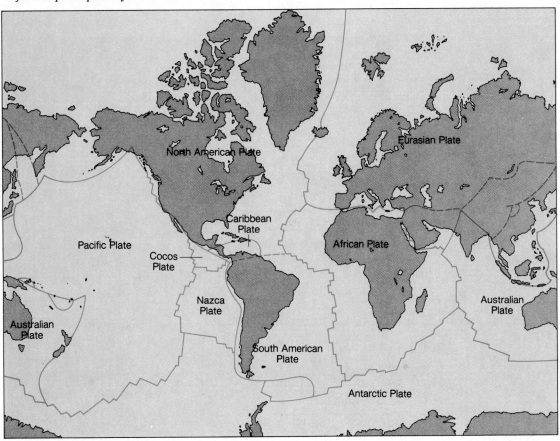

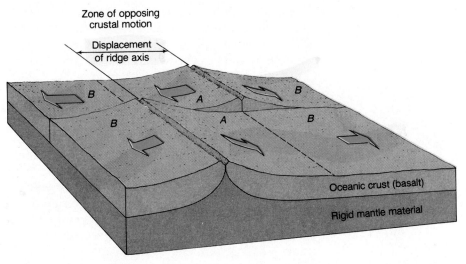

Figure 3.17

Relative motion and displacement of the crust along a transform fault. Motion at A is in opposite directions. Motion at B is in the same direction.

land on the east side of the fault. The relative motion is not uniform. Stress accumulates until the friction along the boundary can no longer withstand it. When this occurs, there is a sudden movement and relative displacement along the fault, with resulting severe earthquakes.

The reason for the many transform faults that are found associated with the ridge system is not clearly understood. It is felt, however, that these faults developed because the crust is subject to different degrees of rotational motion related to latitude. Changes in the direction of crustal motion, due to variations in the

Figure 3.18
(a) The San Andreas fault runs north-south along the San
Francisco peninsula. (b) The sunken portion of the fault south of
San Francisco is used to store water for the city's use.

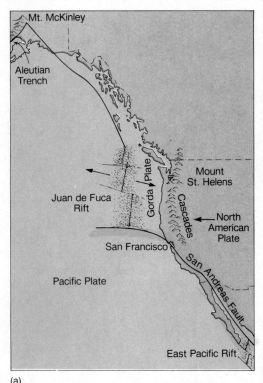

(a)

(b)

strength or location of convection cells or due to collisions between sections of crust, could also result in transform faults.

Rifting

Rift zones, or spreading zones, are regions where the lithosphere splits, separates, and moves apart as new crustal material intrudes along the crack or rift. The major rift zones of the present are the midocean ridge and rise systems (return to fig. 3.8). Rifting is not, however, limited to oceanic areas. It occurs on land and, in the past, has been responsible for breaking up landmasses.

The lithosphere under the continents is 100–150 km (60–90 mi) thick, but under the oceans it varies in thickness from 10 to 100 km (6–60 mi). It is thin where new at the rift zones, and thicker where older and further away from the rift. To crack the lithosphere under the continents requires a mechanism to thin and apply tension to the plate of which the continents are a part. Exactly how this mechanism operates is not known. The crust may be under tension because of subduction occurring at the plate margin, stretching and thinning the

plate to allow magma to rise up beneath it, or a region of ascending magma in a mantle convection cell may heat the overlying plate and bow it upward. This may stretch, thin, and weaken the plate so that it cracks under tension, initially producing a fault zone.

This second case involves a more active role for the mantle processes and appears to be a better explanation of observed features of the earth's crust. For example, production of localized faulting would allow blocks of crust to drop downward, as seen in the Great Rift Valley of Africa stretching from Mozambique to Ethiopia. Such a sunken rift zone is called a **graben.** Continuing divergence along such a fault would result in further thinning of the crust and deepening of the fault, gradually allowing magma to penetrate into the graben, forcing the crack to widen, and forming a rift. If the rift deepens sufficiently, seawater may enter. The Red Sea today is at this intermediate stage, with both oceanic-type basalt and subsided blocks of continental-type crust present on the floor of the rift. As the lithospheric plate cools and moves away from the rift, magma becomes attached to the lower side of the lithosphere, causing the plate to gradually thicken away from the rift. As the crustal plate continues to thin at the rift, magma

Figure 3.19
(a) Rifting is thought to happen when magma rises, causing tension and stretching of the overlying plate, resulting in a graben. (b) As the spreading continues, the fault deepens and cracks, allowing magma to penetrate and eventually form a ridge. Marine sediments accumulate in the fault, smoothing the sea floor.

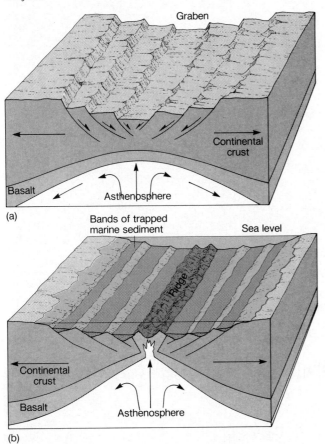

(a)

(b)

wells into the rift at an increasing rate and the landmass is gradually separated into two parts, with a low-lying region of oceanic basalt-type crust between the two sections. A new ocean basin is formed as well as a new ridge system (see fig. 3.19).

Continental Margins

As a plate carrying a continent moves away from a spreading center, the continental margin closest to the midocean ridge is known as the **trailing margin** or **trailing edge.** Basalt-type rock of the seafloor adheres to the trailing edge, and the continent edge slowly subsides as the spreading lithosphere cools and contracts. Along the Atlantic coast, wedge-shaped deposits of sediments eroded from the continent have been carried onto the trailing edge. This edge receives sediments from the land; it is eroded by the action of the waves and the currents, and it may also be modified by organisms building shell or coral reefs. The continental

margin accumulates an orderly series of sediment deposits to a depth of about 3 km (1.8 mi). Turbidity currents move sediments down the slope to the continental rise where thick deposits of graded sediments build up. The trailing edges of continental margins are passive and not greatly modified by tectonic processes; they are often extensive, broad, and shallow with thick sedimentary deposits.

The edge of a continent that is moving toward a deep sea trench or subduction zone is known as the **leading margin** or **leading edge.** The continental margin at a leading edge is modified by tectonic processes when the continental landmass reaches a subduction zone or fault boundary. Continental margins along these plate boundaries are narrow and rapidly modified. Although they receive sediments from the land, these sediments move rapidly down slope to deeper water into trenches or adjacent basins. The active processes at these plate convergence zones do not allow the accumulation of thick sediment deposits from land on the narrow edges of the continental landmass. As one plate collides with and overrides another, the overriding plate edge scrapes, folds, and slices off marine sediments, rapidly carrying them downward. When the descending plate slides past the overriding one, the friction may be great enough to melt the upper part of the downturned slab, including the wet sediments it is carrying. These melted sediments contain water and volatile substances that rise up to form belts of volcanoes along the trench. These volcanoes may form island arc systems when the volcanoes are separated from the continent (see fig. 3.20a). Island arcs may be compressed against a landmass, or volcanoes may arise along the border of the landmass. The California coast range, part of the Alps, and the Andes were formed in this fashion.

When two continents collide, the two landmasses are crumpled and deformed at the colliding edges. Continental material from one plate may override the continental material of the other plate, producing a large thickness of continent. The sediments of the intervening sea floor and the less dense crustal material are not necessarily carried downward into the subduction zone but may be compressed together and piled upward, adding to the mountain range. The Himalayas and the Alps in part are examples of such mountain ranges, and this is why old marine sediments and fossils, including the limestone remains of coral reefs, are found on the summits of peaks in these ranges. This situation is illustrated in figure 3.20b.

If a plate carrying a continent and one carrying an ocean collide, the continental lithosphere is low in density and more buoyant while the oceanic lithosphere is closer to the density of the asthenosphere and is more easily forced down into the mantle. When such

Figure 3.20
Subduction processes produce
(a) island arc systems, volcanic
activity, and (b) additions to a
landmass and explosive volcanoes.

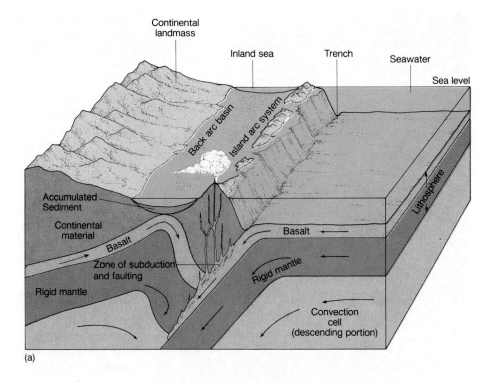

(a)

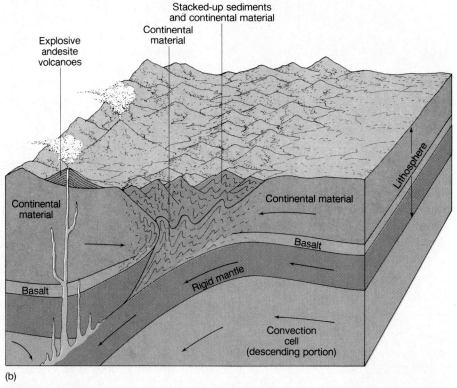

(b)

material becomes molten it moves upward through the crustal fractures adjacent to the subduction zone. The volcanic action that occurs under these conditions is distinctly different from the volcanic action associated with the spreading centers. Volcanic eruptions associated with recycled oceanic crust and silica-rich, water-saturated sediments are explosive and produce a lot of

ash as large amounts of gas and steam are liberated. Volcanic action of this type forms **andesite** volcanoes, named for the type of silicic magma erupted. Such volcanoes are common to the Andes and to the Cascade Mountains of the Pacific Northwest, including Mount St. Helens (see fig. 3.21), and the volcanic peaks that form the island arc chains of the Aleutians, Japan, the Phil-

Figure 3.21

Mount St. Helens erupted violently on May 18, 1980. The mountain lost nearly 4.1 cubic kilometers from its once symmetrical summit, reducing its elevation from 2950 meters to 2550 meters. The force of the lateral blast blew down forests over a 594-square kilometer area. Huge mud floes of glacial melt water and ash flowed down the mountain.

not greatly modified by tectonic processes; they are often extensive, broad, and shallow with thick sedimentary deposits.

often a quiet, smooth outpouring of thick, flowing magma (for example, the Icelandic and Hawaiian volcanoes). See discussion in next section.

Driving Forces for Plate Motion

The actual mechanism that drives the plates apart is still not really understood. The plates could be forced apart at the ridges by the formation of new crust and then pushed downward at the trenches, or the subduction process could be caused by the weight of the thick, dense, older crust and its sediment load sinking into the mantle and dragging the remainder of the plate with it. This latter mechanism would result in tension cracks through which the mantle material escapes upward to form the ridges. In this theory the weight of the dense, cold, descending slab of lithosphere pulls the trailing portion of the plate, and the friction between the moving plate and the asthenosphere helps to drive the convection cell rather than the reverse.

New computer models of this "slab pull" theory lead some researchers to believe that the cold ocean plates can drive plate tectonics without the forces of

seafloor spreading. The situation is probably more complex, combining both mechanisms. Other factors that may be involved include the shape of the earth, changes in the weight of plates resulting from land erosion or the accumulation of sediments, the rate at which magma wells up into the ridges, and the thickening of the older lithosphere.

3.5 The Motion of the Plates

The directions as well as the rates at which pieces of the earth's crust have moved over time have varied. Although we do not understand why such directions have sometimes changed, we are able to trace back these movements and follow paths taken by the continents as they have moved around the earth. Island chains, fixed centers of volcanic activity, and magnetic rocks can help us trace some of these movements.

Rates of Motion

Spreading centers, seafloor spreading, and subduction zones provide the mechanisms for moving the continents. Vine and Matthew's work made it possible to calculate spreading rates, since displacement from the spreading centers is measured against time. Acting like a conveyor belt, the sea floor moves away from the ridge system at a rate of 1 to 10 cm (.5–4 in)/year, and the old crust disappears into the trenches at a comparable rate. Although the rates are small by everyday standards, they produce large changes over geologic time. For example, at the rate of 1.5 cm (0.6 in)/year it takes 100,000 years to move 1.6 km, or 1 mile; therefore, in the 200 million years since the breakup of the Pangaean supercontinent the crust could move 3200 km, or 2000 miles, which is more than half the distance between Africa and South America. Spreading rates moving outward from the ridge system vary in magnitude and with time. In general, a ridge system with a steep profile (like the Mid-Atlantic Ridge) has a slower spreading rate than a ridge system with less steep sides (like the East Pacific Rise). Spreading rates are estimated at 1.25 cm (0.5 in)/year for the Mid-Atlantic Ridge and at 5 cm (2 in)/year for the East Pacific Rise. The largest spreading velocity known is 18.3 cm (7 in)/year off the west coast of South America. It is well to keep in mind that the process of spreading does not occur smoothly and continuously but goes on in fits and starts, with varying time periods between occurrences.

Although seafloor spreading can be observed directly only with great expense and difficulty at sea, there is one place where many of the processes can be seen on land: in Iceland. Iceland is the only large island lying across a mid-ocean ridge, and here geologists can measure seafloor spreading on dry land. Spreading in Iceland occurs at rates similar to those found at the crest of the midocean ridge. Northeastern Iceland had been quiet for 100 years, until volcanic activity began in 1975; this activity has made an 80-km (50 mi) long stretch of the ridge 5m (17 ft) wider in six years. Over one hundred years this spreading rate is 5 cm (2 in)/year, which is within the average range of 1 to 10 cm (.5–4 in)/year. It is also possible that spreading rates may have been more rapid at the time in which the earth's interior was hotter, causing faster convection movements in the mantle.

Hot Spots

Scattered around the earth are approximately forty specific areas of isolated volcanic activity known as **hot spots.** They are found under continents and oceans, in the center of plates, and at the midocean ridges. These hot spots, which channel heat to the surface from deep within the mantle, appear to be fixed at certain locations in the mantle and to erupt periodically. At these sites, a plume of mantle material may force its way through the lithosphere and form a volcanic peak or a seamount directly above, or, if the hot spot does not break through, it may produce a broad swelling of the ocean floor or the continent. Hot spots resupply the asthenosphere, which is constantly cooling and becoming attached to the base of the lithosphere thus thickening the crust. Some people believe the breakup of Pangaea began when the supercontinent came to rest above a chain of hot spots.

Hot spots channel heat to the surface from deep within the mantle. Hot spot plumes of magma are not uniform; they differ in chemistry, suggesting that they come from different depths, and it has been suggested that their discharge rates may also vary. Hot spots may fade away and new ones may form. The life span of a typical hot spot appears to be about 100 million years. Although their positions may change slightly, they tend to remain relatively stable in comparison to the plates, and therefore are useful in tracing plate motions. As the oceanic crust moves over a hot spot, successive eruptions can produce a linear series of peaks or seamounts on the moving crustal plate. In such a series, the youngest peak is above the hot plume and the seamounts increase in age as the distance from the hot spot increases (see fig. 3.22). For example, in the islands and seamounts of the Hawaiian Islands system, the big island of Hawaii is presently located over the hot spot, which feeds its active volcanoes, and in 1981 a new volcano, Loihi, was found to be erupting in this area. Loihi lies 45 km (28 mi) east of Hawaii's southernmost tip.

Figure 3.22

Chains of islands are produced when a crustal plate moves over a stationary hot spot.

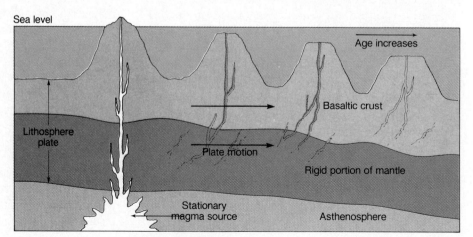

The land features of the island of Hawaii show little erosion, for it is comparatively young. To the west are the islands of Maui, Oahu, and Kauai, which have been displaced by the moving crust. Kauai's canyons and cliffs are the result of erosion over the longer period of time that it has been exposed to the winds and rains. Although these four islands are the most familiar, other islands and atolls attributed to the same hot spot stretch farther west across the Pacific. They are peaks of eroded and subsided seamounts formed by the same hot spot. When the seamounts in tropic areas sank slowly, corals grew upward and coral atolls resulted.

West of Midway Island, the chain of peaks changes direction and stretches to the north, indicating that the crust over the hot spot moved in a different direction some 40 million years ago. This line of peaks is the Emperor Seamount Chain, volcanic peaks that once were above the sea surface as islands but have since eroded and subsided over time, resulting in many flat-topped guyots 1000 m (.6 mi) below the surface. The northern end of the Emperor Seamount Chain is estimated to be 75 million years old, while Midway Island itself may be 25 million years old. Refer to figure 2.12 and colorplate 4 to follow this seamount chain. Notice that the peaks formed by the hot spot get older in the direction in which the plate is moving. The plate is presently moving westward; Midway Island is northwest of the main Hawaiian Islands, and it is also older.

It is possible to check the rate at which the plate is moving by using the distance between seamounts in conjunction with radiometric dating. For example, the distance between the islands of Midway and Hawaii is 2700 km (1700 mi). Midway was an active volcano 25 million years ago, when it was located above the hot spot currently occupied by Hawaii. In other words, Midway has moved 2700 km (1700 mi) in 25 million years, or 11 cm (4 in)/year.

There is another hot spot at 37° 27′S, in the center of the South Atlantic, which is marked by the active volcanic island of Tristan da Cunha. Volcanic activity at this more slowly diverging plate boundary produces seamounts that can be carried either to the east or to the west, depending on the side of the spreading center on which the seamount was formed. The lithosphere is relatively thin here; the hot spot produces a continuous series of seamounts very close together, forming a **transverse ridge:** the Walvis Ridge to the east, between the Mid-Atlantic Ridge and Africa, and the Rio Grande Rise to the west, between the Mid-Atlantic Ridge and South America. Again, refer to figures 2.9 and 2.12.

If a hot spot is located directly underneath a spreading center, the flow of material to the surface under a spreading center is intensified. The crust may thicken and form a platform. Iceland is an extreme example of this process in which the crust has become so thick that it stands above sea level.

Seamount chains, plateaus, and swellings of the sea floor, all products of hot spots, are being used to trace the motion of the earth's plates over known hot spot locations. Recently, hot spot tracks have been used to reconstruct the opening of the Atlantic and Indian Oceans.

Polar Wandering Curves

Magnetic rocks of different ages found on the same continent point to different magnetic pole positions. Records of the angular relationship between the magnetic pole and the geographic pole are locked in these rocks. If the direction of magnetic north recorded in the magnetization of these rocks of various ages is plotted, and if it is assumed that the continent has remained in a fixed position through time, it appears that the earth's magnetic poles have wandered away from their present positions relative to the earth's north-south axis. The path plotted in this way for any continent or region is known as a **polar wandering curve** (see fig. 3.23). Because rocks of the same age from two different continents on different plates point to two different locations for the magnetic pole, and because all evidence points to the magnetic poles remaining close to the geographic poles, even though the magnetic polarity has reversed through

Figure 3.23
The positions of the North Pole millions of years before present (MYBP), judged by the magnetic orientation and the age of the rocks of both North America and Eurasia. The divergence of the two paths indicates the landmasses of North America and Eurasia have been displaced from each other as the Atlantic Ocean opened.

time, it is considered more likely that it is the continents that have moved and not the poles. If the magnetic poles remain close to the geographic poles and the continents move and rotate, then polar wandering curves can be used to track the movements of the continents with time. When the polar wandering curves formed from the changes in the magnetic orientation of North America and northern Europe are superimposed by moving both Europe and North America across the earth's surface, these two landmasses are seen to reconstruct the formation of the two-lobed supercontinent proposed by Wegener, with Laurasia to the north, Gondwanaland to the south, and the Sea of Tethys in between. Based on this method, the present landmasses join nicely at the edges of the continental blocks, at a boundary on the continental slope that is about 2000 m below the present sea level. When the landmasses are moved together, major ancient mountain and fault systems spanning several of our modern continents join; for example, the fault through the Caledonian Mountains of Scotland joins the Cabot Fault extending from Newfoundland to Boston. Also, countries of western Europe and Great Britain that are at present located in temperate and high latitudes were once found in the equatorial zone, which explains their fossil coral reefs and desert-type sand deposits (see fig. 3.6 again).

Current research leads some scientists to believe that the assumption that the rotational polar axis has not varied its location over time may not be correct. The rotating earth possesses tremendous inertia, which keeps it spinning on its axis. However, as the lithospheric plates migrate about on the earth's surface, the earth may tend to roll about its center slightly, trying to adjust so that the majority of the earth's crustal mass is centered about the equator. If this is so, the location of the points marking the earth's rotational axis relative to the crust may also shift. Much more data is required to understand plate motion, assuming fixed positions for the hot spots, and the displacement associated with plates moving relative to the earth's axis of rotation.

3.6 The History of the Continents

The Breakup of Pangaea

About 200 million years ago Pangaea began to break apart. Figure 3.24 traces the plate movements that led to the configuration of the earth's continents and oceans as we know them today. Laurasia (North America and Eurasia) moved away from Gondwanaland, and the South Indian Ocean began to form. India drifted away from Antarctica to move northward toward Asia and its eventual collision with the Asian mainland. Water flooded the spreading center between Africa and South America 135 million years ago; North America and Eurasia were still firmly attached, as were Australia and Antarctica. South America separated from Africa 65 million years ago; Africa moved northward, and the Mediterranean Sea was sealed. India continued to move toward Asia. Australia began to separate from Antarctica, and Madagascar split off from Africa. During the last 16 million years, North America and Eurasia separated, creating the North Atlantic Ocean, and Greenland moved away from Europe. North and South America united, and Australia at last became free of Antarctica. India reached Asia, and because the continental landmass of India was less dense than the basaltic ocean crust, India was not subducted but crumpled up against Asia, resulting in the Himalayas. A similar situation occurred when Italy became detached from Africa and moved across the Mediterranean to collide with Europe and produce the Alps. Twenty million years ago, Arabia moved away from Africa to form the Gulf of Aden and the new and still opening Red Sea.

Before Pangaea

Since the earth is about 4.5 billion years old, there is no reason to believe that Pangaea and its breakup into today's continental configuration was the first or only time during which the sea floor spread or the continents changed position. Scientists are now searching for new and more extensive evidence to indicate the position of the landmasses before the breakup of Pangaea. Because no record of the pre-Pangaean relationships is hidden in the present oceanic crusts, which are only about 200 million years old, scientists must depend on evidence from the continents, where the oldest rocks are found.

Figure 3.24
The movements of the continents from Pangaea to the present.

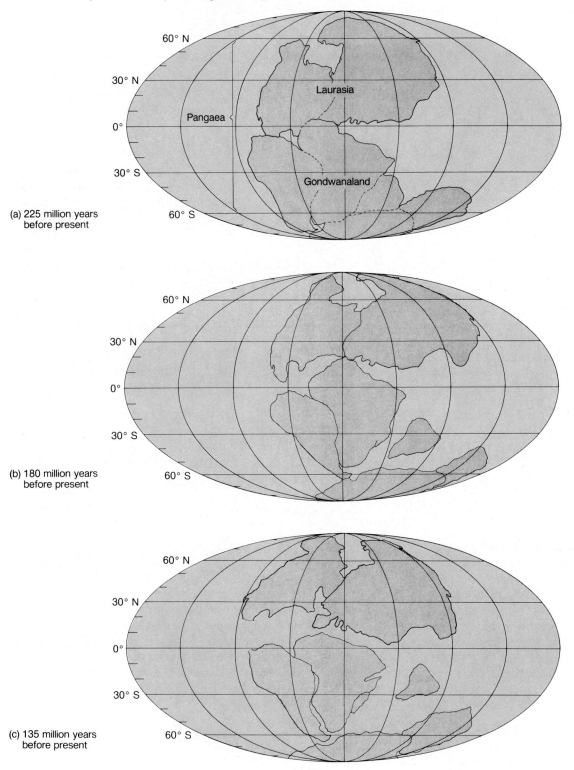

(a) 225 million years
before present

(b) 180 million years
before present

(c) 135 million years
before present

Figure 3.24 continued

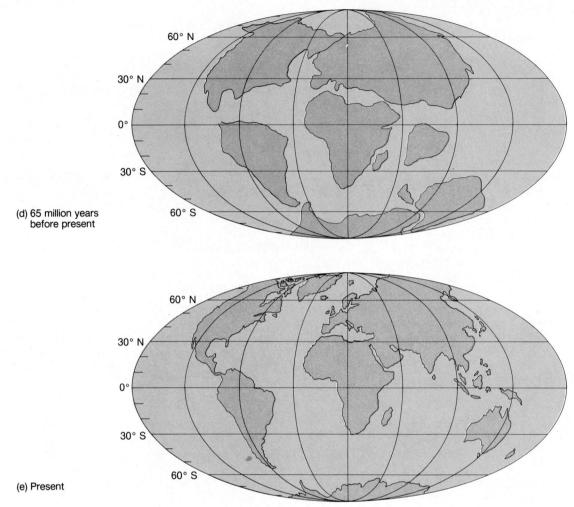

(d) 65 million years
before present

(e) Present

Ancient mountain ranges located in the interiors of today's continents are evidence for collisions of old plates as continents fused together. Seismic evidence has been used to mark old plate edges below the Ural and Appalachian mountains. The magnetic and fossil evidence frozen into the land rocks during the 350 million years prior to Pangaea have been used to propose a series of pre-Pangaea plate movements during the Paleozoic period; which began about 600 million years ago.

Since the earth's polar axis (and therefore its magnetic poles) have held relatively constant positions through time, the magnetic data found in the ancient land rocks (called paleomagnetic data) are the basis for the positioning of the continents during the Paleozoic era. A fixed polar axis also means the earth's climate zones have remained fixed in their latitudinal positions as well. Therefore, any portion of the earth's crust drifting through these climate zones will undergo environmental changes. For example, deposits of mineral

Terranes

Researchers continue to push back the curtain of time and read in the records of landforms, rocks, and fossils the stories of more and more ancient landforms and the ways in which they came to be. Studies of the ancient crustal rocks of the interior of North America indicate that the core of the continent was assembled about 1.8 billion years ago from several large pieces of even older continental land. The large pieces forming the core of the continent are called **cratons.** It is estimated that four or five of these cratons collided and joined over a 100 million-year period about 1.9 billion years ago. Joints along which the pieces fused have been identified and lead to the conclusion that plate tectonics or some form of this process has been operating on earth for at least two billion years.

The boundaries between cratons contain mixtures of sediments from the craton margins and the deep sea floor, the remains of ancient island arc systems, and other small pieces of continental crust trapped between the colliding blocks of continent. The materials in the boundaries became folded, stacked, broken by faulting, and stretched laterally along these joints and fault lines. Radiometric dating has identified boundaries 1.78 billion and 1.65 billion years old between North American cratons, which themselves have ages in excess of 2.5 billion years.

During the 1970s geological studies in Alaska began to show that the geologic pattern found in one area did not necessarily relate to adjacent areas with respect to history, age, structure, or mineral composition; occassionally, an area appeared to be more closely related to an area several hundred miles away. Alaska seemed to be made up of bits and pieces of crustal fragments; some fragments showed the characteristic homogeniety and higher density of oceanic crust while others were highly varied in mineral content and structure and had the lower density associated with continental crust. Crustal fragments associated with cratons, bounded by faults, and with a history distinct from adjoining crustal fragments are known as **terranes.**

Terranes are often elongate, as if produced either by an island arc system that has collided with a craton, or by faulting, which has cut off a sliver of continent similar to the land west of the San Andreas Fault in California. The terranes of Alaska appear to have arrived from the south, rotated clockwise, and been faulted and stretched as the Pacific and North American plates moved relative to each other. For example, a terrane known as Wrangellia originated at or to the south of the equator, moved northward, and crashed into western North America about 70 million years ago. Faulting has spread its fragments throughout Eastern Oregon, Vancouver Island, the Queen Charlotte Islands, and the Wrangell Mountains of South Alaska.

The subcontinent of India is considered by some to be a single, giant terrane; it arrived from Antarctic latitudes and is now firmly attached to the continent of Asia. North of the Himalayas there appear to be terranes which became a part of the continent before the arrival of India. In the western United States the region west of Montana, south into New Mexico, and north into Canada and Alaska appears to be an assemblage of terranes that have been modified by faulting and vulcanism. In Oregon there are volcanic peaks that appear to be basalt seamounts that have moved landward from relatively short distances offshore. In the San Francisco area there are rock formations that are typical of latitudes in the South Pacific. Fossils collected between Virginia and Georgia indicate that a long section up and down the east coast was formed somewhere adjacent to an island arc system; the fossils point to a past European connection.

The more skilled we become at radiometric dating, identifying and relating fossils, and comparative minerology, the more we learn about the diverse nature of the continents. We are gradually beginning to recognize that the continents are not single blocks of land, but are collages made from wandering pieces of land arriving from all corners of our planet. There is much challenging work ahead as geologists follow the history of the earth by charting the movement of terranes, cratons, and crustal plates.

Figure 3.25

Movements of landmasses before Pangaea.

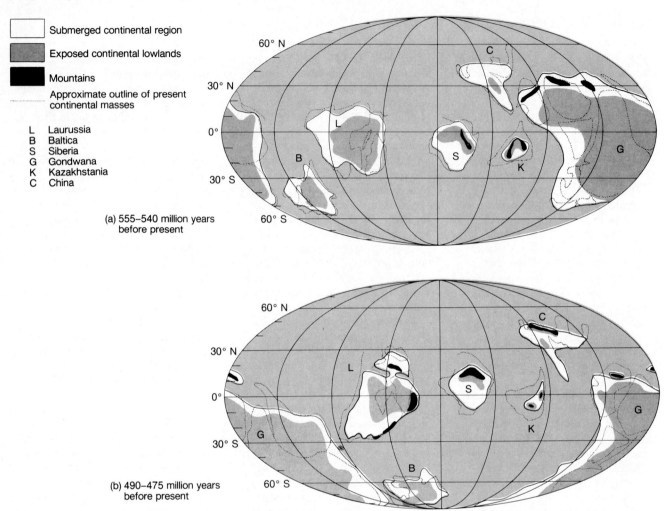

Submerged continental region

Exposed continental lowlands

Mountains

Approximate outline of present
continental masses

L   Laurussia
B   Baltica
S   Siberia
G   Gondwana
K   Kazakhstania
C   China

(a) 555–540 million years
before present

(b) 490–475 million years
before present

salts indicate that an area was once a hot, dry desert, while coal deposits were formed in warm, freshwater swamps; in order for either deposit to have occurred, the land must have been at a latitude favorable to such climatic conditions. Present-day geographic and climatological data are combined with paleomagnetic evidence and paleoclimatology to move the continents back through time. In reconstructing their positions it is assumed that pieces of the earth's crust remained in a climatic zone long enough to collect such a record. The length of the required period varies between 5 million and 15 million years, which is quite reasonable, when we consider that the rate of plate movement is 2 to 8 cm (1–3 in)/year.

Six major continents are recognized from the Paleozoic era: Gondwana (Africa, South America, India, Australia, Antarctica), Baltica (Scandinavia), Laurussia

(North America), Siberia, China, and Kazakhstania. Approximately 540 million to 555 million years ago (fig. 3.25a) these landmasses were strung along the earth's equator; there was no land above 60° N and 60° S, and the polar regions were wide expanses of ocean. Figure 3.25b shows Gondwana and Baltica shifting to the east and south 475 to 490 million years ago. Gondwana's southward motion carried it across the South Pole 430 million to 435 million years ago (fig. 3.25c), and then north again (fig. 3.25d) on the opposite side of the earth, reaching a position that would eventually make it a part of Pangaea about 350 million years ago (fig. 3.25e). This movement completely reversed the north-south orientation of modern Africa and South America. Their present-day southern tips pointed to the north at the beginning of the journey, but as the landmass drifted across the South Pole to the other side of the world, they

Figure 3.25 continued

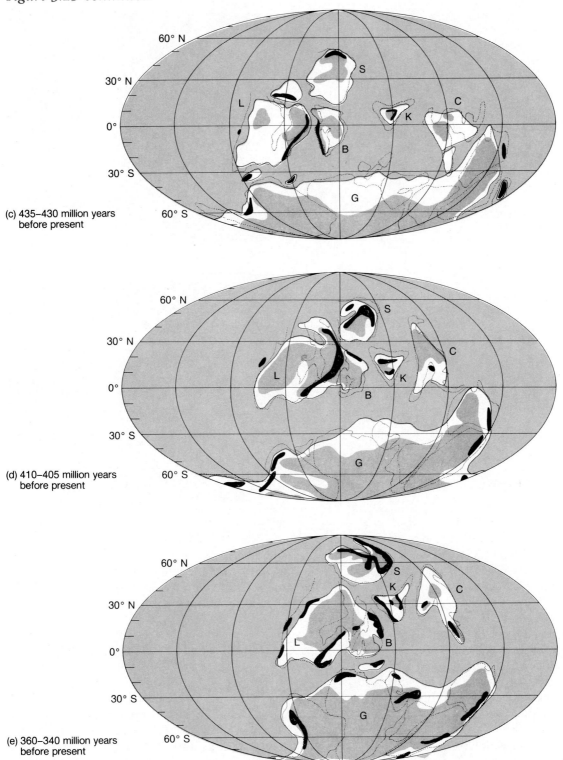

(c) 435–430 million years
before present

(d) 410–405 million years
before present

(e) 360–340 million years
before present

Figure 3.25 continued

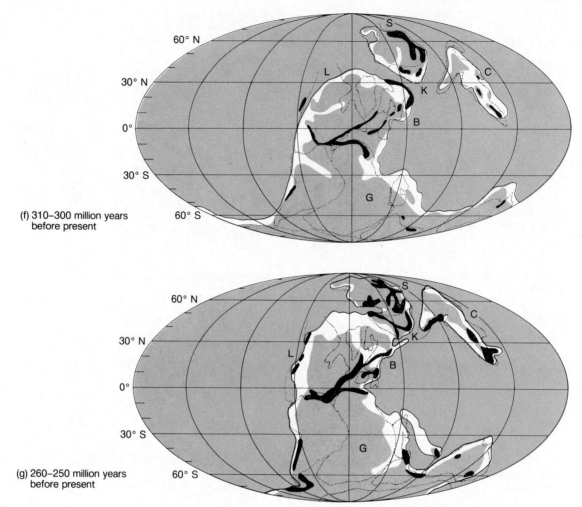

(f) 310–300 million years
before present

(g) 260–250 million years
before present

pointed to the south, as we know them today. About 310 million years ago (fig. 3.25f) the ocean between Gondwana and the other continental fragments began to close. Siberia moved from low to high latitudes and merged with Kazakhstania, while China moved westward (fig. 3.25g). The assembly of Pangaea 250 million to 260 million years ago resulted from a series of collisions that formed great mountain belts that are still detectable on land today.

Because the tectonic processes are continuous, there probably has never been a stable geography of the earth. The modern configuration is no more stable than in the Paleozoic era. They are both steps in an ever-changing pattern. We can even look forward in time to see the Mediterranean Sea being closed and lost as Africa continues to move northward, and the Atlantic and Indian Oceans continuing to expand while the Pacific Ocean narrows as North and South America collide with Asia. Australia will continue to move northward, even-

tually colliding with Eurasia, while Los Angeles and coastal southern California will pass San Francisco on their way toward the Aleutian trench. This is drama on the grandest scale, never to be seen by any one individual, yet it can be appreciated by all.

3.7 Current Research

Drilling Plans

Scientists are currently exploring the ocean bottom with increasingly sophisticated technology as they continue to search for more evidence and a better understanding of present and past events. The new Ocean Drilling Program, guided by the Joint Oceanographic Institutions for Deep Earth Sampling (JOIDES), and using the *JOIDES Resolution* (fig. 3.11) operated by Texas A&M University, plans a ten-year international program of scientific ocean drilling. Objectives of the new program

Figure 3.26
Extruded lava in the rift valley of
the Mid-Atlantic Ridge. This photo
was taken from the submersible
D.S.V. Alvin *during Project FAMOUS.*

include improving our understanding of the nature and evolution of oceanic crust, investigating fossils and the sediments that cover them to create a record of the changing conditions that have occurred on this planet, searching for evidence of global environmental changes, and developing a better understanding of mantle-crust interactions and tectonic processes by recovering rock cores from the seafloor. The ability to drill into the exposed rock of the sea floor will also help us to understand how water circulating through the crust alters the crust's chemical structure and produces the rich mineral deposits associated with divergent plate boundaries (see the section on sulfide deposits in Practical Considerations in chapter 2).

The *JOIDES Resolution* has drilled in the regions of Baffin Bay and the Labrador Sea seeking to determine the tectonic evolution of these two basins. In Caribbean waters, off the island arc of the Lesser Antilles, where the edge of the North American plate moves under the Caribbean plate, cores have been made of the sediments along this plate boundary. Another plate boundary under investigation is along the coast of Peru, where the Pacific plate slides underneath South America. More recently, the *JOIDES Resolution* has drilled in the Antarctic, working to understand the formation of the great ice sheets covering the Antarctic continent. Thousands of meters of core are taken at the drill sites, and it will take many years to complete the analyses of these cores. The results from each cruise will help us to build a more precise and more detailed understanding of the earth, its processes, and its properties.

Project FAMOUS

Modern geological oceanographers and geophysicists have been able actually to descend into the rifts, or cracks, along the ridges where the new crust is produced. During 1972 and 1973 more than twenty U.S., French, Canadian, and British oceanographic vessels mapped a small section of the Mid-Atlantic Ridge. Using photographs, sonar scans, core samples, and dredged samples, this intensive effort produced a detailed map of an area centered at 36°50′N, southwest of the Azores. In this area, in the summer of 1973, French oceanographers aboard the submersible *Archimede* made the first visual study of the rift valley that runs along and splits the ridge, and the following summer (in 1974) the French submersible *Cyana* and the U.S. submersible *Alvin* joined in the dive program. This extraordinary effort was known as the French-American Mid-Ocean Undersea Study, or *Project FAMOUS* (see fig. 3.26). It was conceived to obtain a firsthand, direct, detailed look at the process of seafloor spreading and to increase our understanding of the age and distribution of volcanic activity, lava, and rocks and the location of the plate boundary.

The submersibles descended nearly 3000 m to the floor of the huge rift valley. In a series of 50 dives they took more than 50,000 photographs and made 100 hours of television recording, as well as recovering 150 rock samples and 70 water samples from the bottom. The assumed uniformity of the ocean floor was found to be untrue; the submersibles found an enormous variety of rocks, topography, and sediments. The cracks and fissures running parallel to the direction of the rift indicated a continuous spreading with episodic, localized volcanic eruptions occurring in the central rift area. In the rift valley, most new crust appeared to be formed in a 1-km-wide zone of the inner floor and also in the zone extending out from this area to a total width of about 8 km (5 mi).

Hydrothermal Vents

Another region of intense study is the Galápagos Rift, which lies between the East Pacific Rise and the South American mainland, 964 km (599 mi) west of Ecuador. Measurements of water temperature and chemistry taken in 1972 showed active circulation of seawater through newly formed oceanic crust, which appeared to be the cause of plumes of hot water rising from **hydrothermal vents** along the rift. In 1976 the area's sea floor was mapped in detail in preparation for submersible exploration.

In March 1977, an expedition from Woods Hole Oceanographic Institution, (WHOI) with the submersible *Alvin,* its tender *Lulu,* and the research vessel *Knorr,* arrived to locate and explore the vent areas. The *Alvin* carries two scientists and a pilot. It is equipped with cameras inside and out, as well as baskets for samples taken with its mechanical claws (see fig. 3.27). It dives at a speed of 2 knots to a depth of 3000 m (10,000 ft). It requires three hours for the round trip and *Alvin* spends four to five hours on the bottom. The underwater camera assembly called ANGUS (Acoustically Navigated Underwater Survey System) was slowly towed above the ocean floor to pinpoint the vent area, and the *Alvin* made twenty-four dives to the 2500-m-deep (8000 ft) rift. This process allowed the crew to observe the vents directly, as well as to take still and moving pictures and to collect rock and water samples.

Although the geologic and chemical data in themselves were unique and could be useful in working out the possible mechanisms for mineral deposits around the vents and the heat transfers from the newly formed crust, it was the unexpected and perplexing presence of large communities of animals so far removed from surface food sources that shocked the scientific world. In 1979 the WHOI expedition returned with a full crew of biologists, as well as geologists, from other institutions to explore these dense communities of exceptionally large animals found living in association with the hydrothermal vent areas along the Galápagos Rift.

This time the *Alvin* was equipped with a new color television camera that had a special zoom lens for inspecting the vent animals. Clams, mussels, limpets, tube worms, and crabs were viewed, photographed, and col-

Figure 3.27
The research submersible D.S.V. Alvin *is capable of operating to depths of 3000 meters.*

lected. New species were identified; giant tube worms and clams with red blood and flesh similar to beef were collected for laboratory analysis. The presence of so many large animals at such depths was totally unexpected and immediately brought up the question of their source of food. Such dense communities could not be supported by the fall of organic matter from the sea surface. Instead of being dependent on the sun to provide energy for plant life to produce organic matter to be used as food, these ocean-bottom populations were shown to rely on the hydrogen sulfide and particulate sulfur in the hot vent water, which is utilized by bacteria to provide the energy necessary for life. The other vent animals then feed on the bacteria, which were calculated to reach concentrations as high as 0.1 to 1 gram per liter of vent water. No sunlight is necessary and no food is needed from the surface; the populations are sustained by the vents themselves. In areas where the vents have become inactive the animal communities have died, as

Figure 3.28
Hydrothermal vent with turbid discharge. This "black smoker"
was photographed from Alvin *at the East Pacific Rise.*

their energy source has been removed. Further discussion of these communities, the organisms, and their food supply is found in chapter 16.

Along a section of the Galápagos Rift, the basaltic oceanic crust is new, so new that it shows little sediment cover or chemical change. A spreading rate of 3.5 cm (1.4 in)/year occurs across the rift, while the sedimentation rate is only about 5 cm (2 in)/1000 years. Some of the basalt shows no sediment, indicating its formation within the last 100 years. At the outer edges of the rift valley, about 500 m (1640 ft) from the central rift, sediments are thick enough to cover the new lava, indicating an age of about 10,000 years.

Hot water issuing from the vents has temperatures as high as 17° C, compared to 2° C for the surrounding seawater. The hot water mixes with the cold bottom water to produce shimmering, upward-flowing streams rich in silica, barium, lithium, manganese, hydrogen sulfide, and sulfur. Rocks in the vicintiy of the vents are coated with chemical deposits rich in metals precipitated out of the vent waters.

The Mexican-French-American research program RISE (Riviera Submersible Experiment) investigated ocean floor hot springs near the tip of Baja California in 1979 (see fig. 3.28). RISE's goal was to gain more information about crustal patterns along a spreading

Megaplume

On August 17, 1986, Dr. Edward Baker and Gary Massoth from the Pacific Marine Environmental Laboratory of the National Oceanographic and Atmospheric Administration (NOAA) were making water property measurements aboard the NOAA research vessel *Discoverer.* They were 450 km (260 mi) off the coast of Oregon, just north of the southern end of the Juan de Fuca Ridge. The Juan de Fuca Ridge and the Gorda Ridge to the south were being explored for hydrothermal vents.

Their heat-sensing instruments indicated that they were above a large spherical mass of warm water, 0.25° C above the temperature of the surrounding water and 1300–2000 m (4000–6500 ft) below the surface. Using their heat sensors, they were able to determine that the warm-water mass was about 700 m (2000 ft) thick and 20 km (12 mi) in diameter; its top was about 1000 m (3300 ft)

above the sea floor. The warm water mass was christened a **megaplume.**

Baker and Massoth asked themselves how much heat must have issued from the earth's interior in order to produce such a huge mass of warmed water. They estimated that if the water issuing from the sea floor had been 350° C, the heat that escaped from the earth's crust would have been equal to the amount of heat escaping from about 1000 known hydrothermal vents for an entire year, or the average annual heat supplied from a section of spreading center 50 km (30 mi) long.

Samples of water collected from the megaplume contained gases and large mineral crystals. Because the crystals had not settled out or dissolved, it was concluded that the plume had formed rapidly and was very new when discovered, perhaps less than a week old. The megaplume did not reach the sea surface;

therefore, it is unlikely that it was formed from a small point discharge or a single explosive or violent event. To produce such a large volume of water, it is more likely that the megaplume formed over a great crack in the sea floor with an area larger than that of any known hydrothermal vent.

The discovery of the megaplume has raised more questions than it has answered: How did the large volume of water needed to produce the megaplume seep into the seafloor crust? How was the water stored, and what was the mechanism that released it? Do other megaplumes exist? Was this a special event or are such plumes produced on a regular time cycle? We do not know, but we do know that a plume of this type does not last very long. When the researchers returned to the same place two months later, the megaplume had disappeared.

center. Once again *Alvin* descended to the bottom, 2000 m (6056 ft) down. In this area, mounds and chimney-shaped vents 20 m (65.6 ft) high eject streams of black mud containing particles of lead, cobalt, zinc, silver, and other minerals. More remarkable still are the water temperatures recorded at the mouths of these mounds, up to 350° C. Here, too, the vents are surrounded with rich animal life, including clams up to 40 cm (15.7 in) across, huge tube worms, anemones, and fish.

Since these initial discoveries, expeditions have continued to search for hydrothermal vents along the spreading centers of the world's plates. In September, 1981, researchers first photographed a small section of a rift valley in the ridge system bounding the small Juan de Fuca Plate off the coast of Oregon. The *JOIDES Resolution* identified vents along the Mid-Atlantic Ridge while on its drilling stations in 1986. That same summer

the U.S. Navy submersible, *Sea Cliff,* with scientists from the U.S. Geological Survey, surveyed additional sites on the Gorda Ridge of the Juan de Fuca plate.

3.8 Practical Considerations: The Addition of Waste to the Sea Floor

Using the sea as a dump for trash and garbage is a common practice around the world. Domestic, industrial, and chemical wastes have been poured and dumped into the oceans. After each war, obsolete military hardware and munitions have also been disposed of in this way. The North Atlantic was the dumping ground for toxic gases confiscated from Germany at the end of World War II, and the bays and lagoons of many

South Pacific islands were used for the disposal of jeeps, tanks, bombs, and other items. After the United States left Vietnam, there was a similar unloading of vehicles and explosives in the waters around Southeast Asia.

The production of highly radioactive nuclear waste and hazardous chemical waste products is an unavoidable consequence of nuclear power stations, high-tech industry, modern medicine, and military systems. Today, hazardous chemical wastes and high-level nuclear wastes are usually stored underground or in above-ground repositories. However, it has been suggested that to avoid possible contamination of land areas and freshwater supplies these long-lived wastes should be disposed of away from the continents in the center of the oceanic plate or in the deep-sea trenches of the oceans. Since the trenches are subduction zones, it is pointed out that in time the waste products will sink back into the mantle and so eliminate the problem of long-term hazardous waste storage. Such plans are currently restricted by the 1972 Ocean Dumping Act. Disposal of radioactive wastes requires not only the development of an environmental impact statement but also the approval of both houses of Congress within ninety working days. Because of the latter restriction, the U.S. Navy's plans to dispose of outdated nuclear submarines at sea were abandoned and they are stored on land instead.

The deep ocean basins are the lowest places in the earth's crust, all the products of society and nature ultimately work their way down to these basins. Natural catastrophes, climate changes, and new technologies all leave their mark on the sea floor, to be removed at last by the earth's geologic processes that have been discussed in this chapter.

Summary

The earth is made up of a series of concentric layers: the crust, the mantle, the liquid outer core, and the solid inner core. The evidence for this internal structure comes indirectly from studies of the earth's dimensions, density, rotation, gravity, and magnetic field, and of the remains of meteorites. It also comes from the ways in which seismic waves change speed and direction as they move through the earth. The discontinuity that occurs in the speed of seismic waves between the crust and the mantle is termed the Moho.

Continental crust is formed from granite-type rock, which is less dense than oceanic crust formed of basalt. The top of the mantle is fused to the crust to form the rigid lithosphere. The lithosphere floats on the deformable upper mantle, or asthenosphere. The pressures underneath the elevated continents and depressed ocean basins are kept in balance by vertical adjustments of the crust and mantle, a process known as isostasy.

Alfred Wegener's theory of drifting continents was based on the geographic fit of the continents and the similarity of fossils collected on different continents. His ideas were ignored until the discovery of the midocean ridge system and until the proposal of convection cells in the asthenosphere led to the concept of seafloor spreading. New lithosphere is formed at the ridges, or spreading centers. Old lithospheric material descends into trenches at subduction zones. Seafloor spreading is the mechanism of continental drift. Evidence for lithospheric motion includes the match of earthquake zones to spreading centers and subduction zones, the greater heat flow along ridges, age measurement of seafloor rocks, age and thickness measurements of sediment from deep-sea cores, and the magnetic stripes in the seafloor on either side of the ridge system. Rates of seafloor spreading average 1 to 10 cm/year. Plate tectonics is the unifying concept of lithospheric motion. Plates are made up of continental and oceanic lithosphere bounded by ridges, trenches, and faults.

Rift zones separate ocean basins, and in the past have separated landmasses and produced new ocean basins. Trailing plate margins move away from spreading centers. Leading plate margins move toward subduction zones. Plate collisions can produce island arcs, earthquakes, and volcanic activity. The exact mechanism that drives the plates is unknown.

Plate movements traced over the last 225 million years show the breakup of Pangaea. Paleozoic plate movements in the 225 million years prior to Pangaea have recently been estimated using climatological evidence.

Deep-sea drilling programs are being conducted to investigate tectonic processes, the evolution of oceanic crust, and the sediment records of past conditions. Submersibles have been used to explore the rift valley of the Mid-Atlantic Ridge and the hydrothermal vents and animal communities found along the Galapagos Rift, Baja California, and the Oregon coast.

Dumping waste into ocean waters has been a common practice. Using subduction zones to dispose of hazardous chemical and nuclear wastes has been suggested.

Key Terms

inner core	basalt	seafloor spreading	graben
outer core	Moho	spreading center	trailing margin (edge)
mantle	lithosphere	subduction zone	leading margin (edge)
crust	asthenosphere	polar reversal	andesite
density	isostasy	plate tectonics	hot spot
seismic wave	Gondwanaland	divergent plate boundary	transverse ridge
refraction	continental drift	convergent plate boundary	polar wandering curve
P-waves	Pangaea	fault	hydrothermal vent
S-waves	Laurasia	transform fault	craton
seismic tomography	convection cell	escarpment	terrane
granite	magma	rift zone	megaplume

Study Questions

1. What is meant by the term "polar wandering"? Have the magnetic poles actually wandered?

2. Describe the three types of plate boundaries. What processes take place at each type of boundary? In what direction do the plates move at each boundary?

3. What mechanisms have been proposed to account for plate motion?

4. What is the difference between the leading edge of a plate margin and the trailing edge? A divergent plate boundary and a convergent plate boundary?

5. If the ability of the oceanic crust to transmit heat were uniform, the rate of heat flow through the ocean floor would depend only on the temperature change across the oceanic crust. Under such a condition, how do the heat flow measurements in figure 3.9 indicate the presence of ascending convection cells in the asthenosphere?

6. If the polar wandering curves for North America and Europe are made to coincide, how will these continents move relative to each other?

7. Using the techniques and reasoning employed to discover the properties of the interior of the earth, explain how you would determine what is inside a sealed box (for example, measuring the box, weighing it, spinning it, balancing it on different axes, sampling its exterior). What clue would each of these measurements give you?

8. Why were no further advances made with the ideas of Alfred Wegener until the 1950s?

9. Why does a newly formed midocean volcanic island gradually subside?

10. Explain the formation and symmetry of the magnetic stripes found on either side of the midocean ridge system. What is their significance when the magnetic information is correlated with the age of the crust?

11. Under what conditions will a subduction zone form a mountain range? An island arc system? Why do volcanoes associated with subduction zones usually erupt more explosively than mid-ocean volcanoes associated with hot spots and spreading centers?

Study Problems

1. If a plate moves away from a spreading center at the rate of 5 cm per year, what is the displacement of a land mass carried by that plate after 180×10^6 years?

2. Magnetic stripes with the same magnetic orientation are measured on either side of a ridge crest. The stripe on the west side of the ridge is displaced 11 km from the crest; the stripe on the east side is displaced 9 km from the crest. The age of the rock in both stripes is 4×10^5 years. Calculate the average spreading rate at this ridge.

3. If the north end of the Emperor Seamount Chain is 75 million years old and Midway Island is 25 million years old, what was the rate of movement of the Pacific Plate during the period of the seamount chain's creation? (You will have to use an atlas to determine the distance between the north end of the Emperor Seamount Chain and Midway Island.) What can you deduce about the past direction of the Pacific Plate's movement compared to its present direction of motion from the orientation of the islands and seamounts from Hawaii to Midway and from Midway to the north end of the Emperor Seamount Chain?

The Properties of Water

4

A sudden fog-drift muffled the ocean,
A throbbing of engines moved in it,
At length, a stone's throw out, between the rocks and the vapor,
One by one moved shadows
Out of the mystery, shadows, fishing-boats, trailing each other
Following the cliff for guidance,
Holding a difficult path between the peril of the sea-fog,
And the foam on the shore granite.
One by one, trailing their leader, six crept by me,
Out of the vapor and into it.
The throb of their engines subdued by the fog, patient and cautious,
Coasting all around the peninsula
Back to the buoys in Monterey harbor. A flight of pelicans
Is nothing lovelier to look at;
The flight of the planets is nothing nobler; all the arts lose virtue
Against the essential reality
Of creatures going about their business among the equally
Earnest elements of nature.

Robinson Jeffers,
Boats in a Fog

Water is the most common of substances on our earth and yet it is uncommon in many of its properties. Water is a unique liquid. It makes life possible and, to a very large degree, its properties determine the characteristics of the oceans, the atmosphere, and the land. In order to understand the oceans, it is necessary to examine water as a substance and to learn something of its physical and chemical characteristics. In this chapter we learn about the structure of the water molecule and explore the properties of water.

4.1 The Water Molecule

About a billion years after the formation of the earth and as a consequence of the reorganization of the planet described in chapter 1, water released from the earth's interior and carried to the surface as water vapor condensed to its liquid form. The properties of water have excited scientists for over 2000 years. The early Greek philosophers (500 B.C.) counted four basic elements from which they believed all else was made: fire, earth, air, and water. More than 1800 years later the English scientist Henry Cavendish, in 1783, determined that water was not a simple element but a substance made up of hydrogen and oxygen. Shortly afterward another Englishman, Sir Humphrey Davey, discovered that the correct formula for water was two parts hydrogen to one part oxygen, or H_2O.

The properties that make water such a unique, useful, and essential substance are the result of the chemical properties of the water molecule. The water molecule is deceptively simple. Each water molecule is made up of three atoms: two hydrogen atoms and one oxygen atom. The positively charged ($+$) hydrogen atoms and negatively charged ($-$) oxygen atom attract one another, but the two positively charged hydrogen atoms repel each other. The result is that each water molecule has a specific shape. The central core, or nucleus, of the oxygen atom forms an angle of 105° with the nuclei of the two hydrogen atoms (see fig. 4.1a). This V-shaped molecule has the two hydrogen atoms with their positive charges at one side and the oxygen atom with its negative charge at the other.

The three atoms are linked by **covalent bonds.** Two hydrogen atoms each form a covalent bond with a single oxygen atom; each bond is formed by the sharing of two electrons between the two atoms. The single negatively charged electron of the hydrogen atom is shared with the oxygen atom, which contributes one of its negatively charged electrons to the hydrogen atom (fig. 4.1b). Each V-shaped molecule of water has a neutral charge, but the oxygen atom pulls the shared electrons toward itself and away from the hydrogen atom. In this way, the oxygen atom becomes slightly negatively charged on one side of the molecule and the hydrogen atoms are weakly positively charged on the other. Because the opposite sides of the water molecule have opposite charges, the molecule is termed a **polar molecule.**

Because of the distribution of the charges, when more than one water molecule is present the molecules form bonds between one another. The positively charged side of one molecule attracts and bonds with the negatively charged side of another molecule (see fig. 4.1c). These bonds are known as **hydrogen bonds.** Each water molecule can establish hydrogen bonds with four other water molecules. Any single hydrogen bond is relatively weak (less than one-tenth the strength of the covalent bonds between the hydrogen and oxygen atoms), but as one hydrogen bond is broken, another is formed. As a result, the water molecules cling together.

Figure 4.1

The water molecule. (a) The angle at which the hydrogen atoms bond to the oxygen atom results in a polar molecule. (b) The hydrogen atoms share electrons with the outer ring of the oxygen atom. (c) The positive and negative charges allow each water molecule to form hydrogen bonds with other water molecules.

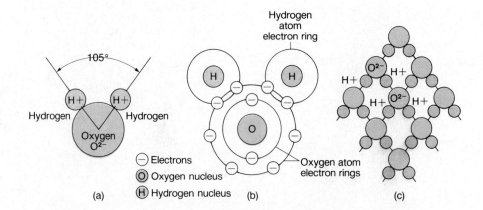

4.2 Changes of State

Water exists in three forms: as a solid, a liquid, and a gas. When it is a solid we refer to it as ice, and when it is a gas we call it water vapor. Pure water ice melts at 0° C and pure water boils at 100° C at standard atmospheric pressure. Pure water is defined as fresh water without suspended particles or dissolved substances, including gases. Standard atmospheric pressure is equal to 76 cm of mercury and is discussed in chapter 7. If you are unfamiliar with the Celsius temperature scale, see the Appendix.

When pure water makes any of the changes among liquid, solid, and gas, it is said to change its state. Changes of state are due to the addition or loss of heat. When enough heat is added to solid water (or ice) to raise its temperature above its freezing point, it melts to form liquid water. When heat is added to liquid water, the temperature of the water rises and some of the water evaporates to form water vapor. When heat is removed from water vapor and the temperature falls below the **dew point,** or temperature of water vapor saturation, the water vapor condenses to liquid. When liquid water loses heat and its temperature is lowered to its freezing point, ice is formed.

Heat is a form of energy and can be measured in **calories.** One calorie is the amount of heat needed to raise 1 gram of water 1° C. Note that 1000 of these calories is equivalent to 1 Calorie, kilocalorie (K-calorie), or food calorie. Do not confuse heat measured in calories and temperature measured in degrees. Heat is energy, while temperature measures only the gain or loss of heat that occurs in a body as heat energy is added or removed. For example, a small pan of fresh water is placed on the stove; the burner adds heat. When enough heat is added to bring the water to 100° C, the water is at the boiling point. If the pan contains 2 cups of water this process will take perhaps 3 to 5 minutes. If a second container of water is then placed on the same burner and this container holds 3 quarts of water, the burner must add considerably more heat, taking 15 to 20 minutes in order to bring the water to the boiling point. The same temperature change occurs in both cases, but the amount of heat required to make that change is different.

To change pure water from its solid state (ice) to liquid water at 0° C requires the addition of 80 calories for each gram of ice. There is no change in temperature; there is a change in the physical state of the water. The reverse of this process is required to change liquid water to ice. For each gram of liquid water that becomes ice, 80 calories of heat must be removed at 0° C. The heat necessary to change the state of water between solid and liquid is called the **latent heat of fusion.** This addi-

tion or loss of heat takes time in nature. A lake does not freeze immediately, even though the surface water temperature is 0° C, nor does it thaw on the first warm day. Time is required to remove or add the heat needed for the change of state.

One gram of liquid water requires 1 calorie of heat to raise its temperature 1° C. Therefore, 100 calories are needed to change the temperature of 1 g of water from 0° to 100°C. In comparison, the 80 calories of heat required per gram to convert ice to and from liquid water at 0° C with no change in temperature is relatively large. Ice is a very stable form of water; it takes a lot of heat to melt it and a large loss of heat to form it.

The change of state between liquid water and water vapor requires even greater quantities of heat. To convert 1 g of water to water vapor at 100° C requires 540 calories of heat. When 1 g of water vapor condenses and returns to the liquid state, 540 calories of heat are liberated. Again, there is no change in temperature; there is only a change in the water's physical state. The heat needed for a change between the liquid and vapor states is the **latent heat of vaporization.**

The energy required to change the state of water from solid to liquid to vapor is shown in figure 4.2. Both the latent heat of fusion and the latent heat of vaporization are presented in this diagram. Note that the calories needed for each change of state and the temperatures at which these changes occur are also included. Water converts from the liquid to the vapor state at temperatures other than 100° C; for example, rain puddles evaporate and clothes dry on the clothesline. This type of change requires slightly more heat to convert the liquid water to a gas at these lower temperatures (see table 4.1).

As water is evaporated from the world's lakes, streams, and oceans, and is then returned as precipitation, heat is being removed from the earth's surface and liberated into the atmosphere. This heat energy is a major source of the energy used to power the earth's weather systems. A hurricane is a dramatic example; in its beginning it is formed over the tropic ocean and extracts large amounts of water vapor, therefore heat, from the sea surface. The vast amounts of heat energy liberated from the condensing water vapor in the atmosphere fuel the storm winds, raising them to destructive levels. A major hurricane contains energy exceeding that of a large nuclear explosion; fortunately, the energy is released much more slowly. The energy generated by a hurricane is about 3×10^{12} watt-hours per day. This amount is the equivalent to the energy in 1 million tons of TNT.

Under certain conditions it is possible (1) to cool liquid water below 0° C and keep it as a liquid, (2) to

Figure 4.2

Heat energy must be added to convert a gram of ice to liquid water and to convert liquid water to water vapor. The same quantity of heat must be removed to reverse the process.

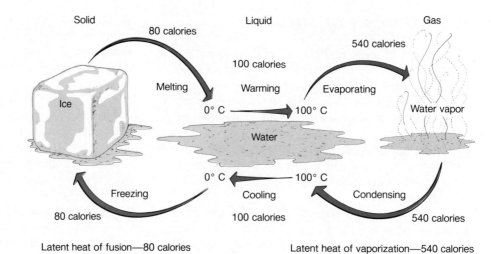

TABLE 4.1.

Energy Required to Convert Water to Water Vapor (Latent Heat of Vaporization)

Temperature of water (° C)	Calories required (Per g of water)
0°	596.0
10°	590.8
20°	585.6
30°	580.4
40°	575.2
50°	568.5
60°	563.2
70°	557.5
80°	551.7
90°	545.8
100°	539.5
110°	532.9
120°	525.7

change ice directly to a gas, a process known as **sublimation,** or (3) to boil water at temperatures below 100° C. Cooling below 0° C is accomplished in the laboratory by controlled cooling of pure water to produce supercooled water. Sublimation is seen in nature when snow or ice evaporates directly under very cold and dry conditions, as on the high desert lands of Utah and Arizona. Anyone living at high altitudes knows that potatoes must be boiled longer and that brewed coffee is cooler there than at sea level because of the decrease in atmospheric pressure and the lower boiling temperature of the water.

The previously described behavior of water is explained by considering processes occurring between water molecules. At the molecular level the addition of heat energy increases the speed with which molecules move while the loss of heat energy decreases their rate of motion. Heat energy is required to break hydrogen bonds between water molecules, and heat is released when hydrogen bonds are formed. The addition of heat to water causes a relatively small change in the temperature of the water, because much of the heat energy is used to disrupt the hydrogen bonds. These bonds must be broken before the molecules are able to move more rapidly. When heat is removed from water, the molecules slow, many additional hydrogen bonds are formed, and considerable energy is released as heat, which prevents any rapid drop in temperature.

Water molecules stay close together because they are attracted to each other. If the molecules are moving fast enough, they overcome their attractions and leave the liquid, entering the air as a gas. Some of the most rapidly moving molecules move into the air even at low temperatures. The greater the addition of heat, the greater the average energy of motion of the molecules, and more water molecules leave the liquid more quickly. Relatively large amounts of heat are needed to evaporate water because hydrogen bonds must first be broken. In the same way, large amounts of heat must be extracted from water to form the hydrogen bonds required to freeze water. On earth, natural temperatures required for boiling are rare and those for freezing are infrequent; therefore, liquid water is abundant.

The addition of salt to water changes its boiling and freezing points. The boiling temperature is raised and the freezing temperature is lowered. The amount of change is controlled by the amount of salt added. The change in the boiling temperature is of little consequence since seawater does not normally reach such high temperatures in nature, but the lowering of the temperature of the freezing point is important in the formation of sea ice. Seawater freezes at about −2° C.

TABLE 4.2

Heat Capacity of Common Materials

Material	Heat capacity (calories/g/° C)
Acetone	0.51
Aluminum	0.22
Ammonia	1.13
Copper	0.09
Grain alcohol	0.23
Lead	0.03
Mercury	0.03
Silver	0.06
Water	1.00

4.3 Heat Capacity

Of all the naturally occurring earth materials, water changes its temperature the least for the addition or removal of a given amount of heat. The ability of a substance to give up or take in a given amount of heat and undergo large or small changes in temperature is a measure of the substance's **heat capacity.** The heat capacity of water is very high compared with that of the land and the atmosphere. For example, summer temperatures in the Libyan desert reach 50° C and temperatures in the Antarctic drop down to −50° C, for a world range of 100° C. Ocean temperatures vary from a nearly constant high of approximately 28° C in the equatorial areas to a low of −2° C in Antarctic waters, for a world range of 30° C. The water in a lake changes its temperature very little between noon and midnight, while the adjacent land and air temperature changes are large during this same time. The high heat capacity of water as well as the ability of water to redistribute heat over depth allow the world's lakes and oceans to change temperature slowly, helping to keep the earth's surface temperature more stable and consequently more liveable.

When the heat capacity of a substance is divided by the heat capacity of pure water the resulting ratio is termed the **specific heat** of the substance. The specific heat of a substance has no units but is numerically equal to heat capacity. The heat capacities of some common materials are given in table 4.2. When salt is added to pure water the changes in heat capacity, latent heat of fusion, and latent heat of vaporization are small. Heat capacity and specific heat, like changes of state, are related to the hydrogen bonding between water molecules. Review the discussion of the topic in the preceding section.

4.4 Cohesion, Surfac[...] and Viscosity

In its liquid state the bonds [...] are quite fragile; they form, [...] frequency. Each bond lasts [...] second. However, at any ins[...] centage of all water molecules [...] neighbors. Therefore, water has [...] other liquids. Collectively, the hyd[...] water together; this is known as **cohes[...]**

Cohesion is related to **surface ten[...]** a measure of how difficult it is to stretch [...] surface of a liquid. At the surface between air [...] water molecules arrange themselves in an [...] system, hydrogen-bonded to each other and to the [...] molecules beneath. This forms a weak elastic m[...]brane at the surface that can be demonstrated by filli[...] a water glass carefully; the water can be made to brim above the top of the glass but not overflow. A steel needle can be floated on water; insects like the water strider walk about on the surface of lakes and streams. All of these things are possible because water has a high surface tension. This property is important in the early formation of waves. A gentle breeze stretches and wrinkles the smooth water surface, enabling the wind to get a better grip on the water and add more energy to the sea surface.

The addition of salt to pure water increases the surface tension. Decreasing the water temperature also increases the surface tension, while increasing the water temperature decreases it.

Liquid water pours and stirs easily. It has little resistance to motion or internal friction. This property is called **viscosity.** Water has a low viscosity when compared to motor oil, paint, or syrup. Viscosity is affected by temperature. Consider pancake syrup. When it is stored in the refrigerator it becomes thick and slow to pour. It has a high viscosity. When the syrup is returned to room temperature or heated, it becomes thin and runny; it has a low viscosity. The same is true of water, but the change in viscosity is much less, so that it is not noticeable under normal temperature changes. Surface water at the equator is warmer and therefore less viscous than surface water in the Arctic. Minute microscopic organisms find it easier to float in the more viscous polar waters; their tropical-water cousins have adapted to the less viscous water by developing spines and frilly appendages to help keep them afloat. The addition of salt to pure water increases the viscosity of the water, but the change is small.

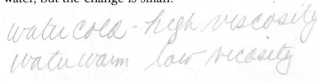

4.5 Compressibility

Pure water is a nearly incompressible substance. The same is true of seawater. Pressure in the oceans increases with depth. For every 10 m of descent in depth, the pressure increases by about 1 atmosphere. One atmosphere is equal to the pressure of a column of mercury 76 cm (30 in) high; in English units, 1 atm equals 14.7 lb/in². Pressure in the deepest ocean trenches, 11,000 m (36,080 ft) deep, is about 1100 atm. These great pressures have only a small effect on the volume of the oceans because of the slight compressibility of the water. A cubic centimeter of seawater at the surface will lose only 1.7% of its volume if it is lowered to 4000 m, with a pressure of 400 atm. The average pressure acting on the total world's ocean volume results in the reduction of ocean depth by about 53 m (175 ft). In other words, if the ocean water were truly incompressible, sea level would stand about 53 m (175 ft) higher than it does at present. The pressure effect is small enough to be ignored in most instances, except when very accurate determination of density is required. Further information on units of pressure is found in the Appendix.

4.6 Density

Density is defined as mass per unit volume of a substance and is measured in grams per cubic centimeter, or g/cm³. Pure water has a density of 1.00 g/cm³ at 4° C. Thus a cube of pure water that is 1 cm high, 1 cm wide, and 1 cm deep has a mass of 1 g. The density of seawater is greater than the density of pure water at the same temperature, because seawater contains dissolved salts. At 4° C the density of seawater of average salinity is 1.0278 g/cm³. The densities of other substances may be considered in the same way (see table 4.3). Less dense substances will float on more dense liquids (for example, oil on water, dry pine wood on water, and alcohol on oil). Note from table 4.3 that ocean water is more dense than fresh water; therefore fresh water floats on salt water.

The Effect of Temperature

Water density is very sensitive to temperature changes. When water is heated, energy is added and the water molecules move apart; therefore the mass per cubic centimeter becomes less because there are fewer water molecules per cubic centimeter. For this reason, the density of warm water is less than that of cold water, so that warm water floats on cold water. When water is cooled it loses heat energy, and the water molecules

TABLE 4.3
Densities of Common Materials

Material	Density (g/cm³)
Ice (pure) 0° C	0.917
Water (pure) 0° C	0.99987
Water (pure) 3.98° C	1.0000
Water (pure) 20° C	0.99823
White pine wood	0.35–0.50
Olive oil 15° C	0.918
Ethyl alcohol 0° C	0.791
Seawater 4° C, 35‰	1.0278
Steel	7.60–7.80
Lead	11.347
Mercury	13.6

slow down and come closer together; there are then more water molecules, or a greater mass, per cubic centimeter. Because cold water is more dense than warm water, it sinks below the warm water. To see these changes try the following experiment: Chill a small quantity of water in the refrigerator; mix it with a little dye such as ink or food coloring to make it easy to see. Allow the water from the tap to run hot and fill a glass half full of the hot water. Now slowly and carefully pour the colored water down the side of the glass on top of the hot water and watch what happens.

Water molecules move closer and closer together as water is cooled to 4° C. At this temperature fresh water is the most dense at 1 g/cm³. As the temperature of the water gets closer to 0° C, the water molecules form a crystal lattice, each molecule bonded to a maximum of four other molecules. Water begins to freeze when the water molecules no longer move enough to break hydrogen bonds with neighboring molecules. The hydrogen bonds keep the molecules far enough apart to make ice about 10% less dense than liquid water at 4° C. At 3° – 0° C the water is less dense than at 4° C (see fig. 4.3) and therefore stays at the surface, where ice is formed with further cooling. As the ice crystals form, the water molecules become widely spaced, resulting in fewer water molecules per cubic centimeter. Therefore ice is less dense than water, and it floats on water.

When ice absorbs enough heat to increase the temperature above 0° C, the hydrogen bonds between molecules are disrupted, and the crystal lattice collapses. The ice melts and the molecules move closer together. Above 4° C, the water begins to expand again, due to the increasing space between the faster moving molecules. If enough heat is added, the energy level of

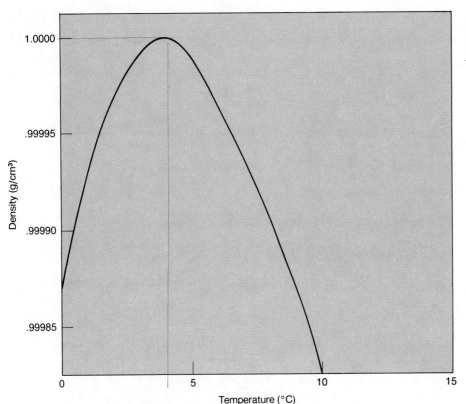

Figure 4.3
The density of pure water reaches its maximum at 3.98° C. Pure water is free of dissolved gases.

the molecules increases until they can overcome the attraction between one another and the vapor pressure of the air at the surface of the water. The molecules can then escape, or evaporate, into the air as water vapor. The density of water vapor is less than that of liquid water; that is, there are fewer water molecules per cubic centimeter. Water vapor is less dense than the mixture of other gases that form the atmosphere. Therefore a mixture of water vapor and dry air is less dense than dry air alone at the same temperature and pressure.

The Effect of Salt

When salts are dissolved in pure water the density of the water increases because the salts have a greater density than water. In other words, they have more mass per cubic centimeter. The typical density of seawater is approximately 1.028 g/cm³, in comparison to 1 g/cm³ for fresh water. Therefore fresh water floats on salt water. To see this happen, take two small quantities of fresh water; add dye to one and some salt to the other. Carefully and slowly add the salt water to the dyed fresh water and watch what happens.

If seawater contains less than 24.7 grams of salt per kilogram of seawater, then the seawater, although it is more dense, will behave much like fresh water. Seawater of this concentration will reach its maximum density before it freezes. Cooling surface water with a salt content of less than 24.7 g/kg causes the surface water to increase in density and sink. This process continues with surface cooling until the temperature of maximum density is reached. Further cooling causes the low-salinity surface water to become less dense and remain at the surface. At a salt content of 24.7 g/kg, the freezing point and the maximum density of seawater coincide at —1.332° C. If the salt content is greater than 24.7 g/kg, freezing occurs before the maximum density is reached. This relationship is shown graphically in figure 4.4. Open ocean water generally has a salt content greater than 24.7 g/kg; therefore cooled surface water sinks continuously as its density increases.

The Effect of Pressure

Density is also affected by pressure, and although the effect is small it is taken into consideration by research oceanographers, because extremely small changes in density are important in creating density-driven vertical water movements. Increasing the pressure increases the density of the water by reducing the volume occupied by a fixed mass of seawater. The reduction in volume results because the water and salt molecules are being crowded together by the applied pressure. The small reduction in volume is the compressibility of the water, which was discussed previously.

Figure 4.4
The initial freezing point and the temperature of maximum density of water decrease as the salt content increases.

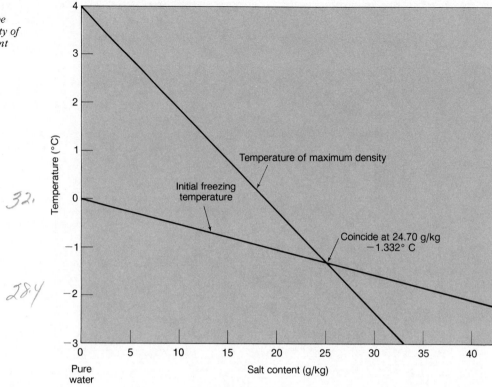

4.7 Dissolving Ability

Water is an extremely good solvent. More substances—solids, liquids, and gases—dissolve in water than in any other common liquid. Water has been called the universal solvent because of its exceptionally good dissolving ability for a wide range of materials.

The excellent dissolving ability of water is also related to its molecular chemistry. For example, common table salt dissolves in water. Common salt is chemically sodium chloride; each molecule is made up of one atom of sodium and one atom of chlorine. When the salt is in its natural crystal form, the positively charged sodium atoms and the negatively charged chlorine atoms are held together in an alternating pattern by their bonds, resulting from the strong attraction force between the two atoms. When the sodium chloride is placed in water these bonds break, and the resulting charged atoms are called **ions.** The positive sodium ions are attracted to the negatively charged oxygen side of the water molecules, while the negative chloride ions are attracted to the positively charged hydrogen side. The salt atoms are surrounded and separated by water molecules (see figs. 4.5a and 4.5b). This ability of water molecules to separate compounds into their ions makes water an excellent solvent.

For millions of years the rainwater washing over the land has been dissolving materials out of the land

Figure 4.5

Salts dissolve in water because the polarity of the water molecule keeps positive ions separated from negative ions. (a) Sodium ions are surrounded by water molecules with their negatively charged portion attracted to the positive ion. (b) Chloride ions are surrounded by water molecules with their positively charged portion attracted to the negative ion.

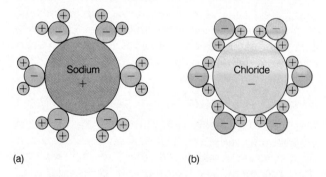

(a)   (b)

and carrying them to the sea via rivers, streams, and groundwater seepage. In this way much of the oceans' salt content has been supplied. Once these dissolved substances reach the ocean basins there is no similar continuous mechanism to return them to the land. The water is recycled to the land by oceanic evaporation followed by precipitation on land to repeat this process, but the salts are trapped in the sea and its bottom sediments. Some salt deposits found on land resulted when ancient seafloor materials were uplifted by geologic processes to become land deposits. Others are the remnants of ancient shallow seas isolated over geologic time and then evaporated, leaving the salt deposits.

4.8 *Transmission of Energy*

Pure water, fresh water, and salt water all transmit some forms of energy. The energy may be in the form of light, heat, or sound. Since the principles of transmission are the same in fresh water and salt water, our discussion will center on energy transmission in the oceans.

Heat

There are three ways in which heat energy may be transmitted: by **conduction,** by **convection,** and by **radiation.** Conduction is a molecular process. When heat is applied at one location the molecules move faster due to the addition of energy; gradually this more rapid molecular motion is passed on to the adjacent molecules and the heat spreads. For example, if the bowl of a metal spoon is placed in a hot liquid, the handle soon becomes hot. In this case, heat has been conducted from the part of the spoon that is in contact with the heat source to the handle, which is not. Certain substances, like metals, are excellent conductors. Water is a poor conductor, and it transmits heat slowly in this way.

Convection is a density-driven process in which a heated fluid material moves and carries heat to a new location. In the older type of heating systems, hot air is supplied through vents at floor level; the hot air rises because it is less dense, and the air carries the heat with it. When this air is cooled it drops down again because it is then more dense. In these energy-conscious days, ceiling fans are being used to force the less dense hot air down, to keep the room warm at the human level rather than accumulating excess heat at the ceiling. Water behaves in the same way, rising when it is heated from below and sinking when it is cooled at the upper surface.

Radiation is the direct transmission of heat from the energy source. Switch on an incandescent light bulb and feel from a foot away the instantaneous heat released. This is radiant energy. The sun provides the earth with radiant energy, and the surface waters of the earth are warmed by this solar radiation. Unlike conduction and convection, which require a medium for the transfer of heat, radiated heat can be transferred through the vacuum of space.

Consider all three processes with respect to a container of water. The surface of the water receives solar radiation; some of this radiation is reflected from the surface and adds no heat to the water, while some of the radiation is absorbed and raises the temperature of the surface water. Since warming the surface water makes it less dense, the warm water floats at the surface. The only way to transmit the heat downward is molecular conduction, unless the container is stirred. Stirring requires that mechanical energy be added from another source. Therefore heating a body of water from the top is not an efficient method of uniformly warming all the water in the container. Compare this situation with that of a similar container heated from below. In this case, the downward molecular conduction of heat is replaced by convection; as the water at the bottom gains heat, decreases in density, and rises, it takes the heat with it and distributes it throughout the water volume. As this warmed water rises it is replaced by colder water, which is warmed in turn. This is a much more rapid and efficient method of distributing heat, since the water circulates on its own without the need for stirring.

The oceans are, of course, heated from above by solar radiation, which is absorbed in the upper surface layers of water. The heat gained at the surface of the ocean is transmitted slowly downward by the natural turbulent stirring action of the wind and the currents, as well as by the slower molecular conduction. The oceans' surface water heated by the sun then loses heat to warm the atmosphere above it. Since the heating of the atmosphere is done from below, the heat is efficiently transferred to the atmosphere by vertical convective motion.

Light

The light striking the ocean surface is one of the many forms of **electromagnetic radiation** the earth receives from the sun. The full range of this radiation may be seen in the **electromagnetic spectrum,** shown in figure 4.6. Note that visible light occupies a very narrow segment of the spectrum. Wavelengths above visible light are shorter and include ultraviolet light, Xrays, and gamma rays. Wavelengths below visible light are longer and include infrared (heat) waves, microwaves, TV, FM, and AM radio waves, all of which are used in modern technology. Visible light may be broken down into the familiar spectrum of the rainbow: red, orange, yellow, green, blue, and violet. Each color represents a range of wavelengths; the longest wavelengths are at the red end of the spectrum, and the shortest are at the blue-violet end. When combined, these wavelengths produce white light.

Seawater transmits the visible light portion of the electromagnetic spectrum. About 60% of the entering light energy is absorbed in the first meter, and about 80% is gone after 10 m (33 ft). Only 1% of the total light available at the surface is left in the clearest water below

Figure 4.6
The electromagnetic spectrum.

Selective absorption of light with depth

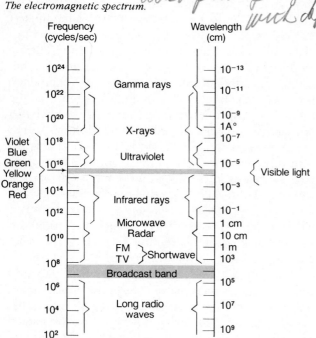

As light passes through the water it is **absorbed** and **scattered** by suspended particles, including silt, single-celled organisms, and the water and salt molecules. It is also absorbed by plants, to be used in their life processes. These various mechanisms combine to decrease the intensity of light over distance; this effect is known as **attenuation.** The clearer the water, the greater the light penetration, and the smaller the attenuation.

Oceanographers use a light meter in a watertight case to measure the intensity of light, first at the sea surface and then at increasing depths. Light meters (fig. 4.9) measure total available light, or they may be equipped with filters to measure only specific wavelengths. The measure of light attenuation with depth allows scientists to estimate the depth at which light limits plant growth. This information is then used to determine the ability of any ocean area to support the plant populations required to feed other living creatures. This ability is of considerable importance to people who control the length of fishing seasons and the size of catches.

The simplest way of measuring light attenuation in surface water is to use a **Secchi disk** (fig. 4.10). This is a white disk about 30 cm (12 in) in diameter, which is lowered on a line to the depth at which it just disappears from view. The measure of this depth can be used to determine the average attenuation of light. Advantages of a Secchi disk include never needing adjustment, low cost, never leaking, and never requiring replacement of electronic components. In waters rich with living organisms or suspended silt, the Secchi disk may disappear from view at depths of 1 to 3 m (3–10 ft) while in the open ocean it may be visible down to 20 to 30 m (65–100 ft).

150 m (492 ft). No light penetrates below 1000 m (3280 ft). Not all wavelengths of the visible spectrum are transmitted equally (see fig. 4.7). The long wavelengths at the red end of the spectrum are lost very rapidly within the upper 10 m (33 ft); the shorter wavelengths of blue-green light are transmitted to the greater depths. Since color perception is due to the reflection of wavelengths of a particular color back to our eyes, the oceans usually appear blue-green because these wavelengths are most available to be reflected from particles within the water. Because all of the wavelengths, or colors, are present to illuminate objects in shallow waters, objects are seen in their natural colors at the surface. Objects in deeper waters usually appear dark in color because they are illuminated by blue light. Coastal waters may appear green or even yellow, brown, or red in color, while deep, open ocean water appears blue. Coastal waters usually contain silt from the rivers and the land, as well as large numbers of microscopic organisms; the color of these particles and organisms is revealed by the light entering the water, so that the color of the water is changed. Open ocean water beyond the influence of land is often clear and blue, which indicates that it is nearly empty of both living and nonliving suspended materials. The clearer the water, the deeper the light penetration, and the less suspended matter present.

When light passes from the air into the water it is bent, or **refracted,** because its speed is faster in air than in water. Because the light rays bend as they enter water from the air, objects seen through the water's surface are not where we believe them to be (see fig. 4.8).

Sound

The sea is a noisy place, because sound is transmitted both farther and faster in seawater than it is in air. Waves break, fish grunt and blow bubbles, crabs snap their claws, and whales whistle and sing. Sound travels in seawater at an average velocity of 1500 m (5000 ft)/sec as compared with 334 m (1100 ft)/sec in dry air at 20° C. The speed of sound in seawater increases with increasing temperature, pressure, and salt content, and decreases with decreasing temperature, pressure, and salt content. High-frequency sounds do not travel as far as low-frequency sounds due to the friction between water molecules, which absorb energy rapidly as the high-frequency sound waves pass through the water, vibrating the water molecules.

Because water conducts sound, sound can be used to locate the objects producing the sounds; because sound is reflected back after striking an object, it can be

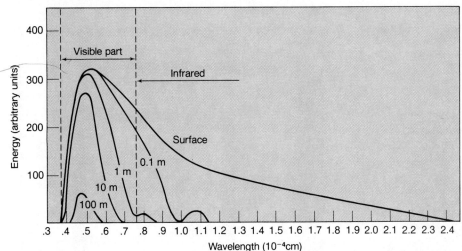

Figure 4.7
Total available solar energy in the sea (shown by the areas under the curves) decreases as depth increases. The longer, red wavelengths are absorbed first. The color peak shifts to the shorter, blue wavelengths as the depth increases.

Shows penetration of vis. light at different depths

highest energy - from sun is in visible light

H₂O absorbed as heat to down to 10-100m

& blue/green light.

Figure 4.8
Objects that are not directly in the individual's line of sight can be seen in water due to the refraction of the light rays. The refraction is caused by the decreased speed of light in water.

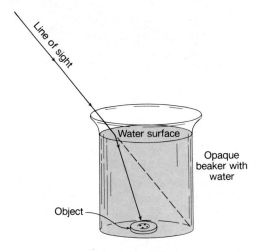

Figure 4.9
An immersion light meter is used to determine the percentage of surface solar radiation at depth. The photo cell on the left is placed on deck to measure solar radiation at the sea surface. The solar cell on the right is lowered in the sea.

Figure 4.10
A Secchi disk.

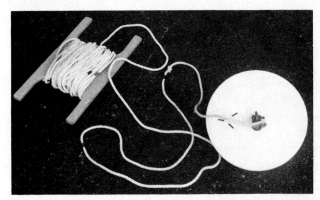

used to find objects, sense their shape, and determine their distance away from the sound's source. If a sound signal is sent into the water and the time required for the return of the reflected sound, or echo, is measured accurately, the distance to the object may be determined. For example, if 6 seconds elapse between the outgoing sound pulse and its return after reflection, the sound has taken 3 seconds to travel to the object and 3 seconds to return. Since sound travels at an average speed of 1500 m/sec in water, the object is 4500 m away, as shown in figure 4.11.

Echo sounding to determine depth uses a narrow sound beam that is directed vertically to the sea bottom. The sound waves pass through nearly horizontal layers of water in which salt content, temperature, and pressure change. The speed of the sound waves continuously changes as they pass from layer to layer, until they

Figure 4.11

Traveling at an average speed of 1500 meters per second, a sound pulse leaves the ship, travels downward, strikes the bottom, and returns. In 4500 meters of water, the sound requires three seconds to reach the bottom and three seconds to return.

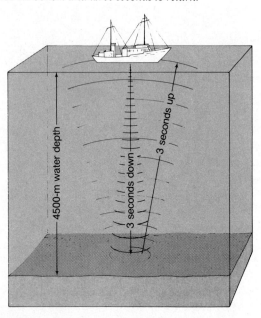

Figure 4.12

A precision depth recorder displaying bottom and subbottom profiles.

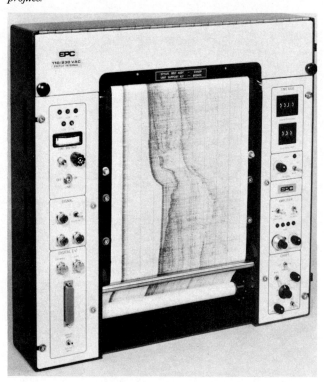

reach the bottom and are reflected back to the ship. Little refraction, or bending, of the sound beam occurs, because the path of the sound beam is perpendicular to the water layers. Echo sounders, or **depth recorders,** are used by all modern vessels to measure the depth of water beneath the ship. Oceanographic vessels record the depths on a chart, producing a continuous reading of depth as the vessel moves along its course. The **precision depth recorder (PDR)** on an oceanographic research vessel uses a narrow sound beam to give detailed and continuous traces of the bottom while the ship is in motion, as shown in figure 4.12.

Geologists can detect the properties of the sea floor by studying the echo charts, because some seafloor materials reflect back a stronger signal than other materials. If the sound beam transmitted is of high intensity, some sound energy may penetrate the sea floor and reflect back from deeper layers in the sediments. In this case, the echo chart will display the layering of the seabed materials.

Porpoises and whales use sound in water in the same way that bats use sound in air. The animal produces a sound, which travels outward until it reaches an object from which it is reflected. The animal is able to judge the direction from which the sound returns, the distance to the reflecting object, and the properties

of that object. Human technology has produced an underwater location system called **sonar (sound navigation and ranging),** which uses sound in a similar way. Sonar technicians direct a sound beam through the water, searching for targets that return echoes. They are then able to determine the distance and direction of the target. An electronic screen is used to display the direction of sound beams relative to the sending vessel. If a reflective target is found, the screen also portrays the distance to the target, using the time difference between the sending of the sound pulse and the return of the echo. After much training and practice, technicians are able to distinguish between the echoes produced by a whale, a school of fish, or a submarine. It is even possible to determine the type and class of a vessel merely by listening to the sounds produced by its propellers and engines.

However, the target may not be at the depth, distance, and angle indicated, because the sound beam may have been bent, or refracted, as it moved through the water. Refraction occurs when the sound beam passes at an angle through water layers of differing density, which is determined by salt content, temperature, and pressure (see figs. 4.13a and 4.13c). Because of this, refraction occurs and targets appear displaced from their actual position (see figs. 4.13b and 4.13d). These dia-

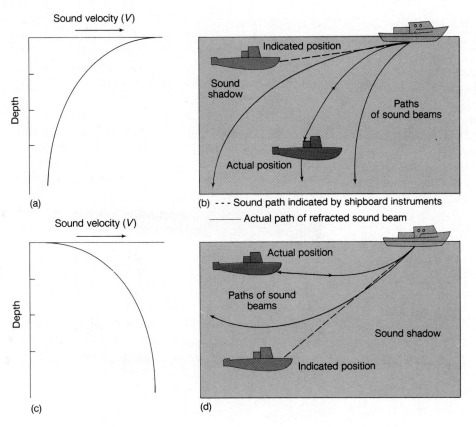

Figure 4.13
Sound waves change velocity and refract as they travel at an angle through water layers of different densities. The angle at which the sound beam leaves the ship indicates a target in the indicated, or ghost, position. To determine the actual position of the target the degree of refraction and change in velocity with depth must be shown.

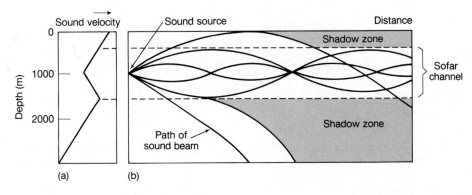

Figure 4.14
(a) The temperature, salinity, and pressure variation with depth combine to produce a minimum sound velocity at about 1000 meters. (b) Sound generated at this depth is trapped in a layer known as the Sofar channel.

grams illustrate areas of the sea into which the sound does not penetrate because it is bent toward regions in which sound travels more slowly and away from regions in which the sound waves travel more rapidly. This can produce areas where the sound beams do not penetrate; these areas are called **sound shadow zones.**

In order to interpret the returning echo and to determine distance and depth correctly, the sonar operator must have information about the properties of the water through which the sound passes. Governments and their navies have conducted intensive research in the field of underwater sound, for in wartime naval operations the winner in an encounter is often determined by who has the best understanding of the water's prop-

erties and its effects on underwater sound. Survival for a surface vessel depends on its accuracy in locating a submarine's position, while the submarine must remain at the correct depth and distance from the sonar detector to remain invisible in the shadow zone.

At depths of about 1000 m (3300 ft) the combination of salt content, temperature, and pressure creates a zone of minimum velocity for sound, the **sofar (sound fixing and ranging) channel** (see fig. 4.14a). Sound waves produced in the sofar channel do not escape from it unless they are directed outward at a sharp angle. Instead, the majority of the sound energy bounces back and forth along the channel for great distances (see fig. 4.14b). Test explosions set off in this channel near

Australia have produced sound heard as far off as Bermuda. It has been suggested that listening stations be established around the world to monitor the sofar channel. Vessels and aircraft in distress could drop small depth charges programmed to explode at the sofar depth and their location could be established by using the difference in arrival time of the sound at listening stations.

Depth recorders can be confused by still another underwater phenomenon. Large numbers of small organisms, including fish, move toward the surface during the night and sink down to greater depths during the day. These organisms form a layer known as the **deep scattering layer (DSL).** This layer reflects a portion of the sound beam energy and creates the image of a false bottom on the depth recorder trace.

Other echoes are returned from mid-depths by fish swimming in schools or by large individual fish. Those who fish learn to recognize these reflections. The depths of the echoes give them the depth to operate their nets, while their location and the appearance of the echo pattern inform them of the fish species.

Note the similarities between the behavior of sound and light in water. Both represent energy in wave form and both are absorbed, reflected, and refracted.

4.9 Practical Considerations: Ice and Fog

Ice and fog are forms of water which are extremely hazardous to ocean navigation and transportation. At polar latitudes in the northern hemisphere, ice prevents the routine use of shipping routes for much of the year. **Sea ice** driven by winds and currents hampers oil drilling, and polar research is also difficult under these conditions. The Arctic is acknowledged to be of critical strategic importance to the military; a knowledge of sea ice is required for operation on or below the sea surface at high latitudes in either hemisphere. Fog restricts visibility. Although electronic eyes (radar) have been developed to help ships under way in foggy seas, safety is compromised, collisions are not infrequent, and many small-boat sailors lose their way. The processes which produce ice and fog at sea are discussed here.

Sea Ice

Ice is present year-round at all latitudes if the elevation is high enough to keep the average earth surface temperature below the freezing point of water. This elevation varies between sea level at the poles and approximately 5000 m (16,400 ft) in elevation at the equator. On land, ice and snow are the result of low temperatures and precipitation. In the sea or in lakes, precipitation is not necessary. Instead the ice is formed when the temperature of the surface water goes below the freezing point of water.

Sea ice is formed at polar latitudes because of the extremely low air and water temperatures. As the seawater begins to freeze, the surface water becomes dull, and clouds of ice crystals are formed. The dissolved salts of the seawater do not fit into the ice crystal structure and are left below in the surface water. As the number of ice crystals increases, a layer of slush is formed, which covers the ocean in a thin sheet of ice. Sheets of new sea ice are broken into "pancakes" by waves and wind (see fig. 4.15a). As the freezing continues the pancakes move about, unite, and form floes (see fig. 4.15b). These floes move with the currents and the wind. They collide with each other, forming ridges and hummocks. As freezing continues, the ice increases in thickness. Snow on the surface may also freeze and contribute to the flat, floating ice masses. Some floes shift constantly, breaking apart and freezing together in response to winds and water motion. Others remain anchored to the landmass. These stationary floes are called **fast ice.**

The water below the ice has an increased salt content because of the salts left behind by the freezing process. This saltier water is very cold and dense, and so it sinks to be replaced by less salty water, which in turn cools to its freezing point. As the ice is formed, some seawater is trapped in the voids between the ice crystals. If the ice forms slowly, most of the trapped seawater drains out and escapes; if the ice forms quickly at subfreezing temperatures, more salt water is trapped. As time passes and the ice continues to age, the salt water slowly escapes through the ice, and eventually the ice becomes pure water, which is quite fresh enough to drink when melted.

In one season ice about 2 m (6.5 ft) thick may be formed at the ocean's surface. The thickness of the ice is limited by the necessity of cooling its surface and extracting sufficient heat through the ice from the underlying water to form new ice beneath the existing layer. The latent heat of fusion must be removed from the

Figure 4.15

Sea ice. (a) Pancake ice, an early stage in sea ice formation.
(b) Ice floes; pressure ridges are formed by colliding floes.
(c) Sections of broken floe ice stack to form a pressure ridge.

(a)   (b)   (c)

water beneath the ice, and it can only be extracted by the process of conduction through the initial ice layer, which is a slow process at polar temperatures. Snow accumulating on the ice surface acts as an insulator that also retards the extraction of heat from the water below.

Sea ice is seasonal around the bays and shores of parts of the northeast United States, Canada, and Alaska (see fig. 4.16 and colorplate 6). The ice exists year-round in the central Arctic and around the Antarctic continent. Sea ice covers the entire Arctic Ocean in the northern winter and pushes far out to sea around Antarctica during the southern hemisphere's winter. Waves, currents, winds, and tides fracture the edges of the ice and cause floes to collide and pile into irregular masses to form pressure ridges (see fig. 4.15c). In summer the ice melts back along its edges. In areas where the ice does not melt entirely during the year, new ice is added each winter, and a thickness of 3 to 5 meters (10–16 feet) can accumulate.

Remember that sea ice is a product of the salt water; it is not to be confused with the freshwater glacial **land ice** that covers Greenland and the Antarctic continent. Pieces of the land ice sometimes break off and fall into the sea. These pieces of freshwater ice are called **icebergs.**

Icebergs

Icebergs are massive, irregular in shape, and float with only about 12% of their mass above the sea surface (see fig. 4.17). They are formed by glaciers, large bodies of ice that begin inland in the snows of central Greenland, Antarctica, and Alaska, and inch their way toward the sea. The forward movement, the melting at the base of the glacier where it meets the ocean, and the waves and wind action cause blocks of ice to break off and float to sea. The production of icebergs by a glacier is called calving, and it occurs mainly during the summer months.

The icebergs produced in the Arctic drift south with the currents as far as New England and the busy shipping lanes of the North Atlantic. It was one of these icebergs that sank the *Titanic.* On its maiden voyage the great luxury liner struck an iceberg and sank with

Figure 4.16

(a) In the winter, sea ice extends out of the Arctic Ocean through the Bering Strait and into the Bering Sea. Sea ice also forms in the Sea of Okhotsk west of the Kamchatka Peninsula. (b) In the summer, the boundary of the Arctic ice pack moves back into the Arctic Ocean.

(a) February sea ice
distribution

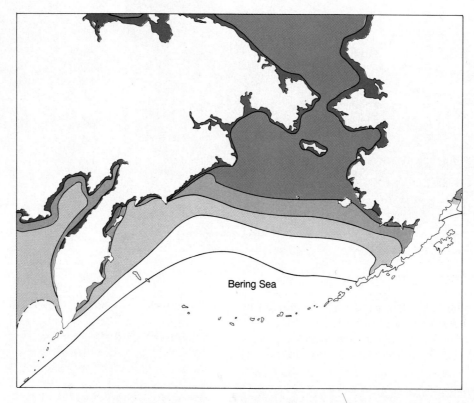

Bering Sea

(b) August sea ice
distribution

Legend:
Average concentration

Open
< 0.1 coverage

Scattered
0.1–0.4 coverage

Broken
0.5–0.7 coverage

Close
0.8–0.9 coverage

Consolidated or fast
1.0 coverage (no water)

Approximate
maximum extent
of 0.1 or greater
concentration

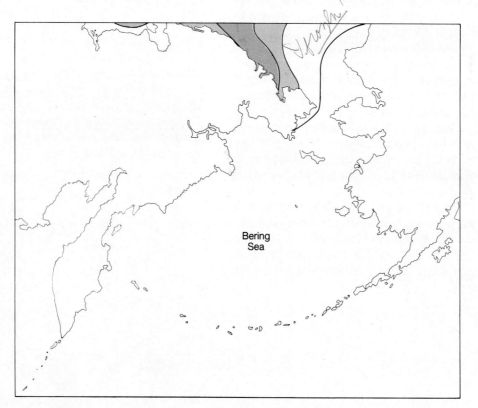

Bering
Sea

Figure 4.17
(a) A castle berg drifting in the Atlantic Ocean just north of the Grand Banks. (b) The edge of a tabular berg in the Arctic Ocean. The dark area at the berg's water line has been cut by the waves. Rocks on the surface indicate the land origin of the ice.

(a)

(b)

a loss of 1517 lives on the night of April 14, 1912. The ice had most probably become detached from a glacier in Greenland and then floated south. Following the sinking, the United States began an iceberg patrol to warn vessels of the location of icebergs. The patrol became an international effort in 1914 and continues to the present.

Icebergs produced in the Antarctic usually stay close to the polar continent, caught by the circling currents, although they have been known to reach latitudes of 40°S to 50°S. Alaskan icebergs are usually released in narrow channels and semienclosed bays; they do not readily escape into the open ocean. Rather than being a hazard to shipping, Alaskan icebergs have become a tourist attraction. Cruise ships visit Glacier Bay, Alaska, to show their passengers the glaciers and icebergs.

Icebergs from valley glaciers are relatively high and narrow, with above-water shapes resembling towers and battlements; these are called **castle bergs.** Icebergs from broad, flat, continental ice sheets are usually much larger than castle bergs. These huge, flat icebergs

are called **tabular bergs** and form the ice islands that are used as bases for polar research. Some of these icebergs have been large enough to accommodate aircraft runways, housing, and research facilities. Such an ice island may be used for years before it eventually breaks up.

Recently, large, floating masses of ice have attracted considerable interest as sources of fresh water. The first of a series of meetings on this possibility was hosted in 1978 by Arab interests. Discussions at the meetings centered on the feasibility of towing icebergs from the Antarctic to the Red Sea. Among the techniques necessary would be those of detaching the ice, shaping it for the long miles of transport, and designing a bridle for towing. Although the expense would be enormous and much of the ice would melt as it was towed into equatorial latitudes, the need for water in the Arab countries is so great that it is believed that the project would be economical and that sufficient ice would remain after the journey to make the project worthwhile.

Figure 4.18
Advective fog obscures the Golden Gate Bridge at the entrance to San Francisco Bay.

Fog

The ability of water to remain as water vapor is a function of temperature: warm air can hold more moisture than cold air. When water vapor in the air is cooled, it condenses around any small particles in the air, forming liquid droplets. Large accumulations of water droplets form clouds. When these clouds form close to the ground they are called **fog.** Fog at sea is a navigation hazard to mariners, but the ocean scientist is interested in fog and in predicting its occurrence because of the role fog plays in transferring water and heat between the atmosphere and the ocean surface.

There are three basic types of fog, each formed by a different process. The most common fog at sea is **advective fog.** Advective fog forms when air warmed by the underlying warm water is saturated with water vapor and then moves over colder water. This advective fog hugs the surface like a blanket. For example, when the warm, moist air over the Gulf Stream flows over the cold water of the Labrador Current, the famous fogs of the Grand Banks result. In northern California, Oregon, and Washington, when offshore warm, moist air moves over the cold coastal waters, particularly in summer, a coastal fog results, which gives moisture to the redwood forests and grassy cliffs and cools the city of San Francisco (see fig. 4.18).

Streamers of **sea smoke** rising from the sea surface on a cold winter day are a spectacular sight. Dry, cold air from the land moves out over the warmer water, and the water warms the air above it. This warmed air picks up water vapor from the sea surface and begins to rise rapidly. As it rises it is cooled below the dew point, and rising ribbons of fog are formed. When sea smoke is formed the vertical convection results in high evaporation from the water's surface, whereas the advective fog returns moisture to the earth's surface by condensation.

Radiative fog is the result of warm days and cold nights. The earth's surface warms and cools, and so does the air just above it. If the air holds enough moisture during the day, it will condense as the earth cools at night, forming the low-lying, thick, white fogs that are often found in river valleys and occasionally in bays and inlets along the coast. These fogs usually disappear during the morning as the sun gradually warms the air, changing the water droplets back to water vapor.

The earth's water, whether found as a liquid filling our lakes, streams, and oceans; as a solid in the land glaciers, snowpacks, and sea ice; or as a gas in the atmosphere, is a remarkable substance. The properties of this water as well as its abundance give the earth its habitable characteristics and make it unique in our solar system.

Measuring with Light

*I*f seawater was completely transparent to the visible spectrum, it would be possible to look down into the seas' depths using a glass-bottom bucket or boat and see all the seafloor features illuminated by sunlight. But water is not totally transparent to light, and salt water is less transparent than fresh water. Light waves are absorbed, reflected, and scattered by salt and water molecules and any suspended particles; the light is first attenuated and then completely absorbed in the first 1000 m of the oceans' depths.

The number of suspended particles per volume, or the abundance of particles, in the water controls, in part, the attenuation of the light. Oceanographers use the light attenuation rate to determine the type and quantity of suspended material in the water. For example, the Secchi disk is used to determine the depth at which light reflected from the disk's surface is indistinguishable from the surrounding light being scattered back to the observer's eye. The Secchi disk's readings can be calibrated to the density of specific suspended particles. Because the degree of opaqueness, size, and abundance of particles from dissimilar sources differs, particles derived from the erosion of soils and silts produce readings that are unlike readings due to microscopic organisms. The Secchi disk can only be used to estimate the abundance of particles when those particles are present at depths less than the depth at which the Secchi disk disappears from view.

A more sensitive instrument to measure particulate scattering and light absorption is made by mounting a light source at one end of a hollow tube and a light-sensitive receptor, or photocell, at the other end of the tube. The tube is flooded with seawater as it is lowered into the ocean. The light source in the tube is focused into a beam by lenses and strikes the photocell. The tube protects the photocell from any outside light sources. The wavelength or quality of the light is controlled by inserting filters between the light source and the photocell. The light that strikes the photocell is the light that remains after absorbtion and scattering have occurred over the path between the light source and the photocell. Photocell readings at the selected depths are then related to the abundance and distribution of kinds of particles in the water column.

If the suspended particles are microscopic organisms, their abundance is used to calculate the distribution of food resources for larger organisms in that water column. Because many toxic substances adhere to inorganic and organic particulate materials, knowing the distribution of particles allows an estimate to be made of toxicants in the water column. The determination of the amounts of suspended particulate material above the sea floor is a measure of the resuspension of particles and their associated toxic materials by currents along the sea floor. The turbid layer above the sea floor is called the **nepheloid layer** and the light-attenuation sensor used to measure its properties is known as a **nephelometer.**

Summary

A water molecule is made up of two positively charged atoms of hydrogen and one negatively charged atom of oxygen. The molecule has a specific shape with oppositely charged sides; it is a dipolar molecule. Because of the distribution of the charges, water molecules interact with each other by forming hydrogen bonds between molecules. As a result of its structure, the water molecule is very stable. The structure of the water molecule is responsible for the properties of water.

Water exists as a solid, a liquid, and a gas. Changes from one state to another require the addition or extraction of heat energy. Water has a high heat capacity; it is able to take in or give up large quantities of heat with a small change in temperature. The surface tension of water, which is related to the cohesion between water molecules at the surface, is high. The viscosity of water is a measure of its internal friction; it is primarily affected by temperature. Water is nearly incompressible. Pressure increases 1 atm for every 10 m of depth in the oceans.

The density, or mass per unit volume, of water increases with a decrease in temperature and an increase in salt content. The effect of pressure on density is small. Less dense water floats on more dense water. Pure water reaches its maximum density at 4° C. Open ocean water does not reach its maximum density before freezing. Water vapor is less dense than air; a mixture of water vapor and air is less dense than dry air.

The ability of water to dissolve substances is exceptionally good. River water dissolves the salts from the land and carries the salts to the sea.

Seawater transmits energy as light, heat, and sound. The long red wavelengths of light are lost primarily in the first 10 m (33 ft); only the shorter wavelengths of blue-green light penetrate to depths of 150 m (495 ft) or more. Light passing into water is refracted. Attenuation, or the decrease in light over distance, is the result of absorbtion and scattering by the water and by particles suspended in the water. Light attenuation is measured by light meters and with a Secchi disk.

The sea surface layer is heated by solar radiation, and heat is transmitted downward by conduction. This is an inefficient process compared to convection, which transfers heat from the earth's surface to the atmosphere.

Sound travels farther and faster in water than in air. Its speed is affected by the temperature, pressure, and salt content of the water. Echo sounders are used to measure the depth of water, and sonar is used to locate objects. Sound is refracted as it passes at an angle through the density layers,

and sound shadows are formed. The sofar channel, in which sound travels for long distances, is the result of salt content, temperature, and pressure in the oceans. The deep scattering layer, formed by small animals moving toward the surface at night and away from it during the day, reflects portions of a sound beam and creates a false bottom on a depth recorder trace.

Sea ice is formed in the extreme cold of polar latitudes. The process is self-insulating, so that large thicknesses of ice do not form each winter. Icebergs are formed from glaciers that break off into the sea.

Fog occurs when water vapor condenses to form liquid droplets. Three types of fog occur: advective fog occurs when warm, water-saturated air passes over cold water; sea smoke occurs when dry cold air moves over warm water; and radiative fog occurs when warm, moist air is cooled at night.

Key Terms

covalent bond	surface tension	scattering	deep scattering layer
polar molecule	viscosity	attenuation	sea ice
hydrogen bond	density	Secchi disk	fast ice
dew point	ions	nepheloid layer	land ice
calorie	conduction	nephelometer	iceberg
latent heat of fusion	convection	echo sounding	castle berg
latent heat of vaporization	radiation	depth recorder	tabular berg
sublimation	electromagnetic radiation	precision depth recorder	fog
heat capacity	electromagnetic spectrum	sonar	advective fog
specific heat	refraction	sound shadow zone	sea smoke
cohesion	absorbtion	sofar channel	radiative fog

Study Questions

1. Discuss the structure of the water molecule.

2. What happens to the arrangement of water molecules when water freezes?

3. What happens to the sea level of the oceans if all the sea ice melts? Assume that the ocean area does not change and ocean temperature stays the same.

4. Compare the heat capacity of the oceans with that of the land and the atmosphere. Compare the climate of the west and east coasts of the United States with that of the central and midwestern states. How is climate related to heat capacity and the direction of the prevailing wind?

5. How do the properties of the water molecule explain (a) surface tension and (b) the dissolving ability of water?

6. Explain why heating the atmosphere from below is an efficient way to distribute heat in the atmosphere, while heating the oceans from above is an inefficient way to distribute heat in the oceans.

7. If there were no scattering of light by seawater, what color would the oceans appear to us? Why?

8. Both advective fog and sea smoke transfer heat between the atmosphere and the oceans. How does the rate and direction of heat transfer differ in each case? Why?

Study Problems

1. If surface cooling and evaporation remove heat from the surface of a deep lake at the rate of 800×10^3 calories per hour, and if the surface temperature of the lake is 0° C, at what rate must deep water be pumped to the surface of the lake to prevent the formation of ice?

2. Determine the freezing point of seawater at 32 g/kg salt content. Use figure 4.4.

3. If an echo sounder measures the depth of the water as 3500 m, but the instrument is shown to have a timing error of ±0.001 second over the total time period of the measurement, what is the error in the depth measurement?

4. If water from a given area of sea surface evaporates at the rate of 1 metric ton per day at 20° C, at what rate is heat being added to the atmosphere above this surface area?

5. If heat is removed from a kilogram of water at a constant rate that lowers the temperature of the water from 25° C to 0° C in thirty minutes, how long will it take to convert 0° C water to ice at this rate of heat removal?

The Salt Water

5

*A*nd truly, though we were at sea, there was much to behold and wonder at, to me, who was on my first voyage. What most amazed me was the sight of the great ocean itself, for we were out of sight of land. All round us, on both sides of the ship, ahead and astern, nothing was to be seen but water—water—water; not a single glimpse of green shore, not the smallest island, or speck of moss anywhere. Never did I realize till now what the ocean was: how grand and majestic, how solitary, and boundless, and beautiful and blue; for that day it gave no tokens of squalls or hurricanes, such as I had heard my father tell of, nor could I imagine how anything that seemed so playful and placid could be lashed into rage, and troubled into rolling avalanches of foam, and great cascades of waves, such as I saw in the end.

*Herman Melville,
from* Redburn

$\mathcal{S}$eawater is salt water. Historically seawater has been valued for its salt. Until relatively recently, salt was enormously important as a food preservative, and at one time it formed the basis for a major commercial trade. Today, although salt is still extracted from seawater, it is the water itself that is becoming increasingly valuable in many areas of the world.

This chapter considers not only the salt in seawater, but also the dissolved gases that are found in seawater. We will investigate the biological and chemical processes that regulate these substances and follow them as they go from the land and atmosphere, to the sea and sea floor, and back to the land and atmosphere.

5.1 The Salts

Ocean Salinities

In the major ocean basins, 3.5% of the weight of seawater, on the average, is dissolved salt and 96.5% is water, so that a typical 1000-g or 1-kg sample of seawater comprises 965 g of water and 35 g of salt. Oceanographers measure the water's salt content, or **salinity,** in grams of salt per kilogram of seawater (g/kg), or parts per thousand (‰). Consequently, average ocean salinity is approximately 35‰.

The salinity of the oceans' surface water is associated with latitude. Latitudinal variations in evaporation and precipitation, as well as freezing, thawing, and freshwater runoff from the land affect the relative amount of salt in seawater. The relationship between evaporation, precipitation, and surface salinity is shown in figure 5.1. Inspection of this diagram shows low surface salinities in the cool and rainy 40°–50°N and S latitude belts, high evaporation rates and high surface salinities in the desert belts of the world centered on 25°N and S, and low surface salinities again in the warm but rainy tropics centered at 5°N.

In coastal areas of high precipitation and river inflow, surface salinities fall below the average. For example, during periods of high flow, the water of the Columbia River lowers the Pacific Ocean's surface salinity to less than 25‰ 20 miles at sea. Also, sailors have dipped up water fresh enough to drink from the ocean surface 50 miles from the mouth of the Amazon River. In subtropic regions of high evaporation and low freshwater input, the surface salinities of nearly landlocked seas are well above the average: 40 to 42‰ in the Red Sea and the Persian Gulf and 38 to 39‰ in the Mediterranean Sea. In the open ocean at these same latitudes the surface salinity is closer to 36.5‰. Surface salinities change seasonally in polar areas, where the surface water forms sea ice in winter, leaving behind the salt and raising the salinity of the water underlying the ice. In summer the water returns as a freshwater layer when the sea ice melts. Sea surface salinities during summer in the northern hemisphere are shown in figure 5.2. Deep-water samples from the mid-latitudes are usually slightly less salty than the surface waters because the deep water is formed at the surface in high latitudes at times of high precipitation. The formation of these deep-water types will be further described in chapter 6.

Figure 5.1
Midocean average surface salinity values match the average changes in evaporation minus precipitation values that occur with latitude.

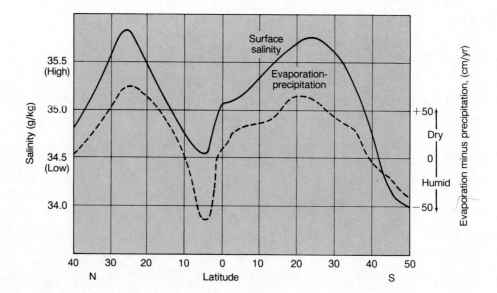

Dissolved Salts

When salts are added to water, the salts dissolve. The salt crystals dissociate (or break apart) into **ions,** which are positively or negatively charged atoms or groups of atoms. An ion with a positive charge is a **cation:** an ion with a negative charge is an **anion.** The salts in seawater are mostly present in their ionic forms as anions and cations. For example, table salt (or sodium chloride) is the most common salt in seawater. When sodium chloride is added to water it is dissolved by the water into positively charged sodium ions and negatively charged chloride ions.

This reaction can be written as:

$$\text{sodium chloride} \xrightarrow{\text{water}} \text{sodium ion} + \text{chloride ion},$$

or it can be written with chemical abbreviations as:

$$NaCl \xrightarrow{H_2O} Na^+ + Cl^-.$$

Six ions make up more than 99% of the salts dissolved in seawater. Four of these ions are cations: sodium (Na^+), magnesium (Mg^{2+}), calcium (Ca^{2+}), and potassium (K^+); and two are anions: chloride (Cl^-) and sulfate (SO_4^{2-}). Table 5.1 lists these six ions and five more, bringing the total to 99.99%. The ions are arranged in order of the amount present in seawater. The ions listed in table 5.1 are known as the **major constituents** of seawater. Note that sodium and chloride ions account for 86% of the salt ions present in seawater.

All the other elements dissolved in seawater are present in concentrations of less than one part per million and are called **trace elements** (see table 5.2). Although many constituents of seawater exist as trace elements, some of these elements are important to organisms living in the water. Certain organisms are able to concentrate these ions in their bodies. For example, long before the presence of iodine could be determined chemically as a trace element of seawater, it was known that shellfish and seaweeds were rich sources for this element. Seaweed has even been harvested commercially for its iodine.

Figure 5.2

Average sea surface salinities in the northern summer, given in parts per thousand (0/00).

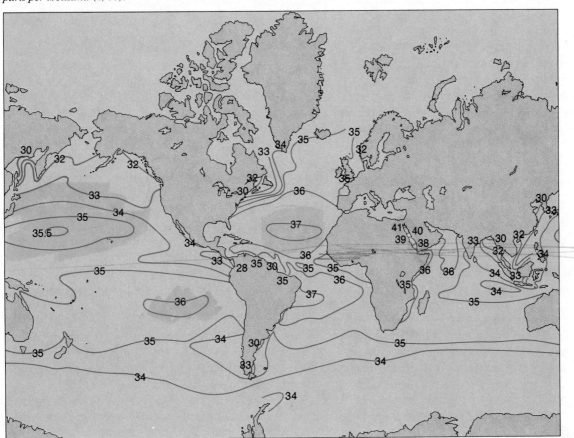

TABLE 5.1

Major Constituents of Seawater

Constituent	Symbol	g/kg in seawater[1]	Percentage by weight
chloride	Cl^-	19.35	55.07
sodium	Na^+	10.76	30.62
sulfate	SO_4^{2-}	2.71	7.72
magnesium	Mg^{2+}	1.29	3.68
calcium	Ca^{2+}	0.41	1.17
potassium	K^+	0.39	1.10
bicarbonate	HCO_3^-	0.14	0.40
bromide	Br^-	0.067	0.19
strontium	Sr^{2+}	0.008	0.02
boron	B^{3+}	0.004	0.01
fluoride	F^-	0.001	0.01
total			99.99

(chloride through potassium braced together: 99.36)

Source: With permission from Riley and Skirrow, *Chemical Oceanography*, vol. 1, 2d ed., New York: Academic Press, Inc., 1975, p. 366. Copyright: Academic Press Inc. (London) Ltd.

[1]Salinity $= 35^0/_{00}$.

Because the major constituents of seawater do not change their ratios to each other with changes in salinity, and because they are not generally removed or added to by living organisms, they are termed **conservative constituents.** Certain of the ions present in much smaller quantities, some dissolved gases, and assorted organic molecules and complexes do change their concentrations with biological and chemical processes that occur in the oceans; these are the **nonconservative constituents.**

Sources of Salt

The original sources of the seas' salts include both the crust and the interior of the earth. As the rains washed over the land they carried dissolved substances down to the ocean basins. The chemical composition of the earth's rocky crust can account for most of the positively charged ions found in seawater. Large quantities of cations are present in **igneous rocks,** which are rocks formed by the crystallization of molten magma from volcanic processes. The physical and chemical weathering of rock over time breaks it into small pieces and dissolves out ions, which are carried to the sea by rivers. Anions are present in the earth's interior and may have been present in the earth's early atmosphere. Some anions may have been washed from the atmosphere by long periods of rainfall, but the more likely source of most of the anions is thought to have been the mantle.

During the formation of the earth, described in chapter 1, gases from the mantle are believed to have carried anions into the newly forming oceans.

Today, anions are released during volcanic eruptions as gases (for example, hydrogen sulfide, sulfur, and chlorine) that are dissolved in rainwater or river water and are carried to the oceans as Cl^- (chloride) and SO_4^{2-} (sulfate). These anions are also dissolved directly in the water formed during a volcanic eruption. River water carrying chloride ions and sulfur-containing ions is acidic and erodes and dissolves the rock over which it flows, helping to liberate the cations. Tests show that the most abundant salts in today's rivers (table 5.3) are the least abundant ones in ocean water and vice versa. This difference results in part because the rivers had previously removed the most easily dissolved land salts, so that they are now carrying the less soluble salts. Exceptions to this pattern are found in rivers used for irrigation purposes. These rivers are flowing through soils that have not lost much of their salt content, because of the arid conditions in these areas. The river water is frequently used several times as it passes through a number of irrigation projects on its way downstream. This usage causes the river water to become increasingly salty and unfit for irrigation purposes toward the end of its run. This situation has produced years of continuous conflict between the United States and Mexico over the waters of the Colorado and Rio Grande rivers, which irrigate much of the agricultural land of the U.S. desert south-

TABLE 5.2
Concentrations of Trace Elements in Seawater[1]

Element	Symbol	Concentration[2]	Element	Symbol	Concentration[2]
aluminum	Al	5.4×10^{-1}	manganese	Mn	3×10^{-2}
antimony	Sb	1.5×10^{-1}	mercury	Hg	1×10^{-3}
arsenic	As	1.7	molybdenum	Mo	1.1×10^{1}
barium	Ba	1.37×10^{1}	nickel	Ni	5×10^{-1}
bismuth	Bi	$\leq 4.2 \times 10^{-5}$	niobium	Nb	$\leq(4.6 \times 10^{-3})$
cadmium	Cd	8×10^{-2}	protactinium	Pa	5×10^{-8}
cerium	Ce	2.8×10^{-3}	radium	Ra	7×10^{-8}
cesium	Cs	2.9×10^{-1}	rubidium	Rb	1.2×10^{2}
chromium	Cr	2×10^{-1}	scandium	Sc	6.7×10^{-4}
cobalt	Co	1×10^{-3}	selenium	Se	1.3×10^{-1}
copper	Cu	2.5×10^{-1}	silver	Ag	2.7×10^{-3}
gallium	Ga	2×10^{-2}	thallium	Tl	1.2×10^{-2}
germanium	Ge	5.1×10^{-3}	thorium	Th	(1×10^{-2})
gold	Au	4.9×10^{-3}	tin	Sn	5×10^{-4}
indium	In	1×10^{-4}	titanium	Ti	$<(9.6 \times 10^{-1})$
iodine	I	5×10^{1}	tungsten	W	9×10^{-2}
iron	Fe	6×10^{-2}	uranium	U	3.2
lanthium	La	4.2×10^{-3}	vanadium	V	1.58
lead	Pb	2.1×10^{-3}	yttrium	Y	1.3×10^{-2}
lithium	Li	1.7×10^{2}	zinc	Zn	4×10^{-1}
			rare earths		$(0.5-3.0) \times 10^{-3}$

[1]Nutrients and dissolved gases are not included.
[2]Parts per billion or μg/kg.
Parentheses indicate uncertainty about concentration.
Source: with permission from Riley and Chester, *Chemical Oceanography*, vol. 8. New York: Academic Press Inc., 1983, p. 172. Copyright: Academic Press Inc. (London) Ltd.

west before becoming available to Mexico. In addition, we know that hot water vents located on the sea floor supply chemicals to the ocean water and also remove them.

Regulating the Salt Balance

The total amount of dissolved material in the world's oceans is calculated to be 5×10^{22} g for ocean water of $36^{0}/_{00}$ salinity. Each year the runoff from land adds about another 2.5×10^{15} g, or 0.000005% of the total ocean salt content. We know from the age of older marine sedimentary rocks that the oceans have been present on the earth for about 3.5 billion years, and chemical and geologic evidence leads researchers to believe that the composition of the oceans has been the same for about the last 1.5 billion years. If we assume that the rivers have been flowing to the sea at the same rate over the last 3.5 billion years, more dissolved material has been added by this process than is presently found in the sea. In order for the oceans to remain at the same salinity, the rate of addition of salt by rivers must be balanced by the removal of salt; input must balance output.

Salts are removed from seawater in a number of ways. Sea spray from the waves is blown ashore, depositing a film of salt on rocks, land, houses, and cars. This salt is later returned to the oceans by runoff from the land. During the earth's history, shallow arms of the sea have become isolated, the water has evaporated, and the salts have been left behind to become land deposits called **evaporites.** Salt ions can also react with each other to form insoluble products that precipitate and

TABLE 5.3

Dissolved Salts in River Water

Ion	Symbol	Percentage by weight
carbonate	CO_3^{2-}	35.15
calcium	Ca^{2+}	20.39
sulfate	SO_4^{2-}	12.14
silica dioxide (nonionic)	SiO_2	11.67
sodium	Na^+	5.79
chloride	Cl^-	5.68
magnesium	Mg^{2+}	3.41
oxides (nonionic)	$(Fe, Al)_2O_3$	2.75
potassium	K^+	2.12
nitrate	NO_3^-	0.90
total		100.00

Source: After C. W. Wolfe et al., *Earth and Space Science,*
D. C. Heath, Lexington, Mass., 1966. Reprinted by
permission of D. C. Heath and Company.

settle to the ocean floor. Biological processes concentrate salts, which are removed if the organisms are harvested. Organisms' excretion products trap ions, which are transferred to the sediments or returned to the seawater. Other biological processes remove Ca^{2+} (calcium) by incorporating it into shells, and silicon is used to form exterior coverings for certain plants and animals. These hard parts are accumulated in the sediments when the organisms die. For a discussion of the sediments produced by biological processes, refer back to chapter 2.

A chemical process known as **adsorption,** the adherence of ions and molecules onto a surface, serves to remove other ions and molecules from the seawater. In this process, tiny clay mineral particles, which were weathered from rock and were brought to the oceans by the rivers, act to adsorb on their surfaces ions such as, K^+ (potassium) and trace metals, carrying them downward and eventually incorporating them into the sediments. Strongly adsorbable ions replace weakly adsorbable ions in a process known as **ion exchange.** If the clay minerals and sediments adsorb and exchange

with one ion more easily than with another, there will be a lower concentration of the more easily adsorbed ion in the seawater. For example, potassium is more easily adsorbed than sodium, and so K^+ is less concentrated than Na^+ in seawater. Nickel, cobalt, zinc, and copper are adsorbed on nodules that form on the ocean floor. The fecal pellets and skeletal remains of small organisms also act as adsorption surfaces. In all of these cases the ions are removed from the water and are transferred to the sediments.

The process of forming the new earth's crust at the ridge system of the deep ocean floor (chapter 3) is also known to participate in the input and output of the ions in seawater. Where molten rock rises from the mantle into the crust, magma chambers are formed. These chambers are found mainly along plate boundaries but have also been identified under volcanic sites in the middle of plates. The Hawaiian islands are an example. Cold water seeps down several kilometers through the fractured crust at a spreading center. The seawater becomes heated by flowing near the magma chamber. By convection, the heated water rises up through the crust and reacts chemically with the rocks of the crust. Magnesium (Mg^{2+}) is transferred from the water to form mineral deposits in the crust. At the same time, hydrogen ions (H^+) are released and the seawater solution becomes more acid. Chemical reactions change sulfate (SO_4^{2-}) to sulfur and then to hydrogen sulfide (H_2S). The acidic water releases and dissolves metals, including copper, iron, manganese, zinc, potassium, and calcium, in the oceanic crust. It has been estimated that a volume of water equivalent to the entire mass of the oceans circulates through the crust at oceanic ridges every ten million years.

The most important process for the removal of most elements from seawater is still the adsorption of ions onto fine particles and their removal to the sediments. Ions deposited in the sediments are trapped there. They are not returned to or redissolved in the seawater; but geologic uplift processes have elevated some sediments from the sea floor to positions now well above sea level. Such upward movements of the earth's crust return sediments and their trapped ions to the land, where they are again eroded and washed down to the sea.

The processes described in this section are summarized in figure 5.3, which you can use as a help in understanding the input and output processes that occur in the oceans.

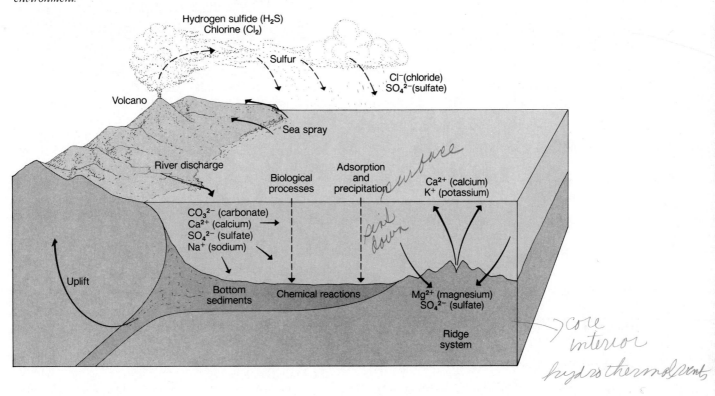

Figure 5.3
The major constituents of the cycle of seawater through the earth's environment.

Residence Time

The relative abundance of the salts in the sea is due in part to the ease with which they are introduced from the earth's crust and in part to the rate at which they are removed from the seawater. Sodium is moderately abundant in fresh water; because it reacts slowly with other substances in seawater, it remains dissolved in the ocean. Calcium, which is more abundant in river water than in seawater, is used to form the shells of marine organisms. It is removed rapidly from seawater, and so is less abundant in the sea.

The average (or mean) time that a substance remains in solution in seawater is called its **residence time.** Aluminum, iron, and chromium ions have short residence times, in the hundreds of years. They react with other substances quickly and form insoluble mineral solids in the sediments. Sodium, potassium, and magnesium are very soluble and have long residence times, in the millions of years (see table 5.4).

TABLE 5.4
Approximate Residence Time of Ions in the Oceans

Ion	Time in years
chloride	80 million
sodium	60 million
magnesium	10 million
sulfate	9 million
potassium	6 million
calcium	1 million
manganese	7 thousand
aluminum	1 hundred
iron	1 hundred

The residence time of an ion may be determined as follows:

1. Calculate the total mass of the ion in the oceans by multiplying the concentration of the ion per kilogram of seawater by the total mass of the oceans:

$$\text{Total amount ion (kg)} = \frac{\text{Ionic concentration (kg)}}{\text{Seawater (kg)}} \times$$

$$\text{Ocean mass (kg)};$$

2. Calculate the input rate by multiplying the concentration of the ion in river water by the volume of river water added to the oceans each year (37,000 km³/yr):

$$\text{Input rate (kg/yr)} = \frac{\text{Ionic concentration (kg)}}{\text{River water (km}^3)} \times$$

$$\frac{\text{River discharge (km}^3)}{\text{Year}};$$

3. Since the ionic concentration of the oceans is nearly constant with time, the input of an ion equals its rate of loss to the sediments, or input equals output:

$$\text{Input (kg/yr)} = \text{Output (kg/yr)};$$

4. Residence time is calculated by dividing the total amount present from step 1 by the output, which is equal to the input as calculated in step 2:

$$\text{Residence time (years)} = \frac{\text{Total amount ion (kg)}}{\text{Input or Output rate (kg/yr)}}.$$

If other input sources of an ion are significant, they must also be considered.

Constant Proportions

The salt water of the oceans is a well-mixed solution; currents at the surface and eddies in deep water, vertical mixing processes, and wave and tidal action all help to stir the pot. Because of the thorough mixing over geologic time, the ionic composition of open-ocean seawater is the same from place to place and from depth to depth. That is to say, the ratio of one major ion or seawater constituent to another remains the same. Whether the salinity is 40‰ or 28‰, the major ions exist in the same proportions. This concept was first suggested by the chemist Alexander Marcet in 1819. In 1865 another chemist, Georg Forchhammer, analyzed several hundred seawater samples and found that these constant proportions did hold true. During the world cruise of the *Challenger Expedition* (1872–1876), seventy-seven water samples were collected from different depths and locations in the world's oceans. When chemist William Dittmar analyzed these samples, he verified Forchhammer's findings and Marcet's suggestion. These analyses led to the **principle of constant proportion** (or **constant composition**), still in use, which states that regardless of variations in salinity, the ratios between the amounts of major ions in open-ocean water are constant. Note that the principle applies to major ions in the open ocean; it does not apply along the shores, where rivers may bring in large quantities of dissolved substances or may reduce the salinity to very low values. The ratios of abundance of minor constituents vary, because some of these ions are closely related to the life cycles of living organisms. Populations of organisms remove ions during periods of growth and reproduction, temporarily reducing the amounts of these ions in solution. Later, the population declines, and decay processes return these ions to the seawater.

Determining Salinity

The great advantage of the major ions' being in constant proportion is that it allows us to determine salinity by measuring the concentration of only one ion. Once the concentration of one major ion is known, then the concentration of all other major ions can be calculated from the known ratios.

Historically, the quantity of chloride ions present in the water sample has been measured to establish the salinity. This measurement is done quite easily and quickly by adding silver nitrate to a seawater sample. The silver combines with the chloride ion; thus if the amount of silver required to react with the chloride ion in a known amount of seawater is measured, then the exact amount of chloride is known. However, the silver also combines with bromine, iodine, and some other trace elements. The chloride concentration measured in this way is termed **chlorinity** (Cl‰); the chlorinity is measured in parts per thousand, or grams of chloride per kilograms of seawater. Chlorinity is defined as the quantity of silver required to remove all the **halogens** from 0.3285 kg of seawater. Halogens are members of a chemical family that includes chlorine, bromine, iodine, and fluorine, which react in similar ways. Chlorinity and salinity are related by the equation

$$\text{Salinity (‰)} = 1.80655 \times \text{Chlorinity (‰)},$$

or

$$S‰ = 1.80655 \times Cl‰.$$

Therefore, if the chlorinity of a sample is determined, we can calculate the salinity or the concentration of other major constituents by using the principle of constant proportion. This is a much faster and easier process than determining the concentration of each of the individual ions.

The Messages in Polar Ice

An intriguing method of looking back into the chemical history of the earth is provided by the earth's polar ice sheets. Studies of ice cores taken from these ice fields allow scientists to look back 100 years to 150,000 years and learn about the earth's previous atmospheric composition, its past temperatures, and its levels of precipitation. Ice cores preserve information in the chemical composition of the ice itself, including a record of soluble substances (sodium, chloride, sulfate, and nitrate), heavy metals (lead, cadmium, and zinc), and particulate impurities carried to the earth's surface by precipitation. Bubbles in the ice contain samples of the atmospheric gases present at the time the ice was formed. The recovery of the ice cores has allowed scientists investigating global pollution to compare the chemical composition of ice layers deposited before and after the Industrial Revolution. Studies show that the concentration of sulfate derived from sulfuric acid has tripled and the concentration of nitrate derived from nitric acid has doubled in Greenland during the last century. Significant increases in the concentration of industrial heavy metals have also been detected. For example, the study of Greenland ice and snow shows that lead, a by-product of leaded fuel combustion, has increased dramatically with the world's increased dependence on gasoline.

Cores have been obtained from Antarctica, from Greenland, Baffin Island, Alaska, and Spitsbergen in the Arctic, and from Peru, Mt. Kenya, and the Tian Shan Mountains of north central China. The oldest ice core was obtained by the Soviets in Antarctica in 1984. This core contains ice estimated to be 150,000 years old.

In order to assure that all determinations of chlorinity and salinity are done in exactly the same way in all oceanographic laboratories around the world, a standard method of analysis and a standard seawater reference is needed. A standard seawater is used to calibrate the silver nitrate solution used in these tests. Since the chemical procedures are all done in the same way and use the same standard seawater, salinity data that were determined in different laboratories can be directly compared.

Until 1975, the Hydrographic Laboratories in Copenhagen, Denmark, were responsible for the production of the standard (or normal) seawater (*eau de mer normale*). This standard seawater had an adjusted chlorinity of 19.4%. It was sealed in glass vials and shipped to oceanographic laboratories around the world. In 1975, the production of standard seawater was taken over by the Institute of Oceanographic Services in Wormly, England. Today's standard seawater is adjusted to constant chlorinity and electrical conductance.

Because of the ions present in solution, seawater conducts electricity; the more ions there are, the greater the conductance. This relationship makes it possible for us to determine salinity by using a salinity (or conductivity) bridge, often called a **salinometer.** The great advantage of the salinometer is that the measurements are made quickly and directly on the water sample with an

Figure 5.4
The salinometer determines salinity by measuring the conductivity of the seawater sample.

electrical probe (see fig. 5.4). It is therefore not necessary to analyze a water sample chemically for its salt content, except to test and calibrate the conductivity instruments. Because the conductivity of seawater is affected by both salinity and temperature, a conductivity instrument must correct for the temperature of the sample if it is calibrated to read directly in salinity units. Other direct measurement techniques are discussed in chapter 6.

TABLE 5.5

Abundance of Gases in Air and Seawater

Gas	Symbol	Percentage by volume in atmosphere	Percentage by volume in surface seawater[1]	Percentage by volume in total oceans
nitrogen	N_2	78.08	48	11
oxygen	O_2	20.99	36	6
carbon dioxide	CO_2	0.03	15	83
argon, helium, neon, etc.	Ar, He, Ne	0.95	1	
totals		100.00	100	100

[1]Salinity = $36^0/_{00}$, temperature = 20° C.

5.2 The Gases

Gases move between the sea and the atmosphere at the sea surface. Atmospheric gases dissolve in the seawater and are distributed to all depths by mixing processes and currents. The most abundant gases in the atmosphere and the sea are nitrogen (N_2), oxygen (O_2), and carbon dioxide (CO_2). Table 5.5 gives the percentage of each of these gases in the atmosphere and in seawater. Oxygen and carbon dioxide play important roles in the ocean because they are necessary to life. Biological activities modify the concentration levels of these gases at various depths. Nitrogen is not used directly by living organisms except for certain bacteria. Gases such as argon, helium, and neon are present in small amounts, but they are inert and do not interact with the ocean water or its inhabitants.

The amount of any gas that can be held in solution without causing the solution either to gain or to lose gas is the **saturation value.** The saturation value changes because it depends on the temperature, salinity, and pressure of the water. If the temperature or salinity decrease, the saturation value for the gas increases. If the pressure decreases, the saturation value decreases. In other words, colder water holds more dissolved gas than warmer water, less salty water holds more gas than more salty water, and water under more pressure holds more gas than water under less pressure.

Distribution with Depth

Plants produce oxygen and use carbon dioxide in the process known as **photosynthesis,** which synthesizes plant tissue. Since plants require sunlight for this process to occur, photosynthesis is confined, on the average, to the top 100 m (328 ft) of the sea, where sufficient light penetrates to energize the process. The photosynthetic process uses carbon dioxide and produces oxygen. By contrast, **respiration,** which breaks down organic substances to provide energy, requires oxygen and produces carbon dioxide and energy. All living organisms respire in order to produce energy in living cells. Respiration occurs at all depths in the oceans. Photosynthesis and respiration are discussed in more detail in chapter 13. Decomposition, the breakdown of nonliving organic material, also requires oxygen and releases carbon dioxide. Decomposition becomes the more important factor in the removal of oxygen from seawater at greater depths because the densities of living populations diminish with depth.

The depth at which the rate of photosynthesis balances the rate of plant respiration is called the **compensation depth.** Above the compensation depth, plants produce oxygen at the expense of carbon dioxide; below it, carbon dioxide is produced at the expense of oxygen. Oxygen can only be added to the oceans at the surface, from exchange with the atmosphere, or in the near-surface water, as a waste product of photosynthesis. Carbon dioxide is also added from the atmosphere at the surface, but it is available at all depths from respiration and decomposition.

Dissolved oxygen concentrations vary from 0 to 10 milliliters per liter of seawater. Very low or zero concentrations occur in the bottom waters of isolated deep basins, which have little or no exchange or replacement of water. Such an area can occur at the bottom of a trench, in a deep basin behind a shallow entrance sill (as in the Black Sea), or at the bottom of a deep fjord (300 to 400 m; 984–1312 ft). The dense, deep water is trapped and becomes stagnant; respiration and decomposition use up the oxygen faster than circulation to this depth can replace it. The bottom water becomes **anoxic,** or stripped of dissolved oxygen; **anaerobic** (or non-oxygen using) bacteria live in such water. This condition is shown in figure 5.5.

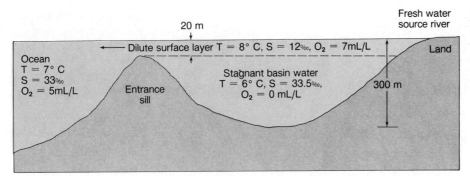

Figure 5.5

High-density ocean water is trapped behind the sill. The trapped water becomes anoxic due to continual respiration and decomposition.

At the surface, where oxygen is produced by the plants and added from the atmosphere, oxygen values are higher. Because oxygen is more soluble in cold water, more oxygen is found in surface waters at high latitudes than at lower latitudes. If the water is quiet, the nutrients and sunlight abundant, and a large population of plants present, oxygen values can rise above the equilibrium (or saturation) value. The oxygen concentration may be 150% or more of the saturation level. This water is **supersaturated.** Wave action will tend to liberate oxygen to the atmosphere and return the condition to the 100% saturation state.

Carbon dioxide levels range between 45 and 54 mL/L of seawater. The concentration is not substantially affected by biological processes, but tends to remain almost constant, controlled by temperature, salinity, and pressure, because of the complex way in which carbon dioxide reacts with seawater. These reactions are discussed in the next section.

Figure 5.6 shows typical oxygen and carbon dioxide concentrations with depth. You may use this diagram to review the relationships of these dissolved gases to the biological processes in the sea and to each other. First, consider oxygen. At the surface, the concentration is high due to the available atmospheric oxygen and to photosynthesis. Below the surface layer, oxygen decreases as animal respiration and the decomposition of organic material remove the oxygen; the **oxygen minimum** occurs at about 800 m (2600 ft). At depths greater than the oxygen minimum depth, the rate of removal of oxygen from the seawater decreases, because the population density of animals and the abundance of decaying organic matter has decreased. The supply of oxygen in water sinking from the surface is sufficient to increase the oxygen concentration above that found at the minimum depth. The carbon dioxide graph in figure 5.6 shows almost the direct opposite effect. The CO_2 concentration at the surface is low, because it is used in photosynthesis. Below the surface layer the concentration increases directly with depth as animal respiration and decomposition continually produce CO_2 and add it to the water.

Figure 5.6

The distribution of O_2 and CO_2 with depth.

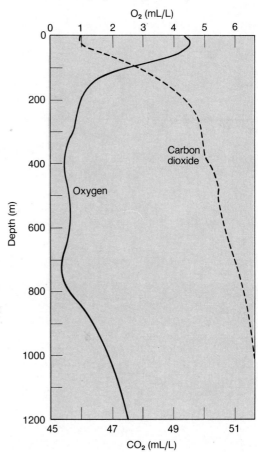

Carbon Dioxide as a Buffer

When carbon dioxide dissolves in seawater, the CO_2 combines with the water to form carbonic acid (H_2CO_3). The carbonic acid dissociates to form bicarbonate (HCO_3^-) and a hydrogen ion (H^+), or carbonate (CO_3^{2-}) and two hydrogen ions ($2H^+$). The CO_2, H_2CO_3, HCO_3^-, and CO_3^{2-} exist in equilibrium with each other

Figure 5.7
The pH scale.

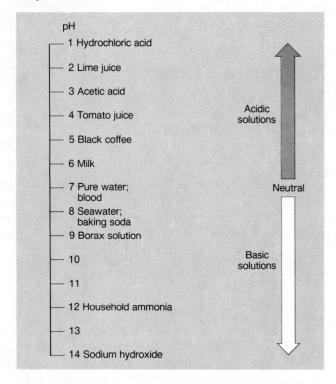

and with the H⁺ ions, as shown below. The arrows indicate that the reactions both produce and take up hydrogen ions:

$$CO_2 + H_2O \leftrightarrows H_2CO_3 \leftrightarrows HCO_3^- + H^+ \leftrightarrows CO_3^{2-} + 2H^+.$$

This series of reactions allows CO_2 to play an additional and extremely important role in the ocean: it acts as a **buffer.** A buffer prevents sudden changes in the acidity or alkalinity of a solution. The buffering capacity of seawater is important to organisms requiring a steady **pH** for their life processes and to the chemistry of seawater, which is controlled, in part, by pH. The pH scale indicates acidity and alkalinity by measuring the concentration of hydrogen ions in a solution. A pH of 7 indicates neutrality, neither acid nor alkaline. The pH values between 1 and 6 are acid while the pH values between 8 and 14 are alkaline, or basic. The pH scale is presented in figure 5.7. The pH range of seawater is between 7.5 and 8.5; an average pH value for the world's oceans is approximately 7.8.

The concentration of hydrogen ions (H^+) in the seawater controls the acidity or alkalinity. An increase in hydrogen ions shifts the reaction series to the left, removing them from solution. If the concentration of hydrogen ions decreases, the reaction moves from carbonic acid (H_2CO_3) to bicarbonate (HCO_3^-), liberating hydrogen ions. Therefore, the removal of CO_2 by photosynthesis or the addition of CO_2 by respiration or decay has little effect on the acid-alkaline balance of seawater.

The Carbon Dioxide Cycle

At present, the annual ocean uptake of carbon as carbon dioxide (CO_2) is calculated to be 2.5 billion metric tons per year. The rate at which the oceans absorb CO_2 is controlled by water temperature, pH, salinity, the chemistry of the ions (presence of calcium and carbonate ions), and biological processes, as well as mixing and circulation patterns of seawater which transfer CO_2 from the surface to the deep ocean (see fig. 5.8).

The transfer of carbon from CO_2 to organic molecules by photosynthesis results in the addition of CO_2 to the deep-ocean water, when the organic material sinks and decays. This process is often called the "biological pump," because biological processes pump carbon to the deep ocean. Enormous quantities of carbon as CO_2 transferred to the ocean bottom in this way are fixed in marine sediments as calcium carbonate ($CaCO_3$) (refer back to the Biogenous Oozes section in chapter 2). The rate at which this occurs is not fully understood. Cold, dense surface water sinking in subpolar regions also transfers CO_2 from the surface to deep water; these transfers are slow, requiring a few hundreds to more than a thousand years. At other latitudes, deep water returns to the surface and provides a return source of atmospheric CO_2. These circulation patterns and the effect of the increased production of CO_2 by the burning of fossil fuels will be discussed further in chapter 6.

The Oxygen Balance

The oceans also play a large role in regulating the oxygen balance in the earth's atmosphere. Photosynthesis in the oceans releases oxygen that is consumed by respiration and decay processes in the same way as on land. However, some organic matter is incorporated into the seafloor sediments, preventing the organic matter's decay and decomposition. Therefore, oxygen is not consumed to balance the oxygen produced in photosynthesis. In this way 300 million metric tons of excess oxygen are released into the atmosphere each year that would result in an excess of oxygen building up in the atmosphere over time. This does not happen because the marine sediments are formed into rocks by the earth's geological processes. Some of these rocks are uplifted onto land and oxygen is eventually consumed in the weathering and oxidation of the materials in the rocks. This balances the atmosphere's oxygen budget. The mechanisms which link and control this process are not well understood.

Figure 5.8
Major carbon dioxide pathways through the earth's environment.

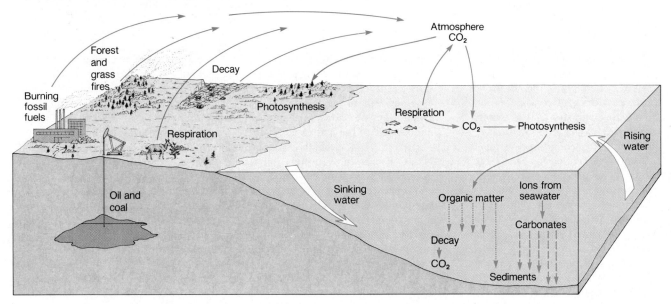

Measuring the Gases

The amount of dissolved oxygen present in seawater samples can be measured by traditional chemical techniques in the laboratory. It is also possible to measure oxygen in the oceans directly, using specialized probes that send a signal back to the ship electronically or store the information in the testing unit. Carbon dioxide is rarely measured directly; rather, it is determined from the pH of the water, which can be measured directly.

5.3 Other Substances

Nutrients

Ions required for plant growth are known as **nutrients**; these are the fertilizers of the oceans. As on land, plants require nitrogen and phosphorus but utilize them in the form of nitrate (NO_3^-) and phosphate (PO_4^{3-}) ions. A third nutrient required in the oceans is the silicate ion (SiO_4^-), which is needed to form the hard outer wall of the single-celled plants called diatoms and the skeletal parts of some protozoans. These three nutrients are among the dissolved substances brought to the sea by the rivers and land runoff. Despite their importance, they are present in very low concentrations (see table 5.6). The nutrients are removed from the water as the plant populations grow and reproduce and are then returned to the water as the organisms die and decay. Nutrients are cycled into the animal populations as the animals feed on the plants. These animals may be eaten in their

TABLE 5.6

Nutrients in Seawater

Element	Concentration $\mu g/kg$[1]
nitrogen (N)	500
phosphorus (P)	70
silicon (Si)	3000

[1]Parts per billion.

turn, and eventually the nutrients are returned to the oceans by way of death and bacterial decomposition. Excretory products from these animals are also added to the seawater, to be broken down and used again by a new generation of plants and animals. The cyclic nature of the nutrient pathways for nitrogen and phosphorus is discussed in chapter 13.

Organics

A wide variety of organic substances is present in seawater. Proteins, carbohydrates, lipids (or fats), vitamins, hormones, and their breakdown products are all present. Some will eventually be reduced to their inorganic components; others are used directly by organisms and are incorporated into their systems. Another portion of the organic matter accumulates in the sediments, where over geologic time it will slowly form deposits of crude oil. In the areas of the ocean that are high in plant and

The Effect of Sunlight on Seawater

*U*nderstanding the many and complex chemical reactions caused by sunlight in the oceans' surface waters is a new area of study in marine chemistry. Photosynthesis is the best known reaction involving sunlight in seawater, but researchers are finding an increasing number of other chemical reactions activated by the absorption of sunlight. When sunlight enters the surface waters, much of the radiation is absorbed by a variety of natural substances. These include living and nonliving particles, dissolved organic matter, and the water itself. Salt ions do not usually absorb sunlight. Some of the substances which are able to absorb the sun's energy become active and initiate a variety of other chemical reactions.

In some cases, the substance which absorbs the sunlight is changed as a direct result of the light energy. In other cases, the absorbing substance causes other substances that have not absorbed light to react. Both possibilities are shown in the figure. An example of the first case is the nitrite ion (NO_2^-), which is changed by the absorption of sunlight to form activated forms of nitrous oxide (NO•) and hydroxide (OH•). The activated compounds then rapidly take part in other chemical reactions. This reaction may affect the availability of nitrogen to marine plants. The second case is illustrated by an organic compound (OC) that absorbs light energy, becomes energized (OC•), and transfers the energy to dissolved oxygen, producing a reactive, excited form of oxygen. The energized organic compound (OC•) is returned to its original form (OC). The excited oxygen passes its energy on to other compounds to be used in reactions thought to be important in the formation of organic molecules and the degradation of some pollutants. Other chemical reactions involving the sun's energy may influence the availability of trace metals (iron, manganese) to marine plants.

There is a great deal more to learn about these processes. The reactions are difficult to follow and to measure because concentrations of substances may be very low. NO• in the surface waters of the equatorial Pacific is only about 2×10^{-9} g/l, and molecules may have a very short lifetime (the half-life of reactive oxygen is 3×10^{-6} second). Marine chemists are just beginning to realize the complexities of these processes and to understand the significance of sunlight's effect on seawater.

Box Figure 5.1
The effect of sunlight on chemicals in seawater.

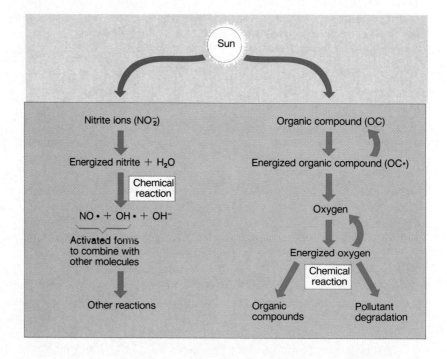

animal life, the surface layer may take on a green-yellow color due to the presence of so many organic substances, which are sometimes called "gelbstoff," or yellow substances.

5.4 Practical Considerations: Salt and Water

Chemical Resources

About 30% of the world's salt is extracted from seawater. In order to keep extraction costs low, the industrially produced energy required to remove the water is kept to a minimum. In warm, dry climates seawater is allowed to flow into shallow ponds; it is then allowed to evaporate down to a concentrated brine solution. More seawater is added, and the process is repeated several times, until a dense brine is produced. Then the evaporation is allowed to continue until a thick, white salt deposit is left on the bottom of the pond. The sea salt is collected and refined to produce table salt. This technique has been recently used in southern France, Puerto Rico, and California (see fig. 5.9).

In cold climate areas, salt has been recovered by freezing the seawater in similar ponds. The ice that forms is nearly fresh; the salts are concentrated in the brine beneath the ice. The brine is removed and heated to remove the last of the water.

Of the world's supply of magnesium, 60% comes from the sea, and so does 70% of the bromine. There are vast amounts of dissolved minerals in the world's

Figure 5.9
The southern end of San Francisco Bay has been diked into shallow ponds, where seawater is evaporated to obtain salt.

Figure 5.10

Solar energy is used to evaporate fresh water from seawater. Solar radiation penetrates the transparent cover of the still and causes evaporation of the seawater contained in the tank.

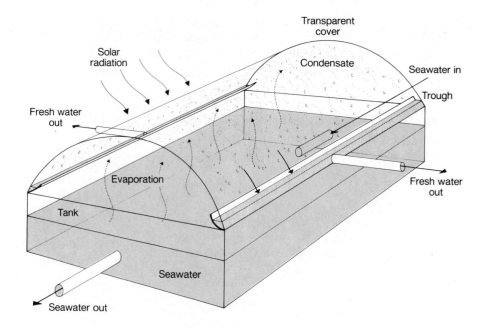

seawater, including 10 million tons of gold and 4 billion tons of uranium, but the concentration is very small (one part per billion or less). In the case of gold, no method has been devised in which the cost of extraction does not exceed the value of the recovered element. Japan, West Germany, and the United States have expressed an interest in the extraction of uranium. Japan set up a land-based test plant in 1986 to produce 10 kg (26 lb) of uranium per year by passing seawater over a uranium absorber. The system, including seawater pumps, absorber costs, and plant operation, has proved to be expensive, and the Japanese are currently looking for methods to use wave energy to run the system. (Some possible ways to harness wave energy are discussed in chapter 8.)

The dense, hot salt brines at the bottom of the Red Sea were discussed in chapter 2. These brines are estimated to contain minerals with values in billions of dollars.

Desalination

Desalination is the process of obtaining fresh water from salt water. The greatest drawback to the production of fresh water from seawater is the high cost, which is due to the energy required. There are several possible desalination methods:

1. **processes involving a change of state of the water** (liquid to solid or liquid to vapor);
2. **processes using a semipermeable membrane,** or **electrodialysis;** and
3. **processes requiring ion exchange.**

The simplest and least expensive process involving a change of state is the solar still. In this process a pond of seawater is capped by a low plastic dome. Solar radiation penetrates the dome, causing evaporation of the seawater. The evaporated water condenses on the undersurface of the dome and trickles down the curved surface to be caught in a trough, where it accumulates and flows to a freshwater reservoir. The rate of collection of fresh water is slow, and a very large system is needed to supply the water requirements of even a small community. Yet the costs for this type of system are low and are principally confined to construction of the facility, since the solar energy is free. This system is illustrated in figure 5.10; it works best in regions with high levels of solar radiation.

When water is distilled by boiling, evaporation proceeds at a rapid rate and large quantities of fresh water are produced, but considerable energy is required to boil a large volume of water. If water is introduced into a chamber that has its pressure reduced, the boiling will occur at a much lower temperature and rapid evaporation proceeds with less expenditure of energy. Saudi Arabia depends on desalinated water for its cities and for its industry. Oil-fired, electrically powered, and evaporative systems are used extensively. The Saudis very clearly express their attitude that their oil is to be used to produce the much needed power and fresh water; its selling price on the world market is of little concern. Compare this attitude to that of those countries with abundant water and less oil.

Producing fresh water from seawater by desalination costs more than obtaining water from groundwater or surface water supplies. Its feasibility is

Figure 5.11
(a) Electrodialysis. A tank is separated into three compartments by membranes A *and* B. *Membrane* A *passes only* Cl⁻. *Membrane* B *passes only* Na⁺. *Electrodes that are placed in the end compartments and are supplied with a direct-current voltage will cause the salt ions to migrate out of the center compartment. Fresh water is recovered from the center compartment; excess salt water is removed from each side compartment. (b) Ion exchange column. Seawater passes through a column of resin particles that exchange* H⁺ *for* Na⁺ *and* Cl⁻ *for* OH⁻ *to produce HOH, or fresh water (H_2O).*

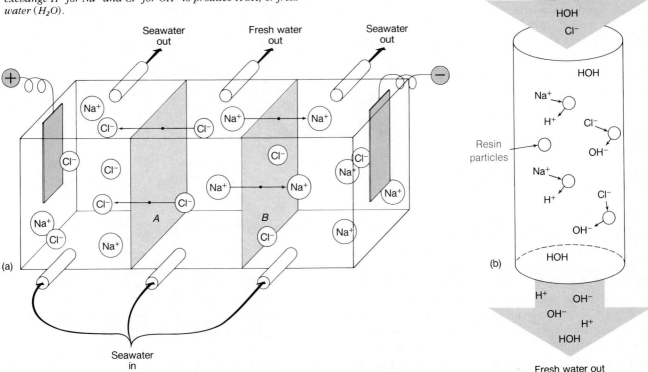

controlled by the degree of need, the uses for the water, the availability of other water sources, and the costs of desalination. This makes the practicality or cost effectiveness of desalination dependent on the situation in which it is required.

Freezing can be used to recover fresh water from seawater. The amount of energy needed is approximately one-sixth of that needed for evaporation, but the mechanical separation of the freshwater ice from the salt brine remains difficult.

Electrodialysis uses an electric field to transport ions out of solution and through semipermeable membranes. This technique works best in low-salinity (or brackish) water. Columns containing ion-exchange resins that extract ions from salt water also work well with water of relatively low salinity, but the resins need to be replaced periodically, which makes the process costly. Neither process produces the large volumes of low-cost fresh water that are required in some areas of the world. These processes are illustrated in figure 5.11.

In areas like Kuwait, Saudi Arabia, Morocco, Malta, Israel, the West Indies, and the Florida Keys, water is a limiting factor for population and industrial growth.

Thousands of cubic meters of fresh water are produced daily in these areas by evaporative desalination. For example, Kuwait produces more than 100,000 m³ each day and is building additional plants powered by waste heat from petroleum-fired power plants. Water costs are low in Kuwait because fuel costs are low, and in any case, there is no other source. Desalination plants in California and Texas produce fresh water at more than twice the cost of water from other sources; it is cheap enough to use for drinking water and for certain industrial applications, but far too expensive for agriculture. Small evaporation-type desalination plants are used on military and commercial vessels, and portable plants are used by the military in arid and Arctic areas.

It is possible to produce both energy and fresh water from a process that exploits the difference in salinity between two bodies of water. If waters of two different salinities are separated by a membrane that is permeable to the water but not to the salt, an **osmotic pressure** difference is produced. **Osmosis** is the movement of water across a semipermeable membrane; the water moves from the side of most concentration (more water) to the side of least concentration (less water).

This means that the water from the high water concentration (or low salinity) side will pass through the membrane, creating a higher pressure on the low water concentration (or high salinity) side of the membrane (see fig. 5.12a). If the salinity on one side of the membrane is 35⁰/₀₀, which is typical of the open ocean, and the salinity on the other side is 0⁰/₀₀ (or fresh water), the fresh water will move through the membrane until the pressure on the salty side builds to 25.84×10^6 dynes*/cm², or 24.5 atmospheres, which is the equivalent of the pressure generated by a column of water about 240 m (787 ft) in height. This pressure difference could be used to generate power, but no practical system has ever been built to accomplish this.

If pressure is applied to the saltwater side of this system, and the pressure is increased above 25.84×10^6 dynes/cm², or 24.5 atmospheres, water molecules will move from the saltwater side to the freshwater side (see fig. 5.12b). By this process, fresh water is squeezed out of the salt water. It requires a pressure of about 103×10^6 dynes/cm², or 101.5 atmospheres, to achieve a reasonable rate of freshwater production. Small units powered by hand pumps are available to produce fresh water for drinking in emergency situations (for example, in lifeboats and rafts). Small commercial units are available for producing fresh water for small vessels at a rate of 400 L (105 gal) per day. Some experiments are being conducted to power pumps to generate the required pressures for this so-called **reverse osmosis** process with wave power. Scale models have been tested and a full-scale test model is being planned that will produce 6000 L (1584 gal) of fresh water daily.

*A dyne is the amount of force required to accelerate a one-gram mass one centimeter per second per second:

$$1 \text{ dyne} = 1 \ \frac{(g)(cm)}{sec^2}.$$

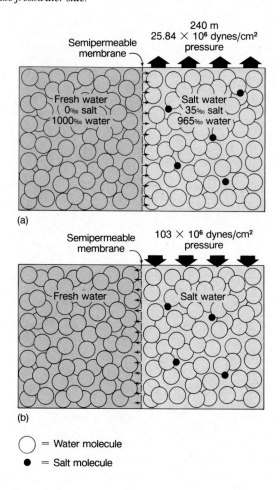

Figure 5.12
(a) Osmosis. Water moves from the freshwater side to the saltwater side. (b) Reverse osmosis. Water moves from the saltwater side to the freshwater side.

○ = Water molecule
● = Salt molecule

Summary

The average salinity of ocean water is 35⁰/₀₀. The salinity of the surface water changes with latitude and is affected by evaporation, precipitation, and the freezing and thawing of sea ice. Soluble salts are present as ions in seawater. Positive ions are cations; negative ions are anions. Six major constituent ions make up 99% of the salt in seawater. Trace elements that are present in very small quantities are particularly important to living organisms.

Most of the positively charged ions come from the weathering and erosion of the earth's crust and are added to the sea by rivers. Gases from volcanic eruptions are dissolved in river water as anions. Since the average salinity of the oceans remains constant, the salt gain must be balanced by the removal of salt; input must equal output.

Salts are removed as sea spray, evaporites, and insoluble precipitates, as well as by biological reactions, adsorption, chemical reactions, and uplift processes. Seawater circulating through magma chambers in the earth's crust deposits dissolved metals and releases other chemicals in solution. The time that salts remain in solution, known as residence time, depends on their reactivity.

The proportion of one major ion to another remains the same for all open-ocean salinities. Ratios may vary in coastal areas and in association with biological processes.

Salinity is determined chemically by the amount of chloride ion present and by the conductivity and temperature of the water.

The saturation value of gases dissolved in seawater varies with salinity, temperature, and pressure. Carbon dioxide is added to seawater at the sea surface from the atmosphere and by respiration and decay processes at all depths; it is removed at the surface by photosynthesis. Oxygen is added only at the surface from the atmosphere and the photosynthetic process; it is depleted at all depths by respiration and decay. Seawater may become supersaturated with oxygen or it may become anoxic. Carbon dioxide levels tend to remain constant over depth. Carbon dioxide has the additional role of buffer in keeping the pH range of ocean water between 7.5 and 8.5. Large quantities of CO_2 are absorbed by the oceans. Biological processes pump carbon as carbon dioxide into deep water, where it is fixed in the marine sediments as calcium carbonate. Atmospheric oxygen is regulated by oceanic processes. The amount of oxygen present in seawater is measured chemically and electronically. Carbon dioxide content is determined from the pH of the water.

Nutrients include the nitrates, phosphates, and silicates required for plant growth. A wide variety of organic products is also present.

Salt, magnesium, and bromine are currently being commercially extracted from seawater. Direct extraction of other chemicals is neither economic nor practical at present. Fresh water is an important product of seawater. Desalination methods include change-of-state processes, movement of ions across semipermeable membranes, and ion exchange. The practicality of desalination is determined by cost and need. Reverse osmosis can produce fresh water, and small processing units are available.

Key Terms

salinity
ion
cation
anion
major constituent
trace element
conservative constituent
nonconservative
  constituent
igneous rock

evaporite
adsorption
ion exchange
residence time
principle of constant
  proportion (constant
  composition)
chlorinity
halogens

salinometer
saturation value
photosynthesis
respiration
compensation depth
anoxic
anaerobic
supersaturation
oxygen minimum

buffer
pH
nutrients
desalination
semipermeable membrane
electrodialysis
osmotic pressure
osmosis
reverse osmosis

Study Questions

1. How is the ocean's salt balance regulated? Make a diagram showing inputs and outputs.

2. Explain the concept of the constant proportions of the major ions in seawater.

3. What is the most common and least expensive method of desalination? Where is it most likely to be used?

4. How does the salinity of midocean surface water change with latitude? What processes produce these changes?

5. List the sources of the salts found in seawater. How is their input regulated?

6. Silicate is a nonconservative constituent of seawater and does not obey the principle of constant proportions. Explain why.

7. Compare the distributions of oxygen and carbon dioxide with depth in seawater. What processes are responsible for the distribution of each gas?

Study Problems

1. If there is 1.4×10^{21} kg of water in the oceans, what is the potential mass of NaCl, sodium chloride, in the oceans? Use table 5.1.

2. If the chloride ion (Cl^-) content of a seawater sample is $18.5^0/_{00}$, what is the concentration of magnesium in the same sample? Express your answer in g/kg.

3. Determine the residence time of calcium using the following information:

Calcium ion present in seawater = 0.41 g/kg
Seawater in the oceans = 1.4×10^{21} kg
Calcium ion = 20.39% by weight of the average dissolved substances in river water
Average salt content of river water = $0.001^0/_{00}$
Annual river runoff = 3.7×10^{17} kg/yr.

4. How many kilograms of seawater would have to be processed to obtain 1 kg of gold? Use table 5.2.

Structure of the Oceans

6

*T*he weeks passed. We saw no sign either of a ship or of drifting remains to show that there were other people in the world. The whole sea was ours, and, with all the gates of the horizon open, real peace and freedom were wafted down from the firmament itself.

It was as though the fresh salt tang in the air, and all the blue purity that surrounded us, had washed and cleansed both body and soul. To us on the raft the great problems of civilized man appeared false and illusory—like perverted products of the human mind. Only the elements mattered. And the elements seemed to ignore the little raft. Or perhaps they accepted it as a natural object, which did not break the harmony of the sea but adapted itself to current and sea like bird and fish. Instead of being a fearsome enemy, flinging itself at us, the elements had become a reliable friend which steadily and surely helped us onward. While wind and waves pushed and propelled, the ocean current lay under us and pulled, straight toward our goal.

Thor Heyerdahl,
from Kon-Tiki

*H*idden below its surface is the ocean's structure. If we could remove a slice of ocean water in the same way we might cut a slice of cake, we would find that, like a cake, the ocean is a layered system. The layers are invisible to us, but can be detected by measuring the changing salt content and temperature of the water from the surface to the ocean bottom. This layered structure is a dynamic response to processes that occur at the surface: the addition and loss of heat, the evaporation and addition of water, and the movement of air forcing the surface water into motion. These surface processes produce a series of horizontally moving layers of water, as well as local areas of vertical motion. In this chapter, we will study both the surface processes and their below-the-surface results in order to understand why the ocean is structured in this way and how the structure is maintained. We will also explore the ways in which oceanographers gather data about this layered system, and we will survey the possibilities of extracting useful energy from it.

6.1 Heating and Cooling the Earth's Surface

The intensity of the solar radiation that is received per unit area of earth surface is greatest at the equator, moderate in the middle latitudes, and least at the poles. If the earth had no atmosphere, the intensity of solar radiation available on a surface at right angles to the sun's rays would be 2 calories per square centimeter per minute (cal/cm²/min); this value is called the **solar constant.** The solar constant is approached only at latitudes between 23½°N (the Tropic of Cancer) and 23½°S (the Tropic of Capricorn), because only between these latitudes can the sunlight strike the earth at right angles. At all other latitudes the earth's surface is always inclined to the sun's rays at a greater angle, because of the earth's spherical shape. Compare the angles at which the sun's rays strike the earth in figure 6.1. Because the sun is so far away from the earth, its rays are parallel when they reach the earth. Where the rays strike the earth at right angles, the same amount of radiant energy strikes each unit area. But as the angle at which the sun's rays strike the earth increases, the unit areas receive less energy. This difference is also shown in figure 6.1.

When the sun stands directly above the equator at noon, during either the vernal or the autumnal equinox, the radiation value is about 1.6 cal/cm²/min. This value is less than the solar constant, because the earth's atmosphere stands between the earth's surface and the incoming solar radiation, and the atmosphere absorbs a portion of the sun's energy. Atmospheric absorption causes the solar radiation per unit of surface area to decrease with increasing latitude in each hemisphere. This additional decrease is due to the increased length of the path the sun's rays must travel through the atmosphere to strike the earth at higher latitudes (see fig. 6.1). The combined effects of atmospheric path length and inclination of the earth's surface cause the earth to receive more heat in the tropics, less at the temperate latitudes, and the least at the poles.

The Heat Budget

In order to maintain its temperature pattern, the earth must lose heat as well as gain it. On the world average, the earth must reradiate as much heat back to space as it receives from the sun. These gains and losses in heat can be represented as a **heat budget.** Just as incoming funds must equal outgoing funds in a balanced mone-

Figure 6.1

Areas of the earth's surface that are equal in size receive different levels of solar radiation as they become more oblique to the sun's rays (see a, b, and c). As latitude increases, the angle between the sun's rays and the earth increases and the solar radiation received on the surface decreases (see d). Note also that the sun's rays must travel an increasing distance through the atmosphere as latitude increases.

tary budget, in the heat budget incoming radiation must equal outgoing radiation. If incoming funds (or deposits) are insufficient, or if outgoing funds (or withdrawals) become too great, serious financial problems arise. In the same way, if less heat were returned to space than is gained, the earth would become hotter, and if more heat were returned than is gained, the earth would become colder. In both cases, the planet would change dramatically.

To follow the deposits and withdrawals in the heat budget, let us assume that 100 units of solar energy are incoming to the earth's atmosphere. Refer to figure 6.2 as you read the following discussion. Thirty-one of these units are reflected back from the earth's atmosphere directly to outer space; no heating of the earth or its atmosphere is accomplished by this reflected radiation. This leaves 69 units of the 100 units, and 4 of these units are reflected by the earth's surface; these 4 units are reflected back through the atmosphere to outer space and also take no part in heating the earth. There are now 65 of the original 100 units left. These 65 units do heat the planet: 47.5 units are absorbed by the earth's surface, and 17.5 units are absorbed by the earth's atmosphere. To balance the budget, 65 units must be lost by **reradiation** of long-wave radiation back to outer space; 5.5 units are reradiated by the earth's surface, and 59.5 units are reradiated from the atmosphere. Our budget is balanced; incoming radiation balances outgoing radiation.

Closer inspection shows that although the earth absorbs 47.5 units, it loses only 5.5 units to space, for a gain of 42 heat units, while the atmosphere absorbs 17.5 units and loses 59.5 units to space, for a loss of 42 units.

In these circumstances, the atmosphere cools and the earth heats. In order to maintain the near-constant average temperatures of the earth's surface and atmosphere, the 42 excess units gained by the earth must be transferred to the atmosphere to make up for the 42 units the atmosphere has lost. This transfer of heat is accomplished in two ways. First, 29.5 of the units are transferred by evaporation processes, which cool the earth's surface and liberate heat when condensation occurs in the atmosphere. The remaining 12.5 units are transferred to the atmosphere in the form of radiation given off by the earth and absorbed by the water content of the atmosphere, and also in the form of sensible heat that raises the temperature of the air. On the world average, the atmosphere is primarily heated from below by heat given off from the earth.

If we consider the heat budget of only a small local portion of the oceans, the following factors must be taken into consideration: total energy absorbed at the sea surface in that area, loss of energy due to evaporation, transfer of heat into and out of the area by currents, warming or cooling of the overlying atmosphere by heat from the sea surface, and heat reradiated to space from the sea surface. These processes vary with time, showing daily and seasonal variations. Measurements made over the earth's surface show that more heat is gained than lost at the equatorial latitudes, while more heat is lost than gained at the higher latitudes (see colorplate 7). Winds and ocean currents remove the excess heat accumulated in the tropics and release it at higher latitudes to maintain the present surface temperature patterns (see fig. 6.3).

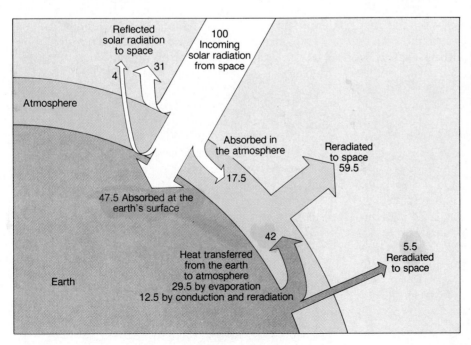

Figure 6.2

The earth's heat budget. Incoming solar energy is balanced by reflected and reradiated energy. The atmosphere's loss of heat is balanced by heat transferred from the earth to the atmosphere by evaporation, conduction, and reradiation.

Figure 6.3
Sea surface temperatures during summer in the northern hemisphere, given in degrees Celsius.

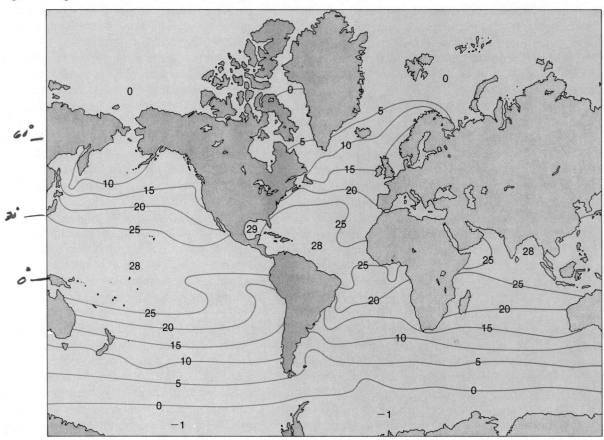

Carbon Dioxide Effects

Atmospheric carbon dioxide is transparent to incoming radiation, but at the same time it absorbs the outgoing long-wavelength back radiation and warms the earth by what is commonly known as the **greenhouse effect.** In the northern hemisphere, the natural cycle of the world's carbon dioxide shows decreasing CO_2 in the spring and summer, when plants increase active photosynthesis and remove CO_2 in greater amounts than that contributed by respiration and decay. In the fall and winter, the photosynthetic activity is reduced, plants lose their leaves, and decay processes release CO_2; atmospheric CO_2 increases. The effects of deforestation and conversion of forest land to agriculture, the burning of fossil fuels, and the growth of human populations has been superimposed on this natural cycle. In preindustrial times, the seasonal cycle was in balance, but current practices have added a huge excess of CO_2 to the equation, currently 18 billion tons per year.

Since 1850, the concentration of carbon dioxide in the atmosphere has increased from 280 parts per million to 345 parts per million. Recently, the average rate of increase has been 1.3 parts per million per year. For more than thirty years scientists have been recording a steady increase in the carbon dioxide concentration in the earth's atmosphere, due to the burning of coal, oil, and other fossil fuels (see fig. 6.4). If the increasing trend continues, the concentration of carbon dioxide will double sometime before the middle of the next century. This will warm the earth's surface and alter the earth's average heat budget by reducing the heat lost to space by long-wave radiation.

Based on this trend of increasing CO_2, predictions of global warming of 2–4° C have been made. A change in temperature of this magnitude in a few decades is comparable to that which has occurred since the last ice age, over 10,000 years ago. It is not possible to predict the extent of the effects brought about by such a change, because other factors could affect this warming trend, including increased evaporation and cloud cover, which decreases incoming radiation. However, if the warming does occur, it is expected to affect the high latitudes, causing melting of polar land ice that would raise the

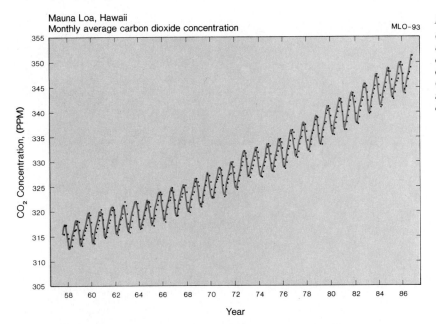

Figure 6.4

Concentration of atmospheric carbon dioxide in parts per million of dry air versus time in years as observed at the Mauna Loa Observatory, Hawaii. The dots indicate monthly average concentrations.

world sea level by about 1 meter; local shifts in weather and climate could influence agriculture.

It is calculated that between 40% and 50% of the CO_2 produced by the burning of fossil fuels now remains in the atmosphere. The majority of the CO_2 missing from the atmosphere enters the oceans and ocean sediments, which serve as a sink for excess CO_2. Refer back to the discussion of carbonate sediments in chapter 2 and dissolved CO_2 in chapter 5. The exact pathways of the excess CO_2 are difficult to trace, and the rates of deposit in oceanic sediments are difficult to measure. The role of the oceans in maintaining the complex cycles and balances of the gases in the atmosphere, and the ocean's ability to cope with human-induced changes are not completely understood; they are the focus of ongoing marine research.

Annual Cycles of Solar Radiation

The intensity of the solar radiation that is available at the earth's surface varies with time. There is an annual cycle of seasonal variations in solar radiation, which is most pronounced at the middle latitudes, for between 40°N and 60°N and between 40°S and 60°S the angle at which the sun's rays strike the earth changes dramatically from summer to winter. At these latitudes, there is a corresponding change in the duration of the daylight hours. This variation is illustrated in figure 6.5. The change in incident radiation, the solar radiation striking the earth's surface, produces seasonal variation in sea-surface temperatures due to heat losses or gains. There are no real seasons in the tropics; the intensity of solar radiation remains fairly constant at tropic latitudes

(between 23½°N and 23½°S) over the year, because the sun's rays are always received at an angle approaching 90°. During the sun's annual migration between 23½°N and 23½°S, its rays cross the intervening latitudes twice. This produces a small-amplitude, semiannual variation in the intensity of solar radiation. The effect is most evident at the equator and can be seen in figure 6.5. Between 70°N and 90°N and between 70°S and 90°S, the sun's rays always strike the earth at an oblique angle. The long duration of daylight hours in summer produces a high level of incident solar radiation averaged over the day (fig. 6.5). However, the intensity of radiation per unit surface area per minute of daylight is much lower than that found at the lower latitudes (see again fig. 6.1). Refer back to figure 1.3 in chapter 1 to review the relationship of the earth, the sun, and the earth's seasons.

Heat Capacity

Land and ocean areas respond differently to the annual cycle of solar radiation. Land has a low heat capacity; it changes temperature rapidly as it gains and loses heat between day and night or summer and winter. The oceans have a high heat capacity; they absorb and release large amounts of heat at the surface with very little change in temperature.

The annual range of surface temperatures for land and ocean are given in figure 6.6. Note that seasonal land temperature ranges at high latitudes are greater than those for the oceans and that the annual range of ocean surface temperatures is greatest at the middle latitudes. Because of the unequal distribution of land between the

Figure 6.5

Average solar radiation values at different latitudes during the year. When it is summer in the northern hemisphere it is winter in the southern hemisphere, therefore the higher values in the northern hemisphere coincide with the lower values in the southern hemisphere. Peak values of solar radiation at the higher latitudes occur during summer in each hemisphere as a result of more hours of sunlight.

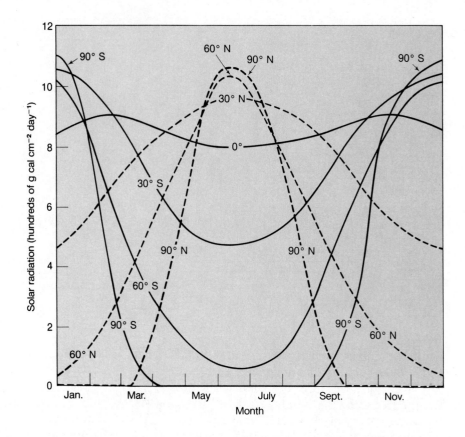

Figure 6.6

The annual range of midocean sea surface temperatures is considerably less than the annual range of land surface temperatures at the higher latitudes. The maximum annual range of sea surface temperatures occurs at the middle latitudes.

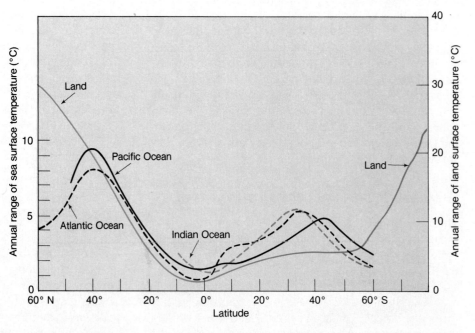

two hemispheres, the summer-to-winter variation in temperature for land at the middle latitudes is much greater in the northern hemisphere than it is in the southern hemisphere. The lack of land in the southern hemisphere means that the oceans, with their high heat capacity, control the earth's annual temperature range at the southern middle latitudes.

Heat that is absorbed at the ocean surface in summer is mixed downward by winds, waves, and currents. Moving the heat downward also reduces the temperature change at the sea surface. The net effect of these processes is the small annual change in midocean surface temperatures: 0 to 2° C in the tropics, 5 to 8° C at the middle latitudes, and 2 to 4° C at the polar latitudes.

An Atmospheric Thermostat

Dimethyl sulfide (DMS) is another gas related to living processes and produced in the oceans. DMS is produced by plants in the oceans' photic layer. Gas exchange between the photic layer and the atmosphere is rapid, measurable in days or weeks, and DMS is present everywhere at the sea surface. It is responsible, in part, for the characteristic smell of the sea one notices when the coast is approached from the land.

Current estimates are that about 60 million tons of DMS or 30 million tons of sulfur are emitted to the atmosphere from the oceans each year. This is equal to one-third to one-half the amount of sulfur entering the atmosphere from the burning of fossil fuels. However, this natural source of sulfur is distributed over the entire earth's surface while sulfur from fossil fuels is concentrated in the industrialized areas of the northern hemisphere. Once DMS is in the

atmosphere, it changes rapidly and one of its main products is sulfuric acid. Sulfuric acid then condenses into small particles that eventually are returned to the oceans and land with the rain. The quantities of sulfuric acid returned to the earth's surface from this process are far below the values associated with the problem of "acid rain," caused by the concentrated emissions from fossil fuels.

Another interesting effect of DMS relates it to global temperature control. This gaseous sulfur compound may help to prevent the overheating of the earth's surface and atmosphere. During the time the sulfur is in the atmosphere, it influences the amount of incoming shortwave radiation by reflecting this radiation back into space, before it can be absorbed to heat the earth's surface. These particles are also

thought to control the density of marine clouds, which in turn affects the clouds' reflective properties, reducing incoming radiation and decreasing the heating of the earth's surface.

The dimethyl sulfide system may act as a feedback mechanism or self-regulating thermostat, helping to control the earth's surface temperature. If a DMS buildup results in an excess of clouds and sulfur, less light strikes the sea surface, surface temperatures drop, and the plant production of DMS decreases. If less DMS is produced, there is less cloud cover and less reflection of shortwave radiation, and the earth's surface warms. Learning about DMS brings additional information to scientists working to understand the complexity of the interactions that occur between the atmosphere and the oceans.

6.2 Evaporation and Precipitation Patterns

The salinity of the midocean surface water is controlled by the distribution of the world's evaporation and precipitation zones (see again figs. 5.1 and 5.2). Salinity of ocean water is measured in grams of salt per kilogram of seawater and expressed in parts per thousand ($^0/_{00}$). The average ocean salinity is considered to be $35^0/_{00}$. In the tropics, the precipitation is heavy on land and at sea. On land, the tropic rain forest is the result; at sea, the surface water has a low salinity, around $34.5^0/_{00}$. At approximately 30°N and 30°S, the evaporation rates are high. These are the latitudes of the world's land deserts and of increased salinity in the surface waters, around $36.7^0/_{00}$. Farther north and south, from 50°N to 60°N and from 50°S to 60°S, precipitation is again heavy, producing cool but less salty surface water, around $34.0^0/_{00}$, and the heavily forested areas of the northern hemisphere. At the polar latitudes sea ice is formed in the winter. During the freezing process, the salinity of

water beneath the ice increases, and in the summer the thawing of the ice reduces the surface salinity once more.

6.3 Density Structure and Vertical Circulation

These variations in the surface temperatures and salinities of the oceans combine to control the **density** of the oceans' surface water. Many combinations of salinity and temperature produce the same density; lines of constant density are displayed in figure 6.7. As the salinity increases, the density increases; as the temperature increases, the density decreases. Salinity may be increased by evaporation or by the formation of sea ice; it may be decreased by precipitation, by the inflow of river water, by the melting of ice, or by a combination of these factors. Changes in pressure also affect density. As the pressure increases, the density increases. Since pressure plays a minor role in determining the density

Figure 6.7

The density of seawater, measured in grams per cubic centimeter, is abbreviated as ρ (rho) and varies with temperature and salinity. The pressure factor is not included. Many combinations of salinity and temperature produce the same density. Low densities are at the upper left and high densities at the lower right.

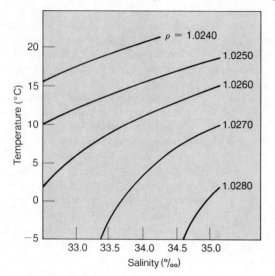

of surface water, its effects will be ignored in the following discussion.

Surface Processes

Less dense water remains at the surface (for example, the warm, low-salinity surface water of the equatorial latitudes). Although the surface water at 30°N and 30°S latitudes is warm, it has a higher salinity. Therefore, it is more dense than the warm, low-salinity equatorial water. This 30° latitude surface water sinks below the equatorial surface water; it extends from the surface at 30°N to below the less dense equatorial layer and back to the surface at 30°S. The combination of salinity and temperature in surface waters at 60°N and 60°S produces a density that is greater than either the equatorial or the 30° latitude surface water. The 60° latitude water therefore extends from the surface in one hemisphere below the other water types to the surface in the other hemisphere. Winter conditions in the polar regions lower the water temperature and increase the salinity as the sea ice forms. The result is a dense water that sinks toward the ocean floor at the polar latitudes. These variations in surface water properties and the resulting density changes produce a layered ocean, for if the density of the seawater is increased at the surface, the water sinks to a level of similar density. This layered system is shown in figure 6.8.

Changes with Depth

The oceans have a well-mixed surface layer of approximately 100 m (330 feet) and layers of increasing density to a depth of about 1000 m (3300 feet). Below 1000 m are found the relatively homogeneous dense waters of the deep ocean. A region between 100 m and about 1000 m where density changes rapidly with depth is known as a **pycnocline** (see fig. 6.9).

Below the 100 m-thick surface layer the temperature decreases rapidly with depth to the 1000 m level. A zone with a rapid change in temperature with depth is called a **thermocline.** Below the thermocline, the temperature is relatively uniform over depth, showing a small decrease to the ocean bottom. A similar situation occurs with salinity. Below the surface water at the middle latitudes, the salinity increases rapidly to about 1000 m; the zone of relatively large change in salinity with depth is called the **halocline.** Beneath the halocline, relatively uniform conditions extend to the ocean bottom. Both a thermocline and a halocline are shown in figure 6.10.

If the density of the water increases with depth, the water column from surface to depth is **stable.** If there is more dense water on top of less dense water, the water column is **unstable.** An unstable water column cannot persist; the more dense water sinks and the less dense water rises. Vertical **overturn** of the water takes place. If the water column has the same density over depth, it has neutral stability and is termed **isopycnal.** A neutrally stable water column is easily mixed in the vertical by wind, wave action, and currents. If the water temperature is unchanging over depth, the water column is **isothermal;** and if the salinity is constant over depth, it is **isohaline.**

Density-Driven Circulation

Processes that increase the water's density at the surface cause vertical water movement, or **vertical circulation,** which ensures an eventual top-to-bottom exchange of water throughout the oceans. Because the density is normally controlled by changes in temperature and salinity, this circulation is known as **thermohaline circulation.** The classic example of thermohaline circulation occurs in the Weddell Sea of Antarctica, where the winter cooling and freezing produce water so dense that it sinks down the continental slope to the sea bottom, forming the densest water found in the open oceans.

At the middle latitudes in the open ocean there is a seasonal effect; this effect is illustrated in figure 6.11. During the summer the surface water is warm, and the water column is stable. In the fall the surface water is cooled. As its density increases with cooling, the water

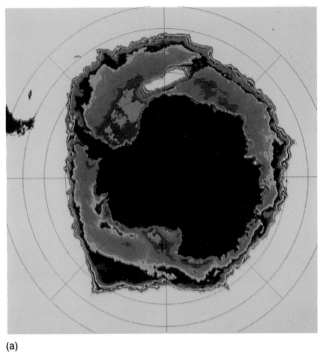

(a)

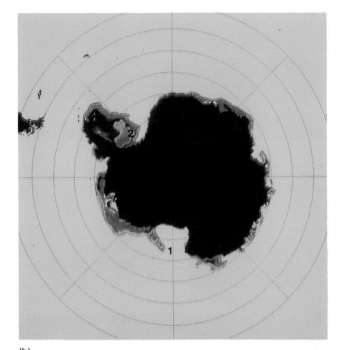

(b)

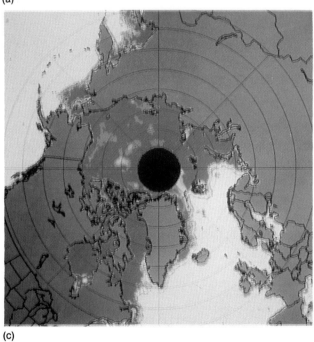

(c)

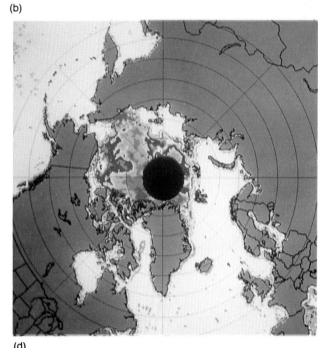

(d)

Colorplate 6
*Sea ice around the Arctic and Antarctic. Data from NASA's
NIMBUS-5 satellite was used to construct seasonal maps of sea ice
coverage at high latitudes. This information is used to help develop
a better understanding of global climate. The Antarctic ice
extends more than 100 km from the land into the Ross (1) and*

*Weddell (2) seas during the winter (a) and is much reduced
during the summer (b). The Arctic Ocean is filled with ice in the
winter (c) but by the next fall the ice cover is much less (d). Short-
term motions of the ice boundaries can be calculated from
repeated satellite images. These show that the ice floes move up to
50 km per day. Purple indicates high ice concentration and blue
indicates open water.*

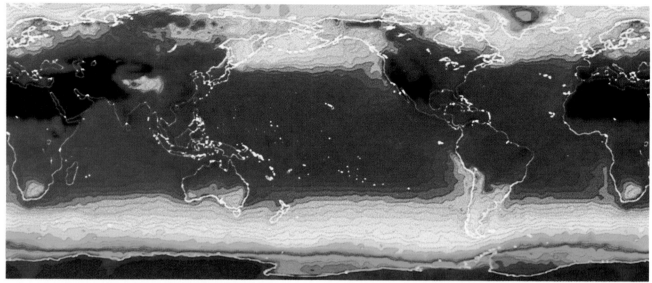

(a)

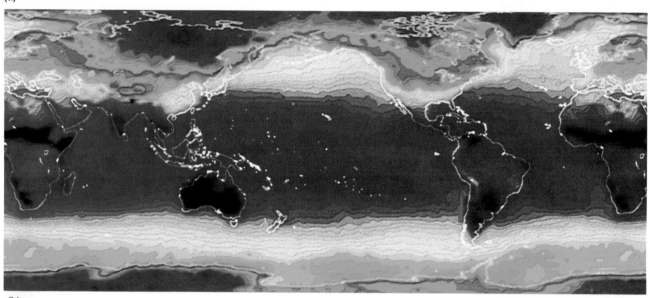

(b)

Colorplate 7

(a) Meteorological satellites, such as NOAA's TIROS, carry high-resolution infrared sensors that measure long-wave radiation emitted from the earth's surface and atmosphere. This radiation is related to the earth's surface temperatures. Green and blue indicate temperatures below 0° C; warmer temperatures are shown in red and brown. In January, Siberia and Canada show surface temperatures near −30° C; at the same time, latitudes between 30° and 50° south show warm summer temperatures.

(b) July data shows the northern hemisphere land masses with considerably warmer temperatures, but the ocean waters do not change dramatically from winter to summer. The oceans' warm surface water moves north and south with the change in the seasons. North-south currents along the coasts of continents are also visible.

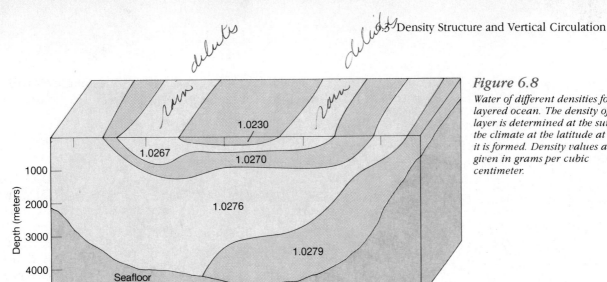

Figure 6.8

Water of different densities form a layered ocean. The density of each layer is determined at the surface by the climate at the latitude at which it is formed. Density values are given in grams per cubic centimeter.

Figure 6.9

Density increases with depth in seawater. The pycnocline is the region in which density changes rapidly with depth. (Based on data from the northeast Pacific Ocean.)

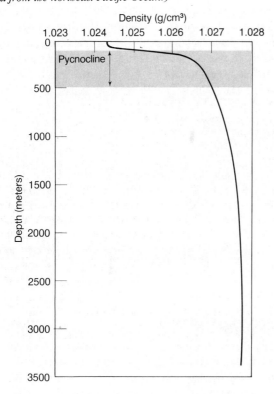

Figure 6.10

Temperature and salinity values change with depth in seawater. Rapid changes in temperature and salinity with depth produce a thermocline and a halocline, respectively. (Based on data from the northeast Pacific Ocean.)

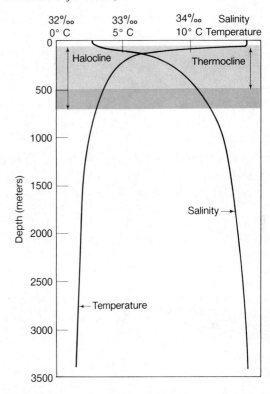

column becomes unstable, and the surface water sinks. Winter storms and winter cooling continue the mixing process. The shallow thermocline formed during the previous summer is destroyed, and the upper portion of the water column becomes isopycnal to greater depths. Spring brings warmth and the thermocline begins to reestablish itself. The water column becomes stable and continues to be stable through the summer.

In the open ocean, temperature is more important than salinity in determining density. For example, the surface water at 30°N and 30°S has the highest salinity of all open-ocean water, but the subtropic water is so warm that it stays at the surface. The water from the North

Figure 6.11

The surface layer temperature structure varies over the year. In the absence of strong winds and wave action in the summer, solar heating produces a shallow thermocline. During the fall and winter the surface cooling and storm conditions cause mixing and vertical overturn, which eliminate the shallow thermocline and produce a deep wind-mixed layer.

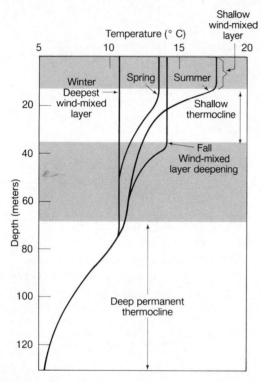

divide the density by 1.000 g/cm³ (the density of pure water) to remove the units:

$$\frac{1.02677 \text{ g/cm}^3}{1.00000 \text{ g/cm}^3} = 1.02677$$

We then subtract 1 and multiple by 1000:

$$(1.02677 - 1) \times 1000 = 26.77$$
$$\sigma_t = 26.77$$

Pressure is disregarded, and σ_t becomes a factor related only to salinity and temperature, as indicated by the symbols σ for salinity and t for temperature.

6.4 Upwelling and Downwelling

When dense water from the surface sinks and reaches a level at which its density is greater than the water above but less than the water below, it moves horizontally in order to make room for the water descending behind it. At the surface, water moves horizontally into the region where sinking is occurring, and elsewhere water rises to complete the cycle. Because water is a fixed quantity in the oceans, it cannot be accumulated or removed at given locations without movement of water between these locations. This concept is called **continuity of flow.** Areas in which water sinks are known as **downwelling zones;** areas of rising waters are **upwelling zones.** Downwelling is a mechanism that transports oxygen-rich surface water to depth, where it is needed for the deep-living animals. Upwelling returns the decay-produced nutrients that have accumulated at depth to the surface waters, where they act as plant fertilizers to promote the production of more oxygen by photosynthesis in the sunlit surface waters.

Upwelling and downwelling are also related to the wind-driven surface currents. When the surface waters are driven together, a surface **convergence** is formed. Water at a surface convergence sinks, or downwells. When surface waters flow away from each other, a surface **divergence** occurs. Water from below is upwelled at a divergence (see fig. 6.12). During the slow movement of water from surface to depth and back, the water continually mixes with adjacent layers of water, gradually exchanging chemical and physical properties.

The upwelling and downwelling water moves at rates of 0.1 to 1.5 m (5 ft)/day. Compare this speed to that of oceanic surface currents, which reach speeds of 1.5 m/sec. Horizontal movement at depth due to the thermohaline flow is about 0.01 cm (.004 in)/sec. Water caught in this slow but relentlessly moving cycle may spend 1000 years at depth before it again reaches the surface. Research carried out by the Geochemical Ocean Sections (GEOSECS) program in 1972–1973 has shown

Atlantic has a lower salinity, but it is cold. This cold water sinks and flows below the salty but warm surface water.

Close to shore, the salinity of seawater becomes more important than temperature in controlling density. This relationship is particularly true in semi-enclosed bays, sounds, and fjords that receive large amounts of freshwater runoff. Here extremely cold but fresh water is added, as the rivers carry precipitation and ice melt to the sea. This water may be only 0 to 1° C, yet its salinity is so low that the water does not sink, but remains instead on the surface as a seaward-moving **freshwater lid.**

Sigma-t

Minute changes in density are responsible for relatively large changes in vertical circulation. For this reason, oceanographers routinely measure density to the fifth decimal place, for example, 1.02677 g/cm³. Because such a number is awkward to handle when performing calculations and processing large quantities of data, a more convenient term is used. Density is converted to a **sigma-t** (σ_t) value. To derive sigma-t from density, we

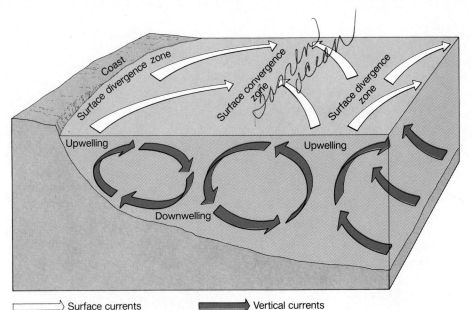

Figure 6.12
Upwelling and downwelling are related to horizontal motions of water at the sea surface. Downwelling occurs where surface currents converge, and upwelling occurs where surface currents diverge.

that radioactive material that was produced by atmospheric testing of nuclear weapons in the late 1950s and early 1960s had already reached the bottom of the North Atlantic, indicating a much faster rate of descent and transport than was previously thought to occur. The 1981 Transit Tracers in the Ocean (TTO) cruises found that small changes had occurred in water properties of the deep North Atlantic water since GEOSECS. This finding is in contrast to earlier research, which indicated that the water properties of the deep ocean remain constant. The rates of formation and properties of this deep water may vary with climate changes at the sea surface.

6.5 *The Layered Oceans*

The layered character of each ocean has been determined by taking salinity and temperature measurements over the depth of the water column from many surface positions. Gradually the accumulation of these data allowed oceanographers to identify layers of water that could be traced back to their sources at the surface, where they received their characteristic values of salinity and temperature. Water that sinks from the surface to spread out at depth eventually rises at another location or slowly mixes with adjacent layers. In all cases, water that sinks to depth displaces an equivalent volume of water upward toward the surface at some other location so that the oceans' vertical circulation may continue.

Structure of the Atlantic Ocean

The structure of the Atlantic Ocean is determined by the properties of the layers of water present under the sea surface. Each layer received its characteristic sa-

linity, temperature, and density at the surface. The water's density controls the depth to which each layer of water sinks. The thickness and horizontal extent of each layer is related to the rate at which it is formed at the surface and the size of the surface region over which it is formed. Figure 6.13 shows the properties of the layers of water making up the Atlantic Ocean.

At the surface in the North Atlantic, water from high northern latitudes moves southward, while water from low latitudes moves northward along the coast of North America and then east across the Atlantic. These waters converge in areas of cool temperatures and high precipitation at approximately 50°N to 60°N, at which areas are located the Norwegian Sea and the boundary of the Labrador Current and the Gulf Stream. The resulting mixed water has a salinity of about 34.9‰ and a temperature of 2 to 4° C. This water, known as **North Atlantic deep water,** sinks and moves southward at depth. North Atlantic deep water from the Norwegian Sea moves down the east side of the Atlantic, while water formed at the boundary of the Labrador Current and the Gulf Stream flows along the western side. Above this water, a lens of very salty (36.5‰) but very warm (25° C) surface water remains trapped by the circular movement of the major oceanic surface currents. Between this surface water and the North Atlantic deep water lies water of intermediate temperature (10° C) and salinity (35.5‰). This water is a mixture of surface water and the upwelled colder, saltier water from the equatorial regions. It moves northward to reappear at the surface south of the convergence in the North Atlantic.

Near the equator, the upper boundary of the North Atlantic deep water is formed by water produced at the convergence centered about 40°S. This is **Antarctic intermediate water.** Since it is warmer (5°C) and less

Figure 6.13

The anatomy of the Atlantic Ocean. Surface and subsurface circulation are related here. The surface currents converge to produce downwelling; water sinks to its density level and flows horizontally. Water at depth rises under zones of surface divergence.

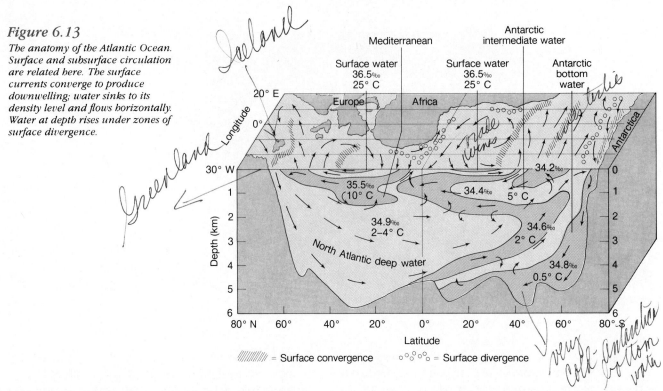

salty (34.4‰) than the North Atlantic deep water, it is less dense and remains above the more dense and saltier water below. Along the edge of Antarctica very cold (−0.5°C), salty (34.8‰), and dense water is produced at the surface during the southern hemisphere's winter. This is **Antarctic bottom water,** the densest water produced in the world's oceans. After it descends to the ocean floor, it begins to move northward. When it meets North Atlantic deep water it creeps beneath it, because it is more dense, and continues to move northward along the coast of South America. Antarctic bottom water is produced in too small quantities to accumulate enough thickness to be able to flow over the midocean ridge system and into the basins along the African coast. It is therefore confined to the deep basins on the west side of the South Atlantic and has been found as far north as the equator.

The North Atlantic deep water, meanwhile, trapped between the Antarctic bottom and intermediate waters, rises to the ocean's surface in the area of the 60°S divergence. As it reaches the surface it splits; part moves northward as **South Atlantic surface water** and Antarctic intermediate water; part moves southward toward Antarctica, to be cooled and modified to form Antarctic bottom water. A mixture of Atlantic deep water and Antarctic bottom water becomes the common water for the circumpolar Antarctic (or Southern) Ocean as it flows around the Antarctic continent.

Warm (25°C), salty (36.5‰) surface water in the South Atlantic is also caught by the circular current pattern at the surface and is centered about 30°S. Below the southern tips of South America and Africa the water

flows eastward, driven by the prevailing westerly winds, which move the water around and around the Antarctic continent.

Because the Atlantic Ocean is a narrow, confined ocean of relatively small volume but great north-south extent, the water types are readily identifiable and their movement can be followed quite easily. Since the bordering nations of the Atlantic have had a long-standing interest in oceanography, the vertical circulation and layering of the Atlantic are both the most studied and the best understood of all the oceans.

Structure of the Pacific Ocean

The vast size of the Pacific Ocean allows waters that sink from relatively small areas of surface convergences to lose their identity rapidly, making the tracking of the weakly defined layers difficult. Antarctic bottom water does form in small amounts along the Pacific rim of Antarctica, but it is quickly lost in the great volume of the Pacific Ocean. The deeper water of the South Pacific Ocean is the common mixture of the circumpolar flow. In the North Pacific, only a small amount of water comparable to North Atlantic deep water can be formed, because of the limited amount of water available from the far northern latitudes, which is due to the blocking of the North Pacific by the continents. Convergence of the southward-flowing cold water from the Arctic and the northward-moving water from the equatorial latitudes occurs primarily in the extreme western North Pacific. The volume of water produced in this convergence area is small and does not result in as dramatic a layering as found in the North Atlantic. Warm, salty surface water is found at subtropic latitudes in each hemisphere, and Antarctic intermediate water is produced, but its influence, too, is small. Deep-water movements in the Pacific are sluggish, and conditions are very uniform below 2000 m (6600 ft).

Structure of the Indian Ocean

Since the Indian Ocean is principally an ocean of the southern hemisphere, there is no counterpart of the North Atlantic deep water. Small amounts of Antarctic bottom water are soon lost; the deeper waters tend to be a fairly uniform mixture of Antarctic bottom water and North Atlantic deep water brought into the Indian Ocean by the Antarctic circumpolar current. A counterpart to the Antarctic intermediate water is identifiable, as is the warm, salty surface water.

Mediterranean Sea and Red Sea Water

Two small but specific water types from bordering seas are readily identifiable, one in the North Atlantic and one in the Indian Ocean. The water from the Mediterranean Sea has a temperature of about 13° C and a salinity of 37.3$\%_{00}$ as it leaves the Strait of Gibraltar. This water mixing with Atlantic Ocean water creates a water mixture that finds its own density in the North Atlantic

at approximately 1000 m (3300 ft) in depth. The influence of Mediterranean water can be traced 2500 km (1500 mi) from the Strait of Gibraltar before it is lost through modification and mixing. In the Indian Ocean very salty (40 to 41$\%_{00}$) water from the Red Sea has been found in a spreading layer at 3000 m in depth more than 200 km (124 mi) south of its source.

Comparing the Oceans

Temperature and salinity distributions as a function of depth are shown in figure 6.14 for each of the three oceans. The distributions of Atlantic Ocean salinity and temperature values form patterns with depth (figs. 6.14a and 6.14b) that are clearly identifiable. Salinity values in the Pacific (figs. 6.14c and 6.14d) identify the high-salinity surface water lenses and the mixture of Antarctic bottom water and North Atlantic deep water, with a salinity of about 34.7$\%_{00}$ and a temperature of about 1° C. A minor intrusion of low-salinity water in the north is the result of the surface convergence in the far northwest Pacific. Note the large extent of deep water that shows a uniform salinity in this ocean. Temperature values also follow a less definite pattern, emphasizing the uniformity of most of this deep water. The Indian Ocean values in figures 6.14e and 6.14f resemble those for the South Atlantic, without the presence of water that is comparable to North Atlantic deep water.

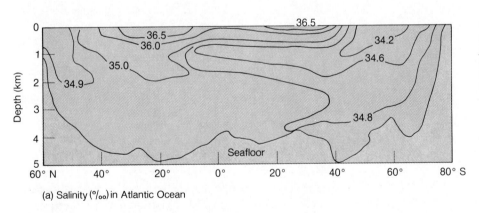

(a) Salinity (°/oo) in Atlantic Ocean

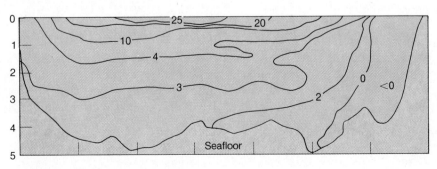

(b) Temperature (° C) in Atlantic Ocean

Figure 6.14
Midocean salinity and temperature profiles of the Atlantic, Pacific, and Indian Oceans. Temperature and salinity are shown as functions of depth and latitude.

Figure 6.14 continued

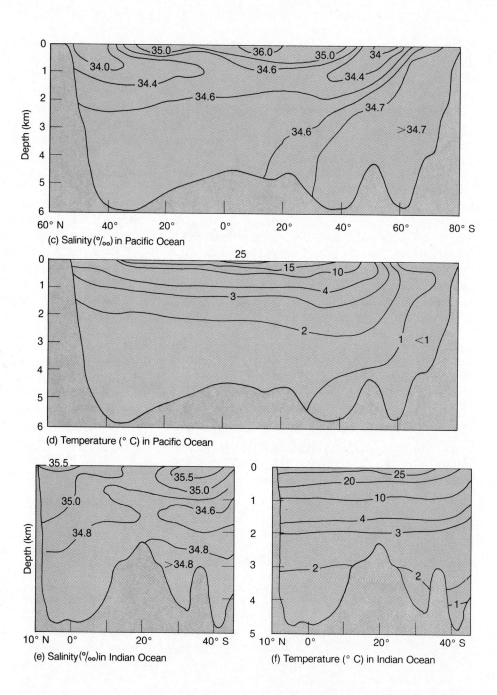

(c) Salinity (‰) in Pacific Ocean

(d) Temperature (° C) in Pacific Ocean

(e) Salinity (‰) in Indian Ocean

(f) Temperature (° C) in Indian Ocean

6.6 T-S Curves

The distribution of salinity and temperature with depth produces distinct patterns in different areas of the world's oceans, as shown by their layered structures. If the salinity and temperature values at each observed depth over the water column are graphed on a temper-ature versus salinity diagram, known as a **T-S diagram**, a curve is produced which is known as a **T-S curve.**

T-S Curves and Water Masses

T-S curves made for large geographic areas of the oceans are similar in shape and fall within narrow zones on a T-S diagram. Figure 6.15 displays specific families of T-S curves which have been given names, and are referred

Figure 6.15
The T-S curves that describe the vertical structure of the recognized oceanic water masses.

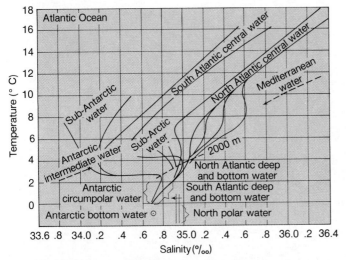

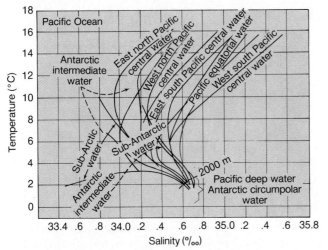

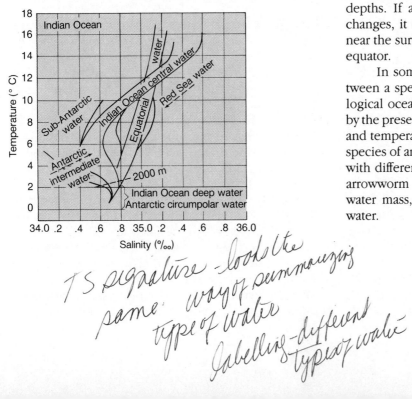

to as **water masses.** Figure 6.16 shows the distribution of these water masses identified by the T-S curves in figure 6.15. Some of the water masses described by these T-S curves have very different characteristics at the surface but are much more uniform with a limited range of salinity and temperature at depth. Where North Atlantic deep water and Antarctic bottom water are formed and sink to depth, the water is not layered, and approximately the same values for salinity and temperature are found at all depths. A water mass with only one temperature value and one salinity value over its entire extent is known as a **water type.** Antarctic bottom water meets the criteria for a water type, as is shown in figure 6.15.

T-S curves are an oceanographer's diagnostic tools. They are used to find errors in data and to indicate the density stability of the water column at different depths. Remember that water density can also be displayed on a T-S diagram (refer back to fig. 6.7). Oceanographers working with this kind of information become so familiar with these distribution patterns that they are able to inspect a set of temperature-salinity values over a series of depths and know in what area of the world's oceans the measurements were made.

Biological Indicators

Salinity and temperature values create different kinds of environments for marine organisms. Organisms that require a warm-water and high-salinity environment are confined to the surface waters of the tropics, but organisms that are adapted to life in temperate and polar waters are often more widespread. Their colder-water, moderate-salinity requirements can be met in both hemispheres at different latitudes and at different depths. If an organism is not restricted by pressure changes, it can find similar environmental conditions near the surface at high latitudes and at depth near the equator.

In some cases, there is a strong relationship between a species and a particular water mass. The biological oceanographer can often identify a water mass by the presence of a species rather than by using salinity and temperature data. For example, the distribution of species of arrowworms, or **chaetognaths,** is associated with different water masses in the North Atlantic. The arrowworm *Sagitta setosa* inhabits only the North Sea water mass, and *S. elegans* is found only in oceanic water.

Figure 6.16

The geographic distribution of water masses that have similar vertical salinity and temperature disributions. Open squares mark the region in which the central and shallow waters are formed; triangles indicate the lines along which the intermediate and deep waters sink; solid circles indicate the formation of bottom water.

From Sverdrup, Johnson, and Fleming, *The Oceans,* © 1942, renewed 1970, p. 741. Reprinted by permission of Prentice-Hall, Inc. Englewood Cliffs, N.J.

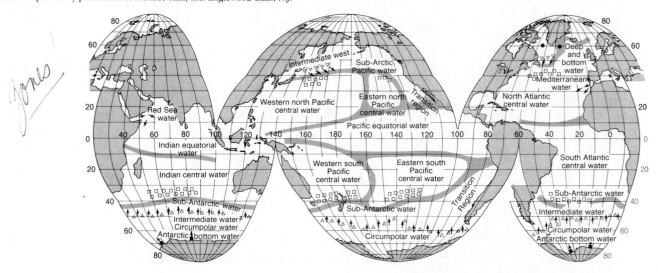

6.7 Measurement Techniques

Water

Specially designed water bottles remove water samples from the oceans for laboratory analysis. Water bottles are attached to a wire rope or cable known as a **hydrowire** (see fig. 6.17). The spacing between the water bottles and the amount of hydrowire let out is measured by a meter wheel. The water bottles are lowered to the desired depth in an open position; a small weight known as a **messenger** is then attached around the hydrowire and is released. The messenger slides down the hydrowire at about 200 m (660 ft)/min. When it strikes the uppermost water bottle it trips a mechanism that causes the water bottle to close, capturing a water sample. When the uppermost water bottle closes, it releases its own attached messenger to slide down and strike the next bottle on the wire. The water bottles are thus tripped in succession. This sequence is shown in figure 6.18.

Some water bottles close by turning over, causing their valves to shut; others stay upright but have spring-loaded valves that seal the bottle. The water bottles have deep-sea reversing thermometers (discussed in the next section) attached, which turn upside down to lock in the temperature of the water at the time the water sample is taken. Water bottles come in different styles, shapes, and sizes; some are metal and others are plastic. Small-

volume samplers of about 1 L (.26 gal) are used if routine chemical analyses are required. If a large water sample is needed the water bottle may be as big as 50 L (13.2 gal). The material that the water bottle is made of becomes important when dealing with trace materials. Metal water bottles are a source of contamination in experiments with trace metals; plastics are a problem if hydrocarbon and organic compounds are being detected.

When all the water bottles have turned over and have taken their samples, the hydrowire is winched up; the water bottles are then removed and are placed in a rack on shipboard. The water samples are carefully drained off into storage bottles and are taken for analysis to a chemical laboratory on the ship or on land.

The process of attaching water bottles to a wire, lowering them, tripping them at the desired depth, and then retrieving them is referred to as taking a **hydrocast.** It is a complicated process and can be a cold, wet, slow, and difficult task at sea under adverse weather conditions. The effort yields data at specific depths in the sea rather than data from instrument sensors that record continuously as they are lowered.

Direct-reading instruments that continuously record salinity measure the electric conductivity of the water, the depth, and the temperature (see fig. 6.19). The conductivity and temperature used to determine salinity are monitored simultaneously, and the data are

Figure 6.17
Hanging a water bottle in the Bering Sea. The hydrowire passes over the meter wheel. The box to the left contains the wire depth indicator.

Figure 6.18
The reversing of a messenger-activated water bottle equipped with thermometers, and the release of a second messenger to activate the next deeper bottle.

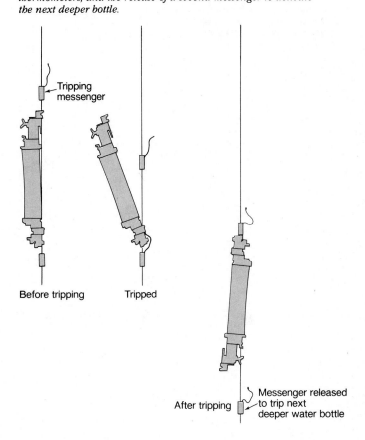

Tripping messenger

Before tripping   Tripped

After tripping   Messenger released to trip next deeper water bottle

returned to the ship as an electric signal. The signal passes through an insulated wire in the cable that suspends the instrument in the water. On board ship the salinity-depth data signal may be fed directly into the computer, recorded on a chart, or made available as numerical data. Large amounts of data can be acquired in this way, whereas water bottles collect only one sample at each depth for later analysis. Sensors can also be left in place, suspended from the surface vessel or buoy, to monitor salinity and temperature changes with time, and the collected data can be stored on magnetic tape for retrieval at another time. Although large quantities of data are collected easily and quickly in this way, these instruments do not measure as many water properties as can be measured by the laboratory analysis of collected samples.

Temperature

Water temperatures are obtained by sending specialized mercury thermometers to the required depth and then retrieving them, because if a water sample taken at a depth below the surface is brought to the ship to have its temperature measured, the temperature of the

Figure 6.19
A conductivity-temperature-depth sensor, or CTD, below a rosette of water bottles. The water bottles are tripped by a signal from the ship when the electronically relayed CTD data indicate an interesting water structure or when water samples are required to calibrate the CTD.

Figure 6.20
(a) Deep-sea reversing thermometers (DSRTs) in set position as they are lowered into the sea. (b) Deep-sea reversing thermometers in reversed position, with the water temperature recorded.

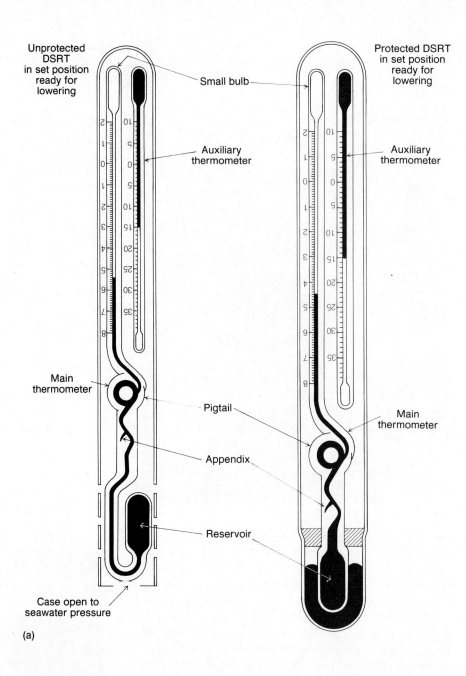

Unprotected DSRT in set position ready for lowering

Protected DSRT in set position ready for lowering

Small bulb

Auxiliary thermometer

Auxiliary thermometer

Main thermometer

Main thermometer

Pigtail

Appendix

Reservoir

Case open to seawater pressure

(a)

sample will change as it passes through the warmer upper layers of water. When the thermometer has come to equilibrium with the temperature at the sample depth, the thermometer must fix this temperature in some way so that the reading will not change as the thermometer is retrieved, passing through layers of different temperatures.

The deep-sea reversing thermometer (DSRT) has been the instrument used through the years to make these measurements. These specially designed thermometers are attached in pairs or groups of three to water sampling bottles and are lowered by cable into the sea. When the desired depth has been reached, the thermometer is allowed approximately five minutes to adjust to the water temperature, then a messenger is used to trip the water bottle, and the thermometers are reversed (or turned upside down) as the valves of the bottles close. This type of thermometer is specially constructed so that the act of turning it over isolates the mercury that registers the temperature from the mercury reservoir. When the thermometer is returned to the ship, the seawater temperature indicated by the isolated mercury is read, and any small expansion of this mercury that has occurred on shipboard is corrected for by using the difference between the indicated seawater temperature and the reading of an auxiliary thermometer. The auxiliary thermometer gives the shipboard temperature of the thermometer case at the time of reading. Figure 6.20 illustrates the DSRT before and after reversal.

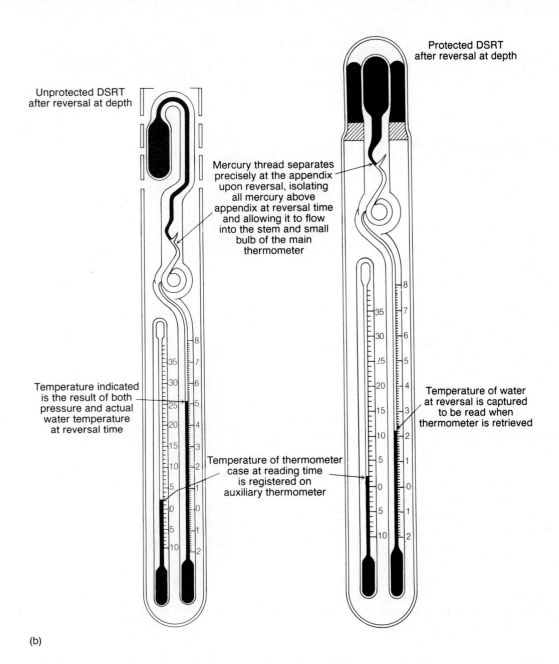

Unprotected DSRT
after reversal at depth

Protected DSRT
after reversal at depth

Mercury thread separates
precisely at the appendix
upon reversal, isolating
all mercury above
appendix at reversal time
and allowing it to flow
into the stem and small
bulb of the main
thermometer

Temperature indicated
is the result of both
pressure and actual
water temperature
at reversal time

Temperature of water
at reversal is captured
to be read when
thermometer is retrieved

Temperature of thermometer
case at reading time
is registered on
auxiliary thermometer

(b)

Reversing thermometers are usually used in pairs so that their readings can be compared as an additional check on their reliability. Their accuracy is about ±0.01° C. The thermometers are specifically referred to as protected when they are totally enclosed by a second glass jacket or covering. Other reversing thermometers are unprotected. Their outside glass jackets are open, so that the seawater exerts pressure directly on the thermometers. This water pressure pushes more mercury up the thermometer tube, about 1°C for each 100 m (330 ft) of depth, resulting in a higher reading on an unprotected thermometer. When an unprotected thermometer is used with a pair of protected thermometers, its elevated reading allows the researchers to calculate the depth at which the temperatures were taken

and to verify that the protected thermometers reversed and recorded the temperature at the desired depth as determined by cable length. Today's DSRTs cost about $900 each and must be frequently calibrated to assure their accuracy.

In order for water bottles and deep-sea reversing thermometers to be used, the ship must be halted, and only one measurement can be made with each thermometer on each lowering and reversal. Modern electronic thermometers respond very rapidly to temperature changes with an accuracy that is comparable to that of deep-sea reversing thermometers. They can be used to record fluctuations in temperatures over time at a single location, or they can be lowered through the water, giving a continuous temperature reading with

Figure 6.21
The mechanical bathythermograph, or BT, is used to record temperature of water versus depth on a coated glass slide.

Figure 6.22
An expendable BT launcher and recorder (a), designed to be used from a cruising vessel or helicopter, shown with the expendable thermal probe (b).

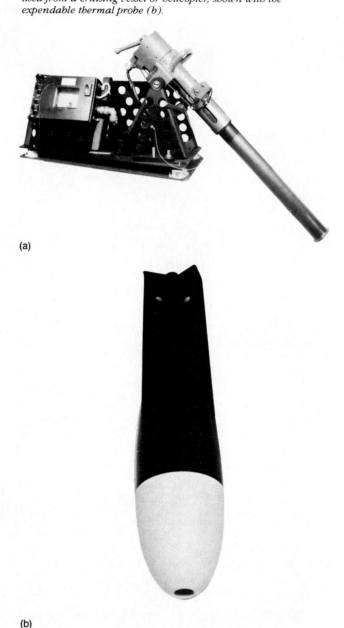

(a)

(b)

depth. The temperature readings may be recorded internally within the instrument, or they may be sent up the suspending wire to the ship, where they may be fed directly to a computer or used to produce a graph of temperature with depth. The internally recorded data are recovered when the instrument is brought back on board. Electronic thermometers can record five or more readings each second. The reversing thermometer obtains one reading for the hour or so necessary to lower it to the desired depth and then retrieve it.

The mechanical **bathythermograph,** or **BT,** was the first instrument developed to measure temperature changes rapidly with depth down to 80 m (260 ft) while a ship was under way (see fig. 6.21). This torpedo-shaped device was used by the navy during World War II to detect the effect of temperature on sound waves. The BT traces a temperature versus depth graph on a small glass slide. The BT is retrieved and the trace is read directly against a grid of temperature versus depth. The mechanical BT has been replaced by the **expendable BT,** or **XBT** (see fig. 6.22). The XBT is a small bomb-shaped device that is dropped into the water and sinks at a known rate. A wire attached to a recorder on the vessel unrolls from the XBT and transmits the temperature as an electric signal directly to the shipboard recorder. The depth is calculated from the elapsed time. The XBT can be dropped from vessels that are stationary or moving at speeds up to 30 knots. It gives the quick temperature-depth profiles to 500 m (1640 ft) that are needed for sonar work (see chapter 4). The XBT, as indicated by its name, is not recovered.

Although the new electronic instruments increase the precision and speed of the measurements as well as the number of measurements that can be made, the simpler older, slower, and less expensive thermometers and water bottles have proved reliable over the years and continue in use for obtaining the control data with which the newer instruments are calibrated and compared.

The choice of instruments used to gather data is determined by the cost of the instruments, the time required to take a sample, the need for detail, accuracy, and speed, the personal preference of the researcher,

and the conditions under which the sampling is done. On any research cruise, data must be obtained under the conditions present at the sampling station on arrival. Despite darkness, wind, waves, heat, or cold, data are taken, for there may not be a second chance. In the laboratory, the researcher controls the experiment and varies individual conditions to see the result of each. At sea, the oceanographer has no control over the environment but tries to measure and interpret a system in which all of the conditions change simultaneously and continually. Nature is in control.

6.8 Practical Considerations: Energy Sources

Temperature and salinity variations with depth represent a potential source of energy and an alternative to the consumption of fossil fuels. Since it is solar radiation that is largely responsible for the temperature and salinity distribution, this energy represents an indirect form of solar energy. Large amounts of solar energy can be stored in the ocean, because the temperature of the water changes very little, as a result of the heat capacity of the water. Therefore, the heat can be extracted independent of daily and seasonal changes in the available solar radiation. Several techniques, some more practical than others, are available to extract this energy.

Salt Fountains

Figure 6.13 shows that between 30°N and 30°S latitudes warm, salty water overlies the colder, less salty water that is found below 1000 m (3300 ft). If a long metallic tube is installed vertically between the warm, high-salinity surface water and the cold, low-salinity water at depth, the water inside the tube is isolated from the water outside the tube, but the heat moves through the tube wall between the inside and outside waters. If cold, low-salinity water is pumped up the pipe the salt content does not change, but the water inside the pipe warms to the temperature of the water outside the pipe. The increasing temperature and the low salinity make the water inside the pipe less dense than the water outside. The water inside the pipe will therefore rise in the pipe without the aid of the pump. Once the flow has been started the pump can be removed and the flow will continue indefinitely, driven by the heat extracted from the surrounding warmer surface water. The density-driven flow is commonly called a salt fountain. The moving water represents energy that can be used to activate a turbine to produce usable energy. Figure 6.23 shows a diagram of such a process. No application of this principle has yet been made.

Osmotic Pressure

Power can also be produced by exploiting the salt differences between two bodies of water. This technique requires two bodies of water with different salt contents (such as river water flowing into the ocean), and it also requires the use of a semipermeable membrane. A semipermeable membrane permits the water molecules to move through the membrane from one side of the membrane to another, but it does not allow the salt molecules to cross the membrane. The water flows from the side on which the water is more concentrated (the

river water side) to the side on which the water is less concentrated (the saltwater side). In a system in which the salt water has the same salt content as the open ocean and the other water source contains no salt, the fresh water moves through the membrane until the pressure on the salt water side is equivalent to a column of water 240 m in height. This pressure difference could be used to produce power by passing the water under pressure through a turbine. The pressure that is formed by water moving through a semipermeable membrane is called **osmotic pressure** (see fig. 6.24 and also the discussion at the end of chapter 5).

Ocean Thermal Energy Conversion

An alternative method of energy extraction depends on the difference in temperature between ocean surface water and water at 600–1000 m (2000–3300 ft) depth. This method of power production is known as **OTEC,** or **ocean thermal energy conversion.** There are two different types of OTEC systems: (1) closed cycle, in which a contained working fluid with a low boiling

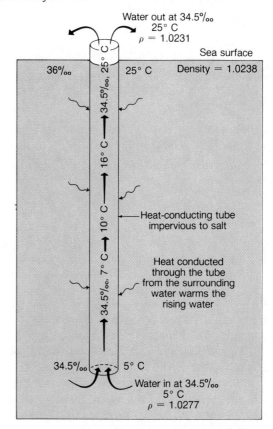

Figure 6.23

Cold, low-salinity water moving up a thermal conducting tube is warmed and exits at the top of the tube with a density that is less than the surface water.

Water out at 34.5‰
25° C
$\rho = 1.0231$

Sea surface

36‰   25° C   Density = 1.0238

34.5‰, 25° C

16° C

10° C

7° C

34.5‰

Heat-conducting tube impervious to salt

Heat conducted through the tube from the surrounding water warms the rising water

34.5‰   5° C

Water in at 34.5‰
5° C
$\rho = 1.0277$

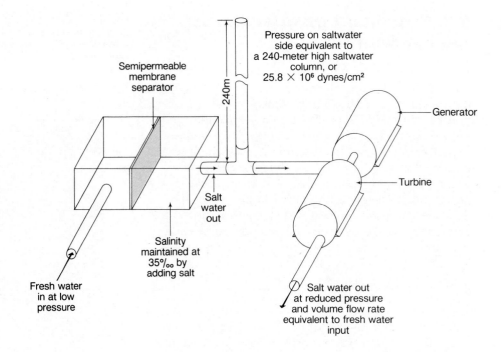

Figure 6.24
The movement of fresh water into salt water through a semipermeable membrane generates pressure on the saltwater side to power a turbine and generator.

point, such as ammonia or Freon, is used; and (2) open cycle, which directly converts seawater to steam.

In a closed system (fig. 6.25b) the warm surface water is passed over the chamber containing the ammonia or Freon; this chamber is the evaporator, and the ammonia or Freon is vaporized by the heat derived from the warm seawater. The vapor builds up pressure in a closed system, and this gas under pressure is used to spin a turbine, which generates power. After the pressure has been released, the ammonia or Freon is passed to a condenser, where it is cooled by the cold water pumped up from depth. Cooling the ammonia or Freon returns it to its liquid state, and it is pumped as a liquid back to the evaporator to repeat the cycle.

In an open system (fig. 6.25c), large quantities of warm seawater are converted to steam in a low-pressure vacuum chamber, and the steam is used as the working fluid. Since less than 0.5% of the incoming water is turned into steam, large quantities of warm water must be used. The steam passes through a turbine and is condensed by using cold water from depth, which cools a condenser. This turns the steam into desalinated water. OTEC plants using either system can be located onshore, offshore, or on a ship that moves from place to place. However, OTEC requires at least a 20° C difference in temperature between surface and depth to generate useful amounts of energy. This requirement means that these power plants would be confined to latitudes between approximately 25°N and 25°S.

In 1979–1980, a floating pilot OTEC plant was installed off Hawaii and was able to produce 50 kilowatts of power while requiring 40 kilowatts to run its pumps, so that a surplus of 10 kilowatts was being produced.

A land-based, closed-cycle, 100-kilowatt pilot plant using Freon was constructed by the Tokyo Electric Power Company in the island republic of Nauru; this plant functioned between 1981 and 1982. In 1981, the Natural Energy Laboratory of Hawaii at Keahole Point on the west coast of the big island of Hawaii built a land-based OTEC plant. This plant is the world's only system using a continuous supply of deep water. Its cold water intake pipe is 0.3 m (1 ft) in diameter and 1500 m (5000 ft) long, and a new pipe of the same length but with a 1-m diameter is being constructed. Plans for closed-cycle plants have been announced by the French for Tahiti, by the Dutch for Bali, and by the British for a floating plant. The problem in all cases is financing. The high capital cost of the plants, which is substantially greater than a conventional steam plant, has delayed all the projects. U.S. cost estimates for a 50-megawatt plant are between $200 million and $500 million, depending on the location, with a marketing price per kilowatt that could be more than twice that from oil-fueled plants.

Research on closed-cycle plants is currently focusing on improving heat exchangers and reducing the problems of condenser corrosion and fouling by marine organisms, which settle and grow on the condenser surfaces. The open system is not as fully developed. Problems include the need for larger turbines for use with the low-density steam, large vacuum chambers, and the large flow rates of water needed (2–4 m³/sec (500–1000 gal/sec)). Another difficulty occurs when the heated seawater allows gases to escape, resulting in the release of large quantities of water low in dissolved oxygen into the local environment, which might cause problems for the organisms present.

Figure 6.25

(a) The proposed Lockheed OTEC Spar, designed for generating electrical energy from the ocean temperature gradient. (b) The simplified working system of an OTEC closed-system electrical generator. (c) The simplified working system of an OTEC open-system electrical generator.

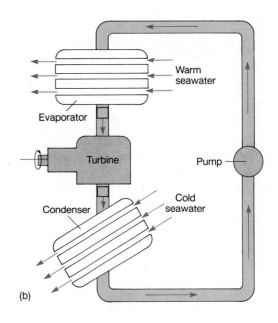

(b)

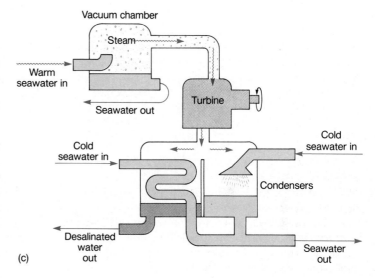

(c)

(a)

However, there are significant advantages associated with an open system. These include using only seawater as the working fluid so that no contamination of the water is possible; direct contact between heat exchangers and the seawater, which is both more efficient and less expensive; the ability to use plastic heat exchangers, which cuts down on fouling and corrosion; and, especially, the desalinated water which can be a by-product of the system.

Because of the costs, the only practical competitive use of power generated by OTEC at the present time is on ocean islands, where fuel oil for power generation must be imported. If open systems are used, plants could produce revenue by selling the fresh water, which might make them competitive in areas lacking both power

and fresh water, for example the Caribbean. Other uses for the cold water include refrigeration and air-conditioning. Because the cold water is pumped from 1000 m (3300 ft) deep, it contains nutrients which can be used as an artificial upwelling to promote fish and seaweed farming. The French consider the potential for fish-farming, fresh water, and refrigeration to be more important than the net power production of their Tahiti plant. Researchers in Hawaii also believe that fish-farming is the greatest incentive for commercial development of OTEC plants in their region.

If, in the future, closed-system, floating OTEC installations capable of producing electric power of 100 to 200 megawatts are considered, the large units will require heat exchangers with many square miles of surface and huge pipes (30 m (100 ft) in diameter) to bring very large volumes of cold, deep water up to the near-surface power unit. Constructing, installing, and maintaining a pipe to 1000 m (3300 ft) depth and miles of thin-walled heat exchange coils present considerable problems.

The power produced at these locations must then be transferred to populated areas with energy needs. At present, there are no electric cables designed to carry underwater the power produced by large non-landbased systems. To get around the necessity of such cables, it is considered possible that the electric power produced by a floating OTEC plant might be used to produce hydrogen from seawater or to produce am-monia gas from the atmosphere. These gases could then be transported by tankers for shore conversion to usable power.

Of the energy extraction methods explained in this section, only OTEC has produced sufficient energy to prove its practical potential, but, as discussed, many problems remain to be overcome before it can produce power at rates that are comparable to hydroelectric, fossil fuel, or nuclear power plants. As the world's ability to produce energy from traditional sources diminishes, however, the costs associated with an OTEC plant will eventually become acceptable, and the technical and engineering problems will be researched and solved. In this century, though, energy conservation will save more power than the systems described here will be able to produce.

Summary

The intensity of solar radiation over the earth's surface varies with latitude. Incoming radiation and outgoing radiation are equal when averaged over the whole earth. Reflection, reradiation, evaporation, conduction, and absorption keep the heat budget in balance with the incoming solar radiation. In any local area of the oceans there is daily and seasonal variation. Winds and ocean currents move heat from one ocean area to another to maintain the surface temperature patterns.

Atmospheric carbon dioxide is responsible for the greenhouse effect which warms the earth's surface. Increases in CO_2 concentration due to the burning of fossil fuels are leading to predictions of global warming.

The oceans gain and lose large quantities of heat, but their temperature changes very little. The heat capacity of the oceans is high compared to that of the land and the atmosphere. Heat that is mixed downward reduces the temperature change at the surface.

The salinity of surface water reflects the pattern of evaporation and precipitation with latitude.

Cooling, evaporation, and freezing increase the density of the sea surface water. Heating, precipitation, and ice melt decrease its density. The surface water changes its density with changes in salinity and temperature that are keyed to latitude. The densities at depth are more homogeneous. Thermoclines and haloclines form where the temperature and salt concentrations change rapidly with depth. If the density increases with depth, the water column is stable; unstable water columns overturn and return to a stable distribution. Neutrally stable water columns are mixed vertically by winds and waves.

Vertical circulation that is driven by changes in surface density is known as thermohaline circulation. In the open ocean, temperature is more important than salinity in determining the surface density. Salinity is the more important factor in areas that are close to shore and to land runoff. Sigma-t is a convenient way of representing density.

Water sinks at downwellings and rises at upwellings. A downwelling occurs at the convergence of surface currents and transfers oxygen to depth. Upwellings bring nutrients to the surface and occur at zones of surface current divergence.

The oceans are layered systems. The layers (or water types) are identified by specific ranges of temperature and salinity. The water types of the Atlantic Ocean are formed at the surface at different latitudes. They sink and flow northward or southward. The water types of the Pacific Ocean lose their identity in the large volume of this ocean; their movements are sluggish. The water types of the Indian Ocean are less distinct than those of the Atlantic, with no water types that correspond to those formed in the northern latitudes. Mediterranean Sea and Red Sea waters enter the Atlantic and Indian oceans at depth as discrete water types that can be tracked for long distances.

Water masses occupy distinct regions of the oceans and have specific patterns of salinity and temperature from surface to depth. Characteristic values of temperature and salinity for these water masses are identifiable when plotted on T-S diagrams. There is sometimes a strong relationship between a certain species and a water mass or water type. Such species are used as indicators of the presence of that water mass.

Water samples are taken with water bottles that turn over and close at the sampling depth. Analysis of these water samples is done in the laboratory. Direct-reading instruments to determine salinity and temperature in the oceans are also available. The temperature of the water at the depth of a water-bottle sampling station is taken with a deep-sea reversing thermometer. Electronic thermometers and bathythermographs can measure temperature while the vessel is moving.

Temperature and salinity variations in the oceans represent possible sources of energy, including salt fountains, osmotic pressure devices, and ocean thermal energy conversion (OTEC).

Key Terms

solar constant
heat budget
reradiation
greenhouse effect
density
pycnocline
thermocline
halocline
stable water column
unstable water column

overturn
isopycnal
isothermal
isohaline
vertical circulation
thermohaline circulation
freshwater lid
sigma-t
continuity of flow
downwelling zones

upwelling zones
convergence
divergence
North Atlantic deep water
Antarctic intermediate
 water
Antarctic bottom water
T-S diagram
T-S curve
water mass

water type
chaetognaths
hydrowire
messenger
hydrocast
deep-sea reversing
 thermometer (DSRT)
bathythermograph (BT)
expendable BT (XBT)
osmotic pressure
ocean thermal energy
 conversion (OTEC)

Study Questions

1. Write an equation for the whole earth's heat budget. Include all the factors for incoming and outgoing energy.

2. What natural processes alter the surface salinity of the oceans? How do these processes work at different latitudes?

3. Describe the changes in water density in the upper ocean layer over the annual cycles at tropic, polar, and temperate latitudes. Indicate when periods of stable and unstable conditions exist in the upper water column.

4. Why are water layers more prominent in the Atlantic Ocean than in the Pacific Ocean?

5. Discuss the use of ocean thermal gradients for the production of energy. Would there be environmental side effects?

6. How can heat be transferred from place to place in the oceans? Why is this heat transfer ignored when considering the world's total heat budget?

7. If the upper layers of an ocean area are homogeneous in salinity, explain why the thermocline coincides with the pycnocline.

8. How does Mediterranean Sea water alter the salinity and temperature of the North Atlantic? Use figure 6.14. At what depth does this change occur?

9. A water sample taken at 4000 m in the Atlantic Ocean has a salinity of $34.8^0/_{00}$ and a temperature of 3°C. At approximately what latitude was this water last at the surface?

10. The following data were taken from a sampling station located at 79°N, 145°W:

Depth (meters)	Temp. (° C)	Salinity ($^0/_{00}$)	Sigma-t
0	1.28	33.29	26.59
50	1.29	33.30	26.59
100	1.36	33.35	26.69
150	1.39	33.55	26.94
200	2.73	33.76	27.01
300	3.07	33.87	27.08
400	3.12	34.03	27.13
500	3.14	34.13	27.21

a. In what ocean region is this station?
b. Is the water column stable or unstable?
c. Does the temperature or the salt content control the density?
d. How deep is the wind-mixed layer?
e. At what time of year were these data obtained? Explain your answer.

Winds and Currents

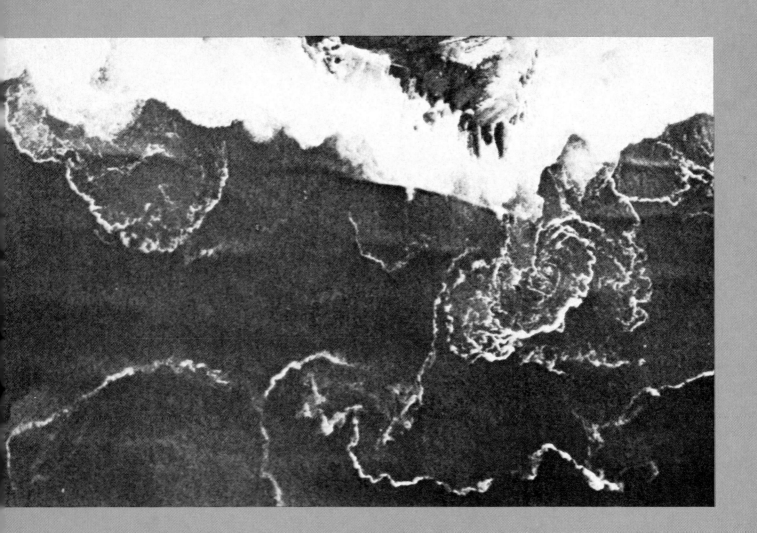

7

T here is a river in the ocean. In the severest droughts it never fails, and in the mightiest floods it never overflows. Its banks and its bottom are of cold water, while its current is of warm. The Gulf of Mexico is its fountain, and its mouth is in the Arctic Seas. It is the Gulf Stream. There is in the world no other such majestic flow of waters. Its current is more rapid than the Mississippi or the Amazon.

Its waters, as far out from the Gulf as the Carolina coasts, are of an indigo blue. They are so distinctly marked, that their line of junction with the common sea-water may be traced by the eye. Often one half of the vessel may be perceived floating in the Gulf Stream water, while the other half is in common water of the sea, so sharp is the line, and such the want of affinity between those waters, and the reluctance, on the part of those of the Gulf Stream to mingle with the common water of the sea.

Mathew Fontaine Maury, from The Physical Geography of the Sea, *1855*

*T*he earth is surrounded by two great, fluid envelopes: the atmosphere and the ocean. Both are in constant motion, driven by the earth's gravity and the density changes occurring in each fluid. Density differences in both the atmosphere and the ocean are produced by the sun. The sun heats the earth, and this heat is exchanged among the land, the atmosphere, and the ocean. In this way, the atmosphere and the ocean are linked; they affect each other. The ocean and land surfaces absorb heat better than the atmosphere. This heat is transferred to the atmosphere from below, setting the air in motion to produce the earth's surface winds. The winds set the sea surface in motion, and wind-driven surface currents result. Both winds and currents carry heat from place to place. As they do so, they further alter the earth's surface temperature pattern. The dynamic interaction between atmosphere and ocean is constantly modifying itself, for as one system drives the other, the driven system in turn acts to alter the properties of the driving system.

In this chapter we will investigate how the pattern of surface temperatures with latitude and with land-water distribution acts to produce the winds, which in turn produce the oceanic surface currents. We will follow these currents and will examine the areas in which they meet, merge, and sink as well as the areas in which they move away from each other, creating regions of water rising from depth to fill the surface voids. We will examine both horizontal and vertical circulation and so will begin to understand that these motions are closely linked to each other as well as to the overall interaction between the atmosphere and the ocean.

7.1 The Atmosphere in Motion

Since the winds drive the major surface currents of the world's oceans, our discussion will begin with an explanation of atmospheric motion and wind systems. Winds are the result of the horizontal movement of air from one location to another. The air moves because at some place the air rises away from the earth's surface while in another place it sinks toward the earth. These actions result in air that flows horizontally along the earth's surface, from an area of accumulation due to air sinking to an area left deficient due to air rising. This process is shown in figure 7.1; note that there are really two horizontal air flows, or wind levels, moving in opposite directions: one at the earth's surface and one aloft. Air circulating in this manner forms a convection cell. The vertical air movements in such a cell are caused by changes in the air's density. Less dense air rises, and more dense air sinks.

Air is a mixture of gases. Table 7.1 lists the gases that compose air. Air becomes less dense when it is warmed, when its water vapor content is increased, and

Figure 7.1
A convection cell is formed in the atmosphere when air is warmed at one location and cooled at another.

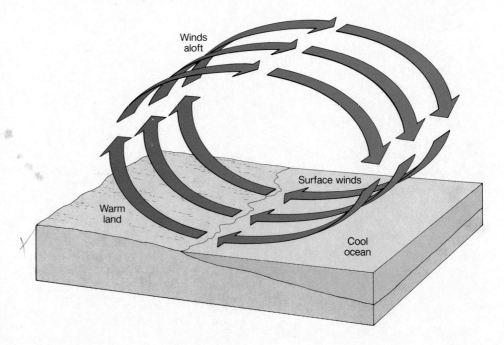

Winds aloft

Surface winds

Warm land

Cool ocean

when the **atmospheric pressure** decreases. Air becomes more dense when it is cooled, when its water vapor content is decreased, and when the atmospheric pressure increases. Because water vapor is a low-density gas, its presence in air decreases the density of the gas mixture. The more low-density water vapor present in any given volume of air, the less dense the air; and the less water vapor present in that volume, the more dense the air.

Atmospheric pressure is the force with which the column of overlying air presses on an area of the earth's surface. The average atmospheric pressure at sea level is 1013.25 millibars (1 bar = 1×10^6 dynes/cm^2), or 14.7 lb/in^2. This standard atmospheric pressure is also equal to the pressure produced by a column of mercury standing 760 mm (29.92 in.) high. Barometers may measure atmospheric pressure in millibars or millimeters or inches of mercury. Pressure can also be recorded in torrs, where 1 torr equals the pressure of a column of mercury 1 centimeter high. Where the density of the air is decreased, atmospheric pressure is below average, and a low pressure zone is formed. Where the density of the air is increased, atmospheric pressure is above average, and a high pressure zone results.

Winds on a Nonrotating Earth

The heating and cooling of air and the gains and losses of water vapor in air are related to the unequal distribution of the sun's heat over the earth's surface, the presence or absence of water, and the variation in temperature of the earth's surface materials in response to heating. Consider a water-covered earth model without continents and rotation, but heated like the natural earth. On this model covered with uniform layers of atmosphere and water, the wind pattern is very simple. At the equator the earth's surface is warm, and the air, warmed from below, rises. Once aloft, the air flows toward the poles. The air above the poles is cooled from below and sinks to flow back toward the equator. As the air moves over the water-covered earth, its water vapor content increases, and as it approaches the equator it warms. The less dense air rises at the equator. As the air rises it cools. The cooling allows the water vapor to condense and rain to form. The air, which is now cool and dry, remains aloft and flows toward the poles. At the poles the air sinks to produce a cold, dry, dense air mass and a zone of high atmospheric pressure. Figure 7.2 shows the air movement on such a model; note that there are two large convective circulation cells, similar to those described at the beginning of this section. In this model, the northern hemisphere surface winds blow from north to south, and the upper winds blow from south to north; in the southern hemisphere the reverse is true, with the surface winds blowing from south to north and the upper winds blowing from north to south.

It is important to remember that winds are named for the direction *from which they blow*. A north wind blows from north to south, while a south wind blows from south to north. In this model, the northern hemisphere surface winds are north winds and the southern hemisphere surface winds are south winds.

The Effects of Rotation

Consider next a model of a rotating water-covered earth. Gravity holds the earth's atmosphere captive, but the atmosphere is not rigidly attached to the earth's surface.

TABLE 7.1
The Composition of Dry Air

Gas	Percentage of mixture by volume
Nitrogen	78.08
Oxygen	20.95
Argon	0.93
Carbon dioxide	0.03
Neon	1.8×10^{-3}
Helium	5×10^{-4}
Krypton	1×10^{-4}
Xenon	1×10^{-5}
Hydrogen	$\leq 1 \times 10^{-5}$
	~100%

Figure 7.2

Heating at the equator and cooling at the poles produces a single large convection cell in each hemisphere on a nonrotating water-covered earth model.

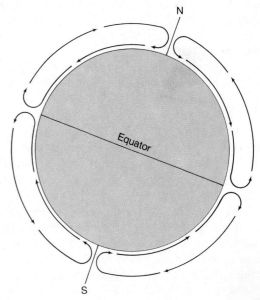

Figure 7.3

Because air moving northward from the equator to point A carries with it its initial eastward velocity, the air is deflected to the right of the initial wind direction in the northern hemisphere. In the southern hemisphere, air moving southward to point B is deflected to the left.

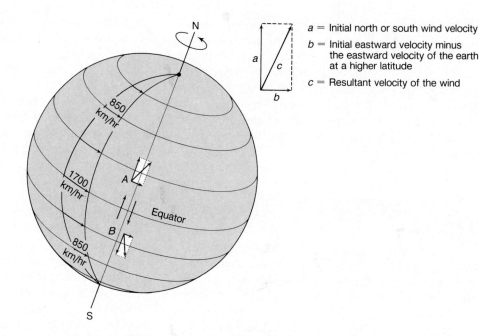

a = Initial north or south wind velocity

b = Initial eastward velocity minus the eastward velocity of the earth at a higher latitude

c = Resultant velocity of the wind

There is little or low friction between the earth's surface and the atmosphere. Because of this feature, the atmosphere moves somewhat independently of the earth's surface. For example, a parcel of air that appears to be stationary above a point on the equator is actually turning with the earth and moving eastward at a speed of about 1700 km(1054 mi)/hr. If a south wind blows this parcel of air due north, it is moved across circles of latitude with progressively smaller circumferences. At these higher latitudes, points on the earth's surface move eastward more slowly. At 60°N, the eastward speed of a point on the earth's surface due to rotation is only half the speed at the equator. For this reason, a parcel of air that was originally moving only northward relative to the earth at the equator carries with it its initial eastward speed, so that it is now moving eastward at a speed that is greater than the eastward speed of the earth's surface at this higher latitude. Therefore the air parcel appears displaced to the east, relative to the earth's surface, as it moves from low to high latitudes; in the northern hemisphere this deflection is *to the right of the direction of the air motion*. This relationship is illustrated in figure 7.3; follow the arrows at *A*.

If the same situation occurs in the southern hemisphere, with a parcel of air moving southward from the equator due to a north wind, the deflection is still to the east relative to the earth's surface. However, this deflection is now *to the left of the direction of the air motion*. See again figure 7.3 and follow the arrows at *B*.

If an air parcel moves southward toward the equator in the northern hemisphere, then it moves from a latitude where a position on the earth has a low eastward speed to a latitude where a position on the earth has a higher eastward speed. These positions move to the east, relative to the air, when the air moves from a

higher to a lower latitude. This relationship causes the air to fall behind the position on the earth, so that the air is moving westward relative to the earth. This pattern is illustrated in figure 7.4. The deflection is still to the right of the direction of motion in the northern hemisphere. A similar pattern in the southern hemisphere shows that the deflection is to the left of the direction of motion.

Deflection due to air moving to the east or to the west is shown in figure 7.5. Air moving eastward is moving eastward faster than the earth beneath it, with reference to the earth's axis. The air is affected by a centrifugal force acting outward from the earth's axis of rotation that is stronger than the centrifugal force that is acting at the earth's surface. This small excess force acting on the air is in part acting against the earth's gravity, and in part acting parallel to the earth's surface directed toward lower latitudes. That part of the centrifugal force acting against the earth's gravity is so small that it has very little effect. That part acting parallel to the earth's surface causes a deflection of the moving air to lower latitudes, or to the right of its motion in the northern hemisphere. Westward-moving air is moving eastward more slowly, relative to the earth's surface. This air is affected by an outward-acting centrifugal force that is weaker than the centrifugal force that is acting at the earth's surface. Because the centrifugal force acting at the earth's surface is greater than the centrifugal force acting on the air, there is a weak force acting toward the earth's axis of rotation. This force is acting in part in the direction of gravity, and in part toward higher latitudes. Note that the deflection continues to be to the right of the initial wind motion in the northern hemisphere. In the southern hemisphere, the deflection of eastward-moving air toward lower latitudes and of westward-

Figure 7.4

Air moving toward the equator passes from a latitude of lower eastward speed to a latitude of higher eastward speed. The result is deflection to the right of the initial wind direction in the northern hemisphere and deflection to the left in the southern hemisphere. There is no deflection at the equator.

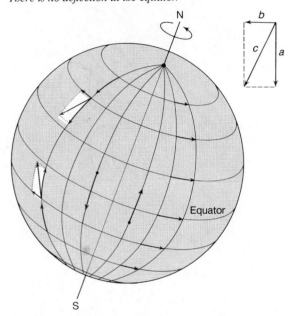

a = Initial north or south wind velocity

b = Initial eastward velocity minus the eastward velocity of the earth at a higher latitude

c = Resultant velocity of the wind

Figure 7.5

The deflection of eastward-moving and westward-moving air. There is no deflection at the equator.

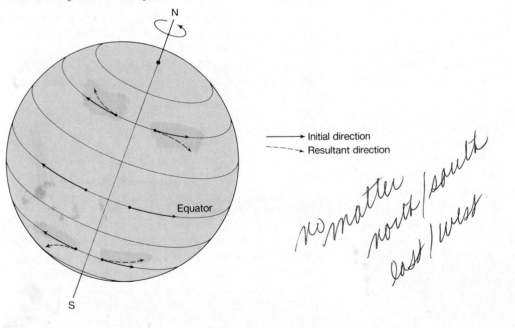

→ Initial direction

--→ Resultant direction

no matter north/south east/west

moving air toward higher latitudes results in a deflection of the air to the left of its original direction of motion. If we were to move from the earth's surface out into space and were able to watch the motions of earth and atmosphere, we would see the atmosphere's in-dependent motion and also see the earth turning out from under the moving parcels of air. Yet we live on a moving earth's surface that we consider stationary, and we make our measurements of air motion relative to the "stationary" surface. Therefore, from the earth's surface

Figure 7.6
The atmospheric convection cell bands on a rotating water-covered earth model.

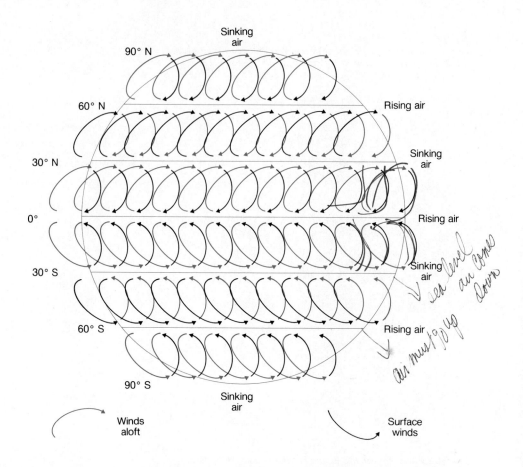

we see the moving air parcels deflected from their paths, deflected to the right in the northern hemisphere and to the left in the southern hemisphere.

The apparent deflection of the moving air relative to the earth's surface is called the **Coriolis effect,** after Gaspard Gustave de Coriolis (1792–1843), who mathematically solved the problem of deflection in frictionless motion when the motion is referred to a rotating body. Because one of the laws of motion in physics tells us that a body set in motion along a straight line continues to move along that straight line unless it is acted upon by a force, such as a push or a pull, the Coriolis effect is often called the Coriolis force. If the term force is used, remember that it is an apparent force, and that a deflection appears only when the motion of an object that is subject to little or no friction is judged against a rotating frame of reference. The magnitude of the Coriolis force increases with increasing latitude, increases with the speed of the moving air, and is dependent on the rotation rate of the earth on its axis.

Wind Bands

Applying rotation, and therefore the Coriolis effect, to the wind system that is found on a water-covered earth model modifies these winds considerably. Refer to figure 7.6 as you read the following description to understand

what happens. The air still rises at the equator and flows aloft to the north and the south, but it cannot continue to move northward and southward without being deflected to the right in the northern hemisphere and to the left in the southern hemisphere. This deflection short-circuits the large, hemispheric, atmospheric convection cells of the stationary-earth model. The air that sinks at 30°N and 30°S moves along the water-covered surface, either back toward the equator or toward 60°N and 60°S. The air that reaches the poles at the upper level sinks over the poles and moves to lower latitudes, warming and picking up water vapor; at 60°N and 60°S it rises again. The result is three convection cells in each hemisphere, rather than the single cell.

Consider the flow of surface air in the three-cell system. Between 0° and 30°N and 0° and 30°S, the surface winds are deflected so their path relative to the earth is from the north and east in the northern hemisphere and from the south and east in the southern hemisphere. This deflection creates moving bands of air, known as the **trade winds:** the northeast trade winds are north of the equator, and the southeast trade winds are south of the equator. Between 30°N and 60°N, the deflected surface flow produces winds that blow from the south and west, while between 30°S and 60°S they blow from the north and west. In both hemispheres these winds are called the **westerlies.** Between 60°N

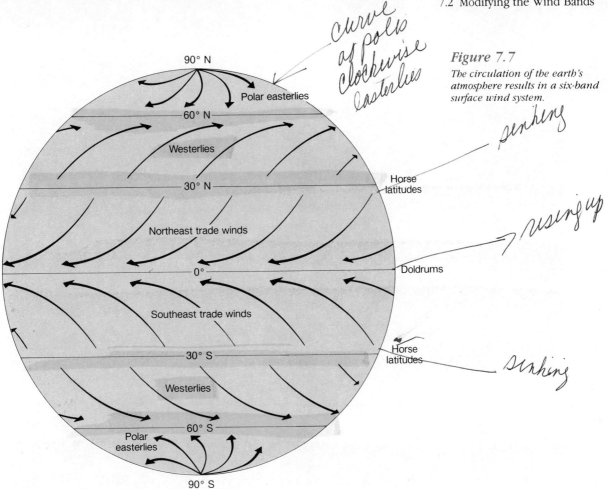

Figure 7.7
The circulation of the earth's atmosphere results in a six-band surface wind system.

[handwritten annotations: "curve of poles clockwise easterlies", "sinking", "rising up", "sinking"]

and the north pole, the winds blow from the north and east, while between 60°S and the south pole, they blow from the south and east. In both cases they are called the **polar easterlies.** The six surface wind bands are shown in figure 7.7.

At 0° and at 60°N and S, the low-density air rises; these are areas of low atmospheric pressure, zones of clouds and rain. The zones of high-density descending air at 30° and 90°N and S are areas of high atmospheric pressure, zones of low precipitation and clear skies. Air flows from regions of high atmospheric pressure to areas of low atmospheric pressure. In the zones of vertical motion, between the wind belts, the surface winds are unsteady. These zones of unsteady and unreliable winds on the model also exist in reality on the earth. They were troublesome areas to the early sailors, who depended on steady winds for propulsion. The area of rising air at the equator is known as the **doldrums,** and those at 30°N and S are known as the **horse latitudes.** In all three areas sailing ships could find themselves becalmed for days. The origin of the word doldrum is obscure, but the horse latitudes are said to have gotten their name from the stories of ships carrying horses that were thrown overboard when the ships were becalmed and the freshwater supply became too low to support both the sailors and the animals.

7.2 Modifying the Wind Bands

Moving from the rotating, water-covered model of the earth to the real earth requires the consideration of two more factors: (1) seasonal changes in the earth's surface temperature, and (2) the addition of the large continental land blocks. Both ocean and land surfaces remain warm at the equatorial latitudes and cold at the polar latitudes all through the year, but the middle latitudes go through seasonal changes. The middle latitudes are warm in summer and cold in winter. Keep in mind that land surface temperatures have a greater seasonal fluctuation than ocean surface temperatures, both because the heat capacity of the ocean water is greater than the heat capacity of land, and because the ocean water has the ability to transfer heat from the surface to depth in summer and from depth to the surface in winter. Land does not transfer heat in this way.

Seasonal Changes

The presence of both land and water in near-equal amounts at the middle latitudes in the northern hemisphere produces average seasonal patterns of atmospheric pressure. During the warm summer months, the

Figure 7.8
Mean sea level pressures expressed in millibars for (a) July and (b) January.

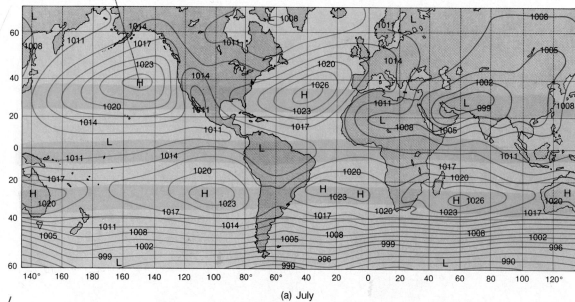

(a) July

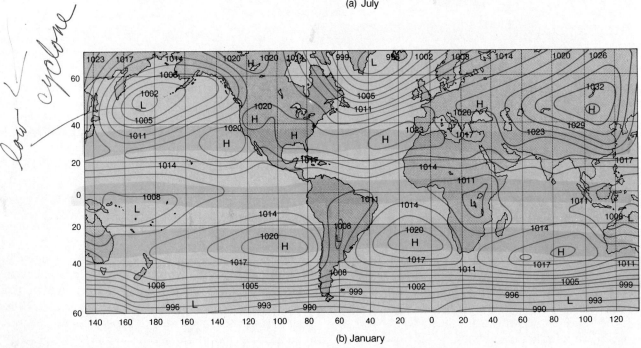

(b) January

land is warmer than the ocean. The air over the land rises, creating a low pressure area, while the air over the sea cools and sinks, producing a high pressure area over the water. In the summer, low pressure zones at 60°N and 0° tend to spread toward each other over the land, cutting off a high pressure zone at 30°N, which expands northward and southward over the ocean. The result at the middle latitudes is the formation of a discrete high pressure cell over the oceans during the summer, rather than latitudinal zones of pressure stretching continuously around the earth. In the winter, the reverse is true

at the middle latitudes. The land becomes much colder than the water, and the air rises over the water and descends over the land, creating a low pressure zone over the ocean. Over the land, the polar high pressure zone spreads toward the high pressure zone at 30°N, cutting off the low pressure belt at 60°N, which is restricted to the sea. This middle-latitude, seasonal alternation of high and low pressure cells breaks up the latitudinal pressure zones and wind belts that are seen in the water-covered model, to produce the situation shown in figure 7.8.

Figure 7.9

(a) Winds flow clockwise about a northern hemisphere high-pressure cell in the summer and (b) counterclockwise about a low-pressure cell in the winter.

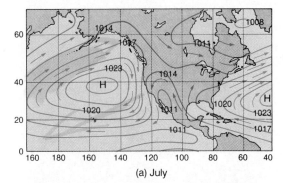

(a) July

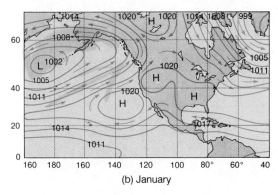

(b) January

During the northern hemisphere's summer, the air in the high pressure cells over the central portion of the North Atlantic and North Pacific oceans descends and flows outward toward the continental low pressure areas. As the descending air moves outward it is deflected to its right, producing winds that spiral in a clockwise direction about the high pressure cells. On the northern side of these high pressure cells are the westerlies, and on the southern side are the north easterlies; the eastern side of the cell has northerly winds, and the western side has southerly ones. In the winter the air circulates counterclockwise about the low pressure cell over the northern oceans, and the prevailing wind directions reverse. Figure 7.9 shows the wind directions related to these pressure cells. Along the Pacific coast of the United States, the northerly winds cool the coastal areas in the summer, and the southerly winds warm them in the winter. The New England coast receives warm, moist air from the low latitudes in the summer, and cold air moves down from the high latitudes in the winter. Although these seasonal changes modify the wind and pressure belts that were developed on the water-covered model, the generalized wind and pressure belts are still identifiable over the earth in the northern latitudes when the atmospheric pressures are averaged over the annual cycle. Rotation of the air flow about high and low pres-

sure air cells is reversed in the southern hemisphere. At the middle latitudes in the southern hemisphere, there is little land to separate the water, and so there is not much of a seasonal effect in the southern hemisphere's middle latitudes. Therefore, the wind pattern in the southern hemisphere changes little over the annual cycle and is very similar to that developed for the water-covered model (see figs. 7.7 and 7.8).

The Monsoon Effect

The differences in temperature between land and water produce large-scale and small-scale effects in coastal areas. In the summer along the west coast of India and in Southeast Asia, the air rises over the hot land, creating a low pressure air system. The rising air is replaced by warm, moist air carried on the southwest winds from the Indian Ocean. As this onshore airflow rises over the land, the moisture condenses, producing a steady, heavy rainfall. This is the wet, or summer, **monsoon.** In the winter, a high pressure cell forms over the land, and northeast winds carry the dry, cool air southward from the Asian mainland and out over the Indian Ocean. This movement produces cool, dry weather over the land, known as the dry, or winter, monsoon. For years, coastal traders in the Indian Ocean who were dependent on sailing craft planned their voyages so that they sailed to their destination on one phase of the monsoon and returned on the next, since the easiest sailing was with the wind. This seasonal reversal in wind pattern is shown in figure 7.10.

On a lesser local scale, the same effect is seen along a coastal area or along the shore of a large lake. During the day, the land is warmed faster than the water. The air rises over the land, and the air over the water moves in to replace it, creating an **onshore** breeze. At night, the land cools rapidly, and the water becomes warmer than the land. Air rises over the water, and the air from the land replaces it, this time creating an **offshore** breeze. Such a local diurnal (once-a-day) wind shift is referred to as a land-sea breeze (see fig. 7.11). The onshore breeze reaches its peak in the afternoon, when the temperature difference between the land and the water is at its maximum. The offshore breeze is strongest in the late night and early morning hours. Sometimes it is possible to smell the land 20 miles at sea, as the offshore wind carries the odors seaward. This daily wind cycle helps fishing boats that depend on only their sails to leave the harbor early in the morning and return late in the afternoon or early evening. Another example of this phenomenon can be seen in San Francisco, California, during the summer. The area east of San Francisco Bay heats up during the day, and the air above it rises. Marine air flows in through the Golden Gate. When the warm marine air from offshore reaches

Figure 7.10

The seasonal reversal in wind patterns associated with (a) the winter (dry) monsoon and (b) the summer (wet) monsoon. The isolines of pressure are given in inches of mercury.

standard
29.92
atm pressure

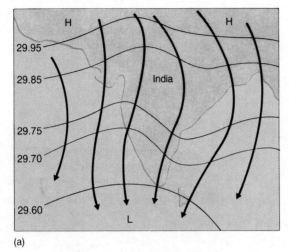

(a)

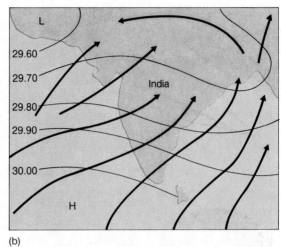

(b)

Figure 7.11

Differences between day and night land-sea temperatures produce an onshore breeze during the day and an offshore breeze at night.

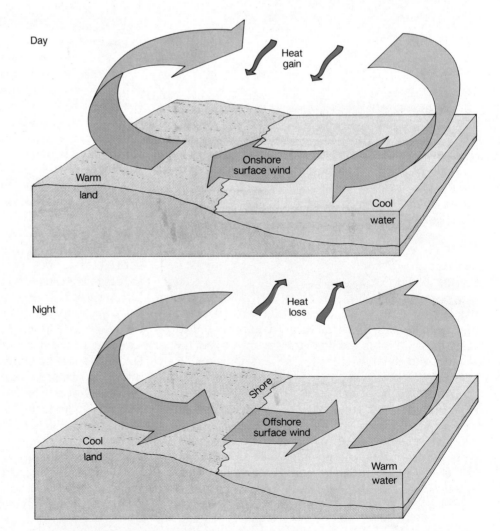

the cold coastal waters, a fog is formed that pours in through the Golden Gate, obscuring first the Golden Gate Bridge and then the city (refer to fig. 4.18 in chapter 4). When the air reaches the east side of the bay, it warms, and the fog dissipates. At night the flow is reversed, and the city is swept free of the fog.

The Topographic Effect

Because the continents rise high above the sea surface, they affect the winds in another way. As the winds sweep across the ocean, they reach the land and are forced to rise to continue moving across the land, as shown in figure 7.12. The upward deflection cools the air, causing rain on the windward side of islands and mountains; on the leeward (or sheltered) side, there is a low precipitation zone, sometimes called a **rain shadow.** For example, the windward sides of the Hawaiian mountains have high precipitation and lush vegetation, while the leeward sides are much drier and require irrigation. On the west coast of Washington State, the westerlies moving across the North Pacific produce the Olympic Rain Forest as the air rises to clear the Olympic Mountains. On the western side of the Olympic Mountains the rainfall is as much as 5 m (200 in)/yr; 100 km (60 mi) away, in the rain shadow on the leeward side of the mountains, it is 40 to 50 cm (16–20 in)/yr. The west side of the mountains of Vancouver Island in British Columbia has a high rainfall, while the eastern side of the island is dry and sunny. The adjacent region of channels and islands is noted for its sunshine and scenic cruising. The southeast trade winds on the east coast of South America sweep up and across the lowlands and then rise to cross the Andes Mountains, producing rainfall, large river systems, and lush vegetation on the eastern side of the Andes and a desert on their western slopes. The control of precipitation patterns due to elevation changes is called the **orographic effect.**

7.3 Wind-Driven Surface Currents

When the winds blow over the oceans they set the surface-water layer in motion. The surface layer is pushed and dragged by the prevailing wind. Except for the seasonal changes that appear at the middle latitudes in the northern hemisphere, the wind blowing over the open ocean drives the large-scale surface currents in a nearly constant pattern. Because the friction coupling between the water and the earth's surface is small, the moving water is deflected by the Coriolis effect in the same way that moving air is deflected. But because water moves more slowly than air, the time required for water to move the same distance as wind is much longer. During this longer time period, the earth rotates farther out from under the water than from under the wind. Therefore, the slower-moving water is deflected to a greater degree than the overlying air. The surface water layer is deflected to the right of the driving wind direction in the northern hemisphere and to the left in the southern hemisphere. In the open sea, the deflection of the surface-water layer is at a 45° angle from the wind direction, as shown in figure 7.13.

Figure 7.13

A wind-driven surface current moves at an angle of 45° to the direction of the wind; this angle is to the right in the northern hemisphere.

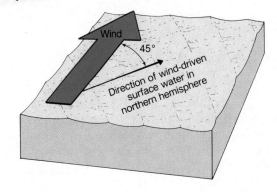

Figure 7.12

Moist air rising over the land loses its moisture on the mountains' windward side. The descending air on the leeward side is dry.

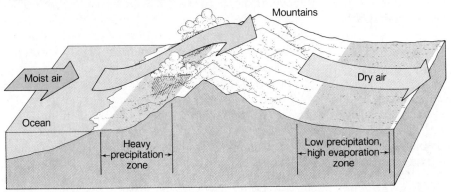

Figure 7.14
The major surface currents of the world's oceans, with principal zones of surface convergence and divergence shown.

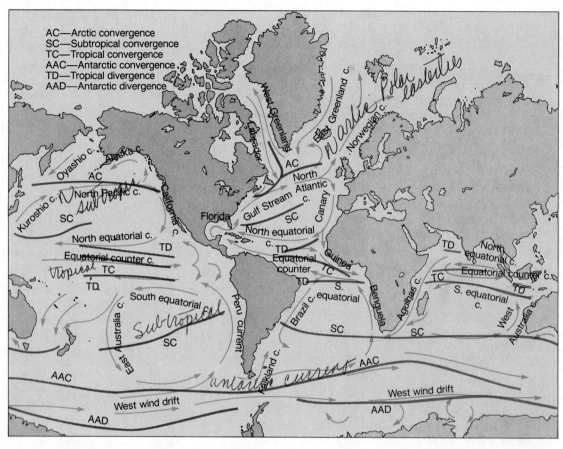

The combined effect of the wind on the surface and the deflection of the water creates the characteristic pattern of the wind-driven surface currents in each of the oceans. The major surface currents have been called the rivers of the sea. Although these rivers have no banks to contain them, they maintain their average course. Slight but important changes in their strength and location occur in response to changes in seasonal wind and climate patterns. Refer to figure 7.14 to follow the paths of these currents, as they are discussed in this section.

The Pacific Ocean Currents

In the North Pacific Ocean, the northeast trade winds push the water toward the west and northwest; this is the **North Equatorial Current.** The westerlies create the **North Pacific Current,** or **North Pacific Drift,** moving from west to east. Note that the trade winds move

the water away from Central and South America and pile it up against Asia, while the westerlies move the water away from Asia and push it against the west coast of North America. The water that accumulates in one area must flow toward areas from which the water has been removed. This movement forms two currents: the **California Current,** moving from north to south along the coast of North America, and the **Kuroshio Current,** moving from south to north along the east coast of Japan. Both flow between the North Pacific Drift and the North Equatorial Current. The Kuroshio and California currents are not completely wind-driven currents; they provide continuity of flow and complete a circular motion centered around 30°N latitude. This circular, clockwise flow of water is called the North Pacific **gyre.** The southern edge of the North Equatorial Current lies, on the average, at 5°N because of the northward displacement of the doldrum belt, which is due to the unequal heating between the northern land hemisphere and southern water hemisphere. Other major North Pacific currents include the **Oyashio,** driven by the polar easterlies, and the **Alaska Current,** fed by water from the North Pacific Current and moving in a counterclockwise gyre in the Gulf of Alaska.

In the South Pacific Ocean, the southeast trade winds move the water to the left of the wind and westward, forming the **South Equatorial Current.** The westerly winds push the water to the east; at these southern latitudes the surface current so formed can move almost continuously around the earth. This current is the **West Wind Drift.** The tips of South America and Africa act to deflect a portion of this flow northward on the east side of both the South Pacific and South Atlantic oceans. As in the North Pacific, continuity currents form between the South Equatorial Current and the West Wind Drift. The **Peru,** or **Humbolt, Current** flows from south to north along the coast of South America, while the **East Australia Current** can be seen moving weakly from north to south on the west side of the ocean. These four currents form the counterclockwise South Pacific gyre.

Note that between the North and South Equatorial currents and below the doldrums there is a current moving in the opposite direction, from west to east. This is a continuity current known as the **Equatorial Countercurrent,** which helps to return accumulated surface water eastward across the Pacific. Under the South Equatorial Current there is a subsurface current flowing from west to east called the **Cromwell Current.** This cold-water continuity current also returns water accumulated in the western Pacific.

The Atlantic Ocean Currents

The North Atlantic westerly winds move the water eastward as the **North Atlantic Current,** or **North Atlantic Drift.** The northeast trade winds push the water to the west, forming the **North Equatorial Current.** The north–south continuity currents are the **Gulf Stream,** flowing northward along the coast of North America, and the **Canary Current,** moving to the south on the eastern side of the North Atlantic. The Gulf Stream is fed by the **Florida Current** and the North Equatorial Current. The North Atlantic gyre rotates clockwise. The polar easterlies provide the driving force for the **Labrador** and **East Greenland** currents.

In the South Atlantic, the westerlies continue the West Wind Drift. The southeast trade winds move the water to the west, but the bulge of Brazil splits the **South Equatorial Current.** Much of it is deflected northward over the top of South America, into the Caribbean Sea, and eventually into the Gulf of Mexico, where it exits as the Florida Current, which joins the Gulf Stream. A portion of the South Equatorial Current slides south of the Brazilian bulge along the western side of the South

Atlantic to form the **Brazil Current.** The **Benguela Current** moves northward up the African coast. The South Atlantic gyre is complete, and it rotates counterclockwise. Because much of the South Equatorial Current is deflected across the equator, the Equatorial Countercurrent appears only weakly in the eastern portion of the Mid-Atlantic. The northward movement of South Atlantic water across the equator means that there is a net flow of surface water from the southern to the northern hemisphere. This flow is balanced by a flow of water at depth from the northern to the southern hemisphere. This deep-water return flow is the North Atlantic deep water, discussed in the previous chapter.

The Indian Ocean Currents

The Indian Ocean is mainly a southern hemisphere ocean. The southeast trade winds push the water to the west, creating the **South Equatorial Current.** The southern hemisphere westerlies still move the water eastward in the West Wind Drift. The gyre is completed by the **West Australia Current** moving northward and the **Agulhas Current** moving southward along the coast of Africa. Since this is a southern hemisphere ocean, the currents move to the left of the wind direction, and the gyre rotates counterclockwise. The northeast trade winds in winter drive the **North Equatorial Current** to the west, and the **Equatorial Countercurrent** returns water eastward toward Australia. With the coming of the wet monsoon season and its west winds, these currents are reduced. The strong seasonal monsoon effect controls the surface flow of the northern hemisphere portion of the Indian Ocean. In the summer the winds blow the surface water eastward, and in the winter they blow it westward. This strong seasonal shift is unlike anything found in either the Atlantic or the Pacific oceans.

The Ekman Spiral and Ekman Transport

Wind-driven surface water sets the water immediately below it in motion. But due to low friction coupling in the water, this next deeper layer will move more slowly than the surface layer and will also be deflected to the right (northern hemisphere) or left (southern hemisphere) of the surface layer direction. The same will be true for the next layer down and the next. The result is a spiral in which each deeper layer moves more slowly and with a greater angle of deflection than the layer above. This current spiral is called the **Ekman spiral,** after the physicist V. Walfrid Ekman, who developed its mathematical relationship. The spiral extends to a depth of approximately 100 to 150 m (330–492 ft), where the much-reduced current will be moving exactly opposite

Figure 7.15
Water is set in motion by the wind. The direction and speed of flow change with depth to form the Ekman spiral. This change with depth is a result of the earth's rotation and the inability of water, due to low friction, to transmit a driving force downward with 100% efficiency. The net transport over the wind-driven column is 90° to the right of the wind in the northern hemisphere.

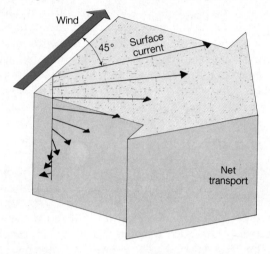

to the surface. Over the depth of the spiral, the average flow of all the water set in motion by the wind, or the net flow (**Ekman transport**), moves 90° to the right or left of the surface wind, depending on the hemisphere in which the motion takes place. This relationship is in contrast to the surface water, which moves at an angle of 45° to the wind direction (see fig. 7.15).

Geostrophic Flow

Due to the Coriolis effect, wind-driven surface layers are deflected toward the center of each current gyre, producing a surface convergence. This convergence creates a mound of surface water that is elevated up to more than 1 m (3 ft) above the equilibrium sea level and also depresses the underlying water. The slope of the mound increases until the force of gravity forcing the water downhill and away from the gyre center equals the Coriolis effect acting to mound the water. At this balance point **geostrophic flow** is said to exist, and no further deflection of the moving water occurs. The currents flow smoothly around the gyre. See figure 7.16 for a diagram of this process. From the subsurface water distribution oceanographers are able to calculate the shape of the surface water mound, and so calculate the volume transport, velocity, and depth of the currents present in the geostrophic flow around the mound.

The Sargasso Sea is the classic example of such a situation. It is located in the central North Atlantic Ocean, and its boundaries are the Gulf Stream on the west, the North Atlantic Current to the north, the Mid-Atlantic Ridge on the east, and the North Equatorial Current to the south. The region is famous for the floating mats of *Sargassum,* a brown seaweed, stretching across its surface. The extent of the floating seaweed frightened the early sailors, who told stories of ships imprisoned by the weed and sea monsters lurking below the surface. The circular motion isolates a lens of clear, warm water 1000 m (3300 ft) deep. The water in this lens is trapped by the geostrophic flow. It is the clearest water in all the oceans. Except for the floating *Sargassum,* with its rich and specialized community of small plants and animals, the area is a near biological desert.

Current Speed

The wind-driven, open-ocean surface currents move at speeds that are about one one-hundredth of the driving wind speed measured 10 m (33 ft) above the sea surface (the water moves between 0.25 and 1.0 knot, or 0.1 to 0.5 m/sec). Currents will flow much faster when a large volume of water is forced to flow through a narrow gap. For example, the northward-flowing South Atlantic Equatorial Current flows into the Caribbean Sea, then into the Gulf of Mexico, and finally out between Florida and Cuba as the Florida Current. The Florida Current's speed may exceed 3 knots, or 1.5 m/sec.

Major currents transport very large volumes of water. For example, the Gulf Stream's transport rate to the northeast is 55×10^6 m³/sec; this rate is more than 500 times the flow of the Amazon River. The flow is distributed over the width and depth of the current. When the current expands its cross-sectional area, it slows down; when it decreases its cross-sectional area, it speeds up. Speed of flow then is not always directly related to surface wind speed, but can be affected by the depth and width of the current as determined by land barriers, by the presence of another current, or by the rotation of the earth, as explained in the next section.

Western Intensification

In the North Atlantic and North Pacific the currents flowing on the western side of each ocean tend to be much stronger and narrower in cross section than the currents on the eastern side. This phenomenon is known as the **western intensification** of currents. The Gulf Stream and Kuroshio currents are faster and narrower than the Canary and California currents, although both the eastern and western currents transport about the same amount of water in order to preserve continuity of flow. Western intensification of currents traveling from low to high latitudes is related to (1) increases in the

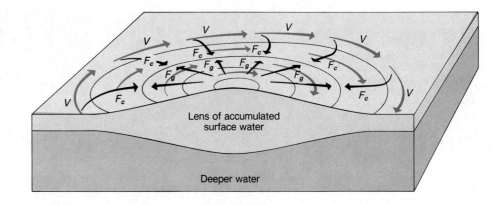

Figure 7.16
Geostrophic flow (V) *exists around a gyre when* F_c, *the deflection due to the Coriolis effect, is balanced by* F_g, *the force due to gravity. The example is of a clockwise gyre in the northern hemisphere.*

Lens of accumulated surface water

Deeper water

Coriolis effect with latitude, (2) the changing strength and direction of the east–west wind field (trade winds and westerlies) with latitude, and (3) the friction between landmasses and ocean water currents. These factors cause a compression of the currents toward the western side of the oceans, where water is moving from lower to higher latitudes. This compression requires that the current speed increase in order to transport the amount of water required in the circulation about the gyre. On the eastern side of the gyre, where currents are moving from higher to lower latitudes, the currents are stretched in the east–west direction. Here the current speed is reduced, but it still transports the required volume of water.

These fast-flowing, western-boundary currents move warm equatorial surface water to higher latitudes. Both the Gulf Stream and the Kuroshio Current bring heat from equatorial latitudes, which is exchanged with the overlying air to moderate the climates of Japan and northern Asia (in the case of the Kuroshio) and the British Isles and northern Europe (in the case of the Gulf Stream, via the North Atlantic and Norwegian Currents). Western intensification is obscured in the South Pacific and South Atlantic, because the continents of Africa and South America deflect a portion of the West Wind Drift and create strong currents on the eastern sides of these oceans.

Formation of Eddies

When a narrow, fast-moving current moves into the open sea, it displaces the quiet water through which it moves by the force of its flow and captures additional water as it does so. The current oscillates, meanders, and develops waves along its boundary, which break off to form **eddies,** or packets of water moving with a circular motion. Meandering waves hundreds of kilometers long may develop along sharp current boundaries. The large eddies that are cut off take with them energy of motion, which gradually dissipates due to friction far away from

the parent current. If the energy of the parent current were not lessened in this way, the current's speed would continue to increase as the winds transfer energy into the water.

As the flow of the Gulf Stream moves away from the coast, it is more likely to develop a meandering path. At times, the western edge of the Gulf Stream develops indentations that are filled by cold water from the Labrador Current side. These indentations pinch off and become eddies, which are displaced to the east and south of the current boundary. The effect is to transfer cold water into warm water. Bulges at the western edge of the Gulf Stream are filled with the warm water from the Sargasso Sea. When these bulges are cut off they drift to the west of the Gulf Stream, into cold water. Figure 7.17 illustrates this process.

The meandering of strong currents (such as the Gulf Stream) and the formation of large eddies that maintain their physical identity for weeks as they wander about the ocean produce surface flow patterns that differ markedly from the uniform current flows shown on current charts. Current charts show the average current flow, not the daily or weekly variations.

Large and small eddies generated by horizontal flows or currents exist in all parts of the oceans (see colorplate 2). These eddies are of varying sizes, ranging from tens to several hundred of kilometers in diameter. Each eddy contains water with specific chemical and physical properties and maintains its identity as it wanders through the oceans. Eddies may appear at the sea surface or be imbedded in waters at any depth.

In the northern hemisphere, most eddies rotate in a clockwise direction similar to atmospheric high pressure systems. These eddies stir the ocean until they gradually dissipate due to fluid friction, losing their chemical and thermal identity and their energy of motion. The water properties of an eddy allow scientists to determine its origin. Small surface eddies found 800 km (500 mi) southeast of Cape Hatteras in the North

Figure 7.17

The western boundary of the Gulf Stream is defined by sharp changes in current velocity and direction. Meanders form at this boundary after the Gulf Stream leaves the U.S. coast at Cape Hatteras. The amplitude of the meanders increases as they move downstream (a and b). Eventually the current flow pinches off the meander (c). The current boundary re-forms, and isolated rotating cells of warm water (W) wander into the cold water, while cells of cold water (C) drift into the warm water (d).

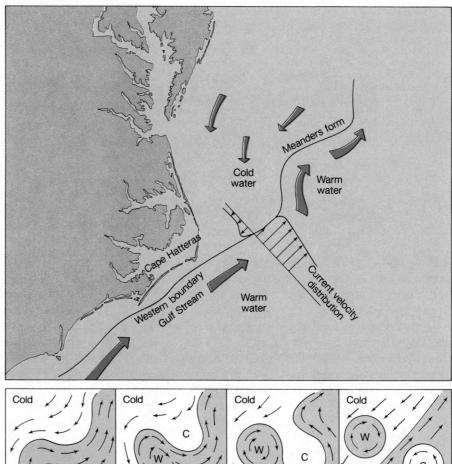

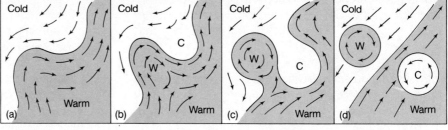

Atlantic have been found with water properties of the eastern Atlantic near Gibraltar, more than 4000 km (2500 mi) away. These eddies from the Strait of Gibraltar are formed from the salty water of the Mediterranean as it spills into the Atlantic and have been nicknamed "Meddies." Deep-water eddies at Cape Hatteras come from the eastern and western Atlantic, the Caribbean, and Iceland. Researchers estimate that these small eddies are several years old; age determination is based on drift rates, distance from source, and biological consumption of oxygen.

Eddies constantly form, migrate, and dissipate. Eddy motion is superimposed on the mean flow conditions of the oceans. If we are to understand the role these swirling masses of water play in mixing the oceans, we need more data and better tracking of eddy positions and size. Satellites are important tools for detecting surface eddies because they can precisely measure the increased elevation of the sea surface and the surface water temperature.

7.4 Convergence and Divergence

When wind-driven surface currents collide with each other or are forced against landmasses, they are said to converge or to produce a surface **convergence.** When surface currents move away from each other or away from a landmass, the currents diverge, and a surface **divergence** is formed. At a convergence, the accumulated surface water sinks; at a divergence, water rises from depth to fill the space left by the diverging flow.

Permanent Zones

There are three major zones of convergence: the **tropical convergence** at the equator, the **subtropical convergences** at approximately 30° to 40°N and S, and the **Arctic** and **Antarctic convergences** at about 50°N and S. Surface convergence zones are regions of downwelling. These areas are low in nutrients and biological

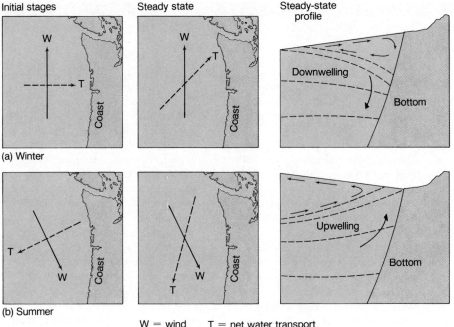

Figure 7.18

*Initially, the Ekman transport is 90°
to the right of the wind along the
northwest coast of North America. As
water is transported (a) toward the
shore in winter or (b) away from
the shore in summer, a sea surface
slope is produced. This slope creates
a gravitational force that alters the
direction of the Ekman transport to
produce a geostrophic balance and
a new steady-state direction.*

W = wind   T = net water transport

productivity. There are two major divergence zones: the
tropical divergence and the **Antarctic divergence.**
Upwelling associated with divergences delivers nutrients to the surface waters to supply the food chains
that support the tropical tuna fisheries and the largely
unexploited but richly productive waters of Antarctica.
These areas of divergence and convergence are shown
in figure 7.14.

The trade winds move the surface waters away from
the western side of major landmasses in the low latitudes, and upwelling occurs nearly continuously
throughout the year (for example, off the west coasts of
Africa and South America, both of which are very productive and yield large fish catches).

Seasonal Zones

Off the west coast of North America, downwelling and
upwelling occur on a seasonal basis as the northern-
hemisphere temperate wind-pattern changes from
southerly in winter to northerly in summer. The downwelling and upwelling occur because of the change in
direction of movement of the wind-driven coastal waters
involved in Ekman transport. Remember that the wind-
driven Ekman transport moves at an angle of 90° to the
right or left of the wind direction, depending on the
hemisphere.

In this area, the wind blows from the north in the
summer, which produces the net movement of the water
layers to the west, or 90° to the right of the wind. This
action results in the movement of the surface water
layers out to sea and the upwelling of deeper water along

the coast to replace them. The upwelling zone is evident from central California to Vancouver Island. In
winter, the winds blow from the south, the wind-driven
surface waters move to the east against the coast, and
downwelling occurs in these waters (see fig. 7.18). The
summer upwelling pattern is what produces the band
of cold coastal water at San Francisco that helps cause
the frequent summer fogs (refer back to chapter 4 and
the previous discussion in The Monsoon Effect section
of this chapter). Because of the lack of land at the middle
latitudes in the southern hemisphere, this type of seasonal upwelling is less common there.

Keep in mind that convergence and divergence of
surface currents result not only in upwellings and
downwellings, but also in the mixing of water from different geographic areas. Waters carried by the surface
currents meet, share properties due to mixing, and form
new water mixtures with specific ranges of temperature
and salinity, which then sink to their appropriate density levels. After sinking, the water moves horizontally,
blending and sharing its properties with adjacent water,
and eventually rises to the surface at a new location. This
process is discussed as thermohaline circulation in the
section on density-driven circulation in chapter 6. Thermohaline circulation and wind-driven surface currents
are closely related. The connections are so close among
formation of water mixtures, upwelling and downwelling, and convergence and divergence, that it is difficult to assign a priority of importance to one process
over another.

El Niño

On the sheltered or lee side of the tropical landmasses, under the trade winds of the Pacific, the upwelling of deep oceanic water is a nearly constant process. For this reason, the surface water along the coasts of Ecuador and Peru is cold and highly productive of living organisms. From time to time, this process falters; the trade winds lose their driving force, the upwelling lessens, and the warm tropical surface water accumulated on the west side of the Pacific moves eastward across the ocean to accumulate along the coast of the Americas. This is the **El Niño**, or Christ Child, named for its frequent coincidence with the Christmas season. The warm water accumulated by the El Niño often results in the death of the cold-water organisms and in the upset of the food chains on which fish, marine mammals, and seabirds depend. In severe cases, the resulting decay of organisms produces quantities of hydrogen sulfide in the surface water; the hydrogen sulfide stained the hulls of fruit company ships painted with white, lead-based paint. Sailors called such events the Callao Painter, after the port of Callao in Peru.

The increase in coastal surface temperatures usually ends by April, but in some years the large quantities of warm water spread north and south, and surface temperatures remain elevated for more than a year. Severe El Niño conditions occurred in 1953, 1957–58, 1965, 1972–73, 1976–77, and 1982–83. During 1982–83, the surface temperatures off Peru were more than 7°C above normal; the widespread effect of this event found tropical species displaced as far north as the Gulf of Alaska. Associated with the 1982–83 occurance were floods along the normally dry Pacific coast of the Americas, severe drought in Australia and in Sahel Africa, and frequent typhoons and heavy rains in Polynesia.

The exact cause of El Niño is still in doubt, but certain processes have been identified with its appearance. One process, in which atmospheric pressure increases on one side of the Pacific, decreases on the other, and then reverses, is known as the Southern Ocean Oscillation. The pressure centers associated with this oscillation lie over Easter Island in the eastern Pacific and Indonesia in the western Pacific. Under normal conditions, there is a high pressure system over Easter Island and a low pressure system over Indonesia, the trade winds are strong and constant, and upwelling occurs along the coast of Peru. When the atmospheric pressure system reverses, the trade winds break down. Westerly winds are formed due to the tropical cyclones (low pressure areas) that develop over Indonesia. Over a period of two to three months, accumulated warm surface water from the western Pacific surges in a wave form across the Pacific to depress the Peruvian upwelling and raise the surface temperatures. The elevated surface temperature of the eastern tropical Pacific is due to both the intrusion of warm water and the inability of cold, upwelled water to reach the surface. The moving mass of warm surface water carries the overlying low pressure zone of rising air and precipitation eastward across the Pacific. This movement causes the westerly winds to move eastward along the doldrum belt, helps the eastward movement of the warm water, and creates high precipitation in normally dry areas.

By midsummer, the El Niño effect lessens; in two to three months the surface water cools. In November and December another slight warming is often observed; the atmospheric-pressure distribution in the Southern Ocean Oscillation reverses again, and the trade winds return to their normal state. The cycle of a severe El Niño is about fifteen months in duration.

Forecasting the El Niño has been attempted, using both the change in the Southern Ocean Oscillation and the surge in strength of the trade winds, which appears to precede their decline and the development of the westerlies. However, in model studies of previous years, observed events were not always predicted, and in some cases, predictions were made of events that did not occur. Researchers continue to search for other factors related to the appearance of this pattern. The location and intensity of tropical cyclones are thought to be related to the appearance of the westerlies in the tropical western Pacific. If these cyclones are related to the beginning of the El Niño, then better methods for predicting the onset of cyclonic winds are needed.

In the winter of 1986–87, an El Niño model predicted its coming. Observations indicated that a warming trend had started, but the trend broke down in midwinter. In late winter, El Niño again began to appear but was not expected to reach the severity of the 1982–83 event. At the present time, we do not know if we will be able to predict the El Niño with 100% confidence, but an increased understanding of El Niño's cause and relationship to other world climate factors appears within our grasp. The impact of El Niño on commerical fisheries will be found in chapter 15.

7.5 *Measuring the Currents*

Direct measurement methods of currents fall into two groups: (1) those that follow a parcel of the moving water, and (2) those that measure the speed and direction of the water as it passes a fixed point. Moving waters may be followed with buoys designed to float at predetermined depths. These buoys signal their positions acoustically to the research vessel, allowing those on board to follow their paths and calculate their speed and displacement due to the current. Surface water may be labelled with buoys or with dye that can be photographed from the air. Buoy positions may also be tracked by satellite. A series of pictures or position fixes makes it possible to calculate the speed and direction of a current.

A variation of this technique uses the **drift bottle.** Thousands of sealed bottles, each containing a postcard, are released at a known position. When the bottles are washed ashore the finders are requested to record the time and location of the find and return the card. In this case only the release point, the recovery points, and the elapsed time are known; the actual path of motion is assumed.

Sensors used to measure current speed and direction are called **current meters.** They consist of a rotor to measure speed and a vane to measure direction of flow (see fig. 7.19). If the current meter is lowered from a stationary vessel, the two measurements can be returned to the ship by a cable or can be stored by the meter for reading upon retrieval. If the current meter is attached to an independent, bottom-moored buoy system, the signals can be transmitted to the ship as radio signals or can be stored on tape in the current meter, to be removed when the buoy and current meter are retrieved.

In order to measure a current passing a location, a current meter must be held fixed in space. Although a vessel can be moored in shallow water so that it does

Figure 7.19

An internally recording Aanderaa current meter. The vane orients the meter to the current while the rotor determines current speed (a). Inside the protective case (b and c), the instrument records coded information from its external sensors. The speed and direction of the current, water temperature, pressure, and conductivity of the water are recorded on magnetic tape (b). A clock (c) controls the frequency at which these data are recorded. The magnetic tape is retrieved when the current meter is picked up from its taut wire mooring.

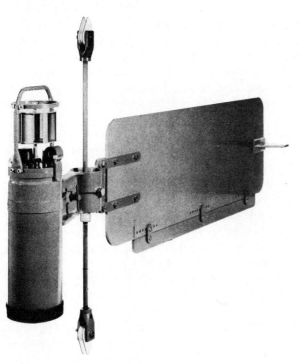

(a)

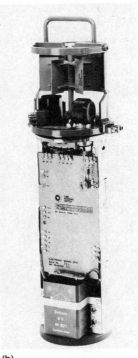

(b)

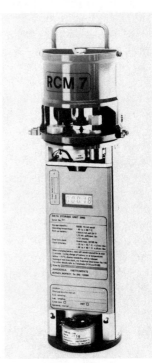

(c)

Figure 7.20
Taut wire moorage. (a) Recovery is accomplished by retrieving the surface buoy and hauling in the wire. If the surface buoy is lost, it is possible to grapple for the ground wire. (b) In this system, a sound signal disconnects the anchor, and the equipment floats to the surface.

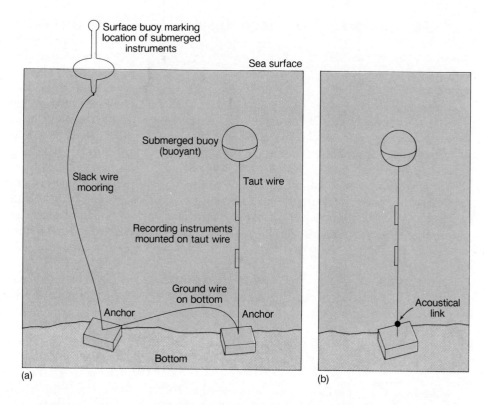

not move, it is very difficult, if not impossible, to moor a ship or surface platform in the open sea so that it will not move and by its motion also move the current meter. The solution is to attach the current meter to a buoy system that is entirely submerged on a taut cable and so not affected by winds or waves. See figure 7.20 for a diagram of this assembly. The string of anchors, current meters, wire, and floats is preassembled on deck and is launched, surface float first, over the stern of the slowly moving ship. As the ship moves away the float, meters, and cable are stretched out on the surface. When the vessel reaches the sampling site the anchor is pushed overboard to pull the entire string down into the water. The floats and meters are retrived by grappling from the surface for the ground wire or by sending a sound signal to a special acoustical link, which detaches the wire from the anchor. The anchor is discarded, and the buoyed equipment returns to the surface. The technique is straightforward, but many problems can occur when launching, finding, and retrieving instruments from the heaving deck of a ship at sea. Whenever oceanographers send their increasingly sophisticated equipment over the side, they must cross their fingers and hope to see it again.

7.6 *Practical Considerations: Energy from the Currents*

The massive oceanic surface currents of the world are untapped reservoirs of energy. Their total energy flux has been estimated at 2.8×10^{14} (280 trillion) watt-hours. Because of their link to winds and surface heating processes, the currents can be considered as indirect sources of solar energy. If the total energy of a current was removed by conversion to electric power, that current would cease to exist; but only a small portion of any current's energy can be harnessed, owing to the current's size. Harnessing the energy from these open ocean currents requires the use of turbine-driven generators anchored in place in the current stream. Large turbine blades would be driven by the moving water, just as windmill blades are moved by the wind; these blades could be used to turn the generators and to harness the energy of the current. (See also the discussion on energy from tidal currents in chapter 9.) Another proposal calls for a barge moored in the current stream with a large cable loop to which parachutes are fastened. The cable would be moved along by the current acting against the open parachutes. When the parachutes reached the end of the loop they would turn the corner and be dragged back against the current while closed. The continuous movement of the cable would be used to turn a generator to produce electricity.

The Florida Current and the Gulf Stream are reasonably swift and continuous currents moving close to shore in areas where there is a demand for power. If ocean currents are developed as energy sources, these currents are among the most likely. But most of the wind-driven oceanic currents generally move too slowly and are found too far from where the power is needed. The cost of constructing, mooring, and maintaining current-driven power generating devices in the open sea makes them noncompetitive with other sources of power at this time.

Summary

The density of air is controlled by the air's temperature, pressure, and water vapor content. The winds are the horizontal air motion in the atmospheric convection cells produced by heating at the earth's surface. Winds are named for the direction from which they blow.

Because of the Coriolis effect, winds are deflected to their right in the northern hemisphere and to their left in the southern hemisphere. This action produces a three-celled wind system in each hemisphere that results in the surface wind bands of the trade winds, the westerlies, and the polar easterlies. Zones of rising air occur at 0° and at 60°N and S; these are low pressure areas of clouds and rain. Zones of descending air at 30° and 90°N and S are high pressure areas of clear skies and low precipitation. Surface winds are unsteady and unreliable at the zones of rising and sinking air, producing the doldrums and the horse latitudes. The doldrum belt is displaced north of the earth's geographic equator.

Seasonal atmospheric pressure changes modify these wind bands and cause coastal winds to change direction seasonally in the northern hemisphere. Differences in temperature between land and water produce the monsoon effect. This seasonal reversal in wind pattern causes the wet and dry monsoons of the Indian Ocean; a similar daily reversal causes the onshore and offshore winds of any coastal area. Winds from the ocean rising to cross the land also produce heavy rainfall.

Winds push the surface water 45° to the right of their direction in the northern hemisphere and 45° to the left in the southern hemisphere. This action creates the large surface current gyres observed in each ocean. Southern hemisphere gyres rotate counterclockwise; northern hemisphere gyres rotate clockwise. The currents of the northern Indian Ocean change with the seasonal monsoons.

Geostrophic flow is produced when the force of gravity balances the Coriolis effect. The water in the Sargasso Sea is isolated by this circular flow.

A current that is transporting a given volume of water flows faster through a smaller area than through a larger one. The speed of a current is affected by the current's cross-sectional area, by other currents, by westward intensification, and by wind speed. Eddies are formed at the surface when a fast-moving current develops waves along its boundary that break off from the parent current. Eddies occur at all depths, wander long distances, and gradually lose their identity.

Downwelling is produced by converging surface currents, while upwelling is produced by diverging surface currents. Upwelling occurs nearly continuously along the western sides of the continents in the trade wind belts, where water diverges from the coast. Seasonal upwellings and downwellings occur in coastal areas that have changing wind patterns and an alternating coastal flow of water onshore and offshore due to the Ekman transport.

A variety of instruments are available to measure currents, either by following the water or by measuring the water's speed and direction as it moves past a fixed point.

Deriving energy from the oceanic current flows is not practical at the present time.

Key Terms

atmospheric pressure	onshore breeze	geostrophic flow	Arctic convergence
Coriolis effect	offshore breeze	western intensification	Antarctic convergence
trade winds	rain shadow	eddy	tropical divergence
westerlies	orographic effect	convergence	Antarctic divergence
polar easterlies	gyre	divergence	El Niño
doldrums	Ekman spiral	tropical convergence	drift bottle
horse latitudes	Ekman transport	subtropical convergence	current meter
monsoon			

Study Questions

1. According to the net transport of the Ekman spiral, wind-driven water is directed toward the center of a large oceanic current gyre. Why does the current not flow to the gyre's center and, instead, flow in a clockwise circular path about a gyre in the northern hemisphere?

2. How is Ekman transport related to coastal upwelling and downwelling?

3. Explain why, on the world average, the earth's atmosphere is unstable. Take into consideration distribution of the earth's surface area with latitude and the variation in heat loss and heat gain with latitude.

4. Why are regions that are noted for their low barometric pressure and rising air famous for their excess precipitation?

5. A frictionless projectile is fired from the North Pole and is aimed along the prime meridian. It takes three hours to reach its landing point, half-way to the equator. Where does it land? If the same projectile is fired from the South Pole under the same circumstances, where does it land?

6. Why does a flow of water that is constant in volume transport per time increase its speed when it passes through a narrow opening?

7. Why does the circulation of the atmosphere depend on its transparency to solar radiation? What would happen to atmospheric circulation if the upper atmosphere absorbed most of the solar radiation?

8. On a map of the world, plot the oceanographic (or meterologic) equator, the six major wind belts, the current system of each ocean, and the main areas of convergence and divergence.

9. Explain why the northeast and southeast trade winds are steady in strength and direction year-round, while the northern hemisphere westerlies alternate from northwest to southwest with summer and winter along the west coast of the United States.

10. Why are the westerlies of the southern hemisphere more consistent than the westerlies of the northern hemisphere?

The Waves

8

None of them knew the color of the sky. Their eyes glanced level, and were fastened upon the waves that swept toward them. These waves were of the hue of slate, save for the tops, which were of foaming white, and all of the men knew the colors of the sea. The horizon narrowed and widened, and dipped and rose, and at all times its edge was jagged with waves that seemed thrust up in points like rocks.

Many a man ought to have a bath-tub larger than the boat which here rode upon the sea. These waves were most wrongfully and barbarously abrupt and tall, and each froth-top was a problem in small boat navigation. . .

A seat in this boat was not unlike a seat upon a bucking broncho, and, by the same token, a broncho is not much smaller. The craft pranced and reared, and plunged like an animal. As each wave came, and she rose for it, she seemed like a horse making at a fence outrageously high. The manner of her scramble over these walls of water is a mystic thing, and, moreover, at the top of them were ordinarily these problems in white water, the foam racing down from the summit of each wave, requiring a new leap, and a leap from the air. Then, after scornfully bumping a crest, she would slide, and race, and splash down a long incline and arrive bobbing and nodding in front of the next menace.

A singular disadvantage of the sea lies in the fact that after successfully surmounting one wave you discover that there is another behind it just as important and just as nervously anxious to do something effective in the way of swamping boats. In a ten-foot dingey one can get an idea of the resources of the sea in the line of waves that is not probable to the average experience, which is never at sea in a dingey. As each salty wall of water approached, it shut all else from the view of the men in the boat, and it was the final outburst of the ocean, the last effort of the grim water. There was a terrible grace in the move of the waves, and they came in silence, save for the snarling of the crests.

Stephen Crane,
from The Open Boat

We have all seen water waves. This up-and-down motion occurs on the surface of oceans, seas, lakes, and ponds, and we speak of waves, ripples, swells, breakers, whitecaps, and surf. Sometimes we observe that they are related to the wind, or to a passing ship, or even to a stone thrown into the water. In all these cases, the smooth surface of the water has been disturbed, and waves are the result.

Waves are of practical significance to us. They may swamp a small boat, or when large, smash and twist the bow of a supertanker. The storm waves of winter crash against our coasts, eating away the shore and damaging offshore structures, such as oil-drilling towers, docks, and breakwaters. Heavy wave action may force commercial vessels to slow their speed and lengthen their sailing time between ports, to the inconvenience and increased cost of the shippers. Special waves associated with seismic disturbances and severe hurricanes have killed thousands of coastal inhabitants and have severely damaged their cities and towns. Surfers search for the perfect wave, and ancient peoples navigated by the patterns that waves form. In order to protect ourselves from destructive waves and possibly to make constructive use of wave energy, we need to know what causes waves, how waves behave, where they come from and where they go, and many other related details.

In this chapter we will describe different kinds of waves, investigate their origin and behavior, and study some of their major characteristics. Our study will be somewhat superficial and simplified, because waves are among the most complex of the ocean's phenomena and because no two waves are exactly the same. Books are written about waves; mathematical models have been devised to explain waves; wave tanks are built to study waves. However, there is much we can learn if we follow the sequence of events beginning at the storm center at sea where a wave is born and ending at last in the spray and surf of a faraway shore.

8.1 How a Wave Begins

Imagine that you are standing on the beach looking out across a perfectly flat, smooth surface of water. In order to produce (or generate) waves, it is necessary to introduce a disturbing force. You throw a stone into the water, or a large, naturally occurring landslide comes down into the water. In both cases, a pulse of energy is introduced, and waves are produced. Such a disturbing force is called a **generating force.** The waves produced by the generating force move (or progress) away from the point of disturbance. Of course, the magnitude of the disturbance and the resulting wave size are very different in each of these examples.

Let us consider the case of the thrown rock in some detail. The rock hits the quiet surface of the water and displaces (or pushes aside) the water surface. As the rock sinks, the displaced water flows back into the space left behind, and as this water rushes back from all sides, its momentum forces it upward, resulting in a higher (or elevated) surface. The elevated water falls back, causing a depression below the surface, which is filled in its turn, with the same result. This action sets up a series of waves (or oscillations) that move outward and away from the point of disturbance. The oscillations continue to move outward from the point of disturbance until they are dissipated through friction among the water molecules.

The generating force in the case we are considering is the rock as it strikes the water surface and introduces a pulse of energy to that surface. The force that causes the water to return to the level of the undisturbed surface is the **restoring force.** If the stone thrown into the water is small, the resulting waves are small, and the restoring force is the **surface tension** of the water surface. (Surface tension, or the elastic quality of the surface due to the cohesive behavior of the water molecules, is discussed in chapter 4.) All very small water waves are affected by surface tension. In the example in which the landslide is the generating force, the much larger oscillations result in waves that are pulled back to the undisturbed surface level of the water by the restoring force of gravity. When waves are of sufficiently large size so that gravity is more important than surface tension as a restoring force, the waves are called **gravity waves.**

The most common generating force for water waves is the moving air, or wind. As the wind blows across a smooth water surface, the friction (or drag) between the air and the water tends to stretch the surface, resulting in wrinkles; surface tension acts on these wrinkles to restore a smooth surface. The wind and the surface tension create small waves, called **ripples,** or **capillary waves** (see fig. 8.1) Often patches of these very small waves can be seen forming, moving, and disappearing as they are driven by pulses of wind. Such patches darken the surface of the water and move quickly, keeping pace with the gusts of wind. Sailors call these fast-moving patches **cat's-paws.** Cat's-paws are capillary waves formed by the wind gusts. These waves die out rapidly due to friction while new ripples form constantly in front of each moving wind gust.

Figure 8.1
Wind-generated capillary waves. Wavelengths are measured in centimeters.

Figure 8.2
Wind waves at sea.

When the wind blows, energy is transferred to the water over large areas, for varying lengths of time, and at different intensities. As waves form, the surface becomes rougher, and it is easier for the wind to grip the roughened water surface and add energy. There is increased frictional drag between the air and the water. As the wind energy is increased, the oscillations of the water surface become larger, and the restoring force changes from surface tension to gravity (see fig. 8.2).

A wave is the result of the interaction between a generating force and a restoring force. Generating forces include any occurrence that adds energy to the sea surface: wind, landslide, sea-bottom faulting or slipping, moving ships, and even thrown objects. The restoring forces are surface tension for capillary waves and ripples, and gravity for larger waves.

8.2 Anatomy of a Wave

In any discussion of waves, certain terms are used. Refer to figure 8.3 during the explanation of these terms. The portion of the wave that is elevated above the undisturbed sea surface is called the **crest;** the portion that is depressed below the surface is called the **trough.** The distance between two successive crests or two successive troughs is the length of the wave, or its **wavelength.** The **height** of the wave is the vertical distance from the top of the crest to the bottom of the trough.

Sometimes the term **amplitude** is used. The amplitude is equal to one-half the wave height, or the distance from either the crest or the trough to the still water, or **equilibrium surface.** The oceanographer characterizes a wave not only by its length and height (or amplitude), but also by its **period.** The period is the time required for two successive crests or two successive troughs to pass a point in space. In other words, if you are standing on a piling and start a stopwatch as the crest of a wave passes and then stop the stopwatch as the crest of the next wave passes, the time you have measured is the period of the wave.

The dimensions and characteristics of waves vary greatly, but the regularity in the rise and fall of the water's surface and the relationship between wavelength and wave period allow mathematical approximations to be made, which give us insight into the behavior and properties of waves. These calculations are done by relating real waves to simple models of waves. Consider figure 8.3a, which reveals a profile of waves at right angles to the crests. This illustration has been drawn to show that real waves tend to have a trough that is flatter than the crest. By contrast, figure 8.3b is that of a symmetrical sine wave. This regular wave, which only approximates a water wave in nature, is one of the wave forms used by physical oceanographers and mathematicians to explore and explain wave motion. The relationships presented below are based on this regular sine wave form.

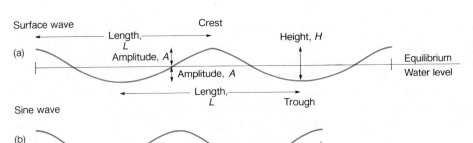

Figure 8.3

The profile of an ideal sea surface wave (a) differs from the shape of a sine wave (b).

Figure 8.4

The moving wave form sets the water particles in motion (note the arrows in the diagram). The diameter of a water particle's orbit at the surface is determined by wave height. Below the surface, the diameter decreases and orbital motion ceases at a depth (D) equal to one-half the wavelength.

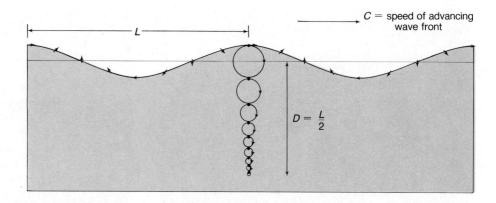

8.3 Wave Motion

As a wave form moves across the water surface, particles of water are set in motion by the energy of the wave. Seaward, beyond the surf and breaker zone, the surface undulates quietly. The water is not moving toward the shore. The ocean wave does not represent a flow of water but rather represents a flow of motion or energy from its origin to its eventual dissipation at sea or loss against the land. In order to understand what is happening during the passing of a wave, let us follow the motion of the water particles as the wave moves through them.

As the wave crest approaches, the surface water particles rise and move forward. Immediately under the crest the particles have stopped rising and are moving forward at the speed of the crest. When the crest passes, the particles begin to fall and to slow in their forward motion, reaching a maximum falling speed and a zero forward speed when the midpoint between crest and trough passes. As the trough advances, the particles slow in falling rate and start to move backward, until at the bottom of the trough they have attained their maximum backward speed and neither rise nor fall. As the remainder of the trough passes, the water particles begin to slow their backward speed and start to rise again, until the midpoint between trough and crest passes. At this point the water particles start their forward motion again as they continue to rise with the advancing crest. This motion (rising, moving forward, falling, reversing direction, and rising again) creates a circular path, or **orbit,** for the water particles (see fig. 8.4).

It is the orbital motion of the water particles that causes a floating object to bob, or move up and down, forward and backward, as the waves pass under it. This motion affects a fishing boat, swimmer, sea gull, or any other floating object on the surface seaward of the surf zone. The surface water particles trace an orbit whose diameter is equal to the height of the wave. This same type of motion is transferred to the water particles below the surface, but less energy of motion is found at each succeeding depth. Therefore, the diameter of the orbits

becomes smaller and smaller as depth increases. At a depth equal to one-half the wavelength, the orbital motion has decreased to almost zero; notice how the orbit decreases in diameter in figure 8.4. Submarines dive during rough weather for a quiet ride, since the wave motion does not extend far below the surface.

The orbits just described are based on the sine wave, as diagrammed in figure 8.3b. Actual ocean waves do not behave exactly the same way. In nature, the water particles move in orbits whose forward velocity at the top of the orbit is slightly greater than the reverse velocity at the bottom of the orbit. Therefore, each orbit made by a water particle does move the water slightly forward in the direction the waves travel. This movement is due to the shape of the real waves, whose crests are sharper than their troughs, as shown in figure 8.3a. This difference means that in nature there is a very slow transport of water in the direction of the waves, but this motion is often ignored.

8.4 Wave Speed

It is possible to relate the wavelength and period of a wave in order to determine the wave's speed. The speed of the wave (C) is equal to the length of the wave (L) divided by the period (T):

$$\text{Speed} = \frac{\text{length of wave}}{\text{wave period}}$$

or

$$C = \frac{L}{T}.$$

The period of the wave is not difficult to measure at sea. The wavelength, however, is hard to measure without a fixed reference point. It is usually determined from its relationship with gravity and wave period. The oceanographer at sea determines the wave period (T) by direct measurement and calculates the wavelength (L) by another equation in order to find the wave speed using this equation.

In deep water, the wavelength is equal to the acceleration due to gravity (g) divided by 2π, times the square of the wave period (T):

$$L = \frac{g}{2\pi} T^2.$$

L in meters $= 1.561\,T^2$; T is measured in seconds. This deep water wave equation, when combined with the equation $C = L/T$, is used to determine wave speed (C) from either wavelength (L) only or wave period (T) only:

$$C = \frac{L}{T} = \frac{g}{2\pi} T \text{ or } C^2 = \frac{L^2}{T^2} = \frac{g}{2\pi} L.$$

8.5 Deep Water Waves

"Deep water" has a precise meaning for the oceanographer studying waves. To be considered as a **deep water wave,** the wave must occur in water that is deeper than one-half the wave's length. For example, a 15-m-long wave must occur in water that is deeper than 7½ m in order to be considered a deep water wave and to behave as the waves described here.

Storm Centers

Most waves observed at sea are **progressive wind waves.** That is to say, they are generated by the wind, they are restored by gravity, and they progress (or travel) in a particular direction. Such waves are formed in local active storm centers or by the steady winds of the trade and westerly wind belts. An active storm may be quite large, covering an area of several thousand square miles. Within the storm area, the winds are not steady but turbulent, and they vary in strength and direction. The winds in the storm area flow in a circular pattern about the low pressure **storm center,** creating waves that move outward and away from the storm in all directions, and the storm center itself may also move. You may gain some idea of the size of storms and their direction of travel by viewing the cloud patterns in the satellite photographs that are presented in televised weather forecasts or printed in the newspaper weather sections.

In the storm area, the sea surface appears as a jumble and confusion of waves of all heights, lengths, and periods. There are no regular patterns. Capillary waves ride the backs of small gravity waves, which in turn are superimposed on still higher and longer gravity waves. This turmoil of mixed waves is called a **sea.** Sailors use the expression, "There is a sea building," to refer to the growth of these waves under storm conditions. Due to variations in the winds of the storm area, energy at different intensities is transferred to the sea surface at different rates, resulting in waves with a variety of periods and heights. The wave periods are a function of the generating force. Once a wave has been created with its period, *the period does not change.* The speed at which the wave moves may change, but its period remains the same. The period is a constant property of the wave until the wave is lost by breaking at sea, through friction, or in the crash of the surf against the shore.

Swell

Among the assortment of waves produced by a storm at sea are those with long periods and long wavelengths. These waves have a greater speed than those with short periods and short wavelengths. These faster, longer waves gradually move through and ahead of the shorter, slower waves. They eventually escape the storm and appear as a regular pattern of undulating crests and troughs moving across the sea surface. Once away from the storm these longer-period, uniform waves are called **swell;** they carry considerable energy, which they lose very slowly. Swell can travel for thousands of miles across the ocean. Groups of large, long-period waves created by individual storms between 40° and 50°S in the Pacific Ocean have been traced across the entire length of that ocean, until they die on the shores and beaches of Alaska.

Dispersion

The movement of faster, longer waves through and ahead of shorter, slower waves is called **sorting** or **dispersion.** Groups of these faster waves move as **wave trains,** or packets of similar waves with approximately the same period and speed. Because of dispersion, the distribution of observed waves from any single storm changes with time. Near the storm center the waves are not yet sorted, while farther away the faster, longer-period waves are out ahead of the slower, shorter-period waves. This process is shown in figure 8.5. The distribution of the waves from a given storm and the energy associated with particular wave periods change predictably with time, allowing the oceanographer to follow wave trains from a single storm over long distances.

Group Speed

Consider again the wave train produced by a rock thrown into the water. In this case the wave group, or train, is seen as a ring of waves moving outward from the point of disturbance. Careful observation shows that waves constantly form on the inside of the train as it moves across the water from its point of origin. As each wave joins the train on the inside of the ring, a wave is lost

Figure 8.5

Dispersion. The longer waves travel faster than the shorter waves. Waves are shown moving in only one direction in the diagram.

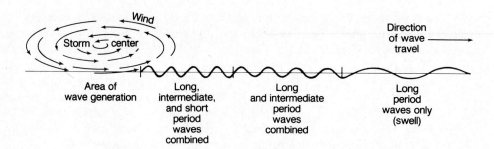

Figure 8.6

Wave trains form interference patterns where they meet. (a) Waves meeting at right angles. (b) Wave crests and wave troughs meet and cancel each other. (c) Wave crests and troughs coincide, building higher waves.

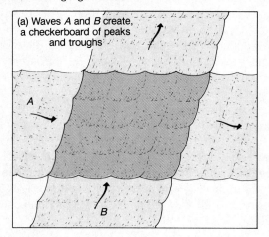

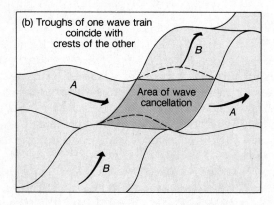

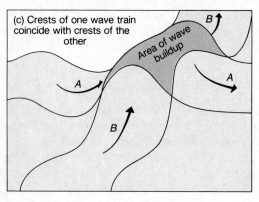

from the leading edge on the outside of the ring. The outside wave's energy is lost in transferring the wave form to the undisturbed water. Therefore, each individual wave actually moves faster than the leading edge of the wave train. The wave train moves outward at a speed one-half that of the individual wave. This speed is called the **group speed,** and it is the speed at which the wave energy is transported away from its source under deep-water conditions:

Group speed = ½ wave speed = speed of energy transport;

or

$$V = \frac{C}{2}.$$

Wave Interaction

Waves that escape, or outrun, a storm and are no longer receiving energy from the storm winds tend to flatten out slightly, and their crests become more rounded. These waves disperse and move across the ocean surface as swell and are likely to meet other trains of swell moving out and away from other storm centers. When the two wave trains meet, they pass through each other and continue on. Wave trains may intersect at any angle, and many possible sea surface patterns (or interference patterns) may result. If the two wave trains intersect each other sharply, then a checkerboard pattern is formed (fig. 8.6a). If the angle is slight and the waves have similar lengths and heights, and if the crests of one train coincide with the troughs of the other train, the wave trains cancel each other (fig. 8.6b). If the crests or the troughs of the two different wave trains coincide, the profiles are additive; the elevation of the crests increases, as does the depression of the troughs (fig. 8.6c). It is possible for two or more wave trains to phase together so that they suddenly develop large waves unrelated to any storms in the crossing area. If these waves become too high, they may break and lose some of their energy. Such waves can take a vessel by surprise and damage it severely.

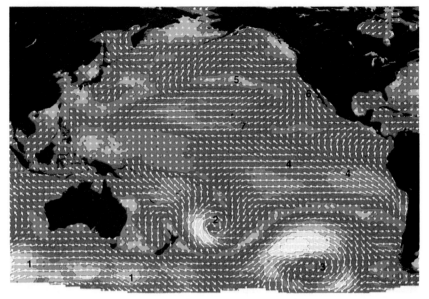

(a)

(b)

(c)

Colorplate 8

(a) *Pacific Ocean wind distributions. The SEASAT satellite obtained the first global measurements of ocean wind data during three days in 1978. Arrows point in the direction the winds blow; longer arrows indicate greater wind speed. Light winds (less then 4 m/sec or 9 mph) are colored blue and strong winds (greater than 14 m/sec or 31 mph) are colored yellow. Wind speeds and directions were calculated from radar reflections from the small wind-driven waves that roughened the sea surface. The approximate accuracy is ± 2m/sec and ± 20°. Winds blowing from the west in the southern hemisphere (1) blow almost continuously around the entire Southern Ocean except near the tip of South America. Intense storms are shown at (2) and (3). The strong and constant southeast trade winds are shown at (4). Summer in the northern hemisphere is characterized by a large high pressure cell in the North Pacific (5). Circulation about this cell produces the winds (6) along the coast that induce upwelling. Winds and squalls are formed at the boundary between the northeast and southeast trade winds (7), the doldrums.*

(b) *Waves at sea observed from a research vessel's hydrocage, the platform used to launch instruments from the deck.*

(c) *Land's End, Cornwall, England is exposed to the full onslaught of the waves roaring in from the North Atlantic.*

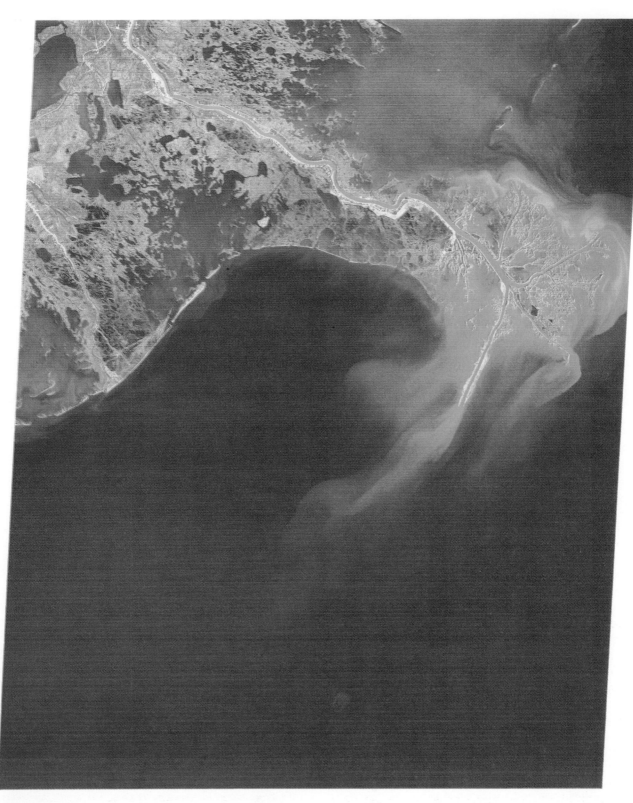

Colorplate 9
The Mississippi River delta is disappearing into the Gulf of Mexico.
Because the river is being kept in its channel by levees, river-
borne sediments are continuing to the river's mouth and being
deposited in deep water instead of spreading out and replenishing
the delta. The loss of sediments also prevents compensation for the
subsidence of the delta (See Going, Going, Gone box in chapter
11).

TABLE 8.1

Properties of Typical Ocean Wind Waves

Wave period (seconds)	Wavelength (m)	Maximum wave height (m)	Wave speed (m/sec)
1	1.56	0.22	1.56
2	6.25	0.89	3.12
3	14.05	2.01	4.68
4	24.98	3.57	6.25
5	39.03	5.58	7.81
6	56.21	8.03	9.37
7	76.50	10.93	10.93
8	99.92	14.27	12.49
9	126.47	18.07	14.05
10	156.13	22.30	15.61
11	188.92	26.99	17.17
12	224.83	32.12	18.74
15	351.29	50.18	23.42

8.6 Wave Height

Table 8.1 presents data for some typical wind waves in deep water. Estimated maximum wave heights are included in this table, but the height of actual waves is produced by the interaction of several factors. The three most important factors are (1) wind speed (how fast the wind is blowing), (2) wind duration (how long the wind blows), and (3) **fetch** (the distance over water that the wind blows in a single direction). The wave height may be limited by any one of these factors. If the wind speed is very low, large waves are not produced, no matter how long the wind blows over an unlimited fetch. If the wind speed is great, but it blows for only a few minutes, again no high waves are produced despite unlimited wind strength and fetch. Also, if very strong winds blow for a long period over a very short fetch, no high waves form. It is when no single one of these three factors is limiting that we find spectacular wind waves at sea.

The area of the ocean in the vicinity of 40° to 50°S latitudes is ideal for the production of high waves. Here, in an area noted for high-intensity storms and strong winds of long duration (what sailors have called the "roaring forties and furious fifties"), there are no landmasses to interfere and reduce the fetch length. The westerly winds blow almost continuously around the earth, adding energy to the sea surface for long periods of time and over great distances, resulting in waves that move in the same direction as the wind. Although this area is ideal for the production of high waves, such waves can occur anywhere in the open sea, given the proper storm conditions.

A typical maximum fetch for a local storm over the ocean is approximately 500 nautical miles (920 km). However, the storm itself may be moving. Keeping in mind that storm winds circulate around a low pressure disturbance, the winds can continue to follow the waves on the side of the storm, along which wave direction is the same as the storm direction. This increases both the fetch and the duration of time over which the wind adds energy to the waves. If the waves move fast enough, their speeds exceed the speed of the moving storm. The waves escape the wind and do not grow larger. Waves 10 to 15 m (33–49 ft) high are not uncommon under severe storm conditions; the lengths of such waves are typically between 100 and 200 m (330–660 ft). This length is about the same as the length of a modern freighter. In such a sea, a vessel of this length encounters hazardous sailing conditions, because the ship may become suspended between the crests of two waves and break its back.

Truly giant to monster waves over 30.5 m (or 100 feet) high are rare. In 1933 the USS *Ramapo*, a Navy tanker, en route from Manila to San Diego, encountered a severe storm, or typhoon. In the course of running the ship downwind to ease the ride, the *Ramapo* was overtaken by waves that, as measured against the ship's superstructure by the officer on watch, were 112 feet (or 34.2 m) high. The period of the waves was measured at

14.8 seconds, and the wave speed was calculated at 27 m (90 ft)/sec. Compare these values with the maximum theoretical values given in table 8.1. Other storm waves in this size category have been reported, but none have been as well documented. It is also probable that ships confronted with such waves do not always survive to report the incidents.

Episodic Waves

Large waves can suddenly appear at sea unrelated to local conditions. These waves are called **episodic waves.** An episodic wave is an abnormally high wave that occurs due to a combination of intersecting wave trains, changing depths, and currents. We do not know a great deal about these waves, as they do not exist for long and they can and do swamp ships, often removing any witnesses. They occur most frequently near the edge of the continental shelf, in water about 200 m (660 ft) deep, and in certain geographic areas with particular prevailing wind, wave, and current patterns.

One region that is noted for such waves is the area where the Agulhas Current, sweeping down the east coast of South Africa, meets the storm waves arising in the Southern Ocean. Storm waves from more than one storm may combine and run into the current and against the continental shelf, producing occasional episodic waves. This area is also one of the world's busiest sea routes, as supertankers carrying oil from the Middle East ride the Agulhas Current on their trip southward to round the Cape of Good Hope en route to Europe and America. The situation is an invitation to disaster, and tankers have been damaged and lost in this area (see fig. 8.7). In the North Atlantic, strong northeasterly gales send large storm waves into the edge of the northward-moving Gulf Stream near the border of the continental shelf, resulting in the formation of large waves. The shallow North Sea also seems to provide suitable conditions for extremely high episodic waves during its severe winter storms.

Researchers studying these waves describe them as having a height equal to a seven- or eight-story building (20–30 m) and moving at a speed of 50 knots, with a wavelength approaching a half mile (0.9 km), making for a truly mountainous mass of water. If such a wave were to topple onto a vessel that had dropped its bow into a trough, there could be no escape from the thousands of tons of water that would crash on its deck. It would be the ultimate "non-negotiable wave." In the North Sea maximum wave height for an episodic wave is calculated as 33.8 m (or 111 feet), but 22.9 m (or 75 feet) is the highest that has been observed. In the Agulhas-Current area, researchers who searched the storm records of the past 20 years calculated a possible wave height of about 57.9 m (or 190 feet).

Figure 8.7
A giant wave breaking over the bow of the ESSO Nederland *southbound in the Agulhas Current. The bow of this supertanker is about 25 meters above the water.*

There are many disappearances of vessels for which episodic waves are now suspected of being the chief cause. It seems that many of the casualties are tankers or bulk carriers of ore, grain, and the like. These vessels are susceptible not only because so many are at sea at any one time, but also because of their design. Bulk carriers comprise a bow section and an aft (or rear) section that includes the engine and the crew accommodations, these sections are separated by a series of flat-bottomed boxes, or storage tanks, which make up the majority of the vessel's length. Since these vessels are about 305 m (1000 ft) long, they are subject to great wrenching forces if a large portion of the hull is left unsupported or only partially supported as it is suspended on a wave crest. These large bulk carriers do well on flat water but pound frightfully in a heavy sea. More traditional and smaller hull forms are stronger and ride more easily, and are therefore less likely to be destroyed by these severe waves.

Wave Energy

Waves carry with them considerable energy per each unit width of their crest. Since the length of a wave is measured in meters, the unit width is one meter. Remember that the wave form represents a flow of energy. The energy is present as **potential energy,** due to the change in elevation of the water surface, and as **kinetic energy,** due to the motion of the water particles in their

Figure 8.8

Wave energy increases rapidly with increases in wave height. Average wave energy is calculated per unit width of crest and averaged over the wave length (L) *and the depth* (L/2)*.*

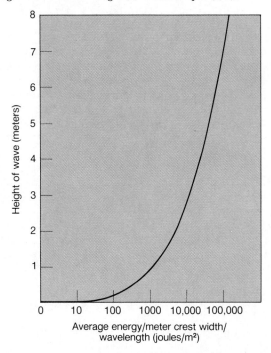

Figure 8.9

Wave steepness. When H/L *approaches 1:7, the wave's crest angle approaches 120° and the wave breaks.*

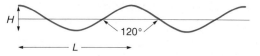

orbits. The higher the wave, the larger the diameter of the water particle orbit, and the greater the speed of the orbiting particle, therefore the greater the kinetic and potential energy. The energy in a deep water wave is nearly equally divided between kinetic and potential energy. Wave energy is calculated as the energy averaged over one wavelength per unit width of crest from sea surface to a depth of $L/2$ and is related to the square of the wave height. Figure 8.8 demonstrates this relationship.

Wave Steepness

There is a maximum possible height for any given wavelength (see table 8.1). This maximum value is determined by the ratio of the wave's height to the wave's length, and it is the measure of the **steepness** of the wave:

$$Steepness = \frac{height}{length}$$

or

$$S = \frac{H}{L}.$$

If the ratio of the height to the length exceeds 1:7, the wave is too steep. Under these conditions the crest angle

is sharper than 120° and the wave is unstable. It will collapse and break. For example, if the wavelength is 70 meters (230 ft), the wave will break when the wave height exceeds 10 meters (33 ft); see figure 8.9 for a pictorial representation of this relationship.

Small unstable, breaking waves are quite common. When wind speeds reach 16 to 18 knots, waves known as whitecaps can be observed. These waves have short wavelengths (about 1 m), and the wind increases their height rapidly. As each reaches the critical steepness and crest angle, it breaks and is replaced by another wave produced by the rising wind.

Long waves at sea usually have a height well below their maximum value. There is rarely sufficient wind energy to force them to their maximum height. If a long wave does attain maximum height and breaks in deep water, tons of water are sent crashing to the surface. The energy of the wave is lost in turbulence and in the production of smaller waves. Rather than breaking, under such conditions it is more likely that the top of a large wave will be torn off by the wind and will cascade down the wave face. This action does not completely destroy the wave, and thus is not considered to be the true collapse, or breaking, of such a wave.

In addition, waves break and dissipate if intersecting wave trains pass through each other in proper phase to form a combined wave with sufficient height to exceed the critical steepness. Waves sometimes run into a strong opposing current, which forces the waves to slow down. Remember that the speed of all waves equals the wavelength divided by the period ($C = L/T$). Once a wave is formed its period does not change; therefore its wavelength must shorten if its speed is reduced by an opposing current. In such a case, the wave's energy is confined to a shorter-length wave, so the wave increases in height to satisfy the direct height-energy relationship. If the increase in height exceeds the maximum allowable height for the shorter wavelength, the wave breaks. Crossing a sandbar into a harbor or river mouth during an outgoing or falling tide is dangerous because the waves moving against the tidal current steepen and break. Entering a harbor or river should be done at the change of the tide or on the rising tide, when the current moves with the waves, stretching their wavelengths, and decreasing their heights.

Universal Sea State Code

*I*n 1806, Admiral Sir Francis Beaufort of the British Navy adapted a wind-estimation system from land to sea use. On land, the clues to wind speed included smoke drift, the rustle of leaves, flags flapping, trees swaying, slates blowing from roofs, and the uprooting of trees. Admiral Beaufort used his knowledge of sea waves to relate observations of the sea-surface state to wind speed, and so designed a wind-speed scale known as the Beaufort Scale. The scale runs from 0 to 12, calm to hurricane, with typical wave descriptions for each level of wind speed. The Beaufort Scale was adopted by the U.S. Navy in 1838, and the scale was extended from 0 to 17. At present, a Universal Sea State Code of 0 to 9 based on the Beaufort scale is in international use for wind speeds and related sea-surface conditions, which the winds produce.

BOX TABLE 8.1

Sea state code	Description	Average wave heights
SS0	Sea like a mirror; wind less than one knot.	0
SS1	A smooth sea; ripples, no foam; very light winds, 1–3 knots, not felt on face.	0–0.3 m 0–1 ft
SS2	A slight sea; small wavelets; winds light to gentle, 4–6 knots, felt on face; light flags wave.	0.3–0.6 m 1–2 ft
SS3	A moderate sea; large wavelets, crests begin to break; winds gentle to moderate, 7–10 knots; light flags fully extend.	0.6–1.2 m 2–4 ft
SS4	A rough sea; moderate waves, many crests break, whitecaps, some wind-blown spray; winds moderate to strong breeze, 11–27 knots; wind whistles in the rigging.	1.2–2.4 m 4–8 ft
SS5	A very rough sea; waves heap up, forming foam streaks and spindrift; winds moderate to fresh gale, 28–40 knots; wind affects walking.	2.4–4.0 m 8–13 ft
SS6	A high sea; sea begins to roll, forming very definite foam streaks and considerable spray; winds a strong gale, 41–47 knots; loose gear and light canvas may be blown about or ripped.	4.0–6.1 m 13–20 ft
SS7	A very high sea; very high, steep waves with wind-driven overhanging crests; sea surface whitens due to dense coverage with foam; visibility reduced due to wind-blown spray; winds at whole gale force, 48–55 knots.	6.1–9.1 m 20–30 ft
SS8	Mountainous seas; very high-rolling breaking waves; sea surface foam-covered; very poor visibility; winds at storm level, 56–63 knots.	9.1–13.7 m 30–45 ft
SS9	Air filled with foam; sea surface white with spray; winds 64 knots and above.	13.7 m and above 45 ft and above

8.7 Shallow Water Waves

A deep water wave at last begins to approach the shore and shallow water. As it passes into water that is shallower than one-half its wavelength, the reduced depth begins to affect the shape of the orbits made by the water particles. The orbits become flattened circles, or ellipses (see fig. 8.10). The wave begins to "feel" the bottom, and the resulting friction and compression of the orbits reduces the forward speed of the wave.

Remember that (1) the speed of all waves is equal to the wavelength divided by period and (2) the period of a wave does not change. Therefore, if the wave slows as it "feels bottom," there must be an accompanying reduction in the wavelength, resulting in increased height and steepness.

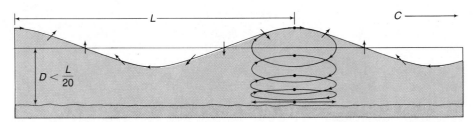

Figure 8.10
Shallow water wave particles move in elliptic orbits. The orbits flatten with depth.

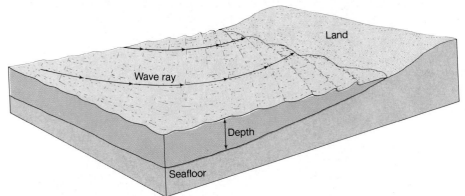

Figure 8.11
Waves moving inshore at an oblique angle to the depth contours are refracted. One end of the wave reaches a depth of L/2 or less and slows while the other end of the wave maintains its speed in deeper water. Wave rays drawn perpendicular to the crests show the direction of wave travel and the bending of the wave crests.

When the wave finally enters water with a depth of less than one-twentieth the wavelength, the wave becomes a **shallow water wave.** While the length and speed of a deep water wave are determined by the wave period, the shallow water wave's length and speed are controlled only by the water depth. Here the speed is determined by the square root of the product of the acceleration due to gravity (g) and depth (D):

$$C = \sqrt{gD} \ .$$

When the condition for a shallow water wave is met, the orbits of the water particles are elliptic; they become flatter with depth until at the sea floor only a back-and-forth oscillatory motion remains (see fig. 8.10). Note that the horizontal dimension of the orbit remains unchanged in shallow water. When the water depth is between $L/2$ and $L/20$, the speed of the wave is also slowed. Waves in this depth range are called intermediate waves. There is no simple algebraic equation to determine the speed of these waves.

Refraction

Waves are refracted, or bent, as they move from deep to shallow water and begin to feel the bottom. When waves from a distant storm center approach the shore, it is likely that they will not approach the beach straight on, but at an angle. One end of the wave crest comes into shallow water and begins to feel bottom, while the other end is in deeper water. The shallow water end moves more slowly than that portion of the wave in deeper water. The result is that the wave crests bend, or **refract,** and tend to become oriented parallel to the shore.

This pattern is shown in figure 8.11. The **wave rays** drawn perpendicular to the crests show the direction of motion of the wave crests. Note that the refraction of water waves is similar to the refraction of light and sound waves, described in chapter 4.

When waves approach an irregular coastline, which has both headlands jutting into the ocean and bays set back into the land, certain refraction patterns occur. Since the depth contours offshore tend to follow the shoreline pattern, a submerged ridge often continues seaward from the headland, and a depression is located in front of the bay. When shallow water waves approach this coastline, the portion of the wave crest over the ridge slows down in the shallower water, while the wave crest on either side continues to move at greater speed. The crests therefore wrap around the headland. This pattern is shown in figure 8.12. This refraction pattern results as the waves shorten their wavelengths and focus their energy on the point of the headland. Since wave energy is proportional to the wave height, the waves gain in height as their wavelength is shortened over the submerged ridge. More energy is thus expended on a unit length of shore at the point of the headland than on a unit length of shore elsewhere.

At the mouth of the bay, the opposite situation exists. The central area at the mouth of the bay is usually deeper than the areas to each side, and so the advancing waves slow down more on the sides than in the center. Because the wavelengths remain long in the center and shorten on each side, the wave crests bulge toward the center of the bay. Since the waves in the center of the bay do not shorten up as much, they have less height and also less energy to expend. This pattern is shown

Figure 8.12

Waves refracted over a shallow submerged ridge focus their energy on the headland. The converging wave rays show the wave energy being crowded into a smaller volume of water, increasing the energy per unit length of wave crest as the height of the wave increases.

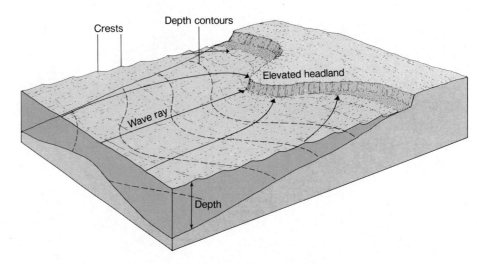

Figure 8.13

Waves refracted by the shallow depths on each side of the bay deliver lower levels of energy inside the bay. The diverging wave rays show the energy being spread over a larger volume of water, decreasing the energy per unit length of wave crest as the wave height decreases.

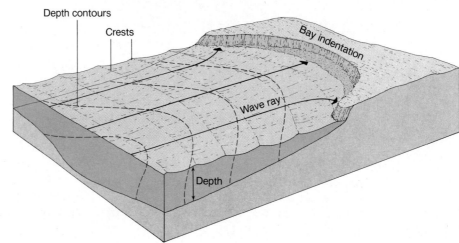

Figure 8.14

The wave from a ship's wake is reflected from a concrete wall. The principal wave is seen at the lower center of the photograph. The reflected wave is to the left. A checkerboard pattern is formed as the principal wave and reflected wave move through each other at approximately right angles.

in figure 8.13. Therefore, the bay is an environment of low wave energy and provides sheltered water. Overall, this unequal distribution of wave energy along the coast results in a wearing down of the headlands and a filling in of the bays, as the sand and mud settles out in the quieter water. If all coastal materials had the same resistance to wave erosion, this process would lead in time to a straightening of the coastline; but the rocky structure of many headlands resists wave erosion, and therefore the cliffs remain.

Reflection

A straight, smooth, vertical barrier in water deep enough to prevent waves from breaking will **reflect** the waves (fig. 8.14). The barrier may be a cliff, steep beach, breakwater, bulkhead, or other structure. Waves approaching such a barrier at an angle reflect from the barrier at the same angle. The reflected waves pass through the incoming waves to produce an interference pattern.

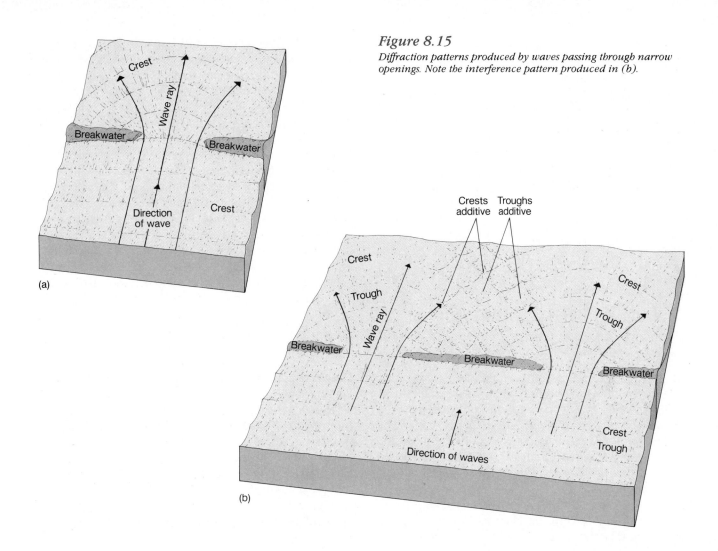

Figure 8.15
Diffraction patterns produced by waves passing through narrow openings. Note the interference pattern produced in (b).

Since the wave forms are additive, steep, choppy seas often result. If the waves reflect at a right angle or back on themselves, the resulting waves appear to stand still, to rise and fall in place.

Waves also reflect from curved vertical surfaces. If the curve of the surface spreads the reflected wave rays, the wave energy is dispersed, but if the surface curves so that the reflected wave rays converge, the energy is focused. This situation may be compared to the reflection of light from curved mirrors. Great care must be taken in the designing of wave protection barriers to be sure that the reflected waves do not focus their energy in another area, with resulting damage.

Diffraction

Another phenomenon that is associated with waves as they approach the shore or other obstacles is **diffraction.** Diffraction is caused by the spread of wave energy sideways to the direction of wave travel. If waves move toward a barrier with a small opening (such as two landmasses separated by a channel or an opening in a break-

water), some wave energy will pass through the small opening to the other side. Once through the opening, the wave crests radiate out and away from the gap. The wave heights decrease as the energy also radiates out in the shape of a semicircle. A portion of the wave energy is transported sideways from the original wave direction; it is diffracted. This effect is shown in figure 8.15a. If more than one gap is open to the waves, the patterns produced by the spreading waves from each opening may intersect and form interference patterns as the waves move through each other (see fig. 8.15b).

If the waves approach a barrier without an opening, diffraction can still occur. Energy will be transported at right angles to the wave crests as the waves pass the end of the barrier (see fig. 8.16). Note that the pattern produced is one-half of the radial pattern observed when the waves pass through a narrow opening, as in figure 8.15a. Energy is transported behind the sheltered (or lee) side of the barrier. This effect is an important consideration in planning the construction of breakwaters and other coastal barriers intended to protect vessels in harbors from wave action and possible damage.

Figure 8.16
Diffraction occurs behind the breakwater.

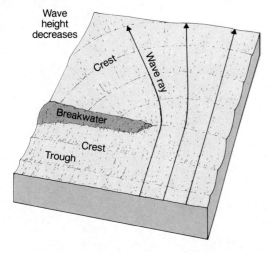

and swell. They learned by living with and observing nature in an area where nature cooperated by behaving in regular and predictable ways.

8.8 The Surf Zone

The surf zone is that area along the coast in which the waves slow, steepen, break, and disappear in the turbulence and spray of expended energy. The width of this zone is variable and is related to both the length of the arriving waves and the changing depth pattern. If shallow depths extend offshore for some distance, the surf zone is wider than it is over a sharply sloping shore. Longer waves, which feel the bottom before shorter waves, become unstable and break farther offshore, in deeper water. Thus longer waves also increase the width of the surf zone.

Navigation from Wave Direction

In those areas of the world where the winds blow steadily and from one direction, as in the trade-wind belts, the waves are very regular in their direction of motion. These regular wave patterns are altered when waves encounter an obstruction or a change in depth. Since waves change speed, shape, and height with water depth, and since waves change direction and pattern due to refraction, diffraction, and reflection from shoals, bars, islands, and coasts, it is possible to deduce the presence of these obstacles from the changes in wave patterns relative to the wind without seeing the obstacles.

Careful direct observation of wave patterns, or the sensing of wave patterns through the motions of a small craft, even at night, allows the detection of a change in the angle between waves and wind direction or a change in the wave pattern. These changing patterns, although subtle, can be detected many miles from land and can be used to bring a small vessel to shore when no landmarks are visible.

The Polynesians of the past lived with the sea; it was their home and they were acutely sensitive to its variations and changes. They made long voyages in their small canoes using their knowledge of wave patterns and how these patterns were associated with winds, shores, and islands. They had no theory to explain these patterns, but they knew what they meant. This information, plus their knowledge of star positions and their observations of cloud forms over land and sea and of bird flights, allowed them to sail many hundreds, even thousands, of miles across the open ocean to reach their destinations. No navigational tools were required other than charts constructed from twigs and shells, portraying island positions relative to stars, wind direction,

Breakers

Breakers are formed in the surf zone because the water particle motion at-depth is affected by the bottom, slowed down, and compressed vertically. However, the orbit speed of water particles near the crest of the wave is not slowed as much, and so these particles move faster toward the shore than the wave form itself. This relationship results in the curling of the crest and the eventual breaking of the wave. It is possible to distinguish different types of breakers; the two most common types are **plungers** and **spillers.** These two types are pictured in figure 8.17.

Plunging breakers form on steep beach slopes. The curling crest outruns the rest of the wave, curves over the air below it and breaks with a sudden loss of energy and a splash. The more common spilling breaker is found over flatter beaches, where the energy is extracted more gradually as the wave moves over the shallow bottom. This action results in the less dramatic wave form, consisting of turbulent water and bubbles flowing down the collapsing wave face. The spilling breakers last longer than the plungers, because they lose energy more gradually. Therefore, spillers give surfers a longer ride, while plungers give them a more exciting one.

A slow curling-over of the crest may be observed on some breakers. The curl begins at a point on the crest and then moves lengthwise along the wave crest as the wave approaches shore. This movement of the curl along the crest occurs because waves are seldom exactly parallel to a beach. The curl begins at the point on the crest that is in the shallowest water or the point at which the crest height is slightly greater, and it moves along the crest as it approaches the beach. The result is the "tube" so sought after by surfers (see fig. 8.18).

Figure 8.17
Breaking waves. A spiller (a) loses energy more slowly than a plunger (b). Note the curling crest of the plunger moving along the wave crest from left to right.

(a)

(b)

Figure 8.18
A surfer rides the tube of a large curling wave.

If the waves approaching the beach are uniform in length, period, and height, then they are the swells from some far-distant storm, which have had time and therefore distance in which to sort into uniform groups. For example, the long surfing waves of the California beaches in summer begin their lives in the winter storms of the South Pacific and Antarctic oceans. If the waves are of different heights, lengths, and periods, and break at varying distances from the beach, then unsorted waves have arrived and are probably the product of a nearby local storm superimposed on the swell.

Water Transport

The waves transport water toward the beach in the surf zone. There is a small net drift of water in the direction the waves are traveling (refer to the previous discussion on the water particle orbits of real waves). This drift is intensified in the surf zone, as the motion toward the beach by the water particles at the crest becomes greater than the return particle motion at the trough. Since the crests usually approach the beach at an angle, the surf zone transport of water flows both toward the beach and along the beach. The result is that water accumulates against the beach and then flows along the beach until it can flow seaward again and return to the area beyond the surf zone. This return flow generally occurs in quieter water with smaller wave heights, as in areas that have troughs or depressions in the sea floor.

Regions of seaward return flow may be narrow and some distance apart. Therefore, the flow in these areas must be swift in order to carry enough water beyond the surf zone to balance the slower but more extensive flow toward the beach. Such regions of rapid seaward flow are called **rip currents.** Swimmers who unknowingly venture into a rip current will find themselves carried seaward and unable to swim back to shore against the flow. The solution is to swim parallel to the beach, or across the rip current, and then return to shore. Because of the danger associated with rip currents, swimmers should be on the lookout for indicators of these currents' presence. These include (1) turbid water and floating debris moving seaward through the surf zone, (2) areas of reduced wave heights in the surf zone, and (3) depressions in the beach running perpendicular to the shore.

Wave action on the beach stirs up the sand particles and keeps them suspended in the water. The sand is carried along the beach parallel to the shore until the rip current is reached, where the sand is then transported seaward. Viewed from a height, such as a high cliff or a low-flying airplane, rip currents are seen as

Figure 8.19
Two large rip currents carry turbid water seaward through the surf zone north of Point Arguello, California.

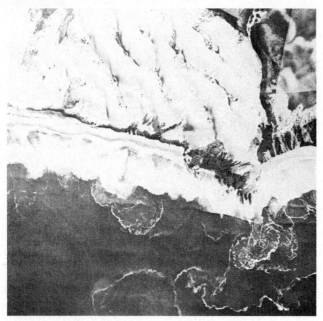

streaks of discolored turbid water extending seaward through the clearer water of the outer surf zone (see fig. 8.19).

Energy Release

Watching the heavy surf pounding a beach from a safe vantage point is an exciting and exhilarating experience. The trick is to determine at what point one is safe. In a narrow surf zone during a period of very large waves, the wave energy must be expended rapidly over a short distance. Under these conditions, the height of the waves and the forward motion of the water particles combine to send the water high up on the beach. The accompanying release of energy is explosive and can result in rocks and debris from the water's edge being hurled high up on the beach by the force of the water. Minot's Lighthouse on the south side of Massachusetts Bay is 30 m (98 ft) high, but it is regularly engulfed in spray. Lonely Tillamook Light off the Oregon coast had to have steel gratings installed to protect the glass that shields the light 40 m (131 ft) above sea level, after the glass had been broken several times by wave-thrown rocks. Waves also move much heavier objects. At Wick Bay in Scotland there is a breakwater built of heavy rocks. Fourteen-meter-high waves have displaced rocks weighing many tons along this structure.

Not all waves expend their energy on the shore. They may break farther seaward on sandbars such as those associated with many river mouths and estuaries. The famous bar at the mouth of the Columbia River is responsible for extremely hazardous conditions for both fishing and commercial vessels, and every winter it produces some casualties. The Coast Guard uses self-righting or rollover surf-rescue boats when it is necessary to cross the bar to render assistance during storm conditions. On an ebbing (or falling) tide, the waves approaching the bar against the current literally stand on end, creating breaking waves up to 20 m (66 ft) high.

The great driftwood logs found stranded on some beaches may be set afloat again at high tide during severe winter storms and can become lethal battering rams when hurled shoreward by the surf. Even with less severe waves, beach logs lying near the water's edge have a dangerous potential. The unsuspecting vacationer sits or plays on such a log, and when the occasional higher wave rolls it over, the person may become trapped under the log and crushed or drowned by the succeeding waves. Logs in or near the surf represent a great danger for the unwary. Where are you safe? You must be the judge.

8.9 Tsunamis

Sudden movements of the earth's crust produce seismic disturbances, or earthquakes. They may also produce **seismic sea waves,** or **tsunamis,** which are commonly but incorrectly called tidal waves. Since a seismic sea wave has nothing to do with tides, oceanographers have adopted the Japanese word tsunami to replace the misleading term tidal wave. It has since been pointed out that tsunami means tidal wave in Japanese, but the term is now accepted as a synonym for seismic sea wave.

If a large area, maybe several hundred square kilometers, of the earth's crust below the sea surface is suddenly raised or lowered, this displacement may cause a sudden rise or fall in the sea surface level above it. In the case of a rise, gravity causes the suddenly elevated water to return to the equilibrium surface level; if a depression is produced, gravity forces cause the surrounding water to flow into it. Both cases result in the production of waves with extremely long wavelengths (100 to 200 km) and long periods as well (10–20 minutes). Since the average depth of the oceans is about 4000 m (or 4 km), this depth is less than one-twentieth the wavelength of these waves. Tsunamis therefore behave like shallow water waves. Their speed is dependent on depth. These seismic waves move out from the point of the seismic disturbance at a speed determined by the ocean's depth ($C = \sqrt{gD}$) and move across the oceans at about 200 m/sec (400 mph). They may be refracted, diffracted, or reflected in midocean, since they are shallow water waves.

A tsunami is not likely to be detected in the open ocean after it leaves its point of origin. It may have a height of 1 to 2 m, but this height is distributed over its many-kilometer wavelength and is not easily seen or felt when superimposed on the other distortions of the sea's surface. Therefore, a vessel in the open ocean is in little or no danger if a tsunami passes. The danger occurs only if the vessel has the misfortune to be directly above the area of the original seismic disturbance.

The energy in a tsunami is distributed from the ocean surface to the ocean floor and over the length of the wave. This wave energy moves rapidly. When the path of the wave is blocked by a coast or island, the wave behaves like any other shallow water wave, and the energy is compressed into a smaller water volume as the depth rapidly decreases. This rapid and sudden increase in energy density causes the wave height to build rapidly. The loss of energy is also rapid, and a tremendous amount of moving water races up over the land. This destructive surge destroys buildings, docks, and trees. It can also deposit large vessels well up on the beach, leaving them stranded when the water recedes. The rising and falling of the water level in some bays, however, may not be as destructive as the extreme currents at the entrance to the bay. The large volume of water entering and leaving causes currents that damage and destroy structures in the way.

The leading edge of the tsunami is either a crest or a trough. If the initial crustal disturbance was an upward motion, a crest is formed first by the rise in sea level; if the crustal motion was downward, a trough is formed. If a trough arrives before the first crest, the result is a rapid drop in sea level, exposing sea plants and animals, all too attractive to the unwary. People have drowned after following the receding water to inspect the marine life, for they find, too late, that it is followed by a wall of advancing water that they cannot outrun.

Tsunamis are most likely to occur in ocean basins that are tectonically active. The Pacific Ocean, ringed by crustal faults and volcanic activity, is the birthplace of most tsunamis. They have also appeared in the Caribbean Sea, which is bounded by an active island arc system, and in the Mediterranean Sea as well. The spectacular havoc and destruction caused by these waves is well recorded. In August, 1883, the then East Indian, now Indonesian, island of Krakatoa erupted and was nearly destroyed in a gigantic steam explosion that hurled several cubic miles of material into the air. A series of tsunamis followed; these waves had unusually long periods of 1 to 2 hours. The town of Merak, 33 miles away and on another island, was inundated by waves over 30.5 m (or 100 feet) high, and a large ship was carried nearly 2 miles inland and left stranded 9.2 m (or 30 feet) above sea level. Over 35,000 people died as a result of these enormous waves. As the waves moved across the oceans, water level recorders were affected as far away as Cape Horn (12,500 km, or 7800 miles) and Panama (18,200 km, or 11,400 miles). The waves were traveling with a speed calculated at about 200 m/sec, or 400 miles per hour.

On April 1, 1946, a disturbance of the crust in the area of the Aleutian Trench off Alaska produced a series of tsunamis. These waves heavily damaged Hilo, Hawaii, killing more than 150 people. Not only the portion of the island facing the oncoming waves was damaged; the waves bent around the island by refraction (and by diffraction when passing between islands), producing high waves that struck the lee side of the island as well. The tsunami waves from the 1946 earthquake were apparently highest in the area of the Aleutian Islands, where they destroyed a concrete lighthouse 10 m (33 ft) above sea level at Scotch Cap and killed the crew. A radio mast in the same area mounted 33 m (108 ft) above sea level was also destroyed. In 1957, Hawaii was hit again, with waves higher than the 1946 series; but no lives were lost, due to early warnings and the evacuation of people along the shore. In 1964, the Alaska earthquake that severely damaged Anchorage and Alaskan coastal towns produced tsunamis that selectively hit areas on the west coast of Vancouver Island and the northern coast of California.

The frequency with which these waves occur in the Pacific has led to the development of a tsunami prediction and warning center located in Hawaii. Whenever an earthquake that might trigger such waves occurs, the time required for a wave to reach population centers is calculated. Areas close to the disturbance report visible wave crests, if any, and monitor tide and water level gauges to detect any rise and fall in sea level. If seismic waves have been observed, then populated areas are alerted to a possible tsunami and its predicted arrival time. This warning allows ships to move out of harbors to the open sea and permits inhabitants of low-lying coastal areas to evacuate and move inland.

8.10 Storm Surges

Periods of excessive high water due to changes in atmospheric pressure and the wind's action on the sea surface are known as **storm surges** or **storm tides.** These storm surges, combined with normal high tide conditions, can spell disaster for low-lying coastal areas.

Intense storms at sea, such as hurricanes and typhoons, are centered about intense low pressure systems in the atmosphere. Under the low pressure area in the center of the storm, the sea surface rises up into a dome, or hill, while the surface is depressed farther away from the center, where the atmospheric pressure is greater. This dome of water travels across the sea under the storm center and raises the water level at the shore when the storm approaches the coast. Such a buildup of water produces extensive flooding of shallow coastal areas. Also, during a storm the drag of the wind on the sea surface not only produces waves but also creates a current, which pushes surface water in the approximate direction of the wind. When this water reaches the shore it may, if the water is deep, sink and return seaward at some depth below the wind-driven surface layer, or it may, if the coastal area is shallow, pile up along the shore, resulting in abnormally high water levels and further flooding.

Storm surges are not typical waves; they are not oscillating motions of the sea surface. They do share with waves the characteristics of a curving sea surface and motion in a direction governed by the storm motion. They are considered here because they produce many of the same effects associated with tsunamis along low-lying shallow coasts and may be misinterpreted as tsunamis.

Along the shallow areas of the East and Gulf coasts of the United States, these elevations of water have caused considerable damage. In 1900, a storm surge caused the "great Galveston flood." It produced water depths of 4 m (13 ft) in downtown Galveston, destroying the city and killing 2000 people. Hurricane Camille, in 1969, was one of the strongest storms ever to hit the Gulf Coast. The high water levels caused severe property damage, and several hundred people were killed. Greater loss of life was averted by early warnings and mass evacuation of low-lying coastal areas.

Storms over the shallow Bay of Bengal spawned storm surges that took 100,000 lives in 1876 and an estimated 300,000 lives in 1970. Another 10,000 people were killed in 1985 in the same area. A storm over shallow regions of the North Sea in 1953 produced a 3.3 m (11 ft) rise in sea level, which combined with a high tide to break through the protective dike system and flood the Netherlands coast, killing 1800. To prevent a repetition of this event, the Dutch are walling off the inlets at the south end of the Rhine delta from the North Sea. They are constructing barrier dams with gates that can be closed when abnormally high tides threaten their coastal lowlands. The project is to be completed by the year 2000.

8.11 Internal Waves

Waves on the ocean surface occur at the boundary between the air and the water. Both air and water are classified as fluids, but vary greatly in their density. Therefore, it can be said that the surface waves form along a boundary between two fluids of different density. Beneath the surface of the ocean there are also fluid layers of varying density. The density differences are small, but waves form along the boundary between two fluids of different density here, too. Waves that occur along these density boundaries are called **internal waves.**

Because the density differences are small, the disturbance needed to begin an internal wave need not be large. Likewise, the restoring force is also small. The requirement of small generating forces plus the viscosity of the water results in large-amplitude, slow-moving waves with long wavelengths. It is believed that internal waves move as shallow water waves, with the thickness of the layers playing a role similar to the water depth in normal shallow water waves.

Internal waves may be caused when the surface layer flows over a more static lower layer in the same way the wind produces surface waves. When the sea surface slopes, as happens in the production of an elevated sea surface under wind stress or a low pressure storm center, the subsurface boundary may tilt also, but in the opposite direction. The tilting of the subsurface boundary occurs as the waters seek to reestablish a pressure balance at depth. When the distorting force is removed, the air-sea boundary returns to its level position and may oscillate several times. The same thing happens at the subsurface boundary, but the oscillation has a longer period. Other possible triggers of internal waves include seismic disturbances, long-term wind changes, and tidal forces.

When the surface layer is relatively shallow, a wave moving along an internal boundary may be high enough to have its crest break the surface. This action creates moving bands of smooth water called **slicks,** each representing a crest position. Slicks are most likely to occur in coastal water, estuaries, or fjords, where the surface layer diluted by freshwater runoff is shallow.

Since internal waves occur below the surface and are usually not visible to the eye, instruments are used to detect their presence. Because they cannot be observed easily and are not as common as surface waves, our knowledge concerning them is still limited.

8.12 Standing Waves

Deep water waves, shallow water waves, and internal waves are all progressive waves; they have a speed and a direction. There are waves, however, that do not progress; these are **standing waves,** which appear as an alternation between a trough and a crest at a fixed position. Consider a dishpan or an aquarium partially filled with water. Lift one end slowly and then rapidly but gently return it to a level position. The surface oscillates about

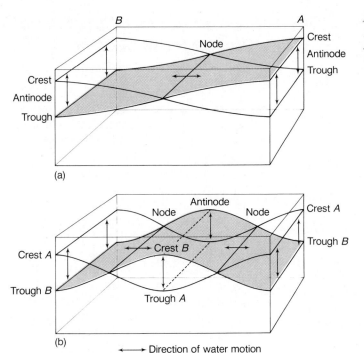

Figure 8.20

Standing waves oscillate about a single node in basin (a) and about two nodes in basin (b).

a point at the center of the container, the **node,** and there is an alternation of low and high water at each end, the **antinodes** (see fig. 8.20a). A standing wave is actually a progressive wave reflected back on itself, the reflection cancelling out the forward motions of the initial and reflected waves. If different-sized containers are treated the same way, we see that as the length of the container increases or as the depth of the water decreases, the period of oscillation increases.

Careful observation of figure 8.20a shows that the single-node standing wave in a basin contains one-half of a wave form. The crest is at end *A* of the container and the trough is at end *B*. Then a trough replaces the crest at *A,* and a crest replaces the trough at *B*. Note that one wavelength, the distance from crest to crest or from trough to trough, is twice the length of the container. By rapidly tilting the basin back and forth at the correct rate, it is possible to produce a wave with more than one node, as shown in figure 8.20b. In the case of two nodes, there will be a crest at either end of the container and a trough in the center. This arrangement alternates with a trough at each end and the crest in the center. The two nodes are one-quarter of the basin's length from each end. In this case, note that the wavelength is equal to the basin's length. The oscillation period of this wave with two nodes is one-half that of the wave with a single node.

Standing waves in bays or inlets with an open end behave somewhat differently than standing waves in closed basins. A node is usually located at the entrance

Figure 8.21

A standing wave oscillates about the node located at the opening to a basin. The antinodes produce the rise and fall of water at the closed end of the basin.

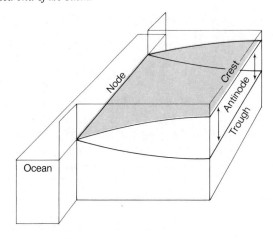

to the open-ended bay, so only one-quarter of the wavelength is inside the bay. There is little or no rise and fall of the water surface at the entrance, but a large rise and fall occurs at the closed end of the bay (see fig. 8.21). Multiple nodes may also be present in open-ended basins.

Standing waves that occur in natural basins are called **seiches,** and the oscillation of the surface is called seiching. In natural basins, the length dimension usually greatly exceeds the depth. Therefore, a standing wave of one node in a natural basin has a very great length in comparison to water depth. Because of this relationship it behaves as a shallow water wave, with the wavelength determined by the length of the basin. In water that is stratified with distinct layers that have sharp density boundaries, standing waves may occur along the fluid boundaries as well as at the air-sea boundary. The oscillation of the internal standing waves is slower than the oscillation of the sea surface.

Standing waves may be triggered by tectonic movements that suddenly tilt the basin, causing the water to oscillate at a period defined by the dimensions of the basin. If storm winds create a change in surface level to produce storm surges, the surface may oscillate as a standing wave in the act of returning to its normal level when the wind ceases. The pulsing of a weather disturbance over a lake may also cause periodic water-level changes, reaching a meter or more in height. If the period of the disturbing force is a multiple of the natural period of oscillation of the basin, the height of the standing wave is greatly increased. To give a parallel example, if a child is riding on a swing, a gentle push timed with each swing period will force the swing higher and higher. The push may be delivered each time the swing passes, every other time, or every third time; all are multiples of the natural period of the swing. In chapter 9, we will learn that repeating tidal forces can produce standing waves in those basins that have natural periods of oscillation approximating the tidal period.

A standing wave in a basin is like a water pendulum. The wave's natural period of oscillation is:

$$T = \left(\frac{1}{n}\right)\left(\frac{L}{\sqrt{gD}}\right),$$

where n is the number of nodes present; D is the depth of water in the basin; g is the earth's acceleration due to gravity, and L is the length of the wave. L equals twice the basin length, l, in a closed basin and four times the basin length in a basin with an open end. This equation is related to the shallow water wave equation when the number of nodes is equal to 1:

$$\frac{L}{T} = n\sqrt{gD}.$$

There is a relationship between progressive and standing waves. A progressive wave reflected back on itself produces a standing wave, because the two waves—original and reflected—are moving at the same speed but in opposite directions. The checkerboard interference pattern produced by two matched wave systems approaching each other at an angle also creates standing waves with crests and troughs alternating with each other in fixed positions (refer back to fig. 8.6a).

8.13 Practical Considerations: Energy from Waves

A tremendous amount of energy exists in ocean waves. The power present is estimated at 2.7×10^{12} watts, which is about equal to 3000 times the power-generating capacity of Hoover Dam. Unfortunately for human needs, this energy is widely dispersed and is not constant at any given location or time. It is, therefore, a difficult task to tap this supply in order to produce power, except in small quantities.

Wave energy can be harnessed in three basic ways: (1) using the changing level of the water to lift an object, which can then do useful work because of its potential energy; (2) using the orbital motion of the water particles or the changing tilt of the sea surface to rock an object to-and-fro; and (3) using the rising water to compress air in a chamber. In each case electric energy may be produced if the wave motion is used either directly or indirectly to mechanically turn a generator.

Consider a large buoyant surface float with a hollow cylinder extending down into the sea (see fig. 8.22). Inside the cylinder is a piston, which is connected to a large-diameter disk (a drag plate) at its bottom end. This disk will cause the piston to resist vertical motion, while the up-and-down motion of the surface float will cause the cylinder to move up and down over the piston. If the cylinder is equipped with inlet and outlet pipes with one-way valves, it will take in water as the surface buoy rises on the crest of the wave, causing the cylinder volume to increase, and will squirt water out as the surface buoy drops with the passing of the trough, causing the cylinder volume to decrease. This pumped water can be used to turn a turbine. Since wave energy is distributed over a volume of water, however, this mechanism does not withdraw much of the passing wave's energy.

Another system, designed by Sir Christopher Cockerell, uses pairs of barges placed stern to stern and hinged at the bottom of their transoms or sterns. Cylinders and pistons connect the two barges at the tops of the transoms (see fig. 8.23). As the wave form passes, the pitching motion of the barges with respect to each other causes the barges to pivot at the hinge, moving the pistons back and forth in their cylinders. These cylinders and pistons pump a fluid in a closed system, forcing it through a turbine, which turns an electric generator. Instead of pairs of barges, the installation could consist of a wedgelike barrier along the shore; the barrier, wide at the top and narrow at the bottom, would be held in a vertical frame so that its sloped surface is exposed to the waves. The wedge would be held so that it can move up and down in the frame. Waves

Figure 8.22
The vertical rise and fall of the waves can be used to power a pump.

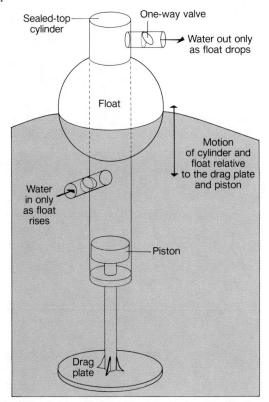

Figure 8.23
Surface-riding Cockerell rafts oscillate about a hinge that is driven by wave motion. This oscillation is used to pump a fluid through a turbine generator to produce electric energy; the fluid returns to the pump.

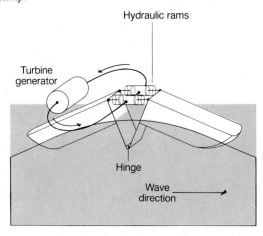

Figure 8.24
Salter's "nodding ducks" device uses the nodding motion of shaped units to turn a common shaft. The shaft's rotation drives a generator.

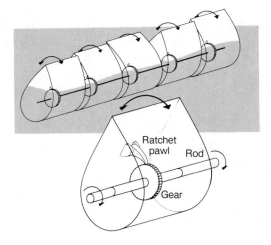

striking the sloped surface force the wedge up, and gravity returns it to a neutral position. This reciprocal motion can be transferred to a crank that turns a generator, or the motion can be used to pump water uphill on land, where it could then fall through a pipe, turning a generator with its momentum. A series of these wedges could be used to supply the water to large, hilltop storage tanks.

The unique "nodding ducks" device, designed by the British engineer S. H. Salter, consists of a series of shapes similar to short teardrop-shaped airfoil sections (see fig. 8.24). A rod runs through these sections; gears and ratchets are installed so that, as each section is moved by a passing wave, it twists the rod in one direction only, resulting in a rotating rod that could be used to turn a generator or to pump fluid through a turbine.

These are examples of systems that might be employed, but all require that many working units be linked together to produce power at useful levels. In addition, the units must be designed to operate within specific limits of wave heights and must be located in an area where waves of the required length or energy are frequent enough to permit the systems to work effectively and continuously.

Along a wave-exposed coast, air traps can be installed so that the crest of a wave moving into the trap

compresses a large volume of air, forcing it through a one-way valve. This compressed air powers a turbine to produce energy. The trough of the wave allows more air to enter the trap, readying it for compression by the following wave crest (see fig. 8.25).

Wave energy expended against the coasts could also be used to produce power. Shores that are continually pounded by large-amplitude waves are most likely to be developed. Great Britain has a coastline with frequent high-energy waves. Average wave power along the British coast is about 5.5×10^4 watts (or 55 kilowatts) per meter of coastline. If the wave energy could be completely harnessed along 1000 km (620 mi) of coast, it would generate enough power to supply 50% of Great Britain's current power needs.

Figure 8.25
Each rise and fall of the waves pumps pulses of compressed air into a storage tank. A smooth flow of compressed air from the storage tank turns a turbine that generates electricity.

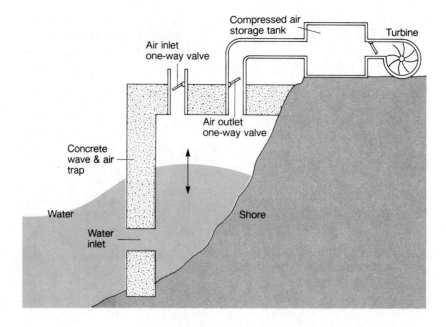

Whether such systems can help us in our current energy dilemma requires the thoughtful consideration of certain questions. If all the wave energy were extracted from the waves in a coastal area, what effect would this action have on the shore area? If the nearshore areas are covered with wave energy absorbers 5 to 10 m (33 ft) apart, what will the effect be on other ocean uses? Since the individual units collect energy at a slow rate, can they collect enough energy over their projected life span to exceed the energy used to fabricate and maintain them? Your answers to these questions will help you understand that the harvesting of wave energy is not without an effect on the environment, that it may not be either cost- or energy-effective, and that its location may present enormous problems for installation, maintenance, and transport of energy to sites of energy use.

Summary

When the water's surface is disturbed, a wave is formed by the interaction between generating and restoring forces. The wind produces capillary waves, which grow to form gravity waves. The elevated portion of a wave is the crest; the depressed portion is the trough. The wavelength is the distance between two successive crests or troughs. The wave height is the distance between the crest and the trough. Wave period measures the time required for two successive crests or troughs to pass a location. The moving wave form causes water particles to move in orbits. The wave's speed is related to wavelength and period.

Deep water waves occur in water deeper than one-half the wavelength. Wind waves generated in storm centers are deep water waves. The period of a wave is a function of its generating force and does not change. Long-period waves move out from the storm center, forming long, regular waves, or swell. The faster waves move through the slower waves and form groups, or trains, of waves. The longer waves are followed by the shorter waves. This process is known as sorting or dispersion. The speed of a group of waves is half the speed of the individual waves in deep water. Swells from different storms cross, cancel, and combine with each other as they move out across the ocean.

Wave height depends on wind speed, wind duration, and fetch. Single large waves unrelated to local conditions are called episodic waves. The energy of a wave is related to its height. When the ratio of the height to the length of a wave, or its steepness, exceeds 1:7, the wave breaks.

Shallow water waves occur when the depth is less than one-twentieth the wavelength. The speed of a shallow water wave depends on the depth of the water. As the wave moves toward shore and decreasing depth, it slows, shortens, and increases in height. Waves coming into shore are refracted, reflected, and diffracted. The patterns produced by these processes helped peoples in ancient times to navigate from island to island.

In the surf zone, breaking waves produce a water movement toward the shore. Breaking waves are classified as plungers or spillers. Water moves along the beach as well as toward it; it is returned seaward through the surf zone by rip currents.

Tsunamis are seismic sea waves. They behave as shallow water waves. A storm surge, or storm tide, is not a wave but a storm-produced elevated sea surface at the shore. Both tsunamis and storm surges produce severe coastal flooding and destruction.

Internal waves occur between water layers of different densities. Standing waves, or seiches, occur in basins as the sea surface oscillates about a node. Alternate troughs and crests occur at the antinodes.

It is possible to harness the energy of the waves by using either the water level changes or the changing surface angle associated with them. Difficulties include cost, location, environmental effects, and lack of wave regularity.

Key Terms

generating force	equilibrium surface	fetch	spiller
restoring force	wave period	episodic wave	rip current
surface tension	water particle orbit	potential energy	seismic sea wave
gravity wave	deep water wave	kinetic energy	tsunami
ripple	progressive wind wave	shallow water wave	storm surge/storm tide
capillary wave	storm center	wave steepness	internal wave
cat's-paw	sea	refraction	slick
crest	swell	wave ray	standing wave
trough	dispersion/sorting	reflection	node
wavelength	wave train	diffraction	antinode
wave height	group speed	breaker	seiche
amplitude		plunger	

Study Questions

1. A surfboard slides downward on the face of a wave. The steepness of the wave face is governed by the decrease of L (wavelength) and the increase of H (wave height) as the wave slows in shallow water. How must the surfer adjust the board in order to stay on the face of the wave as the wave approaches shallow water?

2. Locate a small pond or pool and drop a stone into it. Describe what happens (1) to an individual wave and (2) to the group of waves. Try to determine the group speed and the individual wave speed.

3. Drop two stones into the pond at a short distance from each other. Describe what happens when the wave rings produced by each stone pass through each other. Do the heights of the waves change when they intersect? Do the wave trains pass through each other and continue on?

4. List the forces that act on a smooth water surface to create a fully developed deep water wave.

5. If you were ocean-sailing at night in the trade wind belt, how could you use the waves to keep you on a course of constant direction?

Study Problems

1. Using the equations $C = L/T$ and $L = (g/2\pi)T^2$, show that wave speed can be determined from (a) wave period only and (b) wavelength only.

2. What is the period of a wave moving in deep water at 10 m/sec, if its wavelength is 64 m? When it enters shallow water what will happen to the wave's speed, length, period, and height? How high will the wave have to be to break in deep water?

3. A submarine earthquake produces a tsunami in the Gulf of Alaska. How long will it take the tsunami to reach Hawaii, if the average depth of the ocean over which the waves travel is 3.8 km and the distance is 4600 km?

4. Explain wave dispersion. How far from a storm center will waves with periods of twelve seconds, nine seconds, and six seconds have traveled after twelve hours? If the six-second waves arrive at your beach ten hours after the twelve-second waves, how far away is the storm?

5. Fill a rectangular aquarium or dishpan approximately one-third full of water. Measure the water depth (D). Carefully lift one end of the container and set it down rapidly and smoothly. Time the period between successive high waters at one end (T); this is the wave period. The wavelength (L) is twice the length of the container. Show that C (the wave speed) determined from L/T is equal to C determined from $\sqrt{gD}$.

The Tides

9

*S*pring tides are not the tides of spring as many landsmen suppose. They are the very high and very low tides which occur twice a month, with the new and the full moon, when solar and lunar magnetism pull together to make the circumterrestrial tide wave higher than at other times. The opposite to the spring tide is the neap tide, halfway between these phases of the moon; down East, in Maine, there may be as much as five feet difference in range between springs and neaps.

Spring tides are beloved by all who live by or from the sea. At a spring low, rocky ledges and sandbars which you never see ordinarily are bared; the kinds of seaweed that require air but twice a month appear; sand dollars like tarnished pieces of eight are visible on the bottom. Clam specialists can pick up the big ''hen'' clams or the quahaugs, and with a stiff wire hook deftly flip out of his long burrow the elusive razor clam. Shore birds— sandpiper, plover and curlew—skitter over the sea-vacated flats, piping softly and gorging themselves on the minor forms of life that cling to this seldom-bared shelf.

Samuel Eliot Morison,
from Spring Tides

*B*est known as the rise and fall of the sea around the edge of the land, the tides are caused by the gravitational attraction between the earth and the sun and between the earth and the moon. Far out at sea tidal changes go unnoticed, but along the shores and beaches the tides govern many of our water-related activities, both commercial and recreational. The amount that the tides rise and fall varies from place to place around the world. Early sailors from the Mediterranean Sea, where the daily tidal range is less than 1 m (3.3 ft), ventured out into the Atlantic and sailed northward to the British Isles. Here, to their amazement, they found a tidal range in excess of 10 m (33 ft). The movement of the tide in and out of bays and harbors has been helpful to sailors beaching their boats and to food-gatherers searching the shore for edible plants and animals, but it is also recognized as a hazard by navigators and can produce some spectacular effects when rushing through narrow channels.

This chapter surveys tide patterns around the world and explores the tides in two ways: one is a theoretic consideration of the tides on an earth with no land; the other is a study of the natural situation. The chapter also shows you how to use available tide data to predict water level changes and coastal tidal currents.

Figure 9.1

The three basic types of tides: (a) a once-daily diurnal tide, (b) a twice-daily semidiurnal tide, and (c) a mixed semidiurnal tide with diurnal inequality.

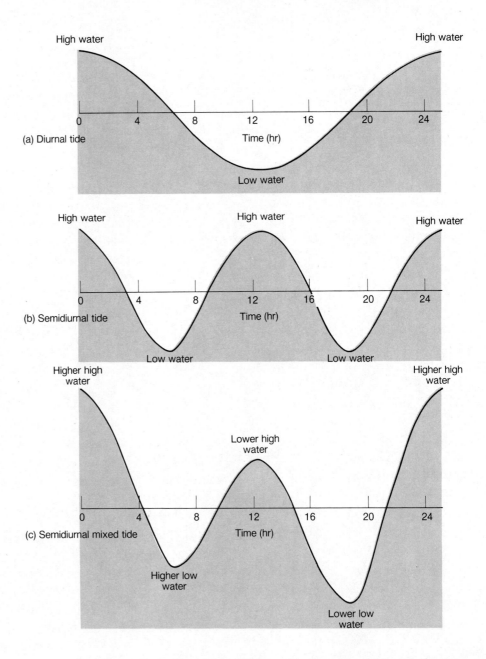

9.1 Tide Patterns

Observations of tidal movements around the world show us that the tides behave differently in different places. In some coastal areas there is a regular pattern of one high tide and one low tide each day; this is a **diurnal tide.** In other areas there is a high water-low water sequence that is repeated twice in one day; this is a **semidiurnal tide.** In a semidiurnal tidal pattern, the high tides reach the same height and the low tides drop to the same level. If there is a semidiurnal tide in which the high tides reach different heights and the low tides drop to different levels, the tide is called a **semidiurnal mixed tide.** This special type of semidiurnal tide has what is called diurnal (or daily) inequality. Diurnal in- equality is created by combining diurnal and semidiurnal tide patterns. The tide curves in figure 9.1 show each type of tide. Figure 9.2 shows curves for typical tides at some United States coastal cities.

9.2 Tide Levels

In a uniform diurnal or semidiurnal tidal system, the greatest height to which the tide rises on any day is known as **high water,** and the lowest point to which it drops is called **low water.** In a mixed tide system, it is necessary to refer to **higher high water** and **lower high water,** as well as **higher low water** and **lower low water** (see fig. 9.1).

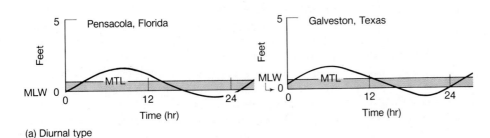

(a) Diurnal type

Figure 9.2

Tide types and tidal ranges vary from one coastal area to another. The zero tide level equals mean low water (MLW) or mean lower low water (MLLW), as appropriate. MTL equals mean tide level. All tide curves are for the same date.

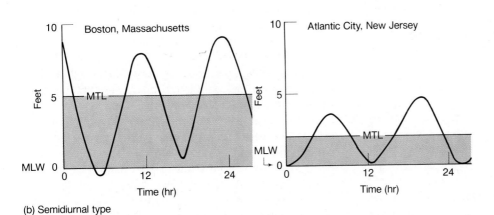

(b) Semidiurnal type

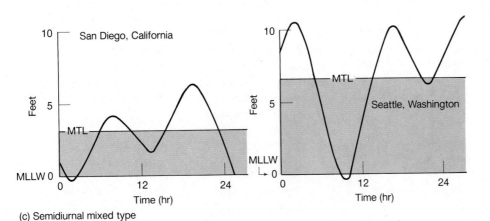

(c) Semidiurnal mixed type

Tidal observations made over long periods of time are used to calculate the **average** (or **mean**) **tide** levels. Averaging all water levels over many years gives the local mean tide level. Averages are also calculated for the high water and low water levels, as mean high water and mean low water. For mixed tides, mean higher high water, mean lower high water, mean higher low water, and mean lower low water are calculated.

Since the depth of coastal water is important to safe navigation, an average low water reference level is established; depths are measured from this level to be recorded on navigational charts. The tide level is added to this charted depth to find the true depth of water under a vessel at any particular time. In areas of uniform diurnal or semidiurnal tide patterns the zero depth reference, or **tidal datum,** is usually established at mean low water. The use of mean low water assures the sailor that the actual depth of the water is, in general, greater than that on the chart. In regions with mixed tides, mean lower low water is used as the tidal datum, for the same reason. Occasionally, the actual low tide level will fall below the mean value used as the tidal datum, producing a **minus tide.** A minus tide can be a hazard to the boater, but it is a cherished joy to the clam digger and the student of marine biology, as it exposes shoreline usually covered by the sea.

As the water level along the shore increases in height, the tide is said to be rising or flooding; a rising tide is a **flood tide.** When the water level drops, the tide is falling or ebbing; a falling tide is an **ebb tide.**

9.3 Tidal Currents

Currents are associated with the rising and falling of the tide in coastal waters. These **tidal currents** may be extremely swift and even dangerous as they move the water into a region on the flood tide and remove it on the ebb. When the tide turns or changes from an ebb to a flood or vice versa, there is a period of **slack water,** during which the tidal currents slow and then reverse. Slack water may be the only time that a vessel can safely navigate a narrow channel with swiftly moving tidal currents that are sometimes in excess of 10 knots (see forward to table 9.4). The relationship of tidal currents to standing wave tides, progressive tides, and tidal current prediction is discussed in later sections of this chapter.

9.4 Equilibrium Tidal Theory

Oceanographers analyze tides in two ways. The tides are studied as mathematically ideal wave forms behaving uniformly in response to the universal laws of physics. This method of study is called **equilibrium tidal theory.** It is based on an earth covered with a uniform layer of water, in order to simplify the study of relationships between the oceans and the tide-raising bodies, the moon and the sun. The tides are also investigated by studying them as they occur in nature. This method is called **dynamic tidal analysis,** and it studies the oceans' tides as they occur, modified by the landmasses, the geometry of the ocean basins, and the earth's rotation.

The effect of the sun's and moon's gravity on tides is most easily explained using the equilibrium theory. Refer to figure 9.3 as you read the following explanation. In this theory, a water-covered earth with its satellite moon orbits the sun. The moon is held in orbit about the earth by the earth's gravitational force (B). There is also a force acting to pull the moon away from the earth and send it spinning into space (B'); this is considered a **centrifugal force** in our discussion. The earth and moon together are held in their orbit by the gravitational attraction between the center of mass of the earth-moon system and the sun (A). A centrifugal force again acts to pull the center of mass of the earth-moon system away from the sun (A'). If the earth and its moon as a system are to remain in orbit about the sun, the gravitational force (A) must equal the centrifugal force (A'). This situation is similar to whirling a weight on a string around your head. The force pulling the weight outward, the centrifugal force, is equal and opposite to the force you exert pulling inward on the string. The result is a circular path (or orbit) of the weight about your head. In the case of the sun and the earth-moon system, the string is replaced by gravitational attraction.

Because the moon also exerts a gravitational force on the earth's center of mass (C), there must be a counterbalancing centrifugal force to keep the earth from moving toward the moon. This centrifugal force (C') is created by the earth's center of mass rotating about the center of mass of the earth-moon system. The concept of centrifugal force is introduced here to help understand how the earth, sun, and moon maintain their orbits about each other and resist the tendency to move together due to gravity.

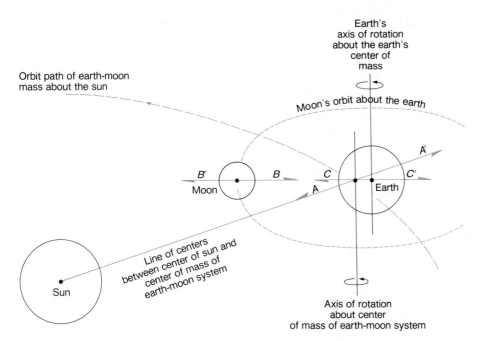

Figure 9.3
Centrifugal and gravitational forces act in pairs to keep the moon in orbit about the earth and the earth-moon system in orbit about the sun. A, B, and C are gravitational forces; A', B', and C' are centrifugal forces.

Sir Issac Newton's universal law of gravitation tells us that any two bodies attract each other. This force of attraction is proportional to the product of the two masses divided by the square of the distance between the centers of the masses:

$$F = G\left(\frac{m_1 m_2}{R^2}\right),$$

where

G = universal gravitational constant, 6.67×10^8 cm^3/g/sec^2;
m_1 = mass of body 1 in grams;
m_2 = mass of body 2 in grams; and
R = distance between centers of masses in centimeters.

If the distribution of gravitational and centrifugal forces is calculated for each gram of material at the surface of the earth and also at its center, the forces at the earth's surface are found to be different from those at its center due to the change in value of R. The earth-moon-sun system is in balance at the earth's center (gravitational forces balanced by centrifugal forces). However, from point to point on the earth's surface these forces are not in balance. At each point, the force difference per unit mass is found to be proportional to $G(M/R^3)$, where G is the gravitational constant, M is the mass of the sun or the moon, and R is the distance between the earth and the sun or between the earth and the moon.

In the earth-moon-sun system, the mass of the sun is very great, but the sun is very far away. By contrast, the moon is small but it is very close to the earth. The result is that when the distribution of these forces is calculated for each water particle at the earth's surface, the moon is found to have a greater attractive effect on the water particles than does the sun. Therefore, because it is more important, the moon's effect will be considered first.

The Moon Tide

The water particles on the side of the earth facing the moon are closest to the moon, and are acted on by a larger moon gravitational force than is present at the earth's center. Because the water covering is liquid and deformable, this force moves the water particles toward a point directly under the moon. This movement produces a bulge in the water covering. At the same time, the centrifugal force of the earth-moon system acting on the water particles at the earth's surface opposite the moon is larger than that present at the earth's center, creating a bulge. This centrifugal force is equal to the magnitude of gravity's tide-raising force and proportional to $-G(M/R^3)$. The minus sign indicates that this centrifugal force is acting opposite to the moon's tide-producing gravity force. If we place the moon opposite the earth's equator and then stop the earth model and its moon in space and time, we see the bulges in the

Figure 9.4

Distribution of tide-raising forces on the earth. Excess lunar gravitational and centrifugal forces distort the earth model's water envelope to produce bulges and depressions.

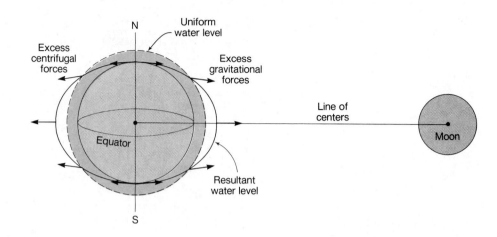

Figure 9.5

The change in water level at point A during one earth rotation through the distorted water envelope. (Refer to fig. 9.4.) Fractions indicate portions of a revolution.

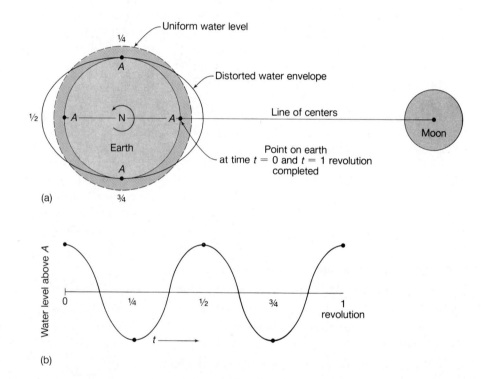

water covering, as shown in figure 9.4. Remember that initially our model earth had a water covering of uniform depth; therefore, as the two bulges are created, an area of low water level is formed between the bulges. We now have a water covering with two bulges and two intervening depressions (or two crests and two troughs, or two high-tide levels and two low-tide levels) distributed around the equator.

The earth makes one rotation in about 24 hours, and the bulges (or crests) in the water covering tend to stay under the tide-producing body as the earth turns. As a result of this movement, a point on earth that is initially at a crest (or high tide) passes to a trough (or low tide), to another high tide, to another low tide, and back to the original high tide as the earth completes

one revolution. You can follow this progess in figure 9.5. The effect created by the motion of the earth as it turns to the east within its water covering when observed from space can also be interpreted as the wave form of the water covering moving westward when observed relative to a fixed position on the earth.

The Tidal Day

While the earth turns upon its axis, the moon is also moving in the same direction along its orbit about the earth. Because the moon has moved along its orbit during a 24-hour period, a point on the earth model that is initially directly under the moon will not be directly under the moon again at the end of the 24 hours. The

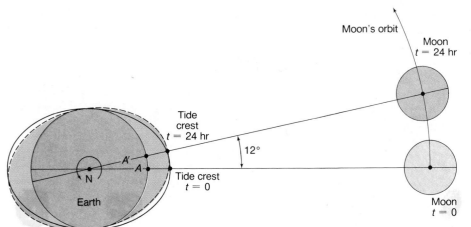

Figure 9.6
Point A *requires twenty-four hours to complete one earth rotation. During this time the moon moves 12° east along its orbit, carrying with it the tide crest. To move from* A *to* A′ *requires an additional fifty minutes to complete a tidal day.*

earth must turn for an additional 50 minutes, about 12°, to bring the earth point back in line with the moon. This relationship is shown in figure 9.6. For this reason, a **tidal day** is not 24 hours long, but 24 hours and 50 minutes. This difference also explains why high tides arrive at any location about one hour later each day.

The Tide Wave

The equilibrium tides produced in this example are semidiurnal, having two highs and two lows each day. The tidal distortion of the model's water covering produces a wave form known as the **tide wave.** The crest of the tide wave is the high water level, and the trough of the tide wave is the low water level. The wavelength in this example is half the circumference of the earth, and the tide wave's period is about 12 hours and 25 minutes.

The Sun Tide

Although the moon plays the greater role in the tide-producing process, the sun is also a participant and produces its own tide wave. Despite the sun's large mass, it is so far away from the earth that its tide-raising force is only 46% that of the moon. Also, the time required for the earth to make a revolution with respect to the sun is on the average 24 hours, and not 24 hours and 50 minutes, as in the case of the earth-moon relationship. For this reason, the tide wave produced by the moon is not only of greater magnitude than that produced by the sun, but it continually moves eastward relative to the tide wave produced by the sun. Because the tidal forces of the moon are greater than those of the sun, the tidal period of the moon is more important; therefore the tidal day is considered to be 24 hours and 50 minutes.

Spring Tides and Neap Tides

During the 29½ days it takes the moon to orbit the earth, the sun, the earth, and the moon move in and out of phase with each other. During the period of the new moon, the moon and sun are lined up on the same side of the earth, so that the high tides, or bulges, produced independently of each other coincide (see fig. 9.7). Since the water level is the result of adding the two wave forms together, tides of maximum height and greatest depression, or tides with the greatest **range** between high water and low water, are produced. These are **spring tides.** The vertical displacement of the tide may also be described as the amplitude of the tide. The amplitude is one-half the range, which is the distance above or below mean sea level.

In a week's time, the moon is in its first quarter, for it has moved along its orbit (about 12° per day) until it is located at approximately a right angle (or 90°) to the line of centers of the earth and sun. Therefore the crest, or bulge, of the moon tide is at right angles to the tide wave created by the sun. Under these conditions, the crests produced by the moon will coincide with the troughs produced by the sun, and the same will be true of the sun's crests and the moon's troughs (see fig. 9.7). The result is that the crests and troughs tend to cancel each other out, and the range between high water and low water is small. These are low-amplitude **neap tides.**

At the end of another week the moon is full, and the sun, moon, and earth are again lined up, producing crests that coincide and tides with the greatest range between high and low waters, or spring tides. These spring tides are also followed by another period of neap tides, produced by the moon in its last quarter when it again stands at right angles to the sun (see fig. 9.7). This four-week progression creates a tidal cycle of changing tidal amplitude, with spring tides occurring every two weeks, and a period of neap tides occurring in between.

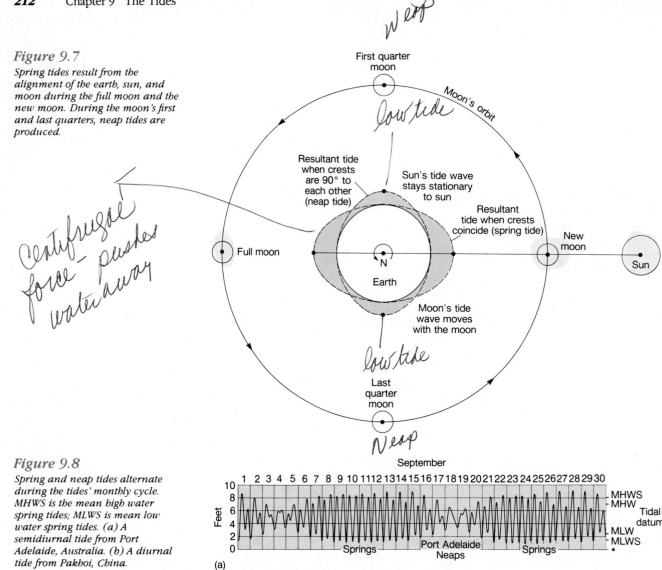

Figure 9.7

Spring tides result from the alignment of the earth, sun, and moon during the full moon and the new moon. During the moon's first and last quarters, neap tides are produced.

Figure 9.8

Spring and neap tides alternate during the tides' monthly cycle. MHWS is the mean high water spring tides; MLWS is mean low water spring tides. (a) A semidiurnal tide from Port Adelaide, Australia. (b) A diurnal tide from Pakhoi, China.

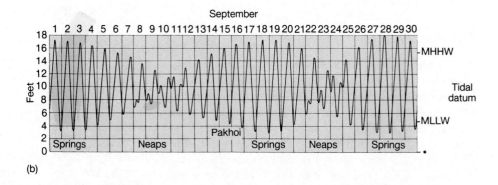

This progression can be seen in the portions of the tide records reproduced in figure 9.8. The effect occurs each lunar month, and is the result of the moon's tide wave moving around the earth relative to the sun's tide wave.

Declinational Tides

If the earth and the moon in the equilibrium tidal model are aligned so that the moon stands north or south of the earth's equator, one bulge (or high water) is in the northern hemisphere and the other is in the southern hemisphere (see fig. 9.9). Under these conditions, a point at the middle latitudes on the earth's surface passes through only one crest (or high) and one trough (or low) during a tidal day. This pattern produces a diurnal tide that is often called a **declinational tide,** because the moon is said to have declination when it stands above or below the equator.

Declinational (or diurnal) tides are influenced by both the moon and the sun. The sun stands above 23½°N at the summer solstice and above 23½°S at the winter solstice. This variation causes the bulge created

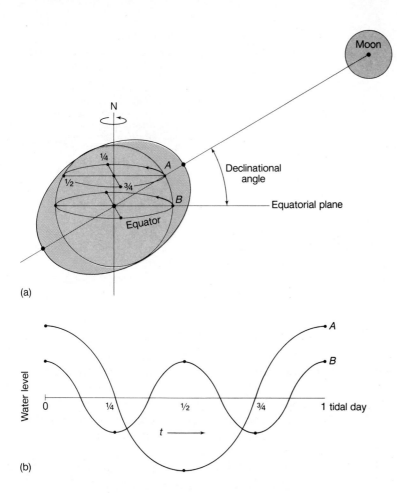

Figure 9.9
The declination of the moon produces a diurnal tide at latitude A and a semidiurnal tide at latitude B. Fractions indicate portions of the tidal day.

by the sun to oscillate north and south of the equator in a regular fashion each year and tends to create more diurnal sun tides during the winter and summer seasons of the year. The moon's declination varies between 28½°N and 28½°S with reference to the earth. Because the moon's orbit is inclined 5° to the earth-sun orbit, it takes 18.6 years for the moon to complete its cycle of maximum declination. Occasionally, the sun's and the moon's declinations coincide; when this happens, both tide waves become more diurnal.

Elliptical Orbits

The moon does not move about the earth in a perfectly circular orbit, nor does the earth circle the sun at a constant distance. These orbits are elliptical, and therefore there are times during an orbit when the earth is closer to a certain tide-raising body. During the northern hemisphere's winter, the earth is closest to the sun, and so the sun plays a greater role as a tide producer in winter than it does during summer.

9.5 Dynamic Tidal Analysis

Although the equilibrium tidal theory gives us an understanding of the distribution of wave-level changes and tide-raising forces in both space and time, it does not explain the tides as observed on the real earth. Return to figure 9.2 and notice the variety of tidal ranges and tidal periods that appear at different locations on the same date. Refer also to figure 9.8, showing different tides at different places occurring during the same time period. The sun-moon-earth relationship for all locations in both figures is the same. Therefore, the equilibrium tidal theory is not sufficient to explain actual or natural tides at any single location. Investigating the actual tides that occur in the discontinuous ocean basins requires the dynamic approach, a mathematical study of tide waves as they occur.

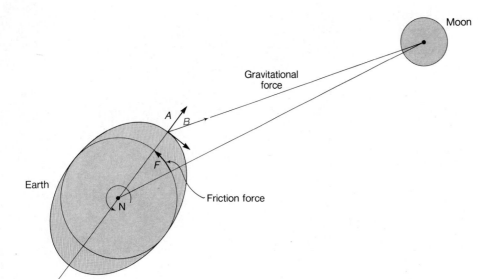

Figure 9.10
*The crest of the tide wave is
displaced eastward until the B
component of the gravitational force
balances the friction (F) between
the earth and the tide wave.
Component A of the gravitational
force is the tide-raising force.
Component B causes the tide wave
to move as a forced wave.*

The Tide Wave

Although the calculated tide-raising forces are the same, the behavior of the tide wave on the earth varies considerably from that on the smooth, water-covered model used to explain the equilibrium theory. Because the continents separate the oceans from each other, the tide wave is discontinuous. The wave starts at the shore, moves across the ocean, and stops at the next shore. Only in the Southern Ocean around Antarctica do the waves move continuously around the earth. The tide wave has a long wavelength as compared to the depth of the oceans; therefore it behaves like a shallow water wave, with its speed controlled by the depth of the water. Because the wave is contained within the ocean basins, it can oscillate in the basin as a standing wave. The tide wave is also reflected from the edge of the continents, refracted by the change in water depth, and diffracted as it passes through gaps between continents. And because the persistence of the tidal motion and the scale on which it occurs are so great, the Coriolis effect plays a role in the water's movement. Together these factors produce tides as they actually occur on the earth. Because the interactions between these factors are complex, it is not possible to understand how they combine to influence the tides unless each is first considered separately.

The tide wave's speed as a **free wave** moving across the water's surface is determined by the depth of the water. The tide wave as a free wave moves at about 200 m/sec (or 400 miles/hour). But at the equator, the earth moves eastward under the tide wave at 463 m/sec (1044 mph), which is more than twice the speed at which the tide can travel freely as a shallow water wave.

Therefore, the tide moves as a **forced wave** that is the result of the moon's attractive force and the earth's rotation. Since the earth turns eastward faster than the tide wave moves freely westward, the tide crest is displaced by friction to the east of its expected position under the moon. This eastward displacement continues until the friction force is balanced by a portion of the moon's attractive force. These two forces, when balanced, hold the real tide crest in a position to the east of the moon rather than directly under it. This process is illustrated in figure 9.10.

Above 60°N or 60°S latitudes, the distance around a latitude circle is less than one-half the distance around the equator. Here, the free propagation speed of the tide wave equals the speed at which the sea floor moves under the wave form. Under these conditions, the crest of the tide stays more easily aligned with the moon, and less friction is generated between the rotating earth and the moving wave form. The friction between the moving tide wave and the turning earth also acts to slow the rotation rate of the earth, adding about 1½ milliseconds per 100 years to the length of a day.

Progressive Tides

In a large ocean basin the tide wave, which moves across the sea surface in the same way that any shallow water wave moves across the water, is a **progressive tide.** Examples of progressive tides are found in the western North Pacific, the eastern South Pacific, and the South Atlantic oceans. **Cotidal lines** are drawn on charts to mark the location of the tide crest at set time intervals, generally one hour apart. The cotidal lines for the world ocean tides are shown in figure 9.11.

amphidromic points

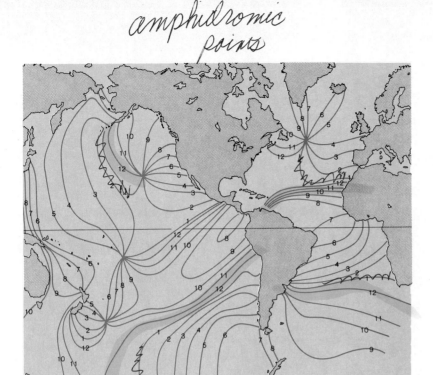

Figure 9.11
Cotidal lines for the world's oceans.
The high tide crest occurs at the
same time along each cotidal line.
Positions of the high tide are
indicated for each hour over a
twelve-hour period for semidiurnal
tides.

Because the tide wave is a shallow water wave, the water particles move in elliptic orbits; the movement extends to the sea floor, and the horizontal component of the motion greatly exceeds the vertical motion. Since the duration of time in which the water particles move in one direction is so long (one-half the tide period), the Coriolis effect becomes important. In the northern hemisphere the water particles are deflected to the right, and in the southern hemisphere they are deflected to the left. This deflection causes a clockwise rotation of the water in the northern hemisphere and a counterclockwise rotation in the southern hemisphere. This circular (or rotary) movement is the oceanic tidal current (refer forward to fig. 9.14).

Standing Wave Tides

In some ocean basins or parts of ocean basins, the tide wave is reflected from the edge of the continents, and a **standing wave tide** is produced. The formation of a standing wave by reflection from the sides of a basin is discussed in chapter 8. Remember that if a container of water is tipped so that the water level is high at one end and low at the other, the water flows to the low end, raising it as the water level at the high end drops. This movement produces a wave having a wavelength that is twice the length of the container, with antinodes at the ends of the basin and a node at the basin's center. This same process occurs in the ocean basins, but with some important modifications.

In the open sea, the time required for the water to flow from the high water side to the low water side of a basin is long, and the Coriolis effect is important. The moving water is deflected to the right in the northern hemisphere. This deflection does not allow the moving water to reach the original low tide end but instead deflects it to some position to the right of its original high tide position. Figure 9.12 illustrates this process. This movement causes the tide crest to be displaced counterclockwise around the basin. Since the process repeats itself continuously, the high tide crest rotates counterclockwise about the basin in which it oscillates, but the tidal current rotates clockwise, because the current is deflected to the right in the northern hemisphere (see fig. 9.12). These directions are reversed in the southern hemisphere.

In the **rotary standing tide wave,** the node becomes reduced to a central point, while the tide crests (shown as cotidal lines) progress around the edges of the basin. (See fig. 9.11 for a demonstration of this pattern in the northeast and southwest Pacific and the north Atlantic.) The central point, or node, for a rotary tide is called the **amphidromic point.** The distance between low and high water levels, or the tidal range, for a rotary tide is shown on a chart by a series of lines decreasing in value as they approach the amphidromic point. The lines of equal tidal range are called **corange lines** (see fig. 9.13). Near the amphidromic point the tidal range

Figure 9.12

Rotary standing tide waves. As water at the high water side of a
basin (shaded areas) begins to flow to the low water side, it is
deflected to its right in the northern hemisphere (a and c). This
deflection causes the tidal current to move in the direction shown
in (b). The tide crest (a, c, e, and g) continues to move
counterclockwise, while the tidal current (b, d, f, and h) moves
clockwise. The movements are summarized in (i) and (j).

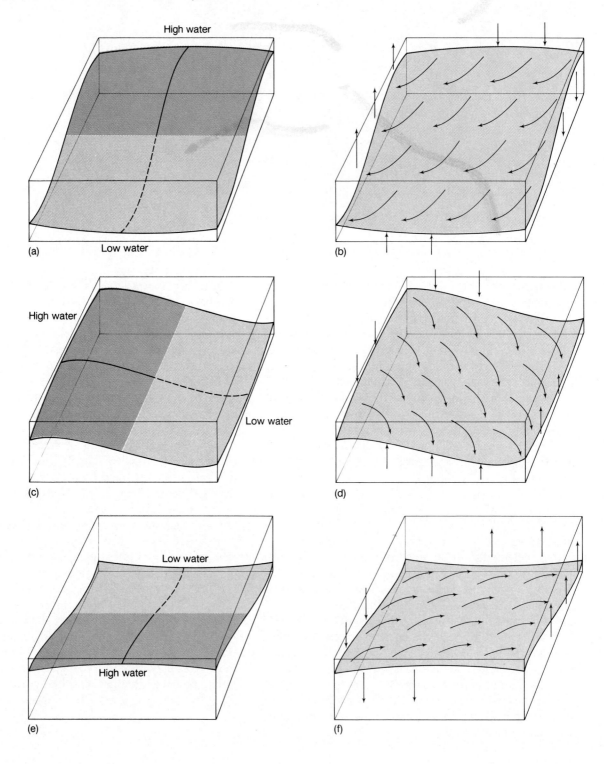

Figure 9.12 continued

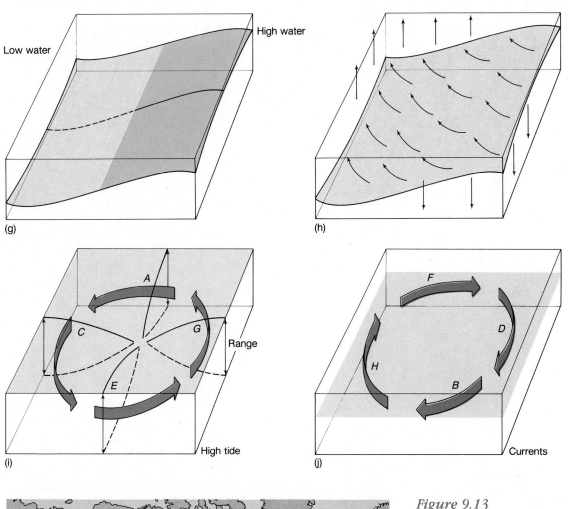

(g) Low water High water

(h)

(i) Range High tide

(j) Currents

Figure 9.13

Corange lines for the world's oceans. Corange lines connect positions with the same spring tidal range. Open-ocean tidal ranges are less well known than nearshore ranges, where tide level recorders are in common use.

Figure 9.14

Ocean tidal currents are rotary currents. The arrows trace the path followed by water particles in a tide wave during a mixed tidal cycle. Two unequal tidal cycles are shown. Numbers indicate hours before and after each tidal stage. The Coriolis effect deflects the horizontal component of the water particles' orbital motion, causing them to move in a circular path.

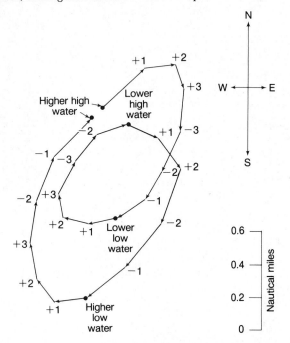

TABLE 9.1

Dimensions of Closed Ocean Basins with Natural Periods Equaling Tidal Periods

Tidal period	Depth (m)	Length or width (km)
Semidiurnal	4000	4428
12.42 hr	3000	3835
44,712 sec	2000	3131
	1000	2214
	500	1566
	100	700
	50	295
Diurnal	4000	8853
24.83 hr	3000	7667
89,388 sec	2000	6260
	1000	4427
	500	3130
	100	1399
	50	590

is small; the farther from the amphidromic point, the greater the range. Since the amphidromic point is located near the center of an ocean basin, many midocean areas have small tidal ranges, while the shores of the landmasses forming the sides of the basins have larger tidal ranges. The value and position of corange lines in midocean are not as well known as they are near shore.

The flow of water from the high water side to the low water side of a standing tide wave produces a rotating tidal current, as shown in figure 9.12. A rotating tidal current is also produced by the orbital motion of water particles in a progressive tide wave (review section 9.3 in this chapter on tidal currents). If the tide wave is diurnal, the water particles travel in one complete circle in a tidal day. A semidiurnal tide causes two circles, and a mixed tide produces two circles of unequal size. Figure 9.14 presents an example of a rotating mixed tidal current recorded at the Columbia River lightship in the North Pacific.

Rotary standing tides occur in basins in which the natural period of the basin approximates the tidal period. Table 9.1 relates different depths, and basin lengths or widths that produce equal tidal periods. Remember from

chapter 8 that the natural period of oscillation of a wave in a basin is $T = \left(\dfrac{1}{n}\right)\left(\dfrac{L}{\sqrt{gD}}\right)$, where L, the wavelength, is twice the basin length, l, in closed basins and four times l in open basins. A comparison of the values given in the table shows that deep-ocean basins must be large in order to support standing waves of tidal period, while shallow basins or seas found at the edge of the oceans may be much smaller in size but may still support a standing tide wave. Most tides are semidiurnal, but in some cases a basin's dimensions cause the basin to resonate with a diurnal tidal period rather than a semidiurnal period. The Gulf of Mexico contains examples of this situation. See the tide curves in figure 9.2 for Pensacola, Florida, and Galveston, Texas. In an open, tidal basin with a mixed tide, the semidiurnal portion of the tide may cause the basin to resonate with a node at the entrance to the basin and another node within the basin. The diurnal portion of the tide may have only a single node at the basin entrance. The result is a diurnal water-level pattern at the second node of the semidiurnal tide and a mixed pattern in all other parts of the basin. The harbor of Victoria, British Columbia, is so located that it alone registers a diurnal tidal pattern in an inlet system of mixed tides.

Ocean tides are the result of combining progressive and standing wave tides with diurnal and semidiurnal characteristics. Different kinds of tides interact with each other along their boundaries, and the results are exceedingly complex.

Figure 9.15
Low tide (a) and high tide (b) at Deer Island in the Bay of Fundy.
The tidal range at the head of the bay exceeds 10 meters.

(a)

(b)

Tide Waves in Narrow Basins

Coastal bays and channels are often long and narrow. Their basins have a length that is considerably greater than their width. In this way they are unlike ocean basins, in which width and length are comparable. These narrow basins have an open end toward the sea, so that the reflection of the tide wave occurs only at the head of the basin. In order to resonate with a tidal period, the open-ended basin's length dimension need only be one-half the length cited for the closed basins in table 9.1. In this type of open basin, the node is at the entrance and only one-fourth of the tide wave's length is present. The presence of only one-quarter of the wavelength, with the node at the entrance to the bay, places an antinode at the head of the bay. This situation is illustrated in chapter 8, figure 8.21. If the open basin is very narrow, oscillation occurs only along the length of the basin; there is no rotary motion because the basin is too narrow. For example, the Bay of Fundy in northeast Canada has a tidal range near the entrance node of about 2 m (6.6 ft), while the range at the head of the bay is 10.7 m (35 ft). This particular bay has a natural oscillation period that is so well matched to the tidal period that every tidal impulse at the Bay of Fundy's entrance creates a large oscillation at the head of the bay (see fig. 9.15). In another bay along another coast, the shape of

the basin may be such that it decreases rather than increases the tide's range. Every naturally occurring basin is different in this regard; each is unique.

9.6 Tidal Bores

In some areas of the world, large amplitude tides approach shallow coasts, bays, or river mouths. Under these conditions, a large amount of water moves in and out of the coastal area in a short period of time. The tide wave is forced to move toward the land at a speed greater than the shallow water wave speed determined by the depth of the water or the speed of the opposing river flow. When the forced tide wave meets those conditions it breaks, forming a spilling wave front that moves into the shallow water or up into the river. This wave front is seen as a wall of turbulent water called a **tidal bore.** The tidal bore produces an abrupt change in water levels as it passes. There may be a single bore formed, or a series of bores may be produced. The bores are usually less than a meter in height but can be as much as 8 m (26 ft) high, as in the case of spring tides on the Tsientang River of northern China. The Amazon, Trent, and Seine rivers all have bores. The fast-rising tide also sends bores across the sand flats of Mont Saint Michel in

Figure 9.16
The fast-rising tide in the Bay of Fundy produces a tidal bore that sweeps across the shallows.

France. The bore in the Bay of Fundy in Canada has been eliminated by the construction of a causeway (see fig. 9.16). Towns in areas having tidal bores often post warnings; their turbulence can be a severe hazard, in that they suddenly flood areas that have been open stretches of beach only minutes before.

9.7 Predicting Tides and Tidal Currents

Because of all the natural combinations of progressive and standing tides and reflection, refraction, and even diffraction of tides, it is not possible to predict the earth's tides from our knowledge of the tide-raising bodies alone; equilibrium tidal theory is not adequate for the task. Our daily tidal predictions are, however, quite accurate, for these tidal predictions are made by combining actual local measurements with known astronomical data.

Water level recorders are installed at coastal sites, and the rise and fall of the tides are measured over a period of years. At a primary tide station, these water level measurements are made for at least 19 years, in order to allow for the long, 18.6 year-period change in the declination of the moon. From these data, mean tide levels are calculated. Oceanographers then use a technique called **harmonic analysis** to separate the tide record into components with magnitudes and periods that match those of the forces produced by the tide-raising bodies. At the end of the procedure, they are left with a record of the effect of the local geography. In order to predict tides for the location, this **local effect** is combined with the astronomer's predicted data. Complex and cumbersome mechanical computers or tide machines were once used to predict the tides, but today modern computers quickly and easily recombine the data and predict the time, date, and elevation of each high water and low water level.

Tide Tables

The National Oceanic and Atmospheric Administration (NOAA) of the U.S. Department of Commerce has the responsibility for determining and publishing an annual predicted tide table in four volumes: for North and South America, Alaska, Hawaii, and the coast of Asia. These tables give the dates, times, and water levels for high and low water at primary tide stations (see table 9.2). There are 196 primary tide stations in these volumes, but many more localities require accurate tide predictions. The data for these other stations are determined by consulting a list of 6000 auxiliary stations and applying corrections for time and height to the appropriate primary station data (see table 9.3). The pocket tide tables provided by sporting goods stores and marine suppliers contain local information selected from the NOAA tables for the area's boaters, beachcombers, and fishing enthusiasts.

Tidal Current Tables

Tidal currents in the open ocean have been explained as rotary currents formed by the passing tide wave form and the deflection of water particles due to the Coriolis effect. Tidal currents in the deep sea are of scientific interest to oceanographers, who are concerned with removing this circular motion from their data to obtain the net flows of the major ocean currents. Tidal currents in harbors and coastal waters are of major interest to commercial vessels and pleasure boaters, since these currents can be very strong and must be taken into account by those wishing to navigate in such waters.

Like the tides, the tidal currents are first measured at selected primary locations in important inland waterways. These current data are studied to determine how the speed and direction of the tidal current are related to the predicted tide level changes. As before, the local effect is determined and is used to predict future tidal currents based on the tide tables.

TABLE 9.2
Times and Heights of High and Low Waters, Aberdeen, Washington, August, 1988

Day	Time	Height		Day	Time	Height		Day	Time	Height	
	(h m)[1]	*ft*	*m*		*(h m)[1]*	*ft*	*m*		*(h m)[1]*	*ft*	*m*
1	0312	9.8	3.0	11	0641	−0.8	−0.2	21	0125	1.1	0.3
M	0945	−1.2	−0.4	Th	1321	8.0	2.4	Su	0730	6.3	1.9
	1558	10.0	3.0		1845	2.8	0.9		1215	4.1	1.2
	2215	0.7	0.2						1905	9.1	2.8
2	0408	9.1	2.8	12	0032	9.6	2.9	22	0233	0.8	0.2
Tu	1027	−0.3	−0.1	F	0720	−0.8	−0.2	M	0845	6.3	1.9
	1643	10.2	3.1		1352	8.3	2.5		1415	4.5	1.4
	2307	0.4	0.1		1928	2.4	0.7		2014	9.3	2.8
3	0507	8.2	2.5	13	0111	9.4	2.9	23	0340	0.2	0.1
W	1112	0.8	0.2	Sa	0756	−0.7	−0.2	Tu	0959	6.7	2.0
	1727	10.2	3.1		1418	8.5	2.6		1539	4.3	1.3
					2007	2.1	0.6		2123	9.6	2.9
4	0005	0.3	0.1	14	0149	9.2	2.8	24	0441	−0.5	−0.2
Th	0611	7.4	2.3	Su	0832	−0.4	−0.1	W	1103	7.4	2.3
	1201	1.8	0.5		1443	8.7	2.7		1645	3.7	1.1
	1815	10.1	3.1		2049	1.8	0.5		2228	10.0	3.0
5	0105	0.3	0.1	15	0225	8.9	2.7	25	0535	−1.1	−0.3
F	0721	6.8	2.1	M	0904	0.1	0.0	Th	1155	8.1	2.5
	1254	2.7	0.8		1510	8.8	2.7		1745	2.8	0.9
	1908	9.9	3.0		2127	1.6	0.5		2326	10.4	3.2
6	0207	0.3	0.1	16	0300	8.5	2.6	26	0623	−1.6	−0.5
Sa	0836	6.5	2.0	Tu	0936	0.8	0.2	F	1239	8.9	2.7
	1355	3.4	1.0		1537	9.0	2.7		1837	1.8	0.5
	2005	9.7	3.0		2206	1.5	0.5				
7	0314	0.2	0.1	17	0339	8.1	2.5	27	0023	10.7	3.3
Su	0956	6.6	2.0	W	1008	1.5	0.5	Sa	0709	−1.7	−0.5
	1503	3.8	1.2		1603	9.1	2.8		1320	9.7	3.0
	2105	9.5	2.9		2245	1.4	0.4		1926	0.9	0.3
8	0415	−0.1	0.0	18	0422	7.6	2.3	28	0116	10.6	3.2
M	1105	6.9	2.1	Th	1034	2.2	0.7	Su	0752	−1.5	−0.5
	1607	3.8	1.2		1635	9.1	2.8		1359	10.2	3.1
	2204	9.5	2.9		2330	1.3	0.4		2015	0.1	0.0
9	0510	−0.4	−0.1	19	0515	7.0	2.1	29	0209	10.3	3.1
Tu	1203	7.3	2.2	F	1058	2.9	0.9	M	0834	−0.9	−0.3
	1705	3.5	1.1		1712	9.1	2.8		1438	10.6	3.2
	2258	9.5	2.9						2104	−0.4	−0.1
10	0559	−0.6	−0.2	20	0022	1.2	0.4	30	0301	9.7	3.0
W	1246	7.7	2.3	Sa	0617	6.6	2.0	Tu	0916	−0.1	0.0
	1759	3.2	1.0		1125	3.5	1.1		1520	10.8	3.3
	2347	9.6	2.9		1800	9.1	2.8		2152	−0.6	−0.2
								31	0357	9.0	2.7
								W	0958	0.9	0.3
									1559	10.7	3.3
									2241	−0.5	−0.2

Time meridian 120°W. 0000 is midnight. 1200 is noon.
Heights are referred to mean lower low water which is the chart datum of soundings.
[1]h.m. = hours, minutes.
Source: U.S. Dept. of Commerce, National Oceanic and Atmospheric Administration National Ocean Survey, 1988.

TABLE 9.3
Tidal Differences and Other Constants, Coast of Washington State, 1988

	Place	Position Lat. N		Long. W		Differences Time High water		Low water		Height High water	Low water	Ranges Mean	Diurnal	Mean tide level
		°	′	°	′	h.	m.	h.	m.	ft	ft	ft	ft	ft
	Washington Coast					(Based on Aberdeen tides)[1]								
919	Long Beach	46	21	124	03	−1	05	−0	59	*0.80	*0.80	6.2	8.1	4.4
	Willapa Bay and River													
921	Willapa Bay entrance	46	43	124	04	−0	37	−0	43	*0.80	*0.80	6.2	8.1	4.4
923	Nahcotta, Willapa Bay	46	30	124	01	+0	20	+0	16	0.0	−0.1	8.0	10.2	5.4
925	Tarlatt Slough, Willapa Bay	46	22	124	00	+0	21	+0	54	*0.91	*0.47	7.9	9.4	4.6
927	Bay Center, Palix River	46	38	123	57	−0	09	+0	04	−1.3	−0.2	6.8	8.9	4.7
929	Toke Point, Willapa Bay	46	42	123	58	−0	36	−0	19	−1.6	−0.2	6.5	8.5	4.5
931	South Bend, Willapa River	46	40	123	48	−0	04	−0	08	−0.3	−0.2	7.8	9.8	5.2
933	Raymond, Willapa River	46	41	123	45	+0	02	−0	03	−0.2	−0.1	7.8	9.9	5.3
935	Grayland	46	49	124	06	−1	05	−0	59	*0.80	*0.80	6.2	8.1	4.4
937	Westport (ocean)	46	53	124	07	−1	05	−0	56	*0.84	*0.84	6.4	8.5	4.6
	Grays Harbor													
939	Point Chehalis	46	55	124	07	−0	32	−0	43	−1.1	−0.1	6.9	9.0	4.8
941	Bay City	46	52	124	04	−0	15	−0	32	−0.9	−0.1	7.1	9.2	4.9
943	Markham	46	54	124	00	−0	19	−0	12	−0.9	−0.2	7.2	9.2	4.9
945	North Channel	46	58	123	57	−0	06	−0	01	−0.4	−0.1	7.6	9.7	5.2
947	ABERDEEN	46	58	123	51		Daily predictions						7.9	10.1	5.4
949	Montesano, Chehalis River	46	58	123	36	+1	21	+1	48	*0.80	*0.53	6.7	8.1	4.1
951	Pacific Beach	47	13	124	12	−1	02	−0	59	*0.85	*0.85	6.5	8.6	4.6
953	Point Grenville	47	18	124	16	−1	02	−0	59	*0.85	*0.85	6.5	8.6	4.6
955	Destruction Island	47	40	124	29	−1	01	−1	03	*0.87	*0.87	6.6	8.7	4.7
957	La Push, Quillayute River	47	55	124	38	−1	00	−0	47	*0.84	*0.84	6.5	8.5	4.6
959	Cape Alava (Flattery Rocks)	48	10	124	44	−0	53	−0	39	*0.81	*0.81	6.0	8.2	4.4

[1] Based on predicted tides for Aberdeen, Washington.
*Predicted Aberdeen tides are to be multiplied by this ratio to obtain local tide level.
Source: U.S Dept. of Commerce, National Oceanic and Atmospheric Administration, National Ocean Survey, 1988.

NOAA publishes tidal current data in a format similar to the tide tables (see table 9.4). The times of slack water, maximum flood currents, and maximum ebb currents, the speed of the currents in knots, and the direction of flow for ebb and flood currents are given. As in the case of the tide tables, auxiliary tidal current stations are listed and are keyed to the primary stations with correction factors to determine current speed, time, and direction at the secondary stations (see table 9.5). This information allows the master of a vessel to decide at what time to arrive at a particular channel in order to find the current flowing in the right direction or how long to wait for slack water before choosing to proceed through a particularly swift and turbulent passage.

TABLE 9.4

Tidal Currents for August 1988, Seymour Narrows, B.C., Canada.

Day	Slack water time	Maximum current Time	Vel.
	h.m.	*h.m.*	*knots*
1 M	0125	0410	10.2F
	0715	1025	11.5E
	1320	1630	12.4F
	1945	2255	12.6E
2 Tu	0205	0500	10.5F
	0810	1115	10.5E
	1410	1715	10.8F
	2020	2340	12.0E
3 W	0250	0555	10.4F
	0910	1210	9.1E
	1505	1805	8.8F
	2100		
4 Th		0025	11.1E
	0340	0650	10.1F
	1015	1310	7.7E
	1610	1900	6.8F
	2145		
5 F		0120	10.1E
	0440	0755	9.8F
	1130	1420	6.7E
	1725	2005	5.2F
	2240		
6 Sa		0220	9.2E
	0540	0905	9.8F
	1245	1545	6.5E
	1850	2120	4.2F
	2345		
7 Su		0330	8.6E
	0645	1010	10.2F
	1355	1700	7.1E
	2010	2230	4.1F
8 M	0055	0440	8.7E
	0745	1115	10.9F
	1455	1800	8.2E
	2115	2335	4.7F
9 Tu	0205	0545	9.1E
	0840	1210	11.6F
	1545	1850	9.4E
	2205		
10 W		0030	5.6F
	0305	0635	9.7E
	0930	1255	12.1F
	1630	1935	10.3E
	2245		

Day	Slack water time	Maximum current Time	Vel.
	h.m.	*h.m.*	*knots*
11 Th		0120	6.5F
	0400	0725	10.2E
	1015	1335	12.3F
	1705	2015	10.9E
	2325		
12 F		0200	7.3F
	0450	0805	10.4E
	1100	1415	12.2F
	1740	2050	11.2E
	2355		
13 Sa		0235	7.9F
	0530	0845	10.3E
	1135	1450	11.7F
	1810	2120	11.2E
14 Su	0030	0310	8.2E
	0610	0920	10.0E
	1215	1520	11.0F
	1840	2155	10.9E
15 M	0100	0345	8.4F
	0650	0955	9.3E
	1250	1555	10.0F
	1910	2225	10.4E
16 Tu	0130	0420	8.4F
	0730	1035	8.5E
	1325	1625	8.8F
	1935	2255	9.8E
17 W	0205	0500	8.2F
	0810	1115	7.5E
	1405	1700	7.5F
	2005	2325	9.0E
18 Th	0240	0540	7.9F
	0900	1200	6.3E
	1455	1745	6.0F
	2035		
19 F		0005	8.3E
	0320	0630	7.6F
	1000	1250	5.3E
	1550	1830	4.5F
	2110		
20 Sa		0050	7.5E
	0410	0730	7.5F
	1110	1405	4.6E
	1710	1935	3.2F
	2155		

Day	Slack water time	Maximum current Time	Vel.
	h.m.	*h.m.*	*knots*
21 Su		0150	6.9E
	0510	0840	7.8F
	1230	1530	4.7E
	1840	2050	2.7F
	2300		
22 M		0300	6.8E
	0620	0950	8.7F
	1340	1645	5.8E
	2000	2210	3.1F
23 Tu	0020	0415	7.5E
	0725	1050	10.0F
	1435	1745	7.4E
	2055	2315	4.3F
24 W	0135	0520	8.7E
	0820	1145	11.4F
	1525	1830	9.1E
	2140		
25 Th		0005	6.0F
	0245	0615	10.2E
	0915	1235	12.7F
	1605	1915	10.8E
	2220		
26 F		0055	7.9F
	0340	0705	11.5E
	1005	1320	13.6F
	1645	1950	12.1E
	2255		
27 Sa		0140	9.6F
	0435	0755	12.5E
	1050	1400	13.9F
	1720	2030	13.0E
	2335		
28 Su		0220	11.1F
	0520	0840	13.0E
	1135	1440	13.6F
	1755	2105	13.5E
29 M	0010	0305	12.1F
	0610	0920	12.9E
	1220	1520	12.8F
	1830	2145	13.5E
30 Tu	0050	0350	12.5F
	0700	1005	12.2E
	1305	1605	11.4F
	1905	2225	13.0E
31 W	0130	0435	12.4F
	0755	1055	10.9E
	1355	1645	9.6F
	1945	2305	12.0E

Time meridian 120°W. 0000 is midnight. 1200 is noon.

F: Flood, Dir. 180° True E: Ebb, Dir. 000° True.

Source: U.S. Department of Commerce, National Oceanic and Atmospheric Administration, National Ocean Survey, 1988.

TABLE 9.5

Current Differences and Other Constants for Georgia Strait Region Based on Seymour Narrows, B.C., Canada.

Tide current station no.	Place	Position Lat. N		Long. W	
		°	′	°	′
	Passages north of Vancouver Island				
	Time meridian, 120°W				
1980	Surge Narrows, Okisollo Channel	50	14	125	10
1985	Hole-in-the-Wall, Okisollo Channel	50	18	125	14
1990	Rapids, near Pulton Bay, Okisollo Channel	50	19	125	16
1995	Arran Rapids, north of Stuart Island	50	25	125	09
2000	Yuculta Rapids, southwest of Stuart Island	50	21	125	09
2010	Godwin Point, Cordero Island	50	28	125	25
2015	Shell Point, Blind Channel	50	26	125	31
2020	Green Point Rapids, Cordero Channel	50	25	125	32
2025	Whirlpool Rapids, Wellbore Channel	50	27	125	47
2030	Shaw Point, Sunderland Channel	50	28	125	56
2035	Root Point, Chatham Channel	50	35	126	12
2040	Littleton Point, Chatham Channel	50	37	126	17
2045	Ripple Bluff, Knight Inlet	50	38	126	31
2050	Owl Island, main entrance to Knight Inlet	50	38	126	41

[1]Current values at Seymour Narrows are multiplied by these ratios to obtain currents at the above secondary tidal current stations.
Source: U.S. Department of Commerce, National Oceanic and Atmospheric Administration, National Ocean Survey, 1988.

9.8 Practical Considerations: Energy from Tides

The possibility of obtaining energy from the tides exists in coastal areas with large tidal ranges or in narrow channels with swift tidal currents. There are two systems for extracting energy from the rise and fall of the tides. Both require building a dam across a bay or estuary so that seawater can be held in the bay at high tide. When the tide ebbs, a difference in water level height is thus produced between the water behind the dam and the ebbing tide. When the elevation difference becomes sufficient, the seawater behind the dam is released through turbines to produce electric power. The reservoir behind the dam is refilled on the next rising tide by opening gates in the dam. This single-action system produces power only during a portion of each ebb tide (see fig. 9.17a). A tidal range of about 7 m (23 ft) is required to produce power using this system.

This same arrangement of dam and turbines can be installed as a double-action system. In this system, power is produced on the ebbing tide. At the end of this power cycle, when the level behind the dam is not suf-

ficient to produce power, the gates are opened and the last of the water in the reservoir is spilled out to sea. The gates are then closed and some of the remaining water in the bay is pumped seaward to further decrease the water level. Although the pumps consume power, this expenditure of energy is worthwhile if the pumping period occurs when power demands are low and if the additional increase in height between water levels allows more power to be produced on the next cycle.

The gates are kept closed and the tide rises on the seaward side of the dam while the reservoir level remains low. When the difference in height is great enough, the incoming water is released through the turbines into the reservoir to produce power. At the end of the rising tide the dam is sealed, and water is pumped into the reservoir to raise the reservoir to its maximum level. A waiting period follows. When the tide drops on the seaward side, the cycle is repeated. This system requires very specialized turbine systems because, unlike a river, the water moves through the turbines in two directions. Therefore, the turbines must be able to generate power on both the flood and ebb tide. Figure 9.17b presents a sketch of the power cycle of such a double-action system.

Slack water before flood		Time differences Flood		Slack water before Ebb		Ebb		Speed ratios[1] Flood	Ebb	Average speeds and direction Maximum flood		Maximum ebb	
h.	*m.*	*h.*	*m.*	*h.*	*m.*	*h.*	*m.*			*knots*	*deg.*	*knots*	*deg.*
Based on Seymour Narrows													
−0	45	−0	45	−0	45	−0	45	0.7	0.7	7.0	140	7.0	320
−0	55	−0	55	−0	55	−0	55	0.8	0.8	7.5	060	7.5	240
−0	50	−0	55	−0	55	−0	55	0.7	0.7	6.5	072	6.5	252
−0	45	−0	45	−0	45	−0	45	0.7	0.7	7.0	065	7.0	245
−0	40	−0	40	−0	40	−0	40	0.5	0.5	5.0	145	5.0	325
−0	55	−0	55	−0	55	−0	55	0.2	0.2	2.2	050	2.2	230
−1	10	−1	10	−1	10	−1	10	0.5	0.5	5.0	170	5.0	350
−1	30	−1	30	−1	30	−1	30	0.5	0.5	5.0	145	5.0	325
−1	50	−1	50	−1	50	−1	50	0.6	0.6	6.0	185	6.0	005
−1	05	−1	05	−1	05	−1	50	0.2	0.2	1.5	060	1.5	240
−1	05	−1	05	−1	05	−1	05	0.6	0.6	5.5	110	5.5	290
−1	05	−1	05	−1	05	−1	05	0.4	0.4	3.5	130	3.5	310
−1	15	−1	15	−1	15	−1	15	0.3	0.3	2.5	105	2.5	285
−1	20	−1	20	−1	20	−1	20	0.3	0.3	2.5	120	2.5	300

This system appears to be a simple and cost-effective method for producing electric power. However, there are few places in the world where the tidal range is sufficient and where natural bays or estuaries can be dammed at their entrances at reasonable cost and effort. Moreover, the appropriate tides and bays are not necessarily located near population centers that need the power. The costs of the installations and the power distribution, in addition to periodic low power production because of the changing tidal amplitude over the tide's monthly cycle, make this type of power expensive in comparison to other sources.

In Europe, two tidal power stations use this method for the production of electric power: a commercial installation on the La Rance River Estuary in France and a small plant in Kislaya Bay in the Soviet Union. The La Rance installation produces 5.4×10^{10} watt-hours per year. Present global energy demands could be satisfied by 250,000 plants of this capacity, but there are only about 255 sites that have been identified around the world with the potential for tidal energy development.

Tidal power has been under consideration for the Bay of Fundy since the 1930s. Canadian and American interest was casual due to the expensive nature of the project, until the rising cost of fossil fuels focused interest on alternative energy sources. Nova Scotia is currently actively pursuing the development of sites in the Minas Basin and Cumberland Basin at the head of the Bay of Fundy. The Minas Basin site would have a capacity of 5300 megawatts and the Cumberland Basin site would produce 1400 megawatts. Ultimately, the development of one or both of these sites will depend upon obtaining financing for the multibillion dollar projects; securing markets in the northeastern United States for excess power; the re-timing of intermittent tidal energy with other, more conventional methods of producing electrical power; and the mitigation of potential environmental effects of the project. On the United States side of the bay, Cobscook Bay and Half-Moon Cove in Maine are being considered as potential power-generating sites. The U.S. Army Corps of Engineers is interested in the Cobscook Bay site, and the Passamaquoddy Indian Tribe is proposing a 12-megawatt development at Half-Moon Cove.

The Province of Nova Scotia commissioned a power station in the tidal estuary of the Annapolis River in 1984 (fig. 9.18). The world's largest straight flow/rim

Figure 9.17
(a) A single-action tidal power system. Power is generated on the ebb tide. (b) A double-action tidal power system. Power is generated on the ebb and flood tides.

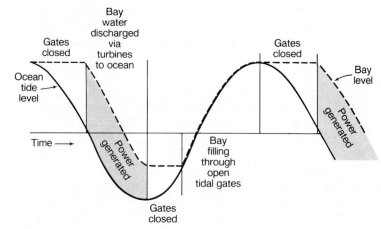

(a) Single-action power cycle; ebb only

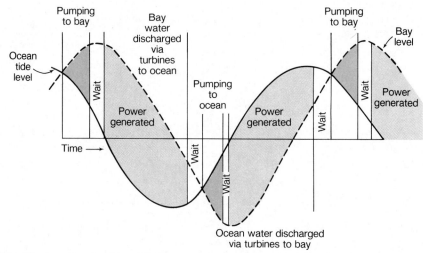

(b) Double-action power cycle; ebb and flood

Figure 9.18
The Annapolis River tidal power project is the first tidal power plant in North America. It was completed in 1984.

type turbine generator has been used to produce power in a single-basin/single-effect scheme. Tidal ranges at the Annapolis site vary from 8.7 m (29 ft) during spring tides to 4.4 m (14 ft) during neap tides. The unit generates up to 20 megawatts of power, from the head of water developed between the upstream basin and the sea level downstream at low tide. Initially, this project was intended as a pilot project to demonstrate the feasibility of a large-scale straight-flow turbine in a tidal setting. The station has now been added to the Province's principal electrical utility's hydro-generating system. Annual production from the unit is between 30,000 and 40,000 megawatt hours. Power availability has been in excess of 95%.

Although tidal power does not release pollutants, it is not without environmental consequences. The dams will isolate the bay from the rivers and estuaries with which it has been previously connected. Damming the head of the bay will effectively decrease the bay's natural period of oscillation by decreasing the bay's length. At present, the natural period of oscillation of the Bay of Fundy is about thirty minutes longer than the tidal period. Although the periods do not match exactly, they are sufficiently alike to allow the water in the bay to resonate with the tides. The result is large-amplitude tides. Shortening the period of oscillation will increase the resonance in the bay. It has been estimated that this change will increase tidal ranges by 0.5 m (20 in) and will increase the tidal currents by 5% along the coast of Maine. Coastal residents fear increased erosion and changes affecting the shellfish populations.

The damming of a bay or estuary has other costs. These installations interfere with ship travel and with port facilities. The dams are barriers to migratory species and alter the circulation patterns of the isolated basin, as well as forming a trap for its deep water and slowing or eliminating its exchange with the outside seawater.

Swift tidal currents in inshore channels represent another possible energy source. Flowing water has been used for several centuries to turn the equivalent of windmills or water wheels for limited power. Since the tidal currents reverse with the tide, these "watermills" must be installed so as to operate with the current flowing in either direction (see fig. 9.19).

Power generated by windmills is dependent on the density of the air, the blade diameter, and the cube of the wind speed. Watermills with a similar design are dependent on the density of the water, the blade diameter, and the cube of the current speed (see table 9.6). Thus, a windmill in a 20-knot wind will produce about the

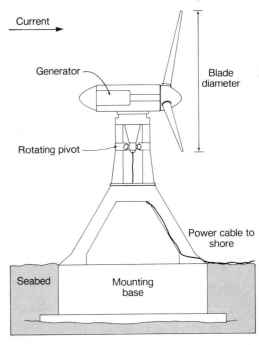

Figure 9.19

A water mill designed to utilize tidal current energy. Power production is related to blade size and current speed.

TABLE 9.6

Estimated Power Generation (in kw) for Watermills at Various Current Velocities

Blade diameter (m)	Current speed		
	5 knots 2.5 m/sec	4.5 knots 2.25 m/sec	3.5 knots 1.75 m/sec
2	8.5	5.5	3.7
5	53	35	23
10	210	140	90
15	480	310	205
20	850	550	370
30	1910	1250	820

same power as a watermill with blades of the same diameter in a 2-knot current, because the density of water is about 1000 times the density of air:

$$0.001 \text{ g/cm}^3 \times (20 Kts)^3 = 1 \text{g/cm}^3 \times (2 Kts)^3$$
$$\text{air density} \times (\text{speed})^3 = \text{water density} \times (\text{speed})^3.$$

The World Ocean Circulation Experiment

*T*he World Ocean Circulation Experiment (WOCE) is an international attempt to globally survey the world's oceans and understand the oceans' physical properties on a global level. This complex project aims to collect data on interactions between the atmosphere and the oceans in order to predict climate over periods of several decades, to predict climate response to natural or human influences, and to understand long-term changes in ocean circulation.

The observation period will begin at the end of this decade and finish in 1995, requiring the world-wide cooperation of the oceanographic and meteorological communities. Major elements of the WOCE observational program include:

- satellite observation systems;
- ship programs to measure salinity, nutrients, oxygen, and density, and to follow geochemical radioactive tracers that detect rates of change in water masses over time and space;
- measurement of surface ocean temperatures from satellites, ships, and buoys;
- a system of selected tide gauges and a sea-level measurement program to coordinate with satellite altimetry;
- floats and buoys to measure interior and surface ocean currents; and
- a program of current measurements and velocity profiles.

Program objectives are:

- to determine how the oceans influence the atmosphere by storing and transporting heat and fresh water;
- to determine how changing surface heat, freshwater content, and surface winds affect ocean circulation and eddies; to understand gyre circulation;
- to understand changes of ocean circulation with time and space; to understand the exchange of water between ocean basins;
- to determine the formation and circulation of water masses and their influence on climate with time scales of from 10 to 100 years;
- to understand the role of the Southern Ocean and its rates of transport; and
- to develop and test ocean models capable of predicting climate change.

The WOCE program builds on data-gathering techniques proven through previous satellite use. However, WOCE is a new program and depends on equipping a new generation of research satellites with improved specialized sensors. In the U.S., these new satellites include the planned Geodetic Satellite (GEOSAT), the Navy Remote Ocean Sensing System (NROSS), the Earth Resources Satellite (ERS-1), the Ocean Topography Experiment Satellite (TOPEX), and the Geopotential Research Mission Satellite (GRM) (see colorplate 1).

This ambitious scientific research effort faces problems other than technical and scientific difficulties. The vast quantities of data produced will require high-capacity computers and new systems of data management. As the detail of measurement increases, the knowledge gained may increase in military significance, meaning that the data may be classified and become unavailable to both national and international research communities. While each system is expensive to create, launch, and monitor, the rate of data gathering is so great that the actual cost per individual measurement is lower than that obtained by conventional means. However, the funding of satellites for exclusive scientific research may become a problem; perhaps scientists will be charged for the data which they receive. Above all, international cooperation between research institutions and nations is required for WOCE to succeed in its objectives.

A 2×10^6 watt generating unit installed on land has a blade diameter of nearly 210 feet mounted on a 120-foot tower. Installation costs are around 1.6 million dollars. The cost of installing and maintaining a similar-sized unit in a tidal channel submerged in seawater would be much greater, and such a unit would still suffer from periodic loss of power production during slack water. These enormous units could be navigation hazards, and channels of sufficient depth with currents of the required speed are few and are not easily accessible. Although the installation of large watermills has been contemplated, economic and practical realities have prevented the construction of even one test unit to date. There is, however, an effort once again to use small watermills to produce limited power to satisfy local needs.

Summary

Diurnal tides have one high tide and one low tide each tidal day; semidiurnal tides have two high tides and two low tides. A mixed tide has two high tides and two low tides, but the high tides reach different heights and the low tides drop to different levels. For diurnal and semidiurnal tides, the greatest height reached by the water is high water, and the lowest point is low water. Mixed tides have higher high water, lower high water, higher low water, and lower low water. The zero depth on charts is referenced to mean low water or mean lower low water; low tides falling below these levels are minus tides. Rising tides are flood tides; falling tides are ebb tides.

Equilibrium tidal theory is used to explain the tides as a balance between gravitational and centrifugal forces. An equatorial tide is a semidiurnal tide, or a tide wave with two crests and two troughs. Because the moon moves along its orbit as the earth rotates on its axis, a tidal day is 24 hours and 50 minutes long. The period of the semidiurnal tide is therefore 12 hours and 25 minutes.

The sun's effect is less than half that of the moon, and the tidal day with respect to the sun is 24 hours. Because the tidal force of the moon is greater than that of the sun, the tidal day is still considered to be 24 hours and 50 minutes.

Spring tides have the greatest range between high and low water; they occur at the new and full moons, when the earth, sun, and moon are in line. Neap tides have the least tidal range; they occur at the moon's first and last quarters, when the moon is at right angles to the sun.

When the moon or sun stands above or below the equator, its tides become more diurnal. Diurnal tides are often called declinational tides. The elliptic orbits of the earth and moon also influence the tide.

The dynamic approach to tides investigates the actual tides as they occur in the ocean basins. The tide wave is discontinuous, except in the Southern Ocean. It is a shallow water wave, which oscillates in some ocean basins as a standing wave, and its motions persist long enough to be acted on by the Coriolis effect. The tide wave is reflected, refracted, and diffracted along its route. Because a point on the earth moves eastward faster at the equator than the tide wave progresses westward as a free wave, the tide wave moves as a forced wave, and its crest is displaced to the east of the tide-raising body. Above 60°N and 60°S latitudes, the crest is more nearly in line with the moon, as the speeds of earth and the tide wave match more closely.

In large ocean basins, the tide wave can move as a progressive wave. The Coriolis effect causes a rotary tidal current. Standing wave tides can also form in ocean basins. They rotate around an amphidromic point as they oscillate. Cotidal lines mark the progression of the tide crest, and regions of equal tidal range are identified by corange lines. The tidal range increases with the distance from an amphidromic point. Standing wave tides in narrow open-ended basins oscillate about a node at the entrance to the bay or basin; the antinode is at the head of the basin, as in the Bay of Fundy. There is no rotary motion when a bay is very narrow.

A rapidly moving tidal bore is caused by a large-amplitude tide wave moving into a shallow bay or river.

Tidal heights and currents are predicted from astronomic data and actual local measurements. NOAA determines and publishes annual tide and tidal current tables.

Single- and double-action dam and turbine systems extract energy from the tides. Tidal power plants are in use in France, the Soviet Union, and Canada. Few places have large enough tidal ranges and suitable locations for tidal dams. Tidal power has environmental drawbacks as well as high developmental costs. Tidal currents are another energy-producing possibility, but installation and service costs are considered high.

Key Terms

diurnal tide	mean tide/average tide	centrifugal force	progressive tide
semidiurnal tide	tidal datum	tidal day	cotidal line
semidiurnal mixed tide	minus tide	tide wave	standing wave tide
high water	flood tide	tide range	rotary standing tide wave
low water	ebb tide	spring tide	amphidromic point
higher high water	tidal current	neap tide	corange line
lower high water	slack water	declinational tide	tidal bore
higher low water	equilibrium tidal theory	free wave	harmonic analysis
lower low water	dynamic tidal analysis	forced wave	local effect

Study Questions

1. Distinguish between the terms in each pair:
   a. Diurnal tide; semidiurnal tide.
   b. Tidal day; tidal period.
   c. Spring tide; neap tide.
   d. Flood tide; ebb tide.
   e. Cotidal lines; corange lines.

2. What is the path of a water particle in the tidal current shown in figure 9.14 if a 1-knot current flowing south is also present?

3. Why is it more efficient to generate power by means of a tidal dam than to erect watermills in tidal currents?

4. Why are standing wave tides produced in small coastal basins as well as in large ocean basins? Use table 9.1.

5. Explain why it is necessary to have both a large tidal range and a relatively large volume of water behind a tidal dam in order to generate electric power from the rise and fall of the tides.

Study Problems

1. How many days will pass before a high tide reoccurs at the same clock time?

2. Using the information in table 9.2 plot the Aberdeen, Wash., tide curves for August 1 and 8. Which is a spring tide? Which is a neap tide? What type of tide is this? Label the water levels on each curve.

3. Choose the best date and time for clam digging during mid-morning hours in the month of August at Aberdeen, Wash. Refer to table 9.2.

4. Use tables 9.2 and 9.3 to correct the Aberdeen tides to tides at Point Grenville for August 20.

5. At what time on August 17 do you arrive at Seymour Narrows to navigate the narrows at slack water between breakfast and dinner time? See table 9.4.

Coasts, Shores, and Beaches

10

The seashore is a sort of neutral ground, a most advantageous point from which to contemplate this world. It is even a trivial place. The waves forever rolling to the land are too far-traveled and untamable to be familiar. Creeping along the endless beach amid the sun-squawl and the foam, it occurs to us that we, too, are the product of sea-slime.

It is a wild, rank place, and there is no flattery in it. Strewn with crabs, horseshoes, and razor-clams, and whatever the sea casts up—a vast morgue, where famished dogs may range in packs, and crows come daily to glean the pittance which the tide leaves them. The carcasses of men and beasts together lie stately up upon its shelf, rotting and bleaching in the sun and waves, and each tide turns them in their beds, and tucks fresh sand under them. There is naked Nature—inhumanly sincere, wasting no thought on man, nibbling at the cliffy shore where gulls wheel amid the spray.

Henry David Thoreau,
from Cape Cod

*T*o most people, the most familiar areas of the oceans are the shores and beaches. People visit the coast to see the ocean, to play along the beach, to enjoy the constant interaction between moving water and what appears to be the stable land. But even the casual visitor senses changes in this area: the tide may be high or low, the logs and drift may have changed position since the last visit, and the dunes and sandbars may have shifted since the previous summer. A visit to a beach farther along the coast or along a different ocean presents a different picture. The sand is a different color or there is no sand at all, the waves break higher or lower, closer or farther away, the slope of the beach is steeper or flatter, and so on.

Coasts and beaches are dynamic, not static. No two regions are exactly the same. People play an important role along coasts and beaches because people and their structures change beaches and coastlines more than they realize. In this chapter we will study the types of coasts and beaches and the natural processes that create and maintain them. We will learn how humans affect these areas, often destroy them, and sometimes try to save them.

10.1 Major Zones

The **coasts** of the world's continents are the areas where the land meets the sea. The terms coast, coastal area, and coastal zone are used descriptively to designate land areas that are associated with the sea and may include areas of cliffs, dunes, beaches, and even hills and plains. The width of the coast, or the distance to which the coast extends inland, varies and may be associated with local geography, climate, vegetation, and even social customs and culture. However, the coast is most generally described as the land area that is or has been affected by such marine processes as tides, winds, and waves, even though the direct effect of these processes may be felt only under extreme storm conditions. The seaward limit of the coast is usually assumed to coincide with the beginning of the beach or shore, but common usage often includes nearby offshore islands.

Coastal areas are regions of change in which the sea acts to alter the shape and configuration of the land. Sometimes these changes are extreme and occur rapidly (for instance, the damage caused by a hurricane occurs in a period of hours). Sometimes the changes are subtle and so slow that they are not perceived by people during their lifetimes, but when considered over long periods of the earth's history these slow changes

Figure 10.1
Driftwood accumulates at the high tide line in areas where timber is plentiful.

are seen to be impressive (for example, the formation of the delta of the Mississippi River or the gradual erosion of Cape Hatteras, North Carolina). In general, coasts that are composed of soft, unconsolidated materials, such as sand, change more rapidly than coasts composed of rock.

The **shore** is that region from just seaward of the lowest tide level to the limit of the waves' direct influence on the land. This limit may be marked by a cliff or an elevation of the land above which the sea waves cannot break. Such features act as barriers to the wave-tossed drift of logs, seaweeds, and human debris. The **beach** is an accumulation of sediment (sand or gravel) that occupies a portion of the shore. The beach is not static, but is moving and dynamic, because the beach sediments are constantly being moved seaward, landward, and along the shore by nearshore wave and current action. Between the high tide mark and the upper limit of the shore there may be dunes or grass flats spotted with occasional drift logs left behind by an exceptionally high tide or severe storm (see fig. 10.1).

10.2 Types of Coasts

Let us first consider the coasts of the world's oceans, how they differ, and why. In our discussion, we will emphasize the different types of coasts, the processes that formed each, and their special characteristics. We will follow the system of the late Francis P. Shepard of the Scripps Institution of Oceanography.

All coastal areas belong in one of two major categories: (1) coasts that owe their character and appearance to processes that occur at the land-air boundary and (2) coasts that owe their character and appearance

(a)

(b)

(c)

(d)

Colorplate 10

(a) The Oregon Coast is famed for its bold rocky headlands and its intervening pocket beaches.

(b) Cape Alava in Washington State is the westernmost conterminous point in the continental U.S. Its wilderness beach is accessible only by hiking the 3.5 miles from the nearest road.

(c) Tide pool environments are plentiful along the protected rocky shores of the San Juan Islands between Washington State and British Columbia, Canada.

(d) These chalky cliffs along the Dorset coast of England were formed from the remains of foraminifera, principally Globigerina.

Colorplate 11

(a) Bayou La Loutre at Ysclosky, Louisiana. The bayou is an old stream course of the Mississippi River. The salt content of its water changes as the wind drives fresh water from the river or salt water from the coast through the delta.

(b) A back water tidal slough near the mouth of the Columbia River. Marshes and mudflats of these sloughs are highly productive nurseries for many marine species.

(c) A quiet bay along the Brittany coast of France during the catastrophic oil spill from the Amoco Cadiz.

(a)

(b)

(c)

to processes that are primarily of marine origin. Classification depends on the large-scale features created by tectonic, depositional, and erosional processes.

In the first category are coasts that have been formed by: (1) erosion of the land by running surface water, wind, or land ice, followed by a sinking of the land or a rise in sea level; (2) deposits of sediments carried by rivers, glaciers, or the wind; (3) volcanic activity, including lava flows; and (4) uplift and subsidence of the land by earthquakes and associated crustal movements. Coasts formed by these processes are **primary coasts,** since there has not yet been time for the sea to substantially alter or modify the appearance given to them by nonmarine processes. The second category includes coasts formed by: (1) erosion due to waves, currents, or the dissolving action of the seawater; (2) deposition of sediments by waves, tides, and currents; and (3) alteration by marine plants and animals. These coasts are **secondary coasts;** their character, even though it may have been originally land-derived, is now distinctly a result of the sea and its processes.

Since a coast that is relatively young in terms of the age of the earth may be rapidly modified by the sea, while a coast that is old in time may retain its land-derived characteristics, the terms primary and secondary may be confusing. Keep in mind that absolute age is not really important, because the classification is based only on whether the characteristics are derived from the land or from the sea, so that it is possible to find both types of features along the same coast.

Primary Coasts

Coasts formed by erosion at the land-air boundary followed by a sinking of the land or a rise in sea level include those that were covered by glaciers during the ice ages. During these glacial periods, sea level was lower than it is at present, because much of the water was held as ice on the continents. The glaciers moved slowly across the land, scouring out valleys as they inched along to the sea. In some cases, the weight of the ice caused the land to subside. When the ice at last began to melt, sea level rose faster than land could rebound upward from its depressed state. In other cases, the glacial troughs were scoured below sea level and filled as the ice receded. These actions resulted in the formation of the fjords of Norway, Greenland, New Zealand, Chile, and southeastern Alaska. Fjords are long, deep, narrow channels, U-shaped in cross section (see fig. 10.2). At their sea end there is often a collection of debris left by the glacier where it met the sea. This debris forms a lip that creates a shallow entrance, or **sill.** In some areas, land that was heavily covered by ice during the last ice age is still slowly rising; in Scandinavia the rate of rise is about 1 to 5 cm (0.4–2 in)/year. Along other coasts,

Figure 10.2
A fjord coast. This type of coast has narrow channels and indentations; glaciers terminate in the sea.

tectonic forces have caused land uplift, for example, the west coasts of North and South America. Along these coasts, old wave-cut terraces have been elevated above sea level.

Glaciers also deposited materials in what is now the coastal area. When a glacier or ice sheet ceases its forward motion and retreats, it leaves a mound of rubble at the point of its greatest extension. Such mounds are called **moraines.** Long Island, off the New York and Connecticut coasts, is a moraine; it acts as a protective barrier to the continental coast and modifies this area of the coastline.

When sea level was lowered during the ice ages, rivers flowed not only across the land but also over the exposed shore to the sea. This process cut typical V-shaped river and stream channels in these areas and in many cases the channels had numerous side branches formed by feeder streams. As the sea level rose, these channels were filled with seawater, producing coastal areas such as Chesapeake Bay and Delaware Bay on the east coast of the United States. A coastal feature of this type is called a **drowned river valley** or **ria coast,** as shown in figure 10.3.

Rivers carrying extremely heavy sediment loads build **deltas** at the edge of the sea. The delta, or deposit of river-borne sediment left at the river mouth, produces a flat, fertile coastal area. Good examples of active deposition are found at the mouths of the Mississippi River and the Amazon River. A similar type of coast is produced when eroded material is carried down from

Figure 10.3
Delaware Bay (middle right) and Chesapeake Bay (center and lower right) are examples of drowned river valleys.

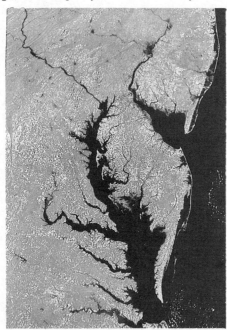

Figure 10.4
Coastal sand dunes near Florence, Oregon.

Figure 10.5
A volcanic crater coast. Hanauma Bay, Hawaii, is a crater that has lost the seaward portion of its rim. Today it is a park and marine preserve.

the hills by surface runoff and many small rivers join together to form an **alluvial plain.** The eastern seacoast of the United States south of Cape Hatteras has been produced in this manner.

It is estimated that during each second the rivers of the world carry 530 tons of sediment to the sea. This rate of removal of material from the land is equal to the erosion of a layer 6 cm (2.4 in) thick from all land above sea level every 1000 years.

Much of this sediment helps to form and maintain the world's beaches. Some of it finally finds its way to the deep-ocean floor; all of it passes through the coastal zone and takes part in coastal processes. Refer back to chapter 2 for sources of terrigenous material. Winds and landslides also deposit material of land origin in the coastal zone. A **dune coast** (fig. 10.4) is a wind-modified depositional coast. In Africa, the Western Sahara is gradually growing westward toward the Atlantic Ocean as the prevailing winds move sand from the inland desert areas to the coast. Along other dune coasts the winds move the sands inland to form dunes, and elsewhere dunes driven by the wind migrate along the shore.

The Hawaiian Islands have excellent examples of coasts formed by volcanic activity, for the islands themselves are the tops of large volcanoes rising from the sea floor. Lava flows extend to the sea, and black sand beaches are formed from the broken lava. These are **lava coasts.** Concave bays have been formed by volcanic ex-

plosions just beneath the sea surface at the water's edge. These volcanic craters that have lost their rims on the seaward side form **cratered coasts** (see fig. 10.5).

Tectonic activity results in faulting and displacement of the earth's crust. When fault movement occurs at the land-ocean boundary, the coast is changed in characteristic ways. The California San Andreas Fault system lies along a boundary where crustal plates are moving parallel to each other along a transform fault. (Plate movements and transform faults are described in chapter 3.) Over long periods of time, faults formed in this area filled with seawater. The Gulf of California, also known as the Sea of Cortez, located between Baja California and the mainland of Mexico, is found at the southern end of the fault system. At the northern end

Figure 10.6

The San Andreas Fault runs along the California coast to the west of San Francisco, separating Point Reyes from the mainland. Bolinas Bay and Tomales Bay are fault bays produced by displacement along the fault line.

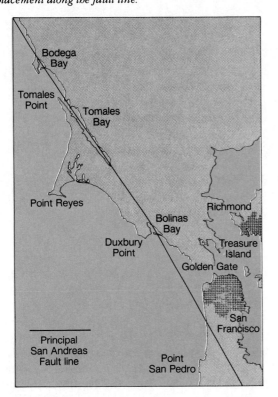

Figure 10.7

(a) Early stage in sea stack formation, Oregon coast. (b) Sea stacks, Point of Arches, Washington coast.

(a)

(b)

lie San Francisco and Tomales Bay, where the fault runs out into the Pacific Ocean. Tomales Bay is a particularly good example of a **fault bay.** See the map of this fault system in figure 10.6. On the other side of the world, the Red Sea provides another example of a **fault coast.**

Secondary Coasts

Secondary coasts have been modified by the sea and owe their present-day appearance to marine processes. As waves batter against the coast, they constantly erode and grind away the shore. Rocks and cliffs are undercut by the wave action and fall into the sea, where they are ground into sand, which, when suspended in the water, adds even greater cutting power to the waves. A coastal area formed of uniform material may have originally been irregular in shape, composed of headlands and bays, but as the waves approach the shore they concentrate their energy on the headlands, wearing them down more rapidly than the straighter shore and cove areas. With time, this action tends to produce a more regular coastline.

If the original coastline is made up of materials that vary greatly in composition and resistance to wave erosion, then the result will be an increasingly irregular coastline. In this case, headlands of hard rock jut into the sea and are separated by sandy coves formed by the erosion of softer materials. The headlands erode irregularly as the shore is cut away. The result is the formation of small rock islands and tall slender pinnacles of resistant rock called **sea stacks** in the nearshore areas (see fig. 10.7a). The sea stacks and rock islands are the more durable remnants of the previous coastline. Headlands still connected to the mainland may be undercut to produce sea caves or even cut through to produce arches or, if the opening is small, a window. Along the wilderness coast of the state of Washington in the Olympic National Park is a rugged stretch of coast containing many sea caves, arches, and windows; this area, known as Point of Arches, is shown in figure 10.7b. Offshore in this area, sea stacks and small islands 20 to 300 feet tall (7–100 meters) with nearly sheer sides stand

Figure 10.8

Sea Island, Georgia, is a barrier island that has been extensively developed. The shallow water between the island and the coast is seen on the left.

Figure 10.9

A spit has formed across the entrance to a small inlet.

lonely and isolated. There are few harbors along this coast, and small fishing boats seeking shelter often anchor at night between the sea stacks and the shore.

In some cases, eroded materials are carried seaward by the waves and currents to areas just off the coast. If sufficient sandy material is deposited in offshore shallows paralleling the beach, **bars** are produced. If still greater amounts of material are added, the bars may grow until they break the surface, helping to form **barrier islands.** Barrier-island formation may also be related to periods of glaciation, when sea level drops to expose offshore bars or when sea level rises to inundate low-lying coastal features. Features of continental margins that help form the continental shelves by trapping sediments, discussed in chapter 2 in the Continental Margins section, also play a role in the complex process of barrier-island formation. Once formed, plants begin to inhabit the islands and the growth of vegetation helps to stabilize the sand and increase the islands' elevation by trapping sediments and accumulating organic matter. The islands become more permanent. A line of these islands along a coast protects the continental coastline from the waves. Such barrier islands are found all along the Gulf and east coasts of the United States (fig. 10.8). Between the islands and the mainland lies a protected water route known as the Intercoastal Waterway, which is ideal for small-boat navigation. Although these barrier islands do protect the mainland from severe storms, the islands themselves sustain the damage that the continental coast is spared. This situation is quite as nature intended, but people who live on these islands look to the government to protect their property from these natural occurrences. Attempts to halt storm-caused erosion by building breakwaters and beach-holding devices have not always been successful and, although well

intentioned, have often resulted in aggravating the loss of material from barrier islands rather than halting it.

Sand spits and **hooks** are bars connected to the shore at one end. They are the result of similar sea forces (see figs. 10.9 and 10.24). They may grow, shift position, wash away in a storm, or rebuild under more moderate conditions. The area between the mainland and these spits and hooks is protected from turbulence and is therefore often the site of beach flats formed of sand or mud. If a spit grows sufficiently to close off the mouth of an inlet, a shallow lagoon is formed. Water percolates through the gravel and sand of the spit, and inside the lagoon the water rises and falls with the tides.

In some parts of the world, **reef coasts** are the result of the activities of sea organisms. The tropic coral reef coast is of this type. In tropic areas, corals grow in the shallow, warm waters surrounding a landmass, and the small animals gradually build a fringing reef, which is attached directly to the landmass. In other places, a lagoon of quiet water may lie between the barrier reef and the land, and, in a few cases, the coral encircles a submerged island to form an atoll. The formation of these reef types is discussed in chapter 2 (see fig. 2.8).

The Great Barrier Reef of Australia, stretching along its northeast coast toward New Guinea, is the most famous of the world's coral reefs. Coral atolls include the Pacific islands of Tarawa, Kwajalein, Eniwetok, and Bikini. The reef-encircled islands of Iwo Jima and Okinawa became familiar as the sites of major battles in the southwest Pacific during World War II.

Other marine animals form reeflike structures as their shells are deposited layer on layer, gradually building up a mass of hard material. There are large reef deposits of oyster shells in the Gulf of Mexico off the

Rising Sea Level

Sea level is once again on the rise around the coasts of the world. Over the past century, sea level has risen about 15 cm (6 in), but recently the rate of rise has increased. It is thought that sea level may gain another 30 cm (1 ft) in the next fifty years. If such an increase occurs, the edge of the shore will move inland from hundreds to thousands of meters along low-lying coasts, causing great damage to homes, towns, ports, farms, and industry. Such an increase in sea level will also rapidly accelerate the erosion of coastal areas. Some believe coastal planners should act now to minimize the damage and loss which will be caused by such a rise. Highly populated shore areas such as Long Island, New York; Atlantic City, New Jersey; Miami Beach, Florida; Ocean City, Maryland; and Galveston, Texas, may be heavily damaged.

The increasing rate of sea level rise has been partly attributed to the greenhouse effect, discussed in chapter 6. The 25% increase in the amount of atmospheric CO_2 since 1900 has increased the surface temperature of the oceans by about 0.5° C and warmed the air over the land by about the same amount. If this trend continues, and if the amount of CO_2 in the atmosphere doubles during the next 100 years, the ocean surface will warm an additional 2–4° C. The greatest warming will occur at the higher latitudes, where the continental ice masses will melt and increase the volume of water in the oceans (see chapter 1, Reservoirs and Residence Time). As the oceans warm, the water will expand, introducing another factor that will affect the sea level.

The rate at which sea level will rise is not clear. The warming process will create changes in cloud cover and precipitation patterns; these will, in turn, change the distribution of incoming solar radiation, which will not only affect ice melt but also wind and storm systems. If the storm patterns that drive the ocean waves change, the rate and location of shore erosion will also change.

Long-term climate cycles, including cold and severe storms alternating with milder climatic periods, play additional roles in water level changes and coastal erosion. In addition, rivers modified by dams and dikes cut off the delivery of sediments to the beaches, and the construction of seawalls, groins, bulkheads, and breakwaters alters the current and wave patterns along the coasts. Again, these are changes that influence coastal erosion rates as the sea level rises. All together, these events and processes weave a complex unity from which it is extremely difficult to isolate any single item as causing the changes that are being forecast.

Whatever the predictions, it is unlikely that people will give up their dreams of a summer cabin on the shore or a home on the beach. However, people are beginning to become more sensitive to the encroaching waters. The 1982 Coastal Barrier Resources Act prohibits federal subsidies for roads, bridges, piers, and flood insurance coverage for homes built after 1983 in certain designated undeveloped barrier-island areas. Western Galveston Island in Texas prohibits new construction and reconstruction of storm-damaged property in some shore areas. States from Maine to North Carolina to Washington have regulated construction of seawalls, jetties, groins, and other permanent structures in the coastal zone. While such regulations do point to an increased awareness of sea level and coastal erosion problems, it may be that human efforts can only be considered too little and too late when confronting these combined natural forces.

coasts of Louisiana and Texas. These reefs are so large that the shells are harvested commercially for lime production. Along the east coast of Florida, large populations of shell-bearing animals have contributed their shells directly to the shore as the naturally occurring small shell fragments that form the sand on the beaches.

Plants as well as animals may modify a coastal area. Along low-lying coasts in warm climates, mangrove trees grow in the shallow, brackish water. Their great roots form a nearly impenetrable tangle, providing shelter for a unique community of other plants and animals (fig. 10.10). Coastal mangrove swamps are found along the Florida coast and in the West Indies. In more temperate

Figure 10.10
Mangrove trees growing along the shore of Guam.

Figure 10.11

A coastal saltwater marsh with tidal drainage channels. Old dikes or raised berms of earth used to exclude seawater from agricultural land are seen parallel to the broad center channel.

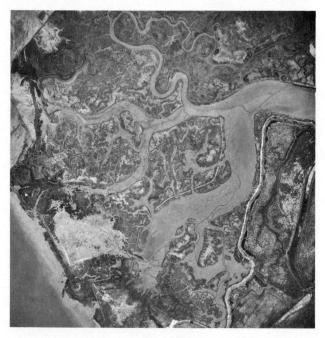

climates, low-lying protected coasts with areas of sand and mud are often thickly covered with grasses, forming another type of plant-maintained environment (fig. 10.11). These **salt marshes** may extend inland a considerable distance if the land is flat enough to permit periodic tidal flooding. Such marshes also form around protected bays and coves that have large changes in tide level. Salt marshes are extremely productive in terms of organic matter and extend the shoreline seaward by trapping sediments. However, these marshes are rapidly disappearing under the pressures of human habitation and the demand for flat industrial sites, harbor space, and recreational property adjacent to the water.

10.3 Anatomy of a Beach

The coast, regardless of the distance it extends inland, is commonly considered to end on the seaward side at the shore. The shore has been defined as the region from just seaward of the lowest tide level to the upper limit of the waves' direct influence on land. The beach is considered to include the sand and sediment lying along the shore, together with the sediments carried along the shore by the nearshore waves and currents. Let us now, however, look only at the beach, along which we may walk or on which we may sunbathe and relax.

Many studies have been made of beach areas around the world, and from these studies there has developed a specific terminology, or vocabulary, applicable to beaches. This standardization of terms allows us to avoid confusion and to describe each area as completely as possible. Therefore, to begin to understand beaches, it is first necessary to understand the words used to describe them. A typical beach profile (fig. 10.12) shows all the features that might appear on any beach. The profile represents a cut through a beach perpendicular to the coast. Although all the possible beach features are shown on this profile, keep in mind that any given beach need not show all these features, and indeed many do not.

The shore is divided into the **backshore,** which is the dry region of beach that is submerged only during the highest tides and severest storms, the **foreshore,** which extends out past the low tide level, and the **offshore,** which comprises the shallow water areas seaward of the low tide level extending to the outer limit of wave action on the bottom.

Terraces appear on beaches in the area where foreshore and backshore meet. These terraces are **berms,** beach features that are formed by the shoreward movement and deposit of materials by the waves. They are recognizable by their slope or rise in elevation (fig. 10.13). When two berms are shown, as in figure 10.12, the berm higher up the beach, or closer to the coastline, is the **winter berm,** formed during severe winter storms, when the waves reach farthest up the beach and pile up material along the backshore. Once this berm is formed, it is not disturbed by the less intense storms of spring, summer, or fall. The seaward berm, the berm closer to the water, is formed by less severe waves that do not reach as far up the beach; this berm is called the **summer berm.** The waves that produce a winter berm erase the lower summer berm. After the storm season is over, a new summer berm is formed by gentler wave action. A berm has a relatively flat top, appearing almost like a terrace, or it has a slight ridge crest that runs parallel to the beach. If there is a ridge, it is called the **berm crest.**

Between the berm and the water level there is a wave-cut **scarp** at the high-water level. The scarp is an abrupt change in the beach slope that is caused by the cutting action of the most frequently occurring waves at normal high tide. Berms and scarps do not form between the normal high and low tide levels, because the continual rise and fall of the water with its wave action does not permit the appearance of any permanent feature. This relatively featureless area is known as the foreshore. The lower portion of the foreshore is often quite flat, forming a **low tide terrace;** the upper portion has a steeper slope and is known as the **beach face.**

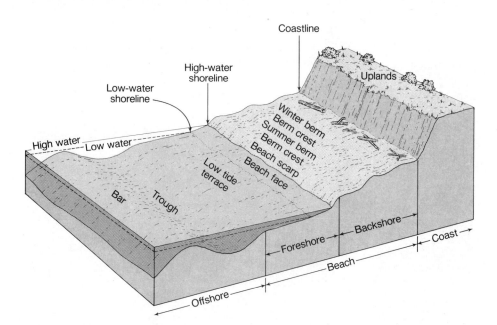

Figure 10.12
A typical beach profile with associated features.

Figure 10.13
Multiple berms on a gravel beach.

Figure 10.14
Rocky headlands separate pocket beaches along the Oregon coast.

Seaward of the low tide level in the offshore region, there may be **troughs** and **bars** that run parallel to the beach. These structures often change seasonally as beach sediments are moved seaward during periods of large waves (winter storms), to enlarge existing bars, and then shoreward during periods of small waves (summer), to diminish the same bars. When a bar accumulates enough sediment to break the surface and is stabilized by vegetation, it becomes an island of the type described for a barrier island coast.

Beaches do not occur along all shores. Along rocky cliffs, no beach area may be exposed between low and high tide levels. However, there may be small pocket beaches between near-vertical headlands, each separated from the next by rocky headlands or cliffs (see fig. 10.14).

10.4 Beach Types

Beaches are also described in terms of (1) shape and structure, (2) composition of beach materials, (3) size of beach materials, and (4) color. In the first category, beaches are described as wide or narrow, steep or flat, and long or discontinuous (as in the case of pocket beaches). A beach area that extends outward from the main beach and then turns and parallels the shore is called a spit. Spits frequently change their shape in response to waves, currents, and storms. In a wave-protected environment, a spit may extend out at an angle from the shore to an offshore island. If this spit builds until it connects the rock or island to the shore, it forms what is called a **tombolo.** Occasionally, a spit extends

Figure 10.15
A spit connected to an offshore island forms a tombolo.

Figure 10.16
An armored beach. Lag deposits of large rocks are left on an eroded beach.

offshore in a wide sweeping arc bending in the direction of the prevailing current. The sediment then moves around the end of the spit and is deposited in the quiet water behind the point. This action may produce a broad point on the spit, and the spit is commonly and descriptively called a hook. See figure 10.15 for an example of a tombolo and figure 10.24 for an example of a hook.

Further identification of the beach is made by reference to the type of material forming the beach, such as shell, coral, rock, lava, and shingle. A shingle beach is composed of flat, circular, smooth stones. The shingles are formed when the beach slope, wave action, and stone size and composition combine to cause the stones to slide back and forth with the water movement. When the stones roll instead of slide, the water action produces round stones, or cobbles. The terms sand, gravel, pebble, cobble, and boulder are used to describe the sediments of a beach. These terms refer to the size of the particles on the beach (see table 2.3 in chapter 2).

Both the composition of the beach material and its size are related to the source of the material and the forces acting on the beach. Land materials are brought to the coast by the rivers or are derived from the local cliffs by wave erosion. Small-sized particles such as sand, mud, and clay are easily transported and redistributed by waves and currents. Larger-sized rocks are usually found in the area close to their source, because they are too large to be moved any distance. The sands of many beaches are the ground-up, eroded products of land materials rich in minerals like quartz and feldspar. The

waves break and erode the coastal cliffs to produce beach materials. The finer soil particles are carried away by the currents, and the larger rocks are left scattered on the beach. In general, a beach littered with large rocks is an eroded beach. The remaining large particles are called **lag deposits** and may accumulate in sufficient quantity to protect a beach from further wave and water erosion. Such a beach is referred to as an **armored beach** (see fig. 10.16).

Some beach materials come to the beach from offshore areas. Coral and shell particles broken by the pounding action of the waves are carried to the beaches by the moving water. A beach covered with uniform small particles of sand or mud is or has been a depositional beach.

Some of the world's beaches have distinctive colors. In Hawaii, there are white sand beaches, derived from coral, and black sand beaches, derived from lava. Green and pink sands can be found in areas where a specific mineral (like olivine) or shell materials are available in large enough quantities.

10.5 Beach Dynamics

A beach formed of smaller-sized particles exists because there is a balance between the supply and the removal of the material that forms it. A beach that does not appear to change with time is not necessarily static (no new material supplied, no old material removed), but is more likely to represent a **dynamic equilibrium.** For a particular beach in this type of system, the supply of material equals the removal of material; the beach is continually changing, but it remains in balance.

Figure 10.17
Seasonal changes of La Jolla Beach: (a) winter and (b) summer.

(a)

(b)

Natural Processes

If we observe a beach for only a single day, we may not see any change other than the rise and fall of the water undergoing the daily tidal cycle. If we continue to watch the same beach over a period of weeks and months, we may note changes in the beach as the weather changes from calm periods, with small waves coming quietly in on the beach, to stormy periods, with great, violent waves pounding high on the beach.

The gentler waves of the summer bring sand to the beach and deposit it there, where it remains until the winter season. The large storm waves of the winter remove the sand from the beach and transport it back offshore, to be deposited in a sandbar. This process occurs alternately, winter and summer, leaving the beach rocky and bare each winter and covered with sand during the summer months (see fig. 10.17). These seasonal changes are fluctuations about the equilibrium state of the beach over a yearly cycle. Other changes also appear in a cyclic fashion, such as the arrival of more sand during periods of late winter and spring river flooding.

Single more drastic and violent events also occur. A great storm may arrive from such a direction that the usual beach current is reversed. A landslide may pile rubble across the beach and some distance into the water, interfering with the transport of sand along the beach. In either case, the beach changes radically due to changes in rates of sediment supply, transport, and removal.

The waves moving toward the beach produce a current in the surf zone that transports sediment to the beach from sandbars and islands off the shore. This landward motion of water is called the **onshore current,** and the beachward transport of sediment suspended in this water is called **onshore transport.** Waves usually do not approach a shore with their crests completely parallel to the beach. Instead, they strike the shore at a slight angle. This pattern sets up a surf zone current that moves down the beach, called the **longshore current.** Figure 10.18 illustrates both onshore and longshore flow. The turbulence that occurs as the waves break in the surf zones abrades the beach material and tumbles it into suspension in the water. The wave-produced longshore current displaces this sediment down the shore in the surf zone. This process is known as **longshore transport** or **littoral drift.**

Each breaking wave's oblique uprush of water, or **swash,** along the beach face moves the sand particles diagonally up and along the beach in the direction of the longshore current. The backwash from the receding wave tends to move directly downslope toward the surf zone, but the backwash is weaker than the swash because much of the receding water percolates down into

Figure 10.18

Waves in the surf zone produce a longshore current that transports sediments down the beach. Arrows indicate onshore and longshore water movement.

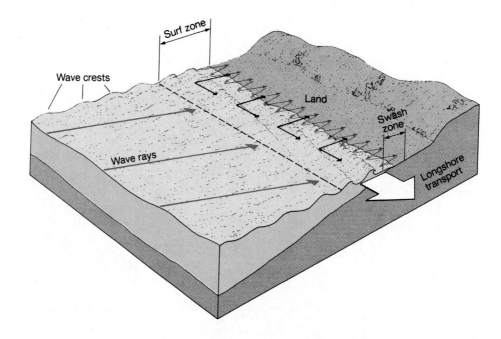

Figure 10.19

A drift sector, or transport cell, extends down the coast from the sediment source to the area of deposit.

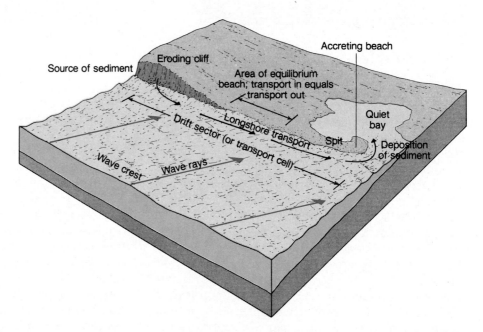

the sand. The combined action of the swash and backwash moves particles in a zig-zag or saw-tooth path along the swash zone of the beach as part of the longshore transport. Along both coasts of the North American continent, the predominant longshore transport steadily moves the sediments in a southerly direction. If a beach receives as much material as it loses, the beach is in equilibrium and does not appear to change.

The path over which beach sediments travel as they are transported from their source to their area of deposition is called a **drift sector** or **littoral cell.** Beaches within the drift sector usually remain in dynamic equilibrium and change little in appearance. Al-though there is often a great flow of materials along the central portion of a drift sector, the supply usually equals the amount removed in this area. At either end of the drift sector, the beaches are seen to change with time. The beaches at the sediment source end of a drift sector are eroded, and those at the other end are depositional. The depositional beaches are growing, or **accreting.** These processes are shown in figure 10.19. The wave-produced longshore current in the surf zone provides the transport mechanism in a drift sector. However, tidal currents and coastal currents associated with large-scale oceanic circulation also affect the sediment transport process in some places.

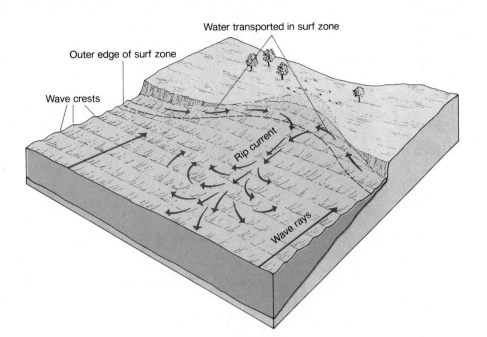

Water transported in surf zone

Outer edge of surf zone

Wave crests

Rip current

Wave rays

Figure 10.20
Rip currents form along a beach in areas of low surf and reduced onshore flow.

Coastal Circulation

Onshore transport accumulates water as well as sand along the beach. This water must flow back out to sea and will do so at some point, where shoreline irregularities or depressions in the seabed create areas where the surf and the onshore current are less intense. Such flows often occur in the form of narrow, fast-moving, seaward-flowing currents called rip currents, which carry sediment seaward through the surf zone (see fig. 10.20). Rip currents are also discussed in chapter 8 (see fig. 8.19).

Just seaward of the surf, the rip current dissipates into an eddy. Some of the sediment is deposited here in quieter deeper water and is lost from the down-beach flow. However, some is returned to shore with the onshore transport on either side of the rip current. This cycle of partial return of sediments to the beach, transport along the beach, transport back to sea in another rip current, and return again to the beach is a part of a littoral cell.

A series of such cells, when linked together along a stretch of coast, forms a major **coastal circulation cell.** In a major coastal circulation cell, the seaward transport of sediment is more dramatic at the cell's end. Here, beach materials are carried and deposited far enough offshore to prevent any recycling. The end point of a major coastal circulation cell is often a submarine canyon. Where a submarine canyon crosses into the continental shelf, the sand transported by the longshore current is deflected seaward, and a large-scale offshore transport of sand frequently occurs. This action removes the sand from its journey along the coast, and sand beaches disappear below this point until a new source

of beach materials contributes sand to form the next major coastal circulation cell.

South of Point Conception in southern California, oceanographers recognize four distinct major littoral cells (fig. 10.21). Each cell begins and ends in a region of rocky headlands, where beaches are sparse and submarine canyons are found offshore. Beaches become wider as river sources contribute their sand to the longshore current. At the end of each cell, the current deposits the sediment in a submarine canyon, where it cascades to the ocean basin floor. Beaches just south of the canyon are sparse, and a new cell begins. Refer back to chapter 2 for a discussion of submarine canyons.

Estimates of the rate at which sediments are moved along a beach are made by observing the rate at which sand is deposited, or accreted, on the upstream side of an obstruction, or by observing the rate at which sand spits migrate. Transport of sand along a section of coast varies between zero transport and several million cubic meters per year. Average values fall between 150,000 m³/yr and 1,500,000 m³/yr. Consider that 150,000 m³ is more than 30,000 dump truck loads. Once the enormous volume of the naturally moving sand is recognized, it becomes apparent why poorly designed harbors fill very quickly, beaches disappear, and spits migrate a considerable distance during a year. It also becomes obvious that the effort required to keep pace with the supply by dredging or pumping operations is enormously expensive and in many cases quite impossible.

The forces that supply the energy for the movement of beach materials come from the waves and the wave-produced currents. These are wind-driven processes. Surface winds supply about 10^{14} watts of power

Figure 10.21

Major coastal sediment circulation cells along the California coast. Each cell starts with a sediment source and ends where beach material is transported into a submarine canyon.

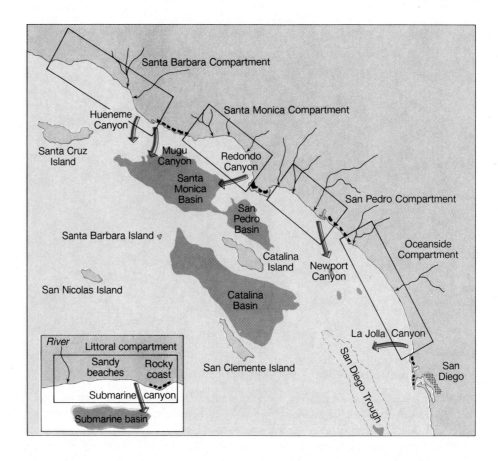

to the ocean surface. There are about 440,000 km (264,000 mi) of coastal zone in the world. Approximately half of this coast is exposed directly to the ocean waves. Ocean waves average 1 m (3.3 ft) in height and produce power that is equivalent to 10^4 watts for each meter of exposed coastline. Under storm conditions, the wave height averages about 3 m (10 ft), and a meter of coastline receives 10^5 watts. It is this supply of energy that causes the beach erosion, the suspension of beach material, and the migration of beach sand along the narrow coastal zone.

Artificial Processes

People also change beaches. Rivers are dammed to control floods and to generate power. When a dam is built, a great still-water lake is formed behind the dam. Sand, gravel, and rock that had once moved down the river to join the sediments at the coast are now deposited in the lake behind the dam. An important source of material critical to the continued balance of a beach has been removed. Although the downstream sediment supply is reduced, the longshore current continues to flow and carry suspended beach material away. The net result is a loss of sediment and sand to the beaches formerly supplied by the river.

Coastal zone engineering projects also result in changes to a beach. **Breakwaters** and **jetties** are built to protect harbors and coastal areas from the force of the waves. The areas behind jetties and breakwaters are quiet; materials suspended by the wave action and carried by the longshore current settle out here. This removal of material from the longshore transport process has the result that less material is available for a beach farther down the coast. As the longshore current continues to flow, the next beach down the coast begins to disappear. Small-scale examples of this process are seen when **groins,** rock, or timber structures are placed perpendicular to the beach to trap sand being carried in the longshore transport (see fig. 10.22). Time and time again, coastal facilities have been constructed in response to the demands of one group, only to find that in doing so a chain of events has been triggered in the dynamic shore zone that results in the requirement for more engineering projects in order to solve newly created problems. Forty-three percent of the shoreline of our national and territorial areas, excluding Alaska, is now losing more sediment than it receives. Our beaches are disappearing into the sea. In many cases, eroding beaches are maintained by supplying the beach with sand and gravel mined from upland areas or dredged from offshore sand bars. Such "beach feed programs" are expensive but necessary to compensate for the natural losses of coastal sediments.

Figure 10.22
The entrance to Grays Harbor, Washington, is protected from southerly swells by a long jetty extending into the ocean. Groins built perpendicular to the land keep the tidal currents from eroding the point. Note the buildup of sand in the angles formed by the groins and the shore.

Figure 10.23
Santa Barbara harbor. The large quantity of sediment that has built up in the protected, quiet water behind the jetty is due to winter storms.

10.6 Practical Considerations: Case Histories of Two Harbors

Designing coastal structures that correct or solve one problem without creating more problems is as much an art as it is a science. Measurements and calculations are made, but more is required. There must be a thorough understanding of the natural processes in an area before the consequences of the new structure can be foreseen. A scale working model is a powerful aid in understanding the present state of a shore area and the results of future changes. Natural processes such as tides, waves, and currents are reproduced on a scale model and then the proposed structure is introduced into the system. Its effect is observed, and design modifications can be made. Models of this type are expensive, but they are much less expensive than the costs of continually correcting the results of poorly designed structures that disturb the natural environment in an undesired manner.

The Santa Barbara Story

A classic example of interference with natural processes in the coastal zone is the Santa Barbara harbor project. The longshore current and sediment transport move down the coast of southern California from north to south. At Santa Barbara, a jetty and breakwater were constructed to form a boat harbor. The jetty at the north side of the harbor juts out into the sea before it turns southward, forming a breakwater that runs parallel to the coast (see fig. 10.23). This situation creates a wave-sheltered area on the jetty's north side and also blocks the longshore current. Sand was deposited on this north side, where the longshore current was blocked, and the beach began to grow. The beaches to the south began

to disappear, because they were starved of the sand that had been deposited to the north.

When the beach to the north had grown until it reached the seaward limit of the jetty, the longshore current could again move beach sediments southward along the ocean side of the breakwater. When the longshore current and the suspended sediment reached the end of the breakwater and the entrance to the harbor, the current formed an eddy that spiraled into the quiet water of the harbor. The sediment settled out and began to fill the harbor, forming a spit connected to the end of the breakwater. The sediment settling in the harbor deprived the beaches farther south of their supply but did not alter the forces acting to remove the sediment from these same beaches.

Today, a dredge pumps the sediments deposited in the harbor by the longshore current from the harbor through a pipe and back into the longshore current on the south side of the harbor. In this way, the harbor remains open and the beaches to the south receive their needed supply of sand. In this case, interference with a natural process has required the expenditure of much time, effort, and money to do the work nature did for nothing.

The History of Ediz Hook

Fifteen hundred miles north of Santa Barbara, another harbor has been altered by interfering with the supply of beach material. The harbor of Port Angeles, located on the Strait of Juan de Fuca in Washington State, is protected from the storm waves in the strait by a naturally occurring 3 ½-mile-long curving spit known as Ediz Hook (fig. 10.24). The hook is composed mainly of sand and gravel, and protects an area large enough and deep enough to accommodate the largest and most modern supertankers. In recent years, however, the hook has

Figure 10.24
Ediz Hook is a long spit forming a natural breakwater at Port Angeles, Washington. The protected harbor behind the spit is deep enough for mooring the largest super tankers.

undergone considerable erosion and the danger of waves breaking through the hook near its base has become severe.

To understand how the hook came to exist at all and how its present condition has come about, it is necessary to go back about 14,000 years, to the time when the glaciers of the last ice age retreated from this area. At that time, sea level was lower than it is now, due to the amount of water on land in the form of ice. The Elwha River west of Port Angeles carried loose glacial deposits to the sea, forming a delta that curved to the east under the influence of local currents and waves in the Strait of Juan de Fuca. As sea level began to rise, the river, currents, and waves continued to carry sediment eastward, and the waves began to erode the cliffs in the area, adding still more sediment to this longshore current. As a result, a hook-type spit was built up at a point where the shoreline makes an abrupt angle with the strait. Over time, Ediz Hook was produced and was kept constantly supplied with sand and gravel from the cliffs and the river.

In 1911, a dam was constructed across the Elwha River to provide a freshwater reservoir. The sediments that had been flowing toward the hook (estimated at 30,000 m³ per year) were now being deposited behind the dam. In 1930, a pipeline was constructed to deliver fresh water from the Elwha Reservoir to Port Angeles. The route chosen was around the cliffs at beach level and required the building of bulkheads along the face of the cliffs in order to protect the pipeline. These bulkheads cut off the source of cliff-eroded sediments, estimated at 380,000 m³ per year. At present, it is believed that only one-seventh of the total sediments once available to feed the spit remain in the longshore current.

Sediment removal from the spit no longer equals sediment supply. Studies by the Army Corps of Engineers show a possible loss of 270,000 m³ of sand each year from the outside of the hook. The hook is now sediment-starved.

Ediz Hook is not merely protection for the harbor. It is also the site of a large pulp mill, a U.S. Coast Guard Station, and a number of small harbor facilities. All are connected by a road running the length of the hook. In the 1950s, the occupants of the hook and the city acted to protect the seaward side of the hook by applying larger boulder armoring called rip-rap as well as steel bulkheads. Cliff material was blasted into the water in hopes of supplying the required sediments. A constant battle between the people and the sea began. Nearly as fast as the armor was applied to the base of the spit, the waves tore away the protective barriers in places and crashed over the spit. Conditions continued to deteriorate.

In 1971 the Corps of Engineers completed a planning study for further corrections. The 1973–1974 winter storms were severe, and the hook was again storm-damaged. Funds were appropriated in 1974, and the work of strengthening the seaward side of the hook began. Over a 50-year period, annual maintenance costs are projected to be about $30 million, while revenue from the harbor is projected at $425 million. Since the benefit:cost ratio is about 14:1, it appears to justify the cost of the project.

Summary

The coast is the land area that is affected by the ocean. The shore extends from the low tide level to the top of the wave zone. The beach is the accumulation of sediment along the shore.

Primary coasts are formed by nonocean processes (for example, land erosion; river, glacier, or wind deposition; volcanic activity; and faulting). Primary coasts include fjords, drowned river valleys, deltas, alluvial plains, dune coasts, lava and cratered coasts, and fault coasts. Secondary coasts

are modified by ocean processes (for example, ocean erosion; deposition by waves, tides, or currents; and modification by marine plants and animals). Erosion produces regular and irregular coastlines. Deposition of eroded materials creates sandbars, sand spits, and barrier islands. Reef coasts are formed by marine organisms; mangrove swamps and salt marshes modify coastlines.

A typical beach has features that include an offshore trough and bar and a beach area comprising a low tide terrace, beach face, beach scarp, and berms. Winter and summer berms are produced by seasonal changes in wave action.

Beaches are described by their shape, the size, color, and composition of beach material, and their status as eroded or depositional beaches.

Beaches exist in a dynamic equilibrium, in which supply balances removal of beach material. The gentle summer waves move the sand toward the shore during onshore transport. In winter, high-energy storm waves scoop the sand off the beach and deposit it in a sandbar during offshore transport. In the surf zone, the breaking waves produce a longshore current. The longshore current moves the sediment down the shore, in a process known as longshore transport. If the beach accumulates as much material as it loses, it is in a state of equilibrium. Drift sectors are defined by the path followed by beach sediments from source to region of deposition.

Rip currents are narrow, fast, seaward movements of water and sediment through the surf zone. A series of small drift sectors forms major transport patterns along the coast, which are known as littoral cells or coastal circulation cells. The volume of naturally moving sand in coastal circulation patterns along a beach is enormous. The energy for the movement comes from the wind-driven waves and the wave-produced currents.

Artificial structures also change beaches (for example, the building of breakwaters, groins, and jetties in coastal waters). Santa Barbara harbor and Ediz Hook are examples of human interference with natural processes in the coastal zone.

Key Terms

coast	cratered coast	berm	onshore current
shore	fault bay	winter berm	longshore current
beach	fault coast	summer berm	onshore transport
primary coast	sea stack	berm crest	longshore transport
secondary coast	bar	scarp	littoral drift
sill	barrier island	low tide terrace	swash
moraine	sand spit	beach face	drift sector
drowned river valley	hook	trough	littoral cell
ria coast	reef coast	bar	accretion
delta	salt marsh	tombolo	coastal circulation cell
alluvial plain	backshore	lag deposit	breakwater
dune coast	foreshore	armored beach	jetty
lava coast	offshore	dynamic equilibrium	groin

Study Questions

1. The processes that form a stretch of flat, uniform coast depend on whether the land is rising or sinking. Discuss the processes that form a flat coastal area under these two conditions.

2. Why are multiple berms more likely to be seen on a beach between March and August than between September and February?

3. Fjord coasts and drowned river valleys, or ria coasts, are primary coasts. Explain why their appearances are distinctly different. Give examples of each.

4. Why is a resort hotel that is located on one of the barrier islands along the southeast coast of the United States a poor long-term investment?

5. Describe the processes required to create a tombolo, or sand spit, between a beach and an offshore island. Consider and discuss (a) the distribution of wave energy, and (b) the longshore transport.

6. What conditions are required to maintain a beach with a constant profile and composition? Consider both a static and a dynamic environment.

7. In order to create a small boat harbor along a wave-exposed sandy beach, a breakwater is built parallel to the shore. What effect will this structure have on the beach behind it?

8. Why does an eroding beach that is supplied with material from cliffs and the banks of an old glacier deposit become an armored beach, while an eroding beach that is supplied by river sediments does not?

9. How does the profile of a sand beach change during alternating seasonal periods of storm waves and more gentle small waves?

10. Sketch a beach section that is stable with respect to the supply and removal of beach sediments. Indicate the source of the sediments and the final deposition of sediments in the system. Add breakwaters and jetties to your beach and indicate what changes each of these structures will produce in your system.

Bays and Estuaries

11

There were many fish moving in through the deep water of the channel that night. They were full-bellied fish, soft-finned and covered with large silvery scales. It was a run of spawning shad fresh from the sea. For days the shad had lain outside the line of breakers beyond the inlet. Tonight with the rising tide they had moved in past the clanging buoy that guided fishermen returning from the outer grounds, had passed through the inlet and were crossing the sound by way of the channel.

As the night grew darker and the tides pressed farther into the marshes and moved higher into the estuary of the river, the silvery fish quickened their movements, feeling their way along the streams of less saline water that served them as paths to the river. The estuary was broad and sluggish, little more than an arm of the sound. Its shores were ragged with salt marsh, and far up along the winding course of the river the pulsating tides and the bitter tang of the water spoke of the sea.

Rachel Carson,
from Under the Sea-Wind

*A*long the coasts of the world, there are embayments, or indentations, where salt water from the oceans and freshwater drainage from the land meet. Sometimes, heavy freshwater runoff floods and dominates these areas; at other times, the seawater is driven landward, submerging the low-lying coasts. In some ways, embayments behave like small oceans, but the frequent addition of fresh water gives them unique and characteristic circulation patterns of their own.

We call embayments bays, harbors, gulfs, inlets, sounds, channels, and straits, names given historically but according to no defined rules. Usually bays and harbors are smaller than gulfs; inlets and sounds are elongate; straits connect two large bodies of water, while channels are narrow water connections between smaller areas. In this chapter, we shall see that it is the properties of the water, the effect of the freshwater drainage, and the ways in which the water mixes and circulates in each that are important. Each coastal embayment is unique, but together they share certain characteristics of circulation which we can use to group them. We will also investigate the relationship between these areas and the increasing populations that live near them and use them.

11.1 Estuaries

An **estuary** is a portion of the ocean that is semi-isolated by land and is diluted by freshwater drainage. Both isolation and dilution are requirements for an estuary. Therefore, all estuaries are coastal embayments, but all embayments are not estuaries unless they are diluted with fresh water. Differences in net circulation and current patterns are determined by tidal and river flow. Circulation and the vertical distribution of salinity, or salt content, are used to place estuaries into a series of categories.

Types of Estuaries

The simplest type of estuary is the **salt wedge estuary.** Salt wedge estuaries occur within the mouth of a river flowing directly into salt water. Here, the fresh water flows rapidly out to sea at the surface, while the denser seawater attempts to flow upstream along the river bottom. The seawater is held back by the flow of the river, which causes a sharply tilted boundary to form between the intruding wedge of seawater and the fresh water moving downstream. This pattern is shown in figure 11.1. The net seaward surface flow is almost entirely river water that moves very fast, because the large discharge of fresh water is forced into a thin layer above the salt wedge. The salt wedge moves upstream on the rising tide or when the river is at a low flow stage; the wedge moves downstream on the falling tide or when the river flow is high. The boundary between the salt water and the overriding river water is kept sharp by the rapidly moving river water. The moving fresh water erodes seawater from the face of the wedge, mixing it upward into the turbulent river water, and raising the salinity of the seaward-moving fresh water. Very little river water is mixed downward into the salt wedge. This one-way mixing process is called **entrainment.** Seawater from the ocean is continually added to the salt wedge to replace the salt water entrained into the seaward-flowing river water. In a salt wedge estuary, the circulation and mixing is controlled by the rate of river discharge; the influence of tidal currents is generally small in comparison to the river flow. Good examples of salt wedge estuaries are found in the mouths of major rivers such as the Columbia and the Mississippi. In the Columbia River, the salt wedge moves as far as fifteen miles upstream at times of high tide and low river flow, but it still maintains its identity with a sharp boundary between the two water types. Salt wedges also occur in the mouths of rivers where the rivers enter other types of estuaries, for example, the mouth of the Sacramento River in San Francisco Bay.

Estuaries that are not of the salt wedge type are divided into three additional categories on the basis of

Figure 11.1

The salt wedge estuary. The high flow rate of the river holds back the salt water, which is drawn upward into the fast-moving river flow. Salinity is given in parts per thousand (°/₀₀).

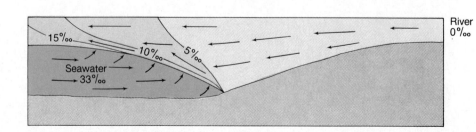

their circulation and the vertical distribution of salinity. These categories are the well-mixed estuary, the partially-mixed estuary, and the fjord-type estuary.

Well-mixed estuaries (fig. 11.2) have strong tidal mixing and low river flow. This pattern creates a slow net seaward flow of water at all depths. Note that the mixing due to strong tidal turbulence is so complete that the salinity of the water is uniform over depth, and decreases from the ocean to the river. Shallow estuaries, including the Chesapeake and Delaware bays, are well-mixed estuaries. There is little or no transport of seawater inward at depth; instead, salt is transferred inward by turbulent diffusion. The nearly vertical lines of constant salinity move seaward on the falling tide or when river flow increases and landward on the rising tide or when river flow decreases.

Partially-mixed estuaries (fig. 11.3) have a strong net seaward surface flow of fresh water and a strong inflow of seawater at depth. The seawater is mixed upward and combined with the river water by tidal current turbulence and entrainment to produce a seaward surface flow that is larger than that of the river water alone. This two-layered circulation acts to rapidly exchange water between the estuary and the ocean. Salt moves into a partially-mixed estuary by turbulent diffusion, but more importantly, by the inflow of seawater at depth. This transport by flowing water is termed **advection.** Examples of partially-mixed estuaries include deeper estuaries, for example Puget Sound and San Francisco Bay.

Fjords or **fjord-type estuaries** (fig. 11.4) are those deep estuaries that have a moderately high river input but little tidal mixing. This pattern occurs in the deep and narrow fjords of British Columbia, Alaska, the Scandinavian countries, and other glaciated coasts. In these estuaries, the river water tends to remain at the surface and move seaward, blending little with the underlying salt water. Most of the net flow is in the surface layer, and there is little influx of seawater at depth. In the fjord-type estuary, advective transport alone supplies salt but the rate is slow. Because of the low inflow at depth and the isolation of the deeper water by the entrance sill, this deeper water may stagnate.

Each of these four types of estuaries is then further analyzed based on the degree of vertical layering, or stratification, measured by salinity changes over depth.

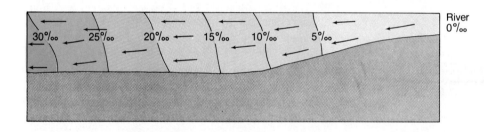

River 0‰

Figure 11.2

The well-mixed estuary. Strong tidal currents distribute and mix the seawater throughout the shallow estuary. The net flow is weak and seaward at all depths. Salinity is given in parts per thousand (‰).

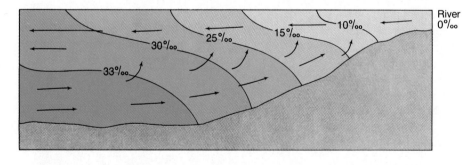

River 0‰

Figure 11.3

The partially mixed estuary. Seawater enters below the mixed water that is flowing seaward at the surface. Seaward surface net flow is larger than river flow alone. Salinity is given in parts per thousand (‰).

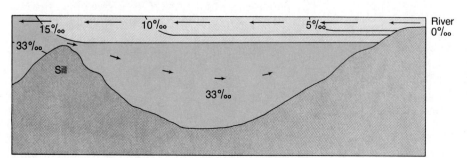

River 0‰

Figure 11.4

The fjord-type estuary. River water flows seaward over the surface of the deeper seawater and gains salt slowly. The deeper layers may become stagnant due to the slow rate of inflow. Salinity is given in parts per thousand (‰).

The salt wedge estuary exists only as a strongly stratified estuary. These estuaries are never very well mixed vertically, except above the boundary between the layer of seaward-moving river water and the salt wedge. The other estuary types have various degrees of vertical stratification. The degree of stratification falls between the highly stratified, or poorly mixed, and the weakly stratified, or well mixed. The processes that govern the degree of vertical mixing and stratification are the strength of the oscillatory tidal currents, the rate of freshwater addition, the roughness of the topography of the estuary over which these currents flow, and the average depth of the estuary. Tidal flow that oscillates with the rise and fall of the tide does not usually result in a preferred directional flow and should not be confused with the net flow in each estuary.

Not all estuaries fit neatly into one of these categories. There are estuaries that fit between the types, and some estuaries change from one type to another seasonally with changes in river flow or weekly as the tides change from springs to neaps. Studying how an individual estuary relates to examples of the different types helps the oceanographer understand the processes that govern an estuary and its water exchange with the ocean.

Circulation Patterns

In this discussion of estuary circulation, we will use as an example the partially-mixed estuary, with its inflow from the ocean at depth and its surface outflow of mixed river and seawater. As the water volume of the estuary increases on the rising tide and decreases on the falling tide, tidal currents are produced. Water in these currents moves back and forth in the estuary, but it does not necessarily leave the system and enter the open sea. However, the surface water moves farther seaward on the falling tide than it moves inward on the rising tide. A parcel of surface water is moved progressively farther seaward on each tidal cycle. In the same way, the water from the sea moves into the estuary on the rising tide farther than it drops back seaward on the falling tide. Averaged over many tidal cycles, this pattern produces a net movement seaward at the surface, and a net movement toward the land at depth.

The **net circulation** of an estuary out (or seaward) at the top and in (or landward) at depth is of great importance. It is this circulation that carries wastes and accumulated debris seaward and disperses them in the larger oceanic system. It is also through this circulation that organic materials and juvenile organisms produced in the estuaries and their marsh borderlands are moved seaward, while nutrient-rich water is brought inward at depth to replenish the estuary.

Understanding the net circulation of an estuary and evaluating the flows of surface water seaward and the underlying ocean water toward land can be a long and expensive process. To directly make this determination requires the installation of recording current meters at various depths and at several cross-channel locations in the estuary. The currents are measured over many tidal cycles for both spring and neap tides, and at times of low, intermediate, and high freshwater discharge. The resulting current records at each depth and channel location are averaged to determine the net current distribution over depth for a given cross section in the estuary. This yields the net flow of the estuary. Data are averaged over one-week periods to determine the importance of spring and neap tides, over monthly cycles to find changes in patterns due to seasonal river and climate fluctuations, and over several years to produce an annual mean pattern. The cost of installing and maintaining equipment, as well as the costs of processing the collected data, make this direct approach a very expensive one.

There is a less expensive, indirect approach. In this method, a simple **water budget** is determined for the estuary. The estuary is assumed to have a constant volume of water when averaged over time. Therefore, all processes that add water to the estuary must equal all processes that remove water from it. A **salt budget** is also assumed; salt added is equal to salt removed. At the entrance to the estuary, measurements of salinity are made and averaged over space and time. In this way, the average salinity of the outflowing surface water, $\overline{S_o}$, and the average salinity of the deeper inflowing seawater, $\overline{S_i}$, are established. The formula

$$\frac{\overline{S_i} - \overline{S_o}}{\overline{S_i}}$$

is used to determine the fraction of river water in the seaward-moving surface layer at the estuary entrance.

If the rate of total river water inflow, R, is known for this same time period, T_o, the volume rate of seaward flow of the surface layer at the estuary entrance can be found by using the following formulas and without direct measurement:

$$\frac{\overline{S_i} - \overline{S_o}}{\overline{S_i}} \, (T_o) = R \ \text{ or } \ T_o = \frac{\overline{S_i}}{\overline{S_i} - \overline{S_o}} \, (R).$$

In a partially-mixed estuary, T_o is always larger than R. This means that both the river inflow, R, and the saltwater inflow, T_i, combine to produce the seaward flow, T_o, and maintain the water budget:

$$T_o = T_i + R.$$

Figure 11.5 shows that both evaporation, E, and precipitation, P, also remove and add water at the surface. If both E and P were large, then R in the above equations becomes $(R - E + P)$.

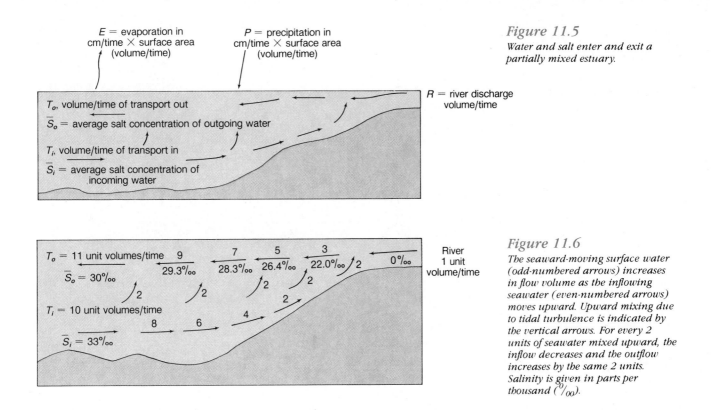

Figure 11.5
Water and salt enter and exit a partially mixed estuary.

Figure 11.6
The seaward-moving surface water (odd-numbered arrows) increases in flow volume as the inflowing seawater (even-numbered arrows) moves upward. Upward mixing due to tidal turbulence is indicated by the vertical arrows. For every 2 units of seawater mixed upward, the inflow decreases and the outflow increases by the same 2 units. Salinity is given in parts per thousand ($^o/_{oo}$).

This method assumes that, over the time period for which the calculations are used, the average water volume and the total salt content of the estuary remain constant.

Temperate Zone Estuaries

In the middle latitudes of North America, most estuaries gain their fresh water from rivers, and the evaporation of water from the estuary surface is minor or is nearly balanced by direct precipitation. The salt concentration of the entering seawater, $\bar{S}_i$, is about $33^o/_{oo}$, and $\bar{S}_o$ is about $30^o/_{oo}$. Using the equation just given, T_o is calculated to be about 11 times the rate of river inflow, or R. T_i in this case is 10 times R. Figure 11.6 shows how the influx of seawater at depth decreases as it moves into the estuary and how the surface flow increases due to mixing and incorporating seawater from below as it moves seaward. While this mixing is occurring, the average salt concentration of the seaward-moving surface flow is also increasing. Ten unit volumes of seawater have combined with the one unit volume of river water to form T_o. These relatively large values of T_o and T_i compared to R make them very important in the exchange of estuary water with ocean water.

In some fjord-like estuaries, the surface freshwater layer is as deep as the shallow sill at the entrance. The average salt content of the surface layer at the entrance is then very low. In this case, T_o is approximately equal to R, and T_i is negligible. In this situation, the deep water of the fjord can stagnate.

11.2 Embayments with High Evaporation Rates

Embayments located near 30°N and 30°S latitudes have low precipitation and high evaporation rates. Embayments subject to these conditions are not estuaries, because they do not experience net dilution. However, they obey the same rules for water and salt budgets as do estuaries. These regions may have rivers, but the contribution of these rivers is minor when compared to the loss of water due to evaporation over the entire basin. The Red Sea and the Mediterranean Sea are examples of such areas. In an evaporative type of sea, the process of evaporation makes the surface water salty and therefore more dense. This water sinks and accumulates at depth. It flows seaward at depth to exit into the ocean at the sea's entrance. The ocean water is less dense than the outflowing deep water and flows into the embayment at the surface. The T_o and T_i flows are present, but they have reversed their depth of occurrence, with T_o at depth and T_i at the surface. Because of this reversal, these embayments are sometimes called **inverse estuaries** (fig. 11.7).

If the excessive evaporation rate removing water from the surface of an evaporative embayment is equal to the rate of river inflow to an estuary, the T_o values for the two embayment types can be compared. In figure 11.7, a typical $\bar{S}_i$ value ($36^o/_{oo}$) is given for ocean surface conditions existing at latitudes of 30°N and 30°S. The

Figure 11.7
The evaporative sea. Evaporation at the surface removes water (1 unit) and river inflow is negligible. Seawater flows inward at the surface ($T_1 = 20$ units), and the seaward flow ($T_o = 19$ units) is at depth. Salinity is given in parts per thousand.

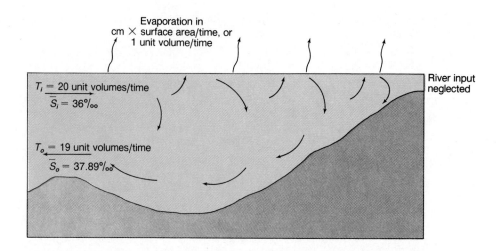

evaporative loss from the embayment is set at 1 unit volume per time. This evaporation results in a T_i flow of 20 units and a T_o flow of 19 units. The resulting $\overline{S_o}$ value is $37.89\%_{00}$. Compare figure 11.7 with the estuary in figure 11.6, where river input of 1 unit volume per time results in T_o and T_i flows of 11 units and 10 units, respectively. In general, if E in an inverse estuary is comparable to R in an estuary, the inverse estuary has a better rate of exchange with the ocean than a true estuary. Not only is T_o likely to be larger, but because the inverse estuary has a continuous overturn as surface water drops to depth and moves seaward, the exchange of water is more complete.

11.3 Flushing Time

If the mean volume of an estuary is divided by T_o, an estimate of the length of time required for the estuary to exchange its water is obtained. This is the **flushing time.** If the net circulation is rapid and the total volume of an estuary is small, the bay flushes rapidly. A rapidly flushing estuary has a high carrying capacity for wastes, because the wastes move rapidly out to sea, where they are diluted by the adjacent ocean water. A slowly flushing estuary risks accumulating wastes and building up high concentrations of land-derived pollutants. Understanding the circulation of our estuary systems is essential to maintaining them as healthy, productive, and useful bodies of water. Population pressures on estuaries are heavy, for these areas are used as seaports, recreational areas, and industrial terminals, and for their fishing resources.

If the waste products are associated only with the fresh water entering the estuary, it is possible to determine the ability of the fresh water to carry these products through the estuary without considering the action of the entire estuary. This requires determining how much fresh water is in the estuary at any one time. The

freshwater volume is estimated from the salinity of the estuary, which is the result of mixing fresh water and ocean water. For example, an estuary with an average salt concentration of $20\%_{00}$ is one-third fresh water if the adjacent ocean salinity is $30\%_{00}$. Dividing the freshwater content by the rate of river water addition yields the flushing time for freshwater and its associated pollutants.

In some instances, there are bays or harbors that are isolated parcels of seawater that do not qualify as estuaries, for they have no freshwater source for dilution. In these cases, any flushing is accomplished by tidal action. On each change of the tide, an amount of water equal to the area of the bay times the water level change between high and low water, called the **intertidal volume,** exits and enters the bay. This volume, when leaving the bay and entering the ocean, may be displaced down the coast by a prevailing coastal current. Then, on the rising tide, an equivalent volume of different ocean water enters the bay. In this case, the flushing time is measured in tidal cycles and is determined by dividing the mean bay volume by the intertidal volume exchanged per tidal cycle. Often this flushing is not complete, because currents along the coast do not always displace the intertidal volume a sufficient distance to prevent some portion of it from cycling back into the bay. Therefore, the number of tidal cycles required for flushing may be greater than the calculated estimate.

The same holds for those estuaries that have a two-layered flow. The seaward-moving surface layer, T_o, may experience some mixing with the T_i flow due to turbulence present at the entrance of the estuary. This mixing may incorporate some of the exiting estuary surface water into the deeper inflowing water, which results in a partial recycling in the estuary. Recycling of this type not only increases the flushing time, but also mixes the surface water to depth, aerating the deeper parts of the estuary.

Occasionally, the net circulation in an estuary is altered by ocean conditions. If oceanic conditions increase the salinity and density of the water present at the entrance to an estuary, for example if coastal upwelling occurs, the inflow of water to the estuary increases with the denser water entering the estuary at depth. This increased inflow accelerates the estuary's circulation and causes the outflow to increase. When upwelling ceases and downwelling occurs, less dense seawater is present at the estuary entrance, and the inflow is reduced. In these circumstances, the estuary's circulation may temporarily reverse as the denser, deeper water flows back to sea and the less dense seawater moves inward at the surface.

11.4 Use Conflicts

Worldwide, major population centers have developed on estuaries and upstream along their rivers. Historically, people depended on access to salt water for trade and transport and to fresh water for their living requirements. In the United States today, over one-half of the population lives within fifty miles of the coasts (including the Great Lakes). This population, with its necessary industries, energy-generating facilities, and waste-treatment plants, has created a tremendous burden on the fringing areas of the ocean and the rivers entering these areas. In the past, these natural waters were assumed to be infinite in their ability to absorb and remove the by-products of human populations. However, too much use and too many wastes discharged into too small an area at too rapid a rate have produced problems that can no longer be overlooked. These discharges can exceed the ability of the natural systems to flush themselves and disperse their wastes into the open ocean.

Recently, the realization has come that changes in the environment of these coastal regions were seriously degrading them, decreasing their aesthetic and economic value, and, in some cases, endangering public health and safety as well as threatening their living resources. The 1960s and 1970s saw a rise in environmental legislative action that produced the laws and agencies that presently monitor and control the uses of these fragile areas. Today, there is a greater concern to act with care and judgment, but today we also have a legacy from past practices that is usually difficult and expensive to correct. The following discussions consider some of the major problems associated with our coastal environment.

Water Quality

Dumping solid waste and pouring liquid pollutants into shore waters are inexpensive solutions to industrial and domestic waste problems. However, these practices have led to environmental degradation, with its hidden costs. The expense of cleaning up and of building plants to process these wastes is high and is increasing. The costs are paid by all of us as increased taxes and increased product prices. In times of high costs and high unemployment, it becomes more difficult to justify the short-term costs required to conserve natural marine resources, no matter how high the potential for eventual economic gain. When living is easy, we are more apt to consider this potential and to pay the costs of protecting the environment.

Off the mouth of the Hudson River lies the New York Bight, between New York and New Jersey. Street sweepings, garbage, dredge spoils, cellar dirt, and waste chemicals have been dumped into this area since the 1890s. Increasing quantities of floating debris found their way back to the beaches until 1934, when laws were passed to prohibit the dumping of "floatables." However, refuse from building and subway construction, toxic wastes from industry, acids, and sewage sludge continued to be dumped. It is estimated that between 1890 and 1971 the volume of solid waste dumped into the water of the New York Bight was 1.4 million m³. M. Grant Gross, in his NOAA monograph on waste disposal in the New York Bight, estimates this waste as the equivalent of a layer covering all of Manhattan Island to the height of a six-story building, or 20 m (66 ft). The dumping in the New York Bight area has degraded the quality of the water and the sediments. Many of the materials added to the water are both toxic and oxygen-demanding. Occasionally, this degraded water upwells along the coast and causes the death of marine organisms in shallower waters. In 1978, the Atlantic coastal waters received 8.3 million tons of material, primarily industrial waste, sewage sludge, and demolition debris.

Although the indiscriminate industrial dumping of materials in estuaries and coastal areas has been substantially reduced in response to recent legislation, there are other routes that pollutants or toxicants follow to reach the coastal zone. Pesticides such as DDT (dichloro-diphenyl-trichloro-ethane), long-lived toxic organic compounds such as polychlorinated biphenyls (PCBs), and heavy-metal ions such as lead, mercury, zinc, and chromium are still making their way through the environment, even though many of these materials are closely controlled or no longer used or manufactured. In the United States, PCB manufacture and sale stopped in 1978 and 1979, but electric transformers charged with PCB are still in use. Although decontamination techniques have been developed to facilitate disposal, PCBs are a widespread pollutant in waters all around the world. Table 11.1 presents data on PCBs in United States waters.

Despite international efforts to tighten controls on manufacturing processes and to reclaim toxic materials, accidents do happen. In November 1986 a fire in Basel,

TABLE 11.1

PCB Concentrations in Water and Sediments

Location	Dissolved in water[1]	In sediments[1]
Fresh waters		
Lake Ontario	0.055	121.0
Lake Michigan		38.0
Hudson River	< 0.01–2800.0	0–6700.0
Connecticut River		0–3500.0
Duwamish River, Wash.	0.03	0–2000.0
Marine waters		
Atlantic Ocean	< 0.0009–0.0036	
Gulf of Mexico	0.0008–0.0041	< 0.2–35.0
California Current	0.0044	
Palos Verdes, Calif.	0.0088	30.0–7900.0
San Francisco Bay		0–1400.0
Puget Sound	0.0067	143.0

Source: From Pavlou, S. P., R. N. Dexter and W. Hom, "Polychlorinated Biphenyl (PCB's) in Puget Sound: Physical/Chemical Aspects and Biological Consequences." 1977; The Use Study and Management of Puget Sound, A Symposium, Washington Sea Grant.
[1]In parts per billion (PPB). EPA standard for chronic level in seawater = 0.030 PPB.

TABLE 11.2

Combined Annual Mass Discharges of Seven Southern California Municipal Wastewater Treatment Facilities[1]

	1971	1972	1973	1974	1975	1976	1977
Flow							
MGD	931	922	955	967	985	1027	966
L day × 10⁹	3524	3490	3615	3360	3728	3889	3658
General constituents (mt/year)							
Total sus. sol.	288,000	279,000	270,000	264,000	287,000	288,000	244,000
5-Day BOD	283,000	250,000	217,000	222,000	237,000[2]	259,000[2]	244,000
Oil and grease	63,500	60,600	57,400	54,700	57,400	59,100	49,000
NH3–N	56,600	39,900	45,900	37,000	36,620	37,350[2]	41,200
Trace metals (mt/year)							
Silver	17.7	21.1	29.0	21.7	25.7	20.2	34.3
Arsenic	ND	ND	ND	20.9[3]	11.9[3]	10.5[3]	14.0
Cadmium	57.3	33.8	49.3	55.4	50.0	45.0	42.4
Chromium	676	673	695	690	580	593.0	366
Copper	559	485	509	575	511	507	412
Mercury	ND	ND	ND	3.1[3]	2.2[3]	2.6[3]	2.8
Nickel	339	273	318	314	234	307	264
Lead	243	226	180	199	196	191	152
Selenium	ND	ND	ND	17.75[2]	16.9[4]	22.0[4]	23.0[4]
Zinc	1880	1210	1360	1320	1142	1064	837
Chlorinated hydrocarbons (kg/year)							
Total DDT	21,700	6600	4120	2120	1989	1673	920
Total PCB	8730	9830	4620	9390	6011	4310	2183

[1]Oxnard included only since 1975. Serra and Encina included since 1982.
[2]Hyperion 7-mile effluent excluded.
[3]Orange County Sanitation Districts data not included.
[4]Total for Hyperion and Joint Water Pollution Control Plant only.
Source: Characteristics of Municipal Wastewater in 1984 and 1985, Southern California Coastal Water Research Project Report, 1986.

Switzerland, consumed the warehouse of a chemical firm. Water used to extinguish the blaze washed 30 tons of herbicides, pesticides, and mercury into the Rhine River. A contaminated slug of river water 25-miles (41.5 km) long moved downstream to the North Sea. Nearly all the plant and animal life over a 185-mile (307 km) stretch of this 820-mile (1361 km) long river was destroyed. Over half a million fish and eels were estimated to have died. Towns dependent on the river for their water supply had to have water trucked in from other sources.

One week later, contaminated water from the same plant spilled into the Rhine again. While checking for the chemicals from these two spills, it was discovered that another, unreported spill of pesticides from a different source had entered the river two days prior to the fire and the initial spill. Ten years of effort to clean and restock the river have been destroyed. It is thought that the river may be able to support fish again in 2–3 years. From the disaster have come new efforts in the nations that border the river to force the reconsideration of laws that regulate the chemical industries of the region.

Surface runoff from agricultural lands finds its way through lakes, streams, and rivers to the estuaries. The runoff supplies pesticides and nutrients, which can poison or overfertilize the waters. In the latter case, the resulting overpopulation of plants eventually dies and decays, removing large quantities of oxygen. Lack of oxygen then kills other organisms, which in turn decay and continue to remove oxygen from the water.

Metropolitan areas have surface runoff via storm sewers that add a wide mix of materials, including hydrocarbons from oil, lead from gasoline, residues from industry, pesticides and fertilizers from residential areas, and coliform bacteria from fecal material. Cities also add treated sewage effluent, which contributes its share of contaminants. Even the chlorine added first to municipal drinking water and then again to the treated sewage effluent as a bactericide may form a complex with organic compounds in the water to produce chlorinated hydrocarbons, which may be toxic in the marine environment. Considering all the pathways, sources, and types of materials that can be classed as pollutants, their management and eventual exclusion from the coastal zone is an extremely difficult and complex problem, since it cannot be solved by sewage treatment alone.

Table 11.2 shows the annual average mass of treated sewage components delivered to the coastal region by seven southern California municipal discharge systems. Because of the area's increasing population, the number of millions of gallons per day

1978	1979	1980	1981	1982	1983	1984	1985
1015	1054	1097	1097	1134	1166	1174	1190
3840	4000	4160	4160	4292	4414	4444	4504
256,000	243,000	233,000	226,000	227,000	247,000	198,000	205,000
237,000[2]	246,000[2]	260,000[2]	264,000[2]	269,000[2]	256,000[2]	230,000	255,000
49,000	45,000	39,000	37,000	31,900	36,300	30,200	34,300
39,500	41,200	42,000[2]	41,000[2]	44,000	40,600[2]	40,800	44,200
32.3	42.2	30.8	27.9	25.9	25.6	24.5	27
14.5	15.4	10.6	12.2	8.7	10.1	18.1	15.8
44.8	42.3	39.5	31.7	21.2	23.6	16.2	16.5
280	237	275	187	203	164	140	110
417	359	336	339	286	247	252	240
1.9	2.5	1.9	1.8	1.2	1.2	1	1
320	256	224	167	169	165	134	120
219	223	175	130	123	98.7	94	120
23.0[4]	7.7[4]	10.5[4]	15.3[4]	9[4]	10[4]	9[4]	13.3
905	724	730	540	549	505	374	377
1110	760	644	474	289	218	306	73
2510	1190	1129	1250	857	1440	1340	891

(MGD) of treated waste effluent has been steadily increasing. However, the mass of sewage components discharged each year has been decreasing, due to increased efforts that prevent harmful materials from entering waste streams at their sources and to increases in the level of wastewater treatment.

Efforts to reduce the discharge of toxic materials into the marine environment are showing some signs of success. Improving the quality of treated sewage discharged into the nation's rivers is improving the levels of dissolved oxygen in the waters around discharge sites. While the increased use of agricultural fertilizers contributes nutrients to the freshwater runoff, improved sewage treatment decreases the nutrient level, and in general nutrient levels appear to be unchanged. The increase in the use of unleaded gasoline appears to be directly related to a reduction in the amount of lead entering the marine environment. Concentrations of lead in the surface waters of the Sargasso Sea have dropped about 30% between 1980 and 1984. Lead carried by the Mississippi River to the Gulf of Mexico has been reduced about 40% in the last decade.

Many toxicants reaching the estuaries do not remain in the water, but become adsorbed onto the small particles of matter suspended in the water column. These particles clump together and settle out due to their increased particle size. Concentrations of toxicants always tend to be higher in the sediment than in the overlying water; compare the PCB values in water and sediments in table 11.1.

Some of the particulate matter adsorbing toxicants has a high organic content and forms a food source for marine creatures. In this way, heavy metals and organic toxicants associated with the particles find their way into the body tissues of organisms, where they may accumulate and be passed on to predators. Throughout the United States, toxic residues are found in estuarine bottom fish. Shellfish concentrate heavy metals at levels that are many thousands of times over the levels found in the surrounding waters. The scallop can elevate levels of cadmium in its tissues about two million times over the water concentration, and oysters can concentrate DDT ninety thousand times over this concentration.

In 1967, a study was made of DDT sprayed on Long Island marshes to control mosquitoes. Although spray concentrations were not directly lethal to fish and birds, table 11.3 shows the effect of this long-lived toxic material as it was concentrated by the food chain. DDT use in the United States has decreased, but there have been increases in the use of other long-lived toxicants such as dieldrin and aldrin, which are also concentrated by fish and shellfish.

An organism's ability to accumulate substances not only increases its concentrations of toxicants to levels that are sometimes injurious or fatal to the accumu-

TABLE 11.3
Food Chain Concentration of DDT

	DDT residues[1]
water	0.00005
plankton	0.04
silverside minnow	0.23
sheepshead minnow	0.94
pickerel (predator)	1.33
needlefish (predator)	2.07
heron (small animal predator)	3.57
tern (small animal predator)	3.91
herring gull (scavenger)	6.00
osprey egg	13.8
merganser (fisheater)	22.8
cormorant (fisheater)	26.4

Source: From Woodwell, et. al. "DDT Residues in an East Coast Estuary" 1967, *Science,* vol. 156 pp. 821–24, 12 May 1967. Copyright 1967 by the American Association for the Advancement of Science.
[1]Given in parts per million (PPM).

lating organism, but also produces organisms that are a hazard to humans, who may use them for food. A tragic example of the results of humans' ingesting organisms that had accumulated a toxin occurred between 1953 and 1960 in Minamata, Japan. Mercury from a local industrial source was released at high levels into the coastal embayment, from which much of the shellfish harvest for a local village was gathered. The mercury formed a complex that was readily taken up by the marine life, which led to severe mercury poisoning and death among those eating the shellfish. The physical and mental degenerative effects of this mercury poisoning were especially severe on children whose mothers had eaten large amounts of shellfish during pregnancy. The condition produced has been named Minamata disease.

The Plastic Trash Problem

Walk any beach in any estuary or along any open coast and see the tide of plastics being washed ashore. It is estimated that every year more than 77 tons of plastic trash is routinely and legally dumped by naval and merchant ships. The National Academy of Sciences estimates that the commercial fishing industry yearly loses or discards about 298 million pounds of fishing gear (nets, ropes, traps, and buoys) made mainly of plastic, and dumps another 52 million pounds of plastic packaging materials. Recreational vessels, passenger vessels, and oil and gas drilling platforms all add their share.

TABLE 11.4

Sources of Oil in the Oceans[1]

	According to U.S. Maritime Administration Tanker Construction Program, 1973	According to National Academy of Sciences, 1973
tankers	1.39	1.85
other vessels	0.85	0.09
offshore production	0.10	0.08
coastal oil refineries	0.30	0.18
industrial waste	0.75	0.27
municipal waste		0.27
urban runoff		0.27
motor vehicles	1.40	
river runoff		1.50
natural seeps	0.10	0.55
atmospheric rainout		0.55
total	4.89	5.61

[1]Amounts are $\times 10^6$ metric tons.
No data indicates source not considered.

Plastics are a worldwide problem, with many sources and effective distribution to even the most remote areas by the currents.

Plastic is inexpensive, strong, and durable; these characteristics make it the most widely used manufacturing material in the world today. These characteristics also make it a major environmental problem. Nobody knows how long plastic stays in the marine environment, but an ordinary plastic six-pack ring could last 450 years. Thousands of marine animals are crippled and killed each year by these materials. More than thirty thousand fur seals are entangled in lost or discarded plastic fishing nets and choke to death in plastic cargo straps. Lost lobster and crab traps made entirely or partially of plastic continue to trap animals; 25% of the 96,000 traps set off Florida's west coast were lost in 1984. Seabirds die entangled in six-pack rings and plastic fishing line. Seabirds and marine mammals swallow plastics. Porpoises and whales have been suffocated by plastic bags and sheeting. Fish are trapped in discarded netting. Sea turtles eat plastic bags and die. Plastics are now considered as great a source of mortality to marine organisms as oil spills, toxic wastes, and heavy metals.

The United Nations' International Maritime Organization has proposed an amendment to the Marine Pollution Convention that would ban dumping of plastic debris at sea. If it is ratified, there is no method of enforcement. Eleven states have passed laws requiring that six-pack rings be made of biodegradable plastic. Manufacturers add light-absorbing molecules that break down the plastic after a few months' exposure to sun-light; there is no evidence that this will solve the problem at sea. Even at the surface, the water keeps the plastic cool and it becomes coated with a thin film of organisms that shade it from the light. As the problem worsens, it may be that only people educated to act in a responsible manner will reduce the tide of plastics rising around the world.

Oil Spills

The twentieth century runs on oil, petroleum, and petroleum products. This dependence requires the bulk transport of oil by sea to bring the crude oil to the land-based refineries and centers of use. This transport process exposes the world's coasts and estuaries to the hazard of oil spills associated with vessel casualties and transfer procedures. Because oil is found below the sea floor, the drilling of offshore wells exposes these areas to the risks of blowouts and accidental spills. Also, because industry, agriculture, and private and commercial transportation require petroleum and petroleum products, oil is constantly being released into the environment, to find its way directly or indirectly down to the sea. Refer to table 11.4 for estimates of the sources of oil entering the oceans.

The transport of oil and oil products on the high seas creates the potential for accidents that release large volumes of oil. Spills far at sea are difficult to assess, because direct visual and economic impact on coastal areas does not occur. Damage to marine life cannot be accurately evaluated in such a case. Spills occurring due

Figure 11.8
The super tanker, Amoco Cadiz, *aground and broken in two off the coast of France.*

Figure 11.9
A catamaran oil skimmer. The vessel cruises at about 3 knots, guiding surface oil between the twin hulls. The oil adheres to a moving belt and is lifted into on-board storage tanks.

to the grounding of vessels or due to accidents in the storing and transferring processes are evaluated in coastal areas, where the environmental degradation and loss of marine life can be readily observed. Spills that occur in estuaries and along coasts affect regions that are oceanographically complex, biologically sensitive, and economically important.

In March 1978 the worst tanker spill to date occurred when the *Amoco Cadiz* lost her steering in the English Channel and broke up on the rocks of the Brittany coast of France. (fig. 11.8). Gale winds and high tides spread the oil over more than 300 km (180 mi) of the French coast; more than 3000 birds died; oyster farms and fishing suffered severely. Of the approximately 2.1×10^5 metric tons of oil spilled, it is considered that 7×10^4 metric tons evaporated and were carried over the French countryside, $3–4 \times 10^4$ metric tons were cleaned up by the army and volunteers, another $3–4 \times 10^4$ metric tons penetrated down into the sand of the beaches to stay until winter storms washed it away, $2–3 \times 10^4$ metric tons (the lighter, more toxic fraction) dissolved in the seawater, and $4–5 \times 10^4$ metric tons sank to the sea floor in deeper water, where it will continue to contaminate the area for an unknown period of time.

There is no technology available to cope with large oil spills, particularly under bad weather and sea conditions, along irregular coastlines, and far away from land-based supplies. Very little of the oil spilled under these conditions is recovered. The technology for oil cleanup at sea includes oil booms and oil skimmers (see fig. 11.9). These devices are useful in confining and recovering small spills in protected waters. A bacterium that is capable of breaking down most of the com-

pounds present in crude oil has also been developed, but before this "superbug" can be used against an oil spill, much more testing is required to determine whether it has other, undesired effects on the environment.

Not all large spills are associated with tankers or transfer procedures. The world's largest and longest-lasting oil spill began in June 1979, when the Petroleus Mexicanos well *Ixtoc 1* blew out and caught fire in the Gulf of Mexico. The well was not capped until March of the next year. Similar but less severe accidents have occurred in the North Sea, the Persian Gulf, and off the United States coast. The bombing of oil facilities in the Persian Gulf during the Iran-Iraq War has produced major spills of unknown volume and consequence.

The immediate damage from a large spill is obvious and dramatic; by contrast, the effect of the small but continuous additions of oil that occur in every port and harbor are much more difficult to assess, because they produce a chronic condition from which the environment has no chance to recover. Refined products such as gasoline and diesel fuel are more toxic to marine life than crude oil, but they evaporate rapidly and disperse quickly. They are less visible and so appear less offensive. Crude oil is slowly broken down by the action of water, sunlight, and bacteria, but the portion that settles on the sea floor moves down into the sediments, which it continues to contaminate for years.

The world's major ports and harbors are usually associated with the world's major bays and estuaries. Since most petroleum is transferred from vessel to shore at these locations, these sensitive, productive, and valuable areas will continue to be particularly vulnerable to oil damage in the foreseeable future.

Figure 11.10

The industrialized estuary of the Duwamish River at Seattle, Washington.

Figure 11.11

The wetlands of Barnegat Bay, New Jersey, were replaced by a housing and recreational complex.

Alteration of Wetlands

Many now recognize the value of estuarine areas as centers of productivity and as nursery areas for the coastal marine environment. Federal and state environmental legislation regulates the development and modification of these areas. Wherever possible, these lands are being managed and rehabilitated to their natural state. Where human use requires the permanent removal of estuarine land, there is now an attempt to mitigate the loss by restoring other degraded aquatic land or to assure that the remaining natural areas be perpetually preserved.

Bordering the estuaries are wetlands, particularly freshwater and saltwater marshes and swamps. These wetlands provide nutrients, food, shelter, and spawning areas for marine species, including such commercially important organisms as crabs, shrimp, oysters, clams, and many fish. In the past, these wetlands have been filled to produce usable ground for industry and port facilities adjacent to the marine waterways. Bulkheads have been built along natural shorelines, and channels have been dredged through these areas to permit the entry of deep-draft vessels into ports and harbors (fig. 11.10). Our desires to live near the shores have created modern Venices of recreation and retirement in what were once marshes bordering estuaries. These projects destroy natural nursery areas for marine organisms by replacing them with waterfront lots and homes that each have their own individual pleasure craft moorage (fig. 11.11). The growth of industry and its diversification, as well as a rising population in the coastal zones, has created enormous pressures on these areas.

It has been estimated that the United States originally had some 127 million acres of natural wetlands. A portion of these wetlands were classified in 1780, showing that 6 million acres were freshwater marshes at the heads of bays, where they were only slightly af-

TABLE 11.5

A Portion of the Natural Wetlands in the United States[1]

	1780	1954	1978
freshwater marshes in marine areas	6	5.98	3.99
salt marshes	4.5	4.49	3.98
mangrove swamps	0.5	0.5	0.5

[1]Values are in millions of acres.

fected by saltwater. Another 4.5 million acres were more coastal and were thus more greatly influenced by seawater; these were the salt marshes. An additional 0.5 million acres of mangrove swamps were associated with warmer coastal waters. Table 11.5 shows what has happened to those originally surveyed acres in the last two hundred years.

Each estuary is different from every other estuary. Although each estuary is unique and must be judged accordingly, the knowledge gained from studying one estuary can be used to understand another system. Understanding the circulation patterns and the processes that form these patterns allows us to judge the degree to which an estuary can be modified and used while still preserving its environmental and economic value. Because estuaries lie at the contact zone between the land and the sea, and because humans are inhabitants of the land, alterations have been made in these areas and probably cannot be entirely avoided in the future. But we owe it to ourselves today and to those who come after us to make the most intelligent and knowledgeable uses possible of these areas, for they represent a great natural resource that can renew itself if treated with care.

Going, Going, Gone

*L*ouisiana is shrinking. Each year 40 to 50 square miles of Mississippi River wetland marsh disappear under the Gulf of Mexico. Twelve thousand years ago, Louisiana's shoreline extended 120 miles further into the Gulf of Mexico. In the mid-1800s, Bailize was a busy river town at the tip of the delta; today, it is under 15 feet of water. The wetlands of coastal Louisiana provide more than 4 million acres of marshes, ponds, lakes, bayous and shallows, habitat for fish, shellfish, birds, and other animals. This great natural resource, important to the economy, biology, and history of the region, is dwindling.

Under natural conditions, the deposition of sediment by a river gradually raises the river bed, decreasing its steepness and speed of flow. During a period of high flow, a river spills over its natural banks or levees and finds a new course where the steepness of the new channel is greater. In this way, a river wanders, leaving its old channel for a new one. This wandering changes the portion of the delta to which sediments and fresh water are being supplied. Older portions of a delta that have lost their sediment supply subside and erode, decreasing in area. The new river channel creates growth of delta wetlands in a new area.

The delta of the Mississippi River has been building over millions of years. Over this time, the sediments deposited by the river have added weight to the earth's crust, causing the crust to sink or subside; at the same time, the sediments have been eroded and dispersed by coastal waves and currents. If the supply of sediments exceeds the rate of subsidence and the rate of erosion, the delta grows and the wetlands increase in area. If the reverse occurs, the wetlands are lost to the Gulf of Mexico.

The Mississippi River has most recently been maintained in a fixed channel by raising the levee heights on each side of the river. This prevents loss of life and property due to flooding, and keeps open the port of New Orleans' access to the river. However, confining the river to a fixed channel cuts off much of the delta from sediments and fresh water. The sediments continue to the river's mouth, where they are discharged into deeper water. The lack of replenishing fresh water allows the water in the sediments to ooze out, and the slow process of soil compaction as well as subsidence results in the sinking of the land. The loss of the sediment prevents the addition of land to subsiding areas and the reduction of fresh water allows the intrusion of seawater over portions of the delta now protected from the river. The result is a gradual loss of marshland and the conversion of vegetation from freshwater- to saltwater-tolerant species. Salt marshes eventually become bays as the sinking and erosion continues.

Other human activities and population growth contribute to the problem. Extraction of freshwater from the soil lowers the water table. Extraction of gas, oil, and sulfur leaves voids into which the land settles. Added weight from construction results in sediments consolidating and subsiding at faster rates than found in natural environments. All act to lower the land level, increasing the flooding. In the past, farming of the rich delta soils was accomplished by constructing extensive drainage systems to remove surplus fresh water. The sinking of these areas converts the drainage channels into access routes for sea water, which ultimately ruins the land for crop plants. Diking and draining or filling for agricultural, industrial, or residential use eliminate natural wetland areas. Marsh soils high in organic material exposed by these processes shrink, decay, and sink.

Erosion quickly follows subsidence, and natural catastrophes such as hurricanes play a role as well. These storms damage barrier islands and drive salt water into the marshes, damaging the freshwater vegetation.

We may be able to save some marshland by rechanneling sediments to sediment-starved areas, building small tidal dams to prevent saltwater flooding of freshwater marshes, and zoning coastal areas to prevent draining and filling. However, we must also recognize that changes upstream that are destructive to the marsh areas have often been designed to protect people and property in that upstream area. The result is a conflict with no simple solution. In addition, subsidence and erosion are natural processes often beyond our control, and change is a natural part of any environment. There will be changes in the Louisiana wetlands with or without the human factor; it is the rate of change that human activities accelerate. (See colorplate 9.)

11.5 *Practical Considerations: Case Histories*

The Development of San Francisco Bay

The San Francisco Bay estuary has a surface area of 1240 km² (480 mi²). Its river systems drain 40% of the surface area of California. When visited by the early Spanish missionaries in 1769, the bay was surrounded by wetlands, where the 10,000–15,000 native peoples gathered much of their food. Development came slowly until 1848, when gold was discovered, and the population of San Francisco increased from approximately 400 to 25,000 in 1850. Within fifty years of the initial gold boom, the marshes were nearly gone, the bay was shallowed, fresh water had been siphoned off for irrigation, and nonnative species of animals had been introduced to the bay's waters, displacing native species.

Early in the bay's development the fisheries resources (salmon, sturgeon, sardines, flatfish, crabs, and shrimp), were heavily exploited to feed the rapidly expanding population. By 1900, many of the fish, shellfish, and wildfowl stocks had been overharvested and depleted. Crabs were fished out in the 1880s, and the fishery was moved offshore, until its collapse in the 1960s. When the transcontinental railroad was completed in 1869, carload after carload of oysters were shipped from the east to mature on the bay's mudflats. The oysters did not become a self-reproducing stock, but many other small marine animals introduced with the oysters did naturalize, including the eastern soft shell clam, the Japanese little neck clam, and the marine pests: the oyster drill and the shipworm. Even the striped bass, the bay's best known sport fish, was introduced from the East Coast in 1879. The only commercially harvested fish left in the bay are herring and anchovy.

Until 1884, hydraulic mining for gold in the surrounding watershed directly washed huge amounts of silt and mud into the streams and rivers. The sands and muds in the rivers destroyed the salmon-spawning areas. Much of the sand and mud reached the bay, where it shallowed some areas, expanded the marshes, and altered the tidal flow. The rivers were flooded with the runoff from the barren land, and the resulting winter and spring floods produced changes in the estuaries, where the rivers enter the bay.

The marshlands of the bay and its river deltas, particularly the deltas of the Sacramento and the San Joaquin rivers, were diked, first to increase agricultural land and then for homes and industries. The conversion of wet land to dry land has reduced the tidal marshes from an area of about 2200 km² (860 mi²) to 125 km² (49 mi²). The increased cropland required increased irrigation, and the state built a series of dams, reservoirs, and canals with a water storage capacity of 20 km³ (.5 mi³). Today, 40% of the Sacramento and San Joaquin rivers' flow is removed for irrigation, and another 24% is sent by aqueduct to central and southern California. By the year 2000, it is expected that San Francisco Bay will receive only 30% of its 1850 freshwater input.

The diversion of water to the fields so reduces the amount of river water during periods of low flow in the summer that the irrigation pumps cause the water to flow upstream from the bay, carrying with it hundreds of juvenile salmon and striped bass, which are drawn into the pumps and die on the fields or in the ditches. It is thought that this has been a major cause of a drop of more than 75% in the population of striped bass since the 1960s. The loss of freshwater flow also changes the distribution of small, floating species on which the juvenile fish feed.

Intensive agricultural practices using fertilizers and pesticides, plus the runoff of water high in salts leached from the irrigated fields has changed the quality of the fresh water entering the bay. In an effort to lessen this effect, run-off water from agriculture was directed into holding reservoirs, which in some cases were also wildlife refuges. In 1982, dead vegetation, deformed wildlife, and depressed reproduction in bird species were discovered at the Kesterson Reservoir. Levels of the element selenium, 130 times that of normal, were found in plant and animal tissues. The selenium was being leached from the irrigated soils and concentrated in the reservoir. Kesterson Reservoir was closed in 1986, but the water from the fields continues to flow into the San Joaquin and on into the bay. Recently, high concentrations of selenium have also been found in south bay-area ducks.

Domestic and industrial wastes from urban areas also enter the bay. Residence time of water in north San Francisco Bay fluctuates between one day during peak river flow in the winter to two months during low summer flow. Winter flows carry fresh water into the south bay and increase its exchange with the central bay, decreasing south bay residence time from months to weeks at this period of the year. Wastewater discharge is expected to be 8% of the total freshwater inflow to the bay by the year 2000.

San Francisco Bay is considered the most modified major estuary in the United States. The city of San Francisco's population is now more than 700,000, in a metropolitan area of 5 million. Despite all the factors involved in this increasingly crowded area, the bay appears to have suffered less in total water-quality degradation than other major estuaries. In part, this is because the greatest urbanization has occurred near the mouth of the bay, allowing wastes to exit quickly. Improvements in sewage treatment since the 1960s have also improved conditions. There are patches of contamination from PCBs, oil, and chemical spills, and at this time the bay does not seem to be suffering from an overproduction of marine plants or a depletion of oxygen. Shellfish collection has recently been permitted for the first time in decades. As the area continues to develop and change, active research, effective monitoring, and intelligent management are needed to understand and protect the Bay system. (After *The Modification of an Estuary*, F. H. Nichols, Cloern, J., Luoma, S., Peterson, D. H. 1986 Science Vol. 231., 567–573.)

The Situation in Chesapeake Bay

The Chesapeake Bay estuary is a shallow, drowned river valley with a surface area of 11,500 km² (7000 mi²) and an average depth of 6.5 m (21 ft). Chesapeake Bay flushes slowly; the residence time of the water is 1.16 years. The Chesapeake estuary has a long history as a major provider of oysters, blue crab, wildfowl, rockfish, and shad. It has also been under continuously increasing population pressure since colonial times, and has been absorbing more and more wastes from the expanding urban centers of Washington D.C., Baltimore, Norfolk, and dozens of towns in Maryland, Virginia, and Delaware. Between 1960 and 1983, the oyster catch dropped by two-thirds, the annual rockfish catch decreased from 6 million to 600,000 pounds, commercial fishing for shad has become nearly extinct in the upper Chesapeake, and the wildfowl population has dropped equally dramatically.

Partially treated and untreated sewage was recognized as the cause for outbreaks of typhoid fever among those drinking water and eating shellfish taken from the system in the early 1900s. Then, decaying untreated sewage robbed the water of oxygen; now plant growth, stimulated by nutrients from the effluent of sewage treatment plants and the runoff from fertilized fields, decays and consumes the dissolved oxygen. Increased plant growth also reduces the clarity of the water and shades the bottom. The low oxygen values and the reduction in bottom plant life due to decreased light together alter the bottom habitat. In addition, wastes from some 5000 factories, from military bases, and from sewage plants located between Virginia and New York find their way into the Chesapeake, adding conventional pollutants and toxic compounds, which are trapped in the bottom sediments.

Efforts to improve water quality in Chesapeake Bay have produced nearly 4000 studies since the 1970s. In 1972, the Federal Clean Water Act provided enforcement standards and a system of permits governing the amount of pollutants that individual dischargers could dump into any body of water. Gradually, industrial and municipal discharges were regulated, but at the same time populations increased and industries expanded, increasing the volume and rate of discharge to the estuary.

At the present time, it is estimated that industries and sewage plants along the shores of Maryland and Virginia discharge about 4 trillion gallons of waste water annually, or about 20% of the water in the bay at any one time. An EPA study costing $27 million over seven years launched the "Save the Bay" campaign in 1983. Scientists working for the campaign reported that industries in Maryland dump more than 2700 tons of heavy metals into the bay each year, and that Virginia industries dump more than 400 tons during the same period. Much of the money has been spent to manage soil and fertilizer runoff from surrounding fields. Critics charge that the program ignores industrial and municipal discharges in excess of the levels allowed under the current permit system. The federal permit system is a self-policing effort, with dischargers setting their own limits depending on available technology and cost. The permit system has not responded to new technologies and enforcement, and until recently has not been independently monitored. Many industries trying to avoid the discharge permit process direct their discharges to municipal sewage systems that are principally equipped to treat domestic and organic wastes.

While trying to understand the effect of human degradation of the Chesapeake, researchers have discovered that the natural changes in rainfall and temperature have large effects as well. At irregular intervals, tropical storms and hurricanes produce flood conditions that have catastrophic effects on the water quality and the organisms. Cyclic variation in survival of eastern oysters and striped bass have been related to such weather changes. In 1972, 1975, and 1979, the annual freshwater inflow was double the average. In 1972 in particular, tropical storm Agnes dropped 25.5 cm (10 in) of rainfall in a short period of time. This sudden surge of freshwater reduced the bay's salinity to about one-fourth its normal minimum level, and rinsed out the small floating organisms on which the larger animals feed. Severe winters, occurring at six- to-eight year intervals, promote heavy ice formation, which depresses the oxygen levels beneath the ice. These irregular but significant events superimposed on the increasingly degrading human influences make it difficult to determine exactly which factors are responsible for which changes in the Chesapeake estuary system.

Summary

The oceanographic classification of coastal embayments is based on the relationships between freshwater input, tidal flow, and net circulation. An estuary is a semi-isolated portion of the ocean that is diluted by fresh water. Estuary types include the salt wedge estuary, in which seawater becomes a sharply defined wedge that moves under the fresh water with the tide. Circulation and mixing in the salt wedge estuary are controlled by the rate of river discharge. In a shallow, well-mixed estuary, there is a net seaward flow at all depths due to strong tidal mixing and low river flow. Salinity is uniform over depth but varies along the estuary. A partially-mixed estuary has strong tidal turbulence, seaward surface flow of mixed fresh water and seawater, and an inflow of seawater at depth. Salt is transported in this system by advection and mixing. Fjord-type estuaries are deep estuaries in which the fresh water moves out at the surface and there is little tidal mixing or inflow at depth.

In a partially-mixed estuary, there is progressive movement of surface water seaward and deep water inward during each tidal cycle. The circulation of the estuary can be measured directly with current meters, which is a long and expensive process, or it can be measured indirectly by determining the water and salt budgets of the estuary.

In temperate latitudes, the volume transport of water between the estuary and the sea is much greater than the addition of fresh water. In fjords, the inward flow is small and the deep water tends to stagnate. In semienclosed seas with high evaporation rates, there is a net removal of fresh water; the seaward flow is at depth and the ocean water enters at the surface.

Estuaries that flush rapidly have a higher capacity for dissipating wastes than those that flush slowly. We can calculate the time period it takes for wastes associated with fresh water to pass through an estuary. In bays and harbors that are flushed only by tidal action, the recycling of resident water and wastes can occur. Partial recycling can also happen in estuaries with two-layered flow, as exiting surface water is mixed with incoming seawater.

Multiple uses of estuaries can overload them with wastes. Water quality is affected by dumping solid waste and liquid pollutants into coastal waters. Examples of practices leading to the degradation of the coastal marine environment include the dumping of solid waste in the New York Bight, the movement of pesticides and long-lived toxicants through the environment, chemical spills, the runoff of fertilizers and pesticides from agricultural lands, and storm sewer runoff from urban areas. Toxicants are adsorbed onto silt particles and become concentrated in the coastal sediments. Organisms concentrate toxicants and pass them on to other members of marine food chains. Plastic trash is an increasing problem to marine animal life. Oil spills are a special problem in inshore waters, for which there is no adequate cleanup technology.

Wetlands border estuaries; they are important as areas of nutrients, food, and shelter for many marine species. Many wetlands have been filled, dredged, developed, and lost. Case histories of San Francisco Bay and Chesapeake Bay are presented.

Key Terms

estuary	partially-mixed estuary	net circulation	inverse estuary
salt wedge estuary	advection	water budget	flushing time
entrainment	fjord	salt budget	intertidal volume
well-mixed estuary	fjord-type estuary		

Study Questions

1. If contaminated sediments are dredged from the floor of a harbor to use in a landfill, what hazards to the environment should be considered during both the dredging and disposal of the contaminated material?

2. Estuaries that receive silt-laden water from rivers and toxicants from urban sources tend to have lower toxicant levels in their waters than estuaries that receive clear river water and urban pollutants. Explain why.

3. Compare the circulation of a semienclosed basin at 30°N with that of an estuary located at 60°N. Which is less likely to accumulate waste products at depth? Why?

4. Estuaries are classified by their net circulation and salt distribution in this chapter. What other features could be used to describe and classify them?

5. Why are estuaries with short flushing times less apt to degrade when used as the receiving water for urban runoff than estuaries with long flushing times?

6. What happens to oil once it has been spilled into coastal waters?

Study Problems

1. Determine the flushing time of an estuary in which $T_o = 9 \times 10^7$ m³/day and the volume of water is 30×10^8 m³.

2. An estuary has a volume of 50×10^9 m³. This water is 5% fresh water, and the fresh water is added at the rate of 6×10^7 m³ day. Why does the rate of addition of fresh water from the estuary to the ocean equal the input of fresh water from the land to the estuary? Consider the average salinity of the estuary as constant. What is the residence time of the fresh water?

3. Water entering an estuary at depth has a salinity of $34.5\%_{00}$. The water leaving the estuary has a salinity of $29\%_{00}$. The river inflow is 20×10^5 m³/day. Calculate T_o, the seaward transport.

4. A bay with no freshwater input can flush only by tidal exchange. If on each tidal cycle (ebb and flood) 10% of the bay's water volume is exchanged with ocean water, how much of the original water will still be in the bay after four tidal cycles?

5. If the flushing time of the bay in Problem 4 is calculated by dividing the volume of the bay by the intertidal volume (10% of the bay volume), the flushing time is ten tidal cycles. Compare this method to that used in Problem 4. In which case does water entering on the flood tide mix completely? In which case does the entering water move to the head of the bay without mixing? Which method best duplicates natural conditions?

Oceans: Environment for Life

12

The river is within us; the sea is all about us;
The sea is the land's edge also, the granite
Into which it reaches, the beaches where it tosses
Its hints of earlier and other creation.
The starfish, the hermit crab, the whale's backbone;
The pools where it offers to our curiosity
The more delicate algae and the sea anemone.
It tosses up our losses, the torn seine,
The shattered lobsterpot; the broken oar
And the gear of foreign dead men. The sea has
many voices,
Many gods and many voices.

T. S. Eliot,
from Four Quartets

A complex variety of environments for living organisms is provided by the oceans and coastal seas of the world. At the surface, conditions range from polar to tropical; over depth, they range from light to total and constant darkness; and around the world, they range from open ocean to sheltered bays. Currents and waves, rising and falling tides, sandy bottoms and rocky shores all contribute to the variation. Sometimes changes are gradual and extend over long distances; at other times changes occur quickly in a small area. Changes also occur with time: light-dark cycles as well as the rise and fall of the tide modify an environment over the day, seasonal cycles modify it over the year, and other cycles take much longer. This chapter introduces the ocean as a habitat by reviewing the properties of the world oceans that act together to produce this special environment for life. Since the topics discussed here are considered in other chapters, their presentation is brief and is keyed to their interaction with the marine organisms.

12.1 *Buoyancy and Flotation*

Organisms that make their home on land require structural strength to support their bodies in air and against gravity. Trees must be able to hold up their canopies of leaves. Animals need skeletons and muscles to give their bodies shape and the ability to move. The salt water surrounding marine organisms has a density similar to the bodies of many of the organisms. The **buoyancy** of the water helps to keep the floating organisms at the surface; it also supports the bodies of the bottom-living creatures and lessens the energy expended by swimmers.

Many organisms have developed ingenious adaptations to help them stay afloat. Some jellyfish-type organisms, for example the Portuguese man-of-war and the by-the-wind sailor (chapter 14, fig. 14.11), are able to secrete gases into a float which enables them to stay at the sea surface. Some algae also secrete gas bubbles and form gas-filled floats, which help them keep their fronds in the sunlit surface waters. One floating snail produces and stores intestinal gases, another forms a bubble raft to which it clings. The chambered nautilus (fig. 12.1a), a relative of the squid, continually adds chambers to its shell and moves to the last chamber as it grows. A specialized tissue removes the ions from the empty chambers, causing water to diffuse out, then the chamber fills with gas, mainly nitrogen, from tissue fluids. The cuttlefish, another relative of the squid, has a soft, porous,

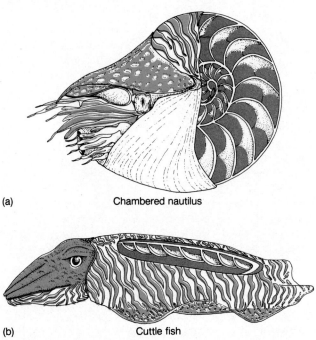

Figure 12.1
(a) Chambered nautilus. (b) Cuttlefish. The chambered shell that provides buoyancy is shown in each organism.

(a)   Chambered nautilus

(b)   Cuttle fish

internal shell or "bone". This animal also regulates its buoyancy by controlling the relative amounts of gas (also mainly nitrogen) and liquid within the shell (see fig. 12.1b).

Many fish have gas-filled swim bladders. The swim bladder allows the fish to become neutrally buoyant. Some fill their swim bladders by gulping air at the surface; others release gas from their blood through a gas gland to the swim bladder. When a fish changes depth, it must adjust the gas pressure in its swim bladder to compensate for the pressure change in the water. This limits its speed of vertical movement. If a fish with a swim bladder is forced suddenly to swim deeper, the increased external water pressure compresses the air, the bladder shrinks, and the fish sinks. If the fish is not able to readjust its system, it tires as it is forced to swim upward continually to compensate for the loss of buoyancy. On the other hand, a deep-swimming fish brought quickly to the surface with a fishing line will show bulging eyes and distended body, as the gas in the swim bladder expands with decreasing pressure. Active, continuously swimming, predatory species such as the mackerel, some tuna, and the sharks do not have swim bladders. Swim bladders are also lacking in bottom fish.

Small members of the floating populations of plants and animals store their food reserves as oil droplets, which decrease their density and retard sinking. Large surface area-to-volume ratios that slow sinking are characteristic of small spheres. Single-celled organisms, particularly plants, take advantage of this by keeping their size small. They have developed spines, ruffles, and feathery appendanges that increase surface area and allow them to remain more easily at or near the sea surface.

Large animals are made up of tissues that are more dense than seawater. To reduce the density of body tissues, the giant squid excludes the more dense ions from its body fluids and replaces them with less dense ions. Another squid species has one pair of its arms filled with low density body fluids. Whales and seals decrease their density and increase their flotation by storing large quantities of blubber, which is mainly low-density fat. Sharks and some other varieties of fish store oil in their liver and muscle.

Seabirds float by using fat deposits in combination with light bones and air sacs developed for flight. Their feathers are waterproofed by an oily secretion called preen, which acts as a barrier to seal air between the feathers and the skin. This is important in keeping the birds warm and it also helps to keep them afloat.

Mechanical strength is not a prerequisite for success in the marine environment. Except where breaking waves are encountered at the surface or along the shore, the mechanical stresses of the water are slight. Many marine organisms are exceedingly delicate and fragile but survive and function well as long as they are surrounded by water.

12.2 Osmotic Processes

Although the salinity of seawater provides increased buoyancy, special problems are posed for living creatures if the salt content of their body fluids differs from the salinity of the water that surrounds them. The body fluids of living plants and animals are separated from the seawater by membrane boundaries that are semipermeable. Semipermeable membranes allow some molecules to move across the membrane boundaries, while other molecules cannot. Molecules that move across these membranes do so along a gradient from a region of high concentration of a substance to a region of low concentration of that substance. This process is called diffusion. Water molecules pass across the membranes in a special type of diffusion known as **osmosis,** which was discussed in chapter 5.

Most fish have body fluids with a salt concentration that is about halfway between fresh water and seawater. In salt water, their tissues tend to lose moisture as the water moves along an osmotic gradient from an internal high water concentration and low salt concentration to the lower water concentration and higher salt concentration of the ocean environment. Fish must constantly expend energy to prevent dehydration by osmosis, which results in an increase in body salt. Fish

Figure 12.2
(a) The salt concentration of the seawater is the same as the salt concentration of the sea cucumber's body fluids (35 %00). The water diffusing out of the sea cucumber is balanced by the water diffusing into it. (b) The salt concentration in the tissues of the fish is much lower (18 %00) than that of the seawater (35 %00). To balance the water lost by osmosis, the fish drinks salt water, from which the salt is removed and excreted.

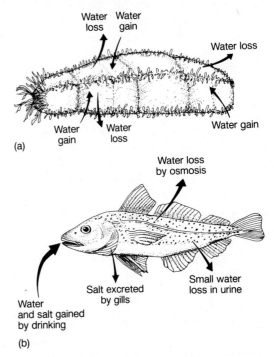

stay in fluid balance by drinking seawater nearly continually and excreting its salt across their gills. Sharks and rays do not have this problem, because their body fluids are at approximately the same salt content as seawater, so that there is no osmotic gradient. These fish maintain a high concentration of urea in their tissues. The urea allows their tissue to retain water, prevents the movement of salt into their bodies, and keeps body fluid at approximately the same salt content as seawater.

Many bottom-dwelling organisms, such as sea cucumbers and sponges, also exist in this condition. Their body fluids are at the same salt concentration as the seawater, existing in a state of equilibrium with the seawater. There is no concentration gradient; the water diffuses equally in both directions across the membranes, and the salt content remains the same on both sides of the membranes. Figure 12.2 compares the fish, which does have to overcome an osmotic gradient, with the sea cucumber, which does not.

Species may be limited in their geographic distribution by changes in salinity, for many organisms can maintain their salt-fluid balance over only limited salinity ranges. Since there is little change in salinity in deep water, species living below the surface layers are dispersed over large areas. The surface-dwelling forms are more likely to find salinity barriers in coastal waters, since the salinity varies in bays and estuaries. Successful estuarine animals are able to stabilize their osmotic processes by regulating the intake of salt ions. Crabs are an example. In some areas, a few species are able to survive in high-salinity lagoons and salt marshes, but they are unlikely to reproduce, and so the population must be replenished by new recruits from the sea. Some animals have an extraordinary ability to adapt to large changes in salinity over their life history. Salmon spawn in fresh water but move down the rivers as juveniles to live their adult lives in the sea. After several years (the time depends on the species), the salmon return to their home streams. The Atlantic common eel reverses this process by migrating downstream to spawn in the Sargasso Sea. The new generation of eels spends one to three years at sea, then returns to fresh water to live for up to ten years before migrating seaward. Other fish and crustaceans use the low-salinity coastal bays and estuaries as breeding grounds and nursery areas for their young, then as adults they migrate farther offshore into higher-salinity waters.

12.3 Temperature

The temperature of the deep oceans is low and nearly constant. At the surface and close to shore the water temperature varies with seasonal climate changes and geographic latitude zones. The temperature of water, like the salinity, affects the density of the seawater. It also affects the water's viscosity. Density and viscosity are discussed in chapter 4. In polar latitudes, the surface water is cold, more dense, and more viscous, so that organisms float more easily. In tropic latitudes, the warm, less dense, less viscous water is home for species with more appendages, larger surface areas, and greater gas-bubble production, for the water is less buoyant and offers less resistance to sinking.

When surface conditions produce a warm, low-density surface layer overlying denser water, the water column is stable. Floating organisms stay in the sunlit upper layers. The floating plant cells increase their rates of photosynthesis and reproduction. When surface waters cool and increase their density, they become unstable and sink. The waters mix, overturn, and take the floating organisms with them into deeper water and away from the sunlight. The photosynthesis and reproduction of plants are decreased.

Plants and marine animals other than birds and mammals do not control their body temperatures. They are cold-blooded and their body temperatures vary with environmental conditions. The physiology of marine organisms is regulated by the temperature of the water, and within limits metabolic processes proceed more rapidly in warm water than in cold water. Cold-water forms frequently grow more slowly, live longer, and attain a larger size. To some species, changes in temperature act as signals to spawn or to become dormant. Although the heat capacity of the water restricts temperature fluctuations, the geographic distribution of seawater temperatures is sufficient to affect the distribution of marine organisms.

Seabirds and mammals are warm-blooded. They maintain nearly constant body temperatures that are well above the temperature of the seawater. Because these animals are less restricted by the temperature of the water, they often have wider geographic ranges. The annual migration of whales between polar and tropical waters is an example. Some fish, although cold-blooded, are able to conserve heat in their swimming muscles, elevating their body temperatures. Because their muscles work more efficiently at higher temperatures, these fish are able to swim rapidly and cruise long distances in the coldest water, making them efficient predators. Fish of this type include some tuna, mackerel sharks, the great white shark, and the dolphin fish.

At the greater depths, the uniformity of temperature with latitude creates an oceanwide environment that is nearly unaffected by seasonal changes. At the sea surface, the temperature changes with latitude much the same as the climate changes on land. Annual changes in surface temperature in the open sea are small at the very high and very low latitudes; at the middle latitudes the annual changes in sea-surface temperature show a seasonal fluctuation. The ocean areas close to land have a still greater change in surface temperature, due to the influence of the greater annual temperature changes over the adjacent landmasses. The seasonal fluctuation in surface temperature is reflected at the middle latitudes in periods of spring and summer reproduction and growth and winter dormancy.

On land, the general pattern of climate zones encountered by approaching the poles is also observed by increasing the altitude. In other words, the climate at sea level in polar zones is similar to the climate at the top of a high mountain peak at a lower latitude. In the ocean, conditions in surface water at polar latitudes are similar to those found at deeper depths at lower latitudes. Some shallow-water species of the polar seas have been found at greater depths at the lower latitudes.

12.4 Pressure

Deep-living organisms such as worms, crustaceans, and sea cucumbers are unaffected by the pressure, because they do not have gas-filled cavities or lungs that must be maintained at high pressure or mechanically protected against the pressure of the overlying water. It is possible that pressure may alter metabolic rates and

TABLE 12.1

Diving Depths and Durations for Some Marine Mammals

Mammal	Diving record (m)	Duration record (minutes)
Weddell seal	600	70
porpoise	300	6
bottle-nosed whale	450	120
fin whale	350	20
sperm whale	> 2000	75–90

growth at greater depths, but little is known about such effects.

When humans descend into the sea, they need either protection from the pressure that will collapse their chest cavities and lungs or air supplied to the lungs at a pressure equal to the outside water pressure. Submarines and submersibles provide the first type of protection, while SCUBA (Self-Contained Underwater Breathing Apparatus) and other commercial diving equipment supply the second type. After breathing gases under pressure, humans may experience the serious problems of decompression sickness and nitrogen narcosis. Divers breathing air under high pressure for extended periods absorb large quantities of gas, particularly nitrogen, into their blood and tissues. If they return to shallow depths of less pressure too quickly, these excess gases form bubbles in the body tissues and blood vessels, causing extreme pain, paralysis, and sometimes death. Excess nitrogen dissolved in the blood stream also has a narcotic effect, which confuses the diver and restricts his or her ability to function normally.

Air-breathing marine mammals make dives of spectacular depth and duration without encountering such difficulties, because of their abilities to adjust their physiology. Table 12.1 gives record diving depths and duration underwater for some of the whales and seals. These animals have a physiology that permits their blood to absorb more oxygen and to tolerate higher concentrations of carbon dioxide than that of land mammals. They also have a vascular shunt system that directs the blood flow through only the brain and heart while underwater. Their muscles store additional oxygen and are able to tolerate the buildup of waste products from exertion to a greater degree than those of other animals. During their dives their lungs collapse completely, forcing the air out and preventing the blood from absorbing compressed gases at high pressure; therefore they do not suffer from diving illnesses and are able to change their depth rapidly.

12.5 Gases

Life in the water requires carbon dioxide and oxygen, as does life on land. Carbon dioxide is required by the plants for photosynthesis; it is contributed by the animals and by decay processes, and it is absorbed by the water from the atmosphere. Because seawater has the capacity to absorb large quantities of carbon dioxide, there is no shortage of carbon dioxide for the plant life. Also, carbon dioxide's role as a buffer limits the ocean's pH range, which keeps it a stable environment for living organisms (see chapter 5).

Oxygen is required by all organisms to liberate energy from organic compounds. Oxygen is available only at the ocean surface as a by-product of photosynthesis and from the atmosphere. Life below the surface depends on the vertical circulation processes (discussed in chapter 6) to replenish the oxygen at depth. The amount of oxygen in the water influences the distribution of organisms and is influenced by the temperature, salinity, and pressure of the water. Shallow tidal pools and bays on warm, quiet days increase in temperature and salinity, decreasing the ability of the water to hold oxygen, forcing motile animals out, and limiting these areas to the organisms that can successfully tolerate these extremes. The bottoms of deep, isolated basins may be so low in oxygen that only nonoxygen-requiring, or **anaerobic,** bacteria can survive there (refer back to chapter 5).

12.6 Nutrients

Nitrate (NO_3^-) and phosphate (PO_4^{-3}) nutrients are required by the sea's plant life. They are the fertilizers of the sea and are stripped from the surface layers by the plants, which incorporate them into their tissues. These nutrients are liberated at depth by the decay of plant as well as animal tissues, or they are returned to the water in the form of waste products of herbivores and carnivores (see fig. 12.3). Vertical circulation and mixing transport the nutrients back to the surface. In these upwelling areas, life is abundant. Estuaries and coastal waters, where nutrients are supplied by land runoff and mixing from the shallow sea floor of the continental shelf, are also rich with organisms. Plant populations are limited by the lack of any essential nutrient; if the concentration of such a nutrient falls below the minumum required, the population's growth ceases until the nutrient is replenished. Nutrients were introduced in chapter 5 and nutrient cycles will be discussed in chapter 13.

Figure 12.3
Nitrate and phosphate distribution in the main basin of Puget Sound, August, 1975. The surface values are the result of nutrient utilization by unicellular marine plants.

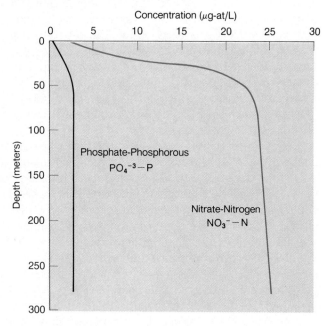

Figure 12.4
The percentage of solar energy available at depth in clear and turbid water.

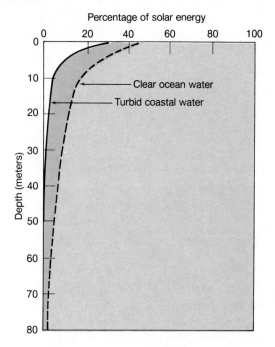

12.7 Light and Color

Sunlight

Without light there is no plant life. Because seawater is relatively transparent to the visible spectrum, plant life exists in the surface layers. Plants are confined to the **photic zone,** where there is sufficient light energy to allow the process of photosynthesis to proceed. The depth of the photic zone is controlled by factors discussed previously in chapters 4 and 6 including (1) the angle at which the sun's rays hit the earth's surface, which is related to latitude and change of season; (2) the different rates at which the wavelengths of light are absorbed, which is determined by the properties of water; and (3) the suspended particulate material present, which affects the rate of absorption. Below the photic zone is the **aphotic zone,** the zone in which there is no photosynthesis.

However, the presence of light does not guarantee plant life. Nutrients must be available. It is because of a lack of nutrients that so much of the open ocean that is exposed to high-intensity sunlight is considered, for all practical purposes, a biological desert. Life is more abundant along the coasts and over the continental shelves because of the larger quantities of dissolved nutrients. Waves and currents of the coastal zone stir the bottom and mix up silt with the nutrients. The silt particles absorb and scatter light, reducing the

depth to which it penetrates. As the single-celled plants reproduce, their increased numbers also act to limit light penetration, so that the photic depth may be reduced to less than 50 m (167 ft). Figure 12.4 compares the penetration of sunlight in clear and turbid seawater.

Bioluminescence

Another source of light is present in the oceans; it is produced by the organisms themselves. On a dark night, the wake of a boat is a glowing ribbon, and disturbed fish leave a trail of light; the water flashes as oars dip; a person's hands glow briefly as a net is hauled in. In each case, the phenomenon producing the light is **bioluminescence;** this cold light is biologically produced by the interaction of the compound **luciferin** and the enzyme **luciferase.** This phenomenon is often incorrectly referred to as phosphorescence; it has nothing to do with phosphorus or with the absorption of radiation, but is a chemical reaction that produces light with a 99% efficiency. The same phenomenon is seen on land in the flashing of a firefly or the ghostly glowing of a fungus in the woods.

In the sea, the agitation of the water disturbs microscopic bioluminescent organisms, causing them to flash and produce glowing wakes and wave crests. Animals that feed on these organisms often concentrate the chemicals in their tissues and also glow. Jellyfish glow in this way and so do one's hands if they come in contact with crushed tissue. Other bioluminescent organisms in the sea include squid, shrimp, and some fish. Many middepth and deep water fish carry light-producing organs, or **photophores.** Some fish have patterns of photophores on their sides, possibly for identification; others show photophores on their ventral surfaces, making them difficult to see from below against the light surface water. Still others have glowing bulbs dangling below their jaws or attached to flexible dorsal spines, acting as lures for their prey. The flashlight fish, found in the reefs of the Pacific and Indian oceans, has a specialized organ below each eye that is filled with light-emitting bacteria. These fish are known to use the light to see, communicate, lure prey, and confuse predators.

Color

Some sea animals are transparent, allowing them to blend with their water background, for example jellyfish and most of the small floating animals in the surface layers. Other animals, particularly the fish, use color in many ways. In the clear waters of the tropics, where light penetrates to greater depths, bright colors play their greatest role. Some fish conceal themselves with bright color bands and blotches. These colors disrupt the outline of the fish and may draw the predator's attention away from a vital area to a less important spot, for example, a black stripe over the eye and an eye spot on a tail or fin. Another use of bright color is to send a warning. Organisms that sting, taste foul, have sharp spines, or poisonous flesh are often striped and splashed with color, for example sea slugs and some poisonous shell fish. Among fish that swim near the surface in well-lighted water, for example herring, tuna, and mackerel, dark backs and light undersides are common. This color pattern allows the fish to blend with the bottom when seen from above and with the surface when seen from below. See figure 12.5.

In temperate regions, coastal waters are more productive, more turbid, and there is less light penetration. Drab browns and grays are concealing against the kelp beds of temperate waters. Cold-water bottom fish are usually uniform in color with the bottom, or speckled and mottled with neutral colors. The flatfish are well known for their ability to change their color. They have skin cells which expand and contract to produce color changes (see fig. 12.6). Their extraordinary color change ability enables them to conceal themselves by matching the bottom type on which they live (see fig. 12.7).

Below the surface layers, some species of shrimp are red when seen in white light. Red pigment below the level of penetration of red wavelengths of light absorbs blue and green light; little is reflected and the animals are dark and inconspicuous. In the deep ocean

Figure 12.5
Viewed from above, the dark dorsal surface of the fish blends with the sea floor; viewed from below, the light ventral surface blends with the sea surface. This type of coloration is known as countershading.

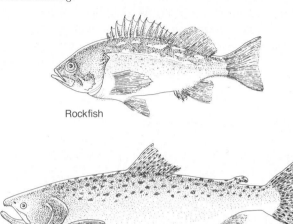

Rockfish

Chinook salmon

Figure 12.6
Pigment cells from a section of fish skin.

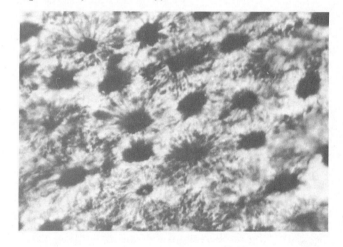

Figure 12.7
The winter flounder resting on a checkerboard pattern shows its use of camouflage.

without light, color is of little importance except perhaps when combined with bioluminesence. Of course, we do not know how the animals see the colors and there may be roles that color plays in the sea we do not know or understand. Color is thought to be important in species recognition, courtship, and possibly in keeping schools of fish together. As we continue to study the oceans we will find more and different roles for color in the marine environment.

12.8 Circulation

The ocean's water is in constant motion. It is moved and mixed by currents (chapter 7), waves (chapter 8), and tides (chapter 9). Below the surface layers, the ocean environment is very uniform, providing marine organisms with similar conditions of temperature and salinity in any ocean at any time. Oceanic circulation brings food and oxygen, replenishes nutrients, and removes waste. It disperses floating organisms and scatters the reproductive stages of swimmers and attached forms alike.

Those plants and animals that drift rather than swim are carried along by the currents and run the risk of being carried out of a suitable habitat by either vertical or horizontal movement. However, analysis of their remains on the sea floor and observations of living populations show that this does not always happen. Populations of drifting animals appear to take advantage of their ability to move in the vertical direction either by swimming or by changing their buoyancy. They are able to maintain their place horizontally in ocean space by moving away from the surface during the day to depths where a current flows in a direction opposite to the surface current. They then move upward at night to be carried back to their starting position. Organisms also appear to maintain their position by adding to their population on the upstream side of a current to balance losses on the downstream side. This pattern is related to the new supply of food and nutrients brought into the population by the current upstream, in contrast to the food-depleted water downstream.

Vertical water motions in the sea are much slower than horizontal motions, but small displacements of organisms in the vertical direction can mean substantial changes in light, salinity, temperature, and nutrient supply. If the vertical flow is upward, it counteracts the tendency of organisms and other particulate matter to sink. In this way, light-dependent organisms are held in the photic zone. The upward motion of the water also supplies nutrients to the photic zone to promote plant growth. At the same time, these upward flows decrease the temperature of the surface waters and return water

with a low oxygen content to the surface, where oxygen is replenished by photosynthesis and atmospheric exchange.

A downward vertical flow under an area of surface convergence accumulates a population of organisms as the surface flows move toward the area of downwelling. If the organisms cannot increase their buoyancy to compensate for the downward current, they are carried down to changes in light, temperature, salinity, nutrients, and gases. Areas of downwelling are usually regions of low plant growth, but at the surface convergence the accumulation of organisms provides a rich feeding ground for carnivores.

12.9 Barriers and Boundaries

In the sea, species and populations are isolated from one another by barriers of different types. The properties of sea water change abruptly at certain boundaries. Near-surface boundaries in the water column may be sharp: for example, rapid changes with depth in temperature (thermocline), density (pycnocline), and salinity (halocline) (refer to chapter 6). Light intensity also changes rapidly in the vertical direction (refer to chapter 4). These boundaries are barriers for marine organisms, because the organisms do not survive if displaced through a boundary. The effect of these barriers decreases at deeper depths, where water properties become more homogeneous.

Similar barriers exist in the horizontal direction, where surface water of one type is adjacent to surface water of another type (refer again to Chapter 6). These boundaries are often associated with zones of surface convergence and divergence in the ocean and also occur in estuaries, between land-derived fresh water and seawater in the coastal zone. When they are associated with a rapidly flowing current of one type of water moving through or adjacent to another type, the boundaries are sharp. For example, populations that do well in the warm, saline waters of the Gulf Stream may suffer if they wander into the cold, less saline Labrador Current water between the coast and the Gulf Stream.

Other boundaries are controlled by the topography of the sea floor (see chapter 2). Ridges that isolate one deep-ocean basin from another prevent the deeper water in the basins from freely exchanging. In this way, water of dissimilar characteristics and populations may exist on either side of a submarine ridge. In other cases, the water and the populations in the two basins may be similar but the populations are not able to move between the basins because the elevation of the ridge that separates them forces the animals to change their depth and pass upward into water with properties that they cannot tolerate. Isolated seamounts with their peaks in shallow water may support specific isolated communities of sea life in much the same way as mountain tops on land support arctic and alpine communities widely separated from each other.

Lateral topographic barriers exist as well. Near-surface and surface species of the tropic regions are prevented from moving between the Atlantic and Pacific oceans by the land barrier of Central America. Tropical surface species such as sea snakes cannot migrate around the continental landmasses of North and South America, because to do so they must pass through regions of much colder water. Africa also acts as a barrier, keeping the tropical species of the Indian Ocean from communicating freely with the tropical species of the Atlantic, because the water south of Africa is too cold for the tropical species of either ocean.

12.10 Bottom Types

Although the properties of the seawater are critical for the survival of marine organisms, for many plants and animals the type of ocean bottom—rock, mud, sand, or gravel—is equally important. A seaweed that requires rock for attachment is unable to live in sand. A burrowing worm from a mud flat or a shrimp from a sand beach cannot survive on a rocky reef. The material of the sea floor, or **substrate,** provides food, shelter, and attachment sites for a vast number of plants and animals. Each substrate type provides suitable living space for a different group of organisms. Substrates show greater variety along the shallow coastal areas; sandbars, mud flats, rocky points, and stretches of gravel and pebble are frequently found along the same strip of coastline. Moving seaward, the substrates tend to become sediment-covered and more uniform. The variety of substrates available declines and so does the diversity of the types of organisms.

Organisms living on the bottom provide food, shelter, and attachment sites for organisms that live above the bottom. For example, great forests of large seaweed attached to rocky bottoms in 20 m (66 ft) of water support populations that are different from the populations found in the eelgrass beds of shallow, quiet, sandy and muddy bays. The bottom also provides a place for the collection of the organic debris that falls from the water environment above. This organic waste is the only food source for many bottom organisms.

12.11 Environmental Zones

Because the marine environment is so large and complex, biological oceanographers, marine biologists, and ecologists interested in the sea divide the marine environment into subunits called zones. The zone classifications used here are based on the system developed by Joel Hedgpeth in 1957. First, the water and seafloor environments are separated. The water environment is referred to as the **pelagic zone** and is separated from the seafloor environment, or **benthic zone.** The pelagic zone is then divided into the coastal or **neritic zone,** which lies above the continental shelf, and the **oceanic zone,** which refers to the deep water away from the influence of land. The oceanic zone is next subdivided by depth. The surface waters to 200 m (660 ft) are the **epipelagic zone.** The **mesopelagic zone** is the twilight zone between 200 m (660 ft) and 1000 m (3300 ft). The **bathypelagic zone** extends to 4000 m (13,200 ft), and the **abyssopelagic zone** extends to the deepest depths. All oceanic subdivisions, except for the epipelagic zone, are aphotic, or without light. The photic zone coincides roughly with the epipelagic zone and also with the neritic zone. Figure 12.8 shows the location of each of these zones.

The sea floor, or benthic environment, has also been subdivided into comparable zones. Tidal fluctuations at the shoreline define the **supralittoral zone,** or **splash zone,** which lies just above the high-water mark and is covered by the sea only during the highest spring tides or by wave spray, and the **littoral zone,** or **intertidal zone,** which lies between high and low water and is covered and uncovered once or twice each day. The **subtidal zone,** or **sublittoral zone,** extends out along the continental shelf. The supralittoral, littoral, and inner portion of the sublittoral zones occupy the same area as the benthic photic zone, because sufficient light is present in these zones to support both floating single-celled plants as well as benthic plants.

Within the aphotic zone are the deeper portions of the subtidal zone: the **bathyal zone,** extending from 200 m (660 ft) to 4000 m (13,200 ft) and coinciding with the continental slope, and the **abyssal zone,** between 4000 m and 6000 m, or roughly the area over the abyssal plain. The **hadal zone** lies below 6000 m (19,800 ft) and is associated with the trenches and deeps (see fig. 12.8).

The properties of the littoral and epipelagic zones are keyed to latitude; keep in mind that these zones differ markedly, depending on whether they are at polar, temperate, or tropic latitudes. Substrate plays a basic role in benthic zones, and there is much less variety in substrates in the deeper zones. Life in all the zones is influenced by variations in temperature, light, dissolved gases, nutrients, and all the factors discussed in this chapter.

Figure 12.8

Zones of the marine environment.

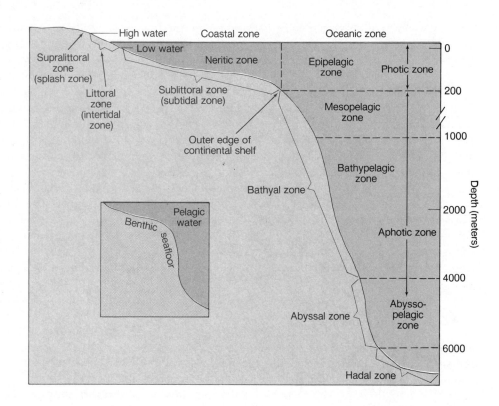

Life at Oil and Gas Seeps

The deep-sea vent environments, with their communities of animals discovered along the rift zones of oceanic ridge systems in the 1970s, depend on **chemosynthesis.** Bacteria in these environments use energy derived from sulfides, particularly hydrogen sulfide, to live and grow. The bacteria may be living inside the tissues of the vent animals, providing their hosts with organic carbon. The relationship between the bacteria and the host is termed **symbiosis.** The discovery of these vent communities is discussed in chapter 3, and the organisms and their chemosynthetic

abilities are explained further in chapter 16. Recently, other animal communities in different environments relying on different types of chemosynthesis have been found.

Along the continental slope of Louisiana and Texas, faulting has fractured the sea bottom and produced environments where oil and gas seep up and onto the surface of the sea floor. Some seeps have been located by the presence of oil at the sea surface, others have been found by identifying gas bubbles in the sediments that interfere with seismic

recording of the sediment structure, and still others by the oil stains in sediment cores. In 1985, clams, mussels, and large tube worms were collected from these sites at depths between 500 to 900 m (1600–3000 ft). Research has shown that the mussels' gills contain bacteria which use methane (the principal component of natural gas) as a carbon source; the mussel is able to use carbon compounds produced by the bacteria. This is the first known example of methane conversion to nutritional organic carbon between a bacterium and its host animal.

12.12 Classification of Organisms

All of these ocean zones together make up the varied marine environment, which is inhabited by a wide variety of organisms uniquely adapted to it. These organisms are divided into groups to promote ease of identification and to increase our understanding of the relationships that exist among them. Classic taxonomic categories are arranged in the manner illustrated in table 12.2. A simpler method is to divide all marine organisms into three groups, based on where and how they live. Plants and animals that float or drift with the movements of the water are the **plankton.** Those that live attached to the bottom or on or in the bottom are the **benthos,** and the animals that swim freely and purposefully in the sea are the **nekton.** Chapters 14, 15, and 16 use this three-part system.

TABLE 12.2

Taxonomic Categories of Some Marine Organisms

	Killer whale	Northern fur seal	Pacific (Japanese) oyster	Giant octopus	Sea lettuce	Giant kelp
kingdom	Animalia	Animalia	Animalia	Animalia	Plantae	Plantae
phylum	Chordata	Chordata	Mollusca	Mollusca	Chlorophyta	Phaeophyta
class	Mammalia	Mammalia	Bivalvia (Pelecypoda)	Cephalopoda	Chlorophyceae	Phaeophyceae
order	Cetacea	Carnivora (Pinnipedia)	Anisomyaria	Octopoda	Ulvales	Laminariales
family	Delphinidae	Otariidae	Ostreidae	Octopodidae	Ulvaceae	Lessoniaceae
genus	*Orcinus*	*Callorbinus*	*Crassostrea*	*Octopus*	*Ulva*	*Macrocystis*
species	*orca*	*ursinus*	*gigas*	*dofleini*	*lactuca*	*pyrifera*

Scientists used submersibles to directly observe the communities. Observations showed dense accumulations of organisms but with some variation. Clams predominate in areas of oil seepage, while mussels are more common at gas seeps. Analysis shows that different sources of carbon are used in each case. While mussels derive their organic carbon from the escaping methane gas, the clams and tube worms acquire their carbon from methane, biologically degraded oil, and from the decay of other organisms. Some species of clams and the tubeworms have been shown to harbor bacteria that are able to utilize hydrogen sulfide. Researchers have also been able to identify carbon compounds from methane sources in animals preying on the shellfish, in this way identifying another pathway for chemosynthetic carbon to enter the general deep-sea animal community.

Thirty-four environments of this type have been identified along the Gulf of Mexico's continental shelf. Thirty-nine samples have been taken at seventeen of these sites; chemosynthetic organisms were found in thirty-five of the samples. The tube worms were the most common organisms found, followed by the mussels. It is not known whether the tube worms are really present in larger numbers or whether the sampling gear is more efficient at catching them.

No one knows how many seeps of this type exist around the world, or how many seeps support populations of organisms in chemosynthetic systems. As our technology allows us to explore the marine environment in more detail, however, we are finding that nature is able to colonize more environments and in more and different ways than was thought possible twenty-five years ago.

12.13 Practical Considerations: Modification and Mitigation

At the same time that we have been acquiring greater understanding of how the physical, chemical, and geological factors discussed in this chapter interact to produce the many and different environments of the ocean, we have been bringing greater and lasting changes to these ocean environments, especially in coastal bays and estuaries, where environments are typically small-scaled and varied. In some cases, using our knowledge of organisms and their preferred habitats, we have altered an area to enhance the populations of organisms we consider desirable over those we consider undesirable. We build artificial reefs using old car bodies, bags of old shells, or chunks of fractured concrete to encourage the growth of organisms that not only do well in a reef's protected nooks and crannies, but are well suited to recreational and sport fishing as well as commercial harvesting.

Development and modification of estuaries and embayments is often done at the expense of wetlands and mud flats (refer back to chapter 11). Our engineering techniques allow us to move sediment, change the slope of the bottom, or alter substrate, resulting in the loss of habitat which translates directly into a loss of organisms. However, the same techniques can be used to create new marshes and tide flats if we wish to do so. We are in a position to choose which habitats and which organisms are to be conserved and which are to be sacrificed.

Our ability to reengineer relatively shallow water environments has resulted in the concept of **mitigation,** which is a coastal management practice in current use. When development projects alter or destroy an environment, management authorities at the local, state, or federal level may choose to enforce mitigation of these effects by requiring the developer to purchase an area of the same type as that to be developed and to arrange for it to be held in its natural state, or the developer may be required to reengineer an area to resemble what has been lost. While mitigation does preserve some habitats and species, it can only approximate the lost environment, not duplicate it. The result is still a shift or a change in habitat produced by human pressures.

Development pressures on deep-sea areas associated with mining and energy projects are still in the future. OTEC plants (chapter 6), nodule mining (chapter 2), and other developments will bring changes to the deep ocean as well. However, because of the vast size of deep-sea areas with common properties, changes in deep water are likely to be less significant than changes in the coastal zone.

Summary

Organisms living in the sea are buoyed and supported by the seawater. Adaptations for staying afloat include low-density body fluids, gas bubbles, gas-filled floats, swim bladders, oil and fat storage, and extended surface areas and appendages.

Most marine fish lose water by osmosis. They drink continually and excrete salt to prevent dehydration. Sharks have the same concentration of salt in their tissues as there is in seawater. They therefore do not have a water-loss problem. Salinity is a barrier to some organisms; others can adapt to large salinity changes.

Temperature affects density, viscosity, and the water's buoyancy, as well as the stability of the water column. The body temperature and metabolism of all marine organisms, except for birds and mammals, are controlled by the sea temperature. Some fish conserve heat in their body muscles and elevate the temperature in these muscles. Temperatures at depth are uniform; sea-surface temperatures change with latitude and seasons.

Changes in pressure affect organisms with gas-filled cavities. Marine mammals have a unique ability to undergo large pressure changes due to their physiology and body chemistry. The swim bladders of fish are affected by pressure changes, and the fish must change depth slowly.

The carbon dioxide-oxygen balance in the oceans influences the distribution of all organisms. The availability of nutrients and light limits plant populations. The depth of light penetration in the oceans is controlled by the angle of the sun's rays, the properties of the water, and the material in the water. Light limits plant life, but nutrients are also required. Some organisms produce chemical light, known as bioluminescence. Animals use color for concealment and camouflage, also to warn predators of poisonous flesh and bitter taste.

Winds, tides, and currents mix the water. Moving water carries food and oxygen, removes waste, and disperses organisms. Floating populations are not necessarily scattered, but keep their place due to movements between surface currents and deeper currents. Upwellings supply nutrients and hold plants in the surface layers. Downwellings are regions of low plant growth.

Barriers for marine organisms include water properties, light intensity, zones of convergence and divergence, seafloor topography, and geography. Different substrates provide food, shelter, and attachment for different groups of organisms.

The marine environment is subdivided into zones. The major environments are the benthic and pelagic zones;

there are numerous subdivisions of each of these zones.

The organisms of the sea are classified for identification and relationship. Organisms are also grouped as plankton, nekton, and benthos.

Development of marine areas with the consequent loss of habitat and therefore populations has led to the concept of mitigation, under which developers are required to preserve or replace habitats in an effort to maintain and preserve species.

Key Terms

buoyancy	substrate	abyssopelagic zone	abyssal zone
osmosis	pelagic zone	supralittoral zone/splash	hadal zone
anaerobic	benthic zone	zone	chemosynthesis
photic zone	neritic zone	littoral zone/intertidal	symbiosis
aphotic zone	oceanic zone	zone	plankton
bioluminescence	epipelagic zone	sublittoral zone/subtidal	benthos
luciferin	mesopelagic zone	zone	nekton
luciferase	bathypelagic zone	bathyal zone	mitigation
photophore			

Study Questions

1. How do so many delicate and fragile organisms exist in the oceans without damage?

2. What will happen to the body fluids of a frog placed in seawater? A sea cucumber placed in fresh water?

3. Discuss the effect of temperature on the distribution of organisms. Consider changes with latitude and with depth.

4. Although seals and whales are mammals, they do not suffer from either decompression sickness or nitrogen narcosis during deep dives of long duration. Explain why.

5. How does the role of bioluminescence differ from the role of sunlight in the sea?

6. Compare the flotation problems of a many-armed organism with those of an organism without arms but of the same density. Consider both organisms in 4° C water and in 20° C water.

7. In what ways does upwelling contribute to increasing the populations of surface organisms? What properties of seawater act as barriers for marine organisms?

8. Explain why the substrate of the sea floor becomes less diversified as one moves from the shore to the deep ocean.

9. What characteristics determine whether a plant or an animal belongs to the plankton, the nekton, or the benthos?

10. What properties of seawater act as barriers for marine organisms?

Production and Life

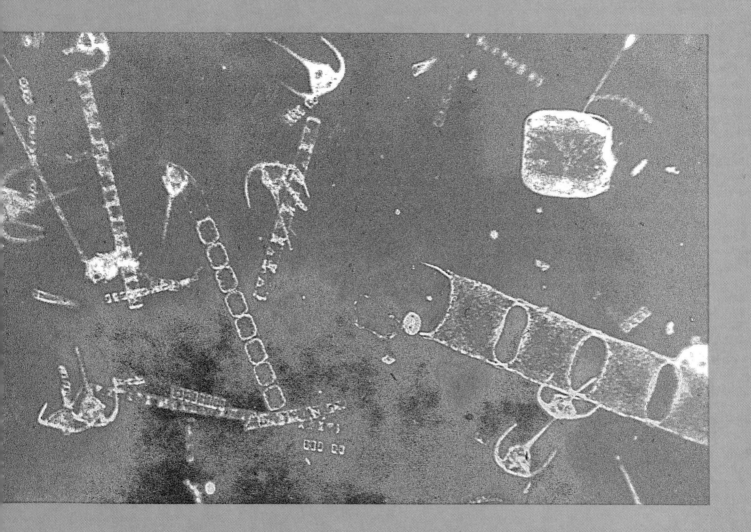

13

*O*ne's first reaction to a close view of the life of the sea is confusion, unease. How can things be so beautiful, with such intricate balance and symmetry in their infinitely varied forms and, at the same time, seem so hostile, so threatening, ready to lunge, to snatch life from each other? So many of the moving parts of sea life seem to be designed, exquisitely tooled, for nothing but destruction and devouring. It is disconcerting, almost as though we have had the wrong idea about beauty and harmony; living things so evidently aimed at each others' throats should not have, as these things do, the aspect of pure, crystalline enchantment.

Perhaps something is wrong in the way we look at them. From our distance we see them as separate, independent creatures interminably wrangling, as a writhing arrangement of solitary adversaries bent on killing each other. Success in such a system would have to mean more than mere survival; to make sense, the fittest would surely have to end up standing triumphantly alone. This, in the conventional view, would be the way of the world, the ultimate observance of nature's law. It was to delineate such a state of affairs that the hideous nineteenth-century phrase "Nature red in tooth and claw," was hammered out.

What is wrong with this view is that it never seems to turn out that way. There is in the sea a symmetry, a balance, and something like the sense of permanence encountered in a well-tended garden.

Lewis Thomas,
from Sensuous Symbionts of the Sea

*I*n all the waters of all the oceans, one organism preys on another: big fish eat little fish, and little fish eat littler fish, and so on. This series of prey-predator relationships is called a **food chain,** and each food chain has a beginning and an end. We are generally familiar with the creatures at the upper end of the chains (for example, salmon, tuna, swordfish and seals), but without the other end (or beginning) of the chain, such large carnivores could not exist. The plankton are at the lower ends of the open-ocean food chains, and the single-celled, microscopic plant plankton, or **phytoplankton,** are the first link. The biological oceanographer devotes much time and study to this first link, for understanding the variation in abundance of the phytoplankton is the key to understanding the ocean's productivity, or the rate at which organic material is produced in the sea.

13.1 Primary Production

Gross and Net

The phytoplankton, like plants on land, have certain requirements for growth: sunlight, **nutrients** or fertilizers, carbon dioxide gas, and water. The nutrients and carbon dioxide are found dissolved in the water that surrounds and supports the plant cells. The plant cells contain the pigment **chlorophyll,** which traps the sun's energy for use in **photosynthesis.** The photosynthetic process converts the carbon dioxide and water to high-energy organic compounds, which form new plant material. The production of new plant material by photosynthesis is termed **primary production.** The total amount, or mass, of organic material produced by photosynthesis is the **gross primary production** of the sea.

Photosynthesis is represented by the equation:

$$6CO_2 \ + \ 6H_2O \ \xrightarrow[\text{chlorophyll}]{\text{solar energy}} \ C_6H_{12}O_6 \ + \ 6O_2$$

$$\begin{matrix} 6 \\ \text{molecules} \\ \text{carbon} \\ \text{dioxide} \end{matrix} + \begin{matrix} 6 \\ \text{molecules} \\ \text{water} \end{matrix} \xrightarrow[\text{chlorophyll}]{\text{solar energy}} \begin{matrix} 1 \\ \text{molecule} \\ \text{sugar} \end{matrix} + \begin{matrix} 6 \\ \text{molecules} \\ \text{oxygen.} \end{matrix}$$

The sugars produced by photosynthesis can be broken down with the addition of oxygen to yield energy, carbon dioxide, and water in the process known as **respiration.** Respiration provides the plant with energy for its life processes.

Respiration is represented by the equation:

$$C_6H_{12}O_6 \ + \ 6O_2 \ \rightarrow \ 6CO_2 \ + \ 6H_2O \ + \ \begin{matrix}\text{life} \\ \text{support} \\ \text{energy}\end{matrix}$$

$$\begin{matrix} 1 \\ \text{molecule} \\ \text{sugar} \end{matrix} + \begin{matrix} 6 \\ \text{molecules} \\ \text{oxygen} \end{matrix} \rightarrow \begin{matrix} 6 \\ \text{molecules} \\ \text{carbon} \\ \text{dioxide} \end{matrix} + \begin{matrix} 6 \\ \text{molecules} \\ \text{water} \end{matrix} + \begin{matrix} \text{life} \\ \text{support} \\ \text{energy.} \end{matrix}$$

The gain in new organic material by a plant population is the population's primary production after the organic material required for respiration has been deducted. This gain is the **net primary production.** The net primary production is available to the food chain for consumption by animals and for decomposition by bacteria.

Net and gross primary production are usually reported for a unit volume of water or the volume of water under a fixed area of the sea surface and over a given period of time. The organic matter produced can be expressed as the number of organisms or as their weight, which is known as **biomass.** Because organic substances are based on the element carbon, primary production is most often given as the mass of dry weight organic carbon in grams, produced under a square meter of sea surface over a time of years, months, weeks, or days, or $gC/m^2/time$, the rate of change of biomass.

Standing Crop

The total amount of plant material (or biomass) under any area of sea surface at any instant in time is known as the **standing crop.** The standing crop is the result of growth, reproduction, death, and grazing. If the standing crop on two successive days shows an increase, then there is net production, but if the standing crop shows no change over the two days, the net production has not necessarily remained at zero. This relationship is shown in figure 13.1, where a population of herbivores is grazing the phytoplankton population at the same rate as the net primary production increases it, resulting in a nearly constant standing crop.

Figure 13.1

Net primary production is balanced by the grazing of herbivores. Both populations increase during the spring; the standing crop remains nearly constant throughout the year.

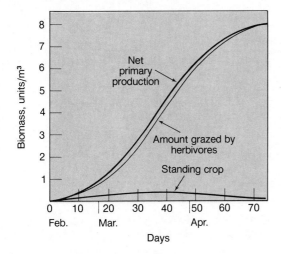

Figure 13.2

Phytoplankton growth cycles vary with the light available at different latitudes. At high latitudes (a), growth is limited to a brief period in midsummer. Seasonal light changes at the middle latitudes (b) increase growth in early spring and continue it through the summer. Sunlight levels vary little at tropic latitudes (c), where phytoplankton growth is nearly uniform throughout the year.

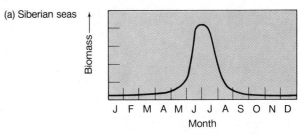

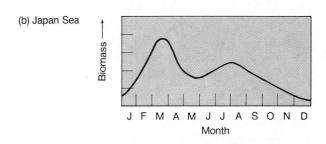

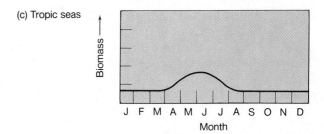

13.2 Factors Controlling Primary Production

Available Light

At the polar latitudes, changes in the duration of daylight and darkness mark the seasonal cycle. During the summer days of nearly constant daylight, the light intensity and the penetration of light below the surface are low, because of the angle at which the sun's rays strike the earth above 66½°N and 66½°S. For the remainder of the year, little sunlight is available. This pattern results in a short period of rapid plant growth during the midsummer. This growth period is shown in figure 13.2a, which presents the phytoplankton biomass with reference to the months of the year.

At the middle latitudes, the intensity of the sunlight and the duration of the daylight periods vary with the seasons. Figure 13.2b demonstrates the extended growing period at these latitudes. The graph shows an initial increase in phytoplankton biomass associated with the lengthening of the available hours of daylight in the spring and a second increase in the population during the midsummer.

Compare figure 13.2a and b with figure 13.2c, which shows the extended but lower-level growth in the tropics. There is a small population increase associated with a slight seasonal increase in sunlight, but the year-round sunlight provides abundant high-intensity solar energy at all times.

In order to take advantage of the available sunlight, the phytoplankton must remain in the ocean's lighted surface layers. In the tropics, the oceans have a warm surface layer of low-density water. This layer exists all year long, and the phytoplankton are not easily displaced into the denser underlying water. At the middle latitudes, a low-density, warm oceanic surface layer is found only during the summer. This layer appears gradually and is dependent on the water's absorption of the increasing solar radiation. If the warm weather is "late," or if it is interspersed with cool periods or strong storms, the formation of this layer is delayed and the active growth of the phytoplankton does not occur until the water is sufficiently stable and stratified to allow the plankton to remain in the upper layers. In the wintertime, the low-density surface layer of the middle latitudes is mixed with deeper water by surface cooling and winter storms. Vertical mixing, as well as the decrease in available sunlight, restricts phytoplankton growth. At polar latitudes, the low level of sunlight during the summer months forms a weakly stable density layering that is often associated with the fresh water that has returned from the melting sea ice. These factors combine with the short growing season to determine phytoplankton growth.

Figure 13.3

Phytoplankton biomass, nutrient supply, and surface water stability respond to solar energy changes at the middle latitudes in the northern hemisphere.

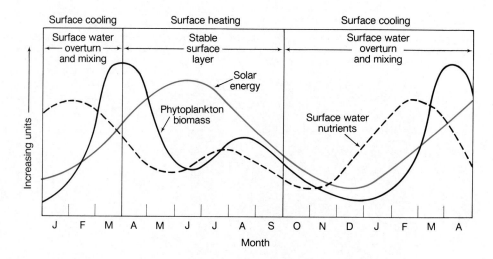

If the annual distribution of solar energy over the earth's surface were the sole factor influencing plant growth, we would expect the level of phytoplankton growth in the tropics (fig. 13.2c) to be maintained year-round, but at biomass levels approaching the summer values shown for the polar and middle latitudes (fig. 13.2a and b), and we would expect the peak values for the polar biomass to occur at much lower levels. However, the nutrient supply in these areas must also be taken into account. Remember that land drainage, mixing, overturn, and upwelling supply nutrients to the surface layers, and that these processes occur seasonally and in specific areas of the oceans. Both light and nutrients must be available if the phytoplankton population is to increase.

Available Nutrients

In the polar seas, the growing season is short and the nutrients rarely fall below the level that is necessary for population increase. Here, the availability of light controls phytoplankton growth. In the tropics, the phytoplankton remain in the surface layers, because of the stability of the water column. The phytoplankton extract the nutrients from the low-density surface waters, but the upwelling and mixing processes for renewing the nutrient supply are weak. Phytoplankton production in the tropics and subtropics is limited by the nutrient supply to the photic layers.

In temperate latitudes, if winter storms are severe, or if the cooling of the surface water produces an overturn, the surface nutrients are replenished. Nutrients are then available to the plant cells when the spring increase in solar radiation warms the surface and increases the stability of the water column. The increase in sunlight sparks a phytoplankton bloom in the nutrient-rich water. Eventually, the bloom uses up the available nutrients in the surface waters, at which time

the population growth rate declines. The rate at which nutrients are recycled or added to the surface layer controls the continued growth of the population. At these latitudes, the phytoplankton growth is limited either by the supply of nutrients or by the level of sunlight.

The interaction between the sunlight, the stability of the surface water, the nutrients in the surface water, and the phytoplankton biomass are presented in figures 13.3 and 13.4. Figure 13.3 demonstrates the situation at the middle latitudes in the northern hemisphere. Note the replenishment of nutrients to the surface during periods of overturn and mixing, and the decrease in nutrients during periods of increased sunlight and plant growth. In figure 13.4, the lack of surface mixing and overturn at low latitudes is reflected in the low levels of nutrients and plant biomass despite the constant high level of solar energy. Grazing by herbivores is also a factor in the reduction of the phytoplankton biomass after it reaches its maximum (refer back to fig. 13.1).

Nutrient Cycles

If a plant or an animal dies naturally, or if portions of an organism remain uneaten, organisms of decay, known as **decomposers** (bacteria and fungi), release the energy within the organism's body to the environment as heat and break down the organic molecules making up its mass to basic molecules such as carbon dioxide, nutrients, and water. Since no significant amount of new matter comes to the earth from space, living systems must recycle inorganic molecules to form the organic compounds of living organisms. Organisms contain many elements, but among the most important are nitrogen and phosphorus.

Nitrogen in the form of nitrate and phosphorus in the form of phosphate are two of the nutrients required for life (refer back to chapter 5). Nitrogen is essential in the formation of proteins, and phosphorus is required in energy reactions, cell membranes, and nu-

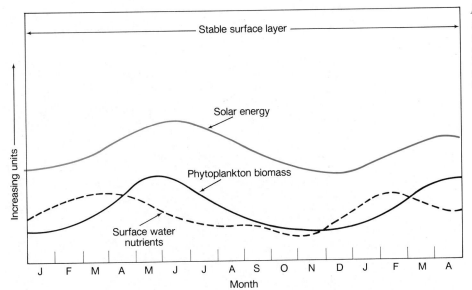

Figure 13.4

Lack of surface mixing and overturn at low latitudes results in a depressed phytoplankton biomass. This pattern is related to solar radiation levels that produce a year-round stable water column.

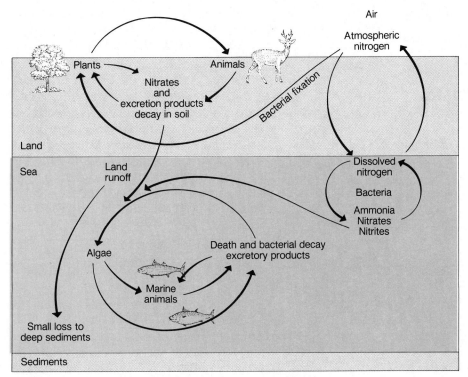

Figure 13.5

The nitrogen cycle. Although nitrogen makes up 78% of the earth's atmosphere, only a few microorganisms are able to change nitrogen gas to the nitrate that is used by land and sea plants, which are in turn eaten and recycled by animals.

cleic acids. Nitrates and phosphates are removed from the water by the primary producers as the plant populations grow and reproduce. Both are cycled into the animal populations as the animals feed on the plants and are returned to the water as the organisms die and decay. Excretory products from the animals are also added to the seawater, to be broken down and used again by a new generation of plants and animals. The cyclic nature of the nutrient pathway for nitrogen and phosphorus is illustrated in figures 13.5 and 13.6.

To cycle nitrogen, a number of different microorganisms must participate. Nitrogen gas is useless to

plants; it must be converted to nitrate to be incorporated into plant tissue. A few species of microorganisms are able to convert nitrogen gas to ammonia, and then other species convert ammonia to nitrite. Still another species is required to convert nitrite to nitrate, the form most easily absorbed by the plants.

It is possible to follow the cyclic path of any molecule required for life (refer back to the hydrologic cycle described in chapter 1). Water is a requirement for life; it is part of all life processes and is cycled over and over through life forms, time, and space.

Figure 13.6
The phosphorous cycle. Dissolved phosphorus is carried to the sea by runoff from the land, where it is taken up by plants and recycled through animals until it is released from dead tissue by bacterial action.

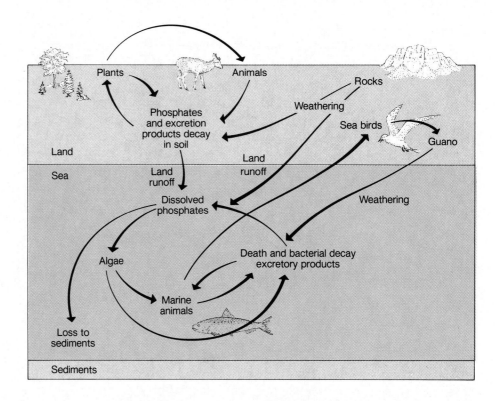

13.3 Global Primary Production

Figure 13.7 shows the geographic distribution of global primary production. There are a few narrow areas of very high productivity scattered around the world. These areas are located against the northwest coasts of North and South America, the west coast of Africa, and along the west side of the Indian Ocean. These special areas are the product of upwellings that transport nutrients to the photic zone, and, as we shall see in chapter 15, these are the areas of the world's greatest fish catches. Coastal areas are generally considered more productive than open ocean, because in the more shallow coastal areas and estuaries, tidal currents cause mixing and turbulence, which supply nutrients to the surface layers. The rivers and the land runoff in coastal areas also supply nutrients. At the same time and despite the turbulence, fresh water from these sources helps to create a stable, low-density surface layer, which keeps the plant cells up where sunlight is plentiful.

Considering equal areas, we can compare the quantities of organic material that are produced by upwelling, the coastal ocean, and the open sea. Upwelling areas are, on the average, three times more productive than coastal areas and six times more productive than the open ocean for the same units of area and time. This relationship is shown in the first column of table 13.1. The second column shows that the total area of each type of oceanic region is inversely related to its primary production. There is about one hundred times more coastal area than upwelling area, and nearly ten times

more open ocean area than coastal area. In the third column, the total primary production of each ocean area type per year is shown. These values demonstrate that most of the organic carbon produced by the oceans' plants is scattered in low concentrations over the large area of the open sea. Smaller total amounts are produced along the coasts and in upwelling regions, but these amounts are concentrated in smaller areas, making these areas very rich in primary production on a per unit area basis. A comparison of gross production in major upwelling areas is given in table 13.2.

The situation described here is currently being challenged by some new ideas on the subject of open ocean productivity; for a summary, see the Primary Production in Oceanic Areas box in this chapter.

To compare primary production on land with that at sea, study table 13.3. It shows that primary production per square meter in the open sea is about the same as that of the deserts on land. The vast areas of open ocean are productive only because of their size. Remember that they cover 63.9% of the earth's surface, or 90% of the ocean's surface. Upwelling areas are comparable to pastureland and lush forestland, while certain estuary systems approach the most heavily cultivated land. Areas of intensively cultivated, fast-growing crops on land produce much more carbon than do most of the regions of the sea; however, on land people put large quantities of time and energy into raising their crops to produce the same high yields that natural processes provide in shallow estuaries. Keep in mind that from the human standpoint, the plants on land are often used directly for food, while those of the sea are not.

Figure 13.7
The distribution of primary production in the world's oceans.

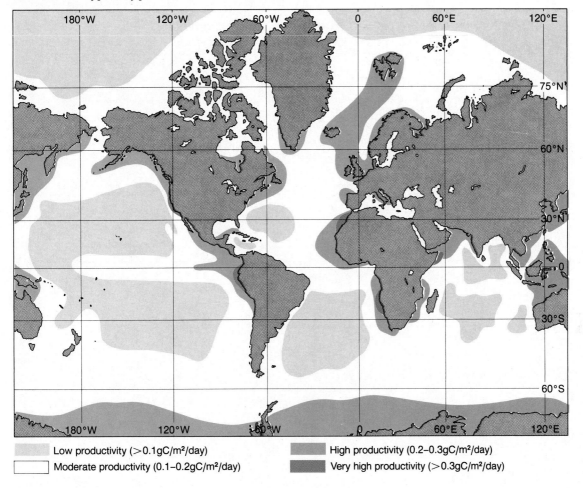

Low productivity (>0.1gC/m²/day)

Moderate productivity (0.1–0.2gC/m²/day)

High productivity (0.2–0.3gC/m²/day)

Very high productivity (>0.3gC/m²/day)

TABLE 13.1

World Ocean Primary Production

Area	Primary production (gC/m²/yr)	World ocean area (km²)	(%)	Total primary production (metric tons of carbon/year)
upwellings	300	0.36×10^6	0.1	0.1×10^9
coasts	100	36×10^6	9.9	3.6×10^9
open oceans	50	325×10^6	90.0	16.3×10^9

Source: After J. H. Ryther, 1969.

TABLE 13.2

Gross Primary Production in Major Upwelling Areas

Amount (gC/m²/yr)	Upwelling areas	Amount (gC/m²/yr)	Upwelling areas
> 360	Peru, Canary Islands, Benguela, South Arabia (Yemen), Vietnam, Gulf of Thailand	36–100	California, Costa Rica, Chile, Benguela, New Guinea, Andaman Islands, Northwest Australia
100–360	Peru, Canary Islands, Benguela, Somalia, Malagasy (Madagascar), Orissa coast, Java, Sri Lanka (Ceylon), Flores, Banda, East Arafura	< 36	open ocean

TABLE 13.3

Gross Primary Production: Land and Ocean

Ocean area	Amount (gC/m²/yr)	Land area	Amount (gC/m²/yr)
open ocean	50	deserts grasslands	50
coastal ocean	25–150	forests common crops pastures	25–150
upwelling zones deep estuaries	150–500	rain forests moist crops intensive agriculture	150–500
shallow estuaries	500–1250	(sugarcane and sorghum)	500–1250

Source: After E. P. Odum, 1963.

13.4 Measuring Primary Production

It is possible to take a water sample, filter out the phytoplankton, count the cells, and multiply the number counted by the average mass per individual cell in order to determine the amount of plant material present. Other less direct but less tedious methods are also available. If the chlorophyll is extracted chemically from a sample of phytoplankton, the concentration of pigment can be determined and used to estimate the total quantity of plant material. Repeated sampling and chlorophyll determination in a volume of water or under a fixed area of sea surface over time yield data on net primary production. Another method exposes the chlorophyll in the phytoplankton cells to certain wavelengths of light, which cause the chlorophyll pigment to fluoresce. The strength or intensity of the fluorescence is read electronically to give a direct measure of the chlorophyll and phytoplankton biomass present in a given volume of water.

When new plant material is produced by the phytoplankton, dissolved carbon dioxide is converted to organic carbon compounds and oxygen gas; nitrogen from nitrate and phosphorus from phosphate (the nutrients dissolved in seawater) are also required. A ratio exists between the oxygen gas and organic carbon produced and the nitrogen and phosphorus removed from the seawater. These fixed ratios by weight are: $O_2 : C : N : P = 109 : 41 : 7.2 : 1$. These ratios are used to estimate primary carbon production by measuring the rate of nutrient uptake by the plant population or the rate of production of dissolved oxygen. If 10 g of phosphorus is removed by the plants from a volume of water in a given time, then 410 g of carbon have been produced and incorporated into the phytoplankton biomass. If the rate at which the nitrogen or phosphorus is carried into a region by upwellings and currents, and the rate at which it is removed by other currents is known, the rate at which it is being incorporated into the phytoplankton can be calculated. Primary production is then determined from these data. This allows the oceanographer to estimate primary production over large areas of the ocean and to associate it with large-scale water movements and chemical cycles.

Notice that 7.2 times more nitrogen than phosphorus is needed to produce plant material. If the mechanisms supplying these nutrients to the photic zone are similar, then the available nitrogen will normally be used up before the phosphorus. Depletion of the seawater's nitrate level tends to be the limiting nutrient factor for the ocean's primary production, because nitrogen is critical to living organisms for the production of proteins and nucleic acids. Without nitrogen from nitrates, phytoplankton growth and reproduction will lag, and primary production will decrease.

A simple technique with light and dark bottles demonstrates the relationship between net and gross primary production in an area over depth. Seawater samples with their natural phytoplankton population are collected at selected depths. The depths are keyed to the percentage of surface solar radiation available: for example, 100% (surface), 75%, 50%, 20%, and 10%. The water sample from each depth is divided into three subsamples. Subsample A is treated immediately to determine the amount of dissolved oxygen in the water at the beginning of the experiment. Subsample B is placed in a tightly sealed **dark bottle** covered with black tape or aluminum foil so that no light can penetrate the bottle. Subsample C is placed in a tightly sealed clear glass bottle, a **light bottle.** The light and dark bottles are then attached to a line and returned to the depth at which they were collected. They are left there for several hours and then brought back to the surface to have the amount of oxygen in each bottle determined as quickly as possible. The changes in the dissolved oxygen content with time can be used to describe the primary production at the sampled depths and light levels by subtracting oxygen values, as follows:

Bottle A − Bottle B_{dark} = measure of oxygen used in respiration

Bottle C_{light} − Bottle A = measure of net oxygen produced by photosynthesis, or net primary production

Bottle C_{light} − Bottle B_{dark} = measure of total oxygen produced, or gross primary production.

The carbon:oxygen ratio for photosynthesis, the volumes of water in the bottles, and the duration of the experiment are used to convert the calculated changes in dissolved oxygen to changes in gC/volume/time.

A light and dark bottle experiment is illustrated in figure 13.8. Note that the primary production varies with

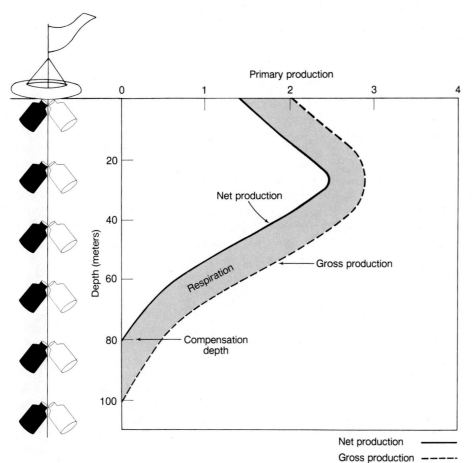

Figure 13.8
A light and dark bottle experiment provides values for respiration, net primary production, and gross primary production with depth. Respiration and production are measured in milligrams of O_2 per liter per six-hour interval. The compensation depth occurs where net production is zero. Light, nutrients, and stability of the water combine to provide conditions for the highest primary production at the depth of 30 meters.

Primary Production in Oceanic Areas

The generally accepted concept of low biological productivity in oceanic areas away from coastlines and upwelling zones is being challenged by recent research. Data taken from the central North Pacific shows two to three times as much organic matter per surface area being created by photosynthesis than had been reported in the past. Some oceanographers think that the open ocean has not been sampled often enough in the past to catch periods of high productivity, and that this has resulted in low productivity assumptions. Others believe that the techniques used may have given erroneous results. One problem may have been the death of relatively large numbers of photosynthesizing cells in sparse populations during the incubation periods of the experiments which measure production rates. High production results in the North Pacific were achieved using ultraclean conditions to avoid poisoning the phytoplankton. Despite this care, microscopic examination showed phytoplankton dying in the water bottles during the photosynthesis experiments. When working in coastal and upwelling areas in which the phytoplankton populations are large, such experimental death rates would not significantly change productivity calculations.

Another approach to measuring productivity is to use ocean-scale volumes of water isolated by natural ocean processes rather than artificially contained water samples. William Jenkins and Joel Goldman of the Woods Hole Oceanographic Institution used data collected over eighteen years near Bermuda in the Sargasso Sea. They calculated levels of production so high in the upper 100m that there was not sufficient nitrate in the water to allow it. Jenkins and Goldman suggest that the nitrate is supplied from below, possibly infrequently during storms. They propose a two-layer system for the area. In the upper layer, samples would show the traditional low levels of production. In the lower layer, just above the nutrient supply, the production would be much higher. Some evidence of a two-layer system has also been found in the Pacific.

If open-ocean production is indeed higher, then this large area of the world is more important on a world scale of total carbon production than is indicated in the text tables. Additional field studies and reassessment of production data continue as scientists work toward a better understanding of the open ocean's ability to produce organic matter.

depth and that the values for respiration remain nearly constant. Where net primary production is zero, the oxygen produced by photosynthesis exactly meets the demand for oxygen by phytoplankton respiration. This is the **compensation depth.**

If it is not convenient to return the light and dark bottles to the sea, the experiments can be carried out by wrapping or covering the clear light bottles with perforated screens to reduce the sunlight penetrating the bottle by the desired percentage and then placing the bottles in a controlled temperature bath and exposing them to light equivalent to that available at the sea surface. In this way, light and dark bottle values can be obtained while the ship is on its way from station to station.

Similar experiments today are usually made with light and dark bottles to which the radioactive isotope carbon-14 is added. After an incubation time the phytoplankton are filtered from the water, and the amount of carbon-14 incorporated into the plant cells is measured; the net primary production is determined by subtracting any radioactivity incorporated into cellular material in the dark bottle sample to account for non-photosynthetic effects.

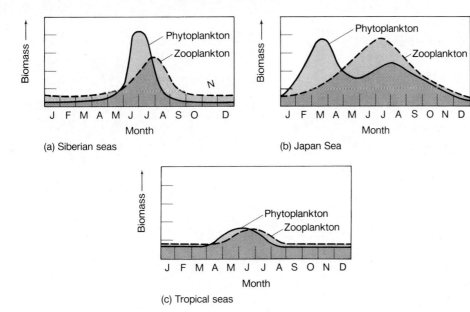

Figure 13.9
The zooplankton biomass varies
directly with the phytoplankton
biomass in (a) polar latitudes,
(b) middle latitudes, and
(c) tropical latitudes.

13.5 Total Production

Food Chains and Food Webs

Primary production forms the first step on the nutritional ladder linking plants, herbivorous animals, and carnivorous animals. Together they make up the total production of the sea. Where there is high primary production by phytoplankton, seaweeds, or other marine plants, there are large populations of animals. Animals are consumers; they feed on the primary producers or on other consumers. The **herbivores** eat plant life directly; the **carnivores** feed on the herbivores or on other carnivores. Notice that in either case the carnivores are still linked to the primary producers. The most numerous and the greatest biomass of herbivores are the plant-eating **zooplankton** or animal plankton. These animals are the primary consumers that convert plant tissue to animal tissue; in turn they become the food for the next rung on the ladder, the flesh-eating zooplankton, which are the carnivores, or secondary consumers. See figure 13.9 for the relationship between herbivorous zooplankton and their phytoplankton food

source. Again, the response varies with latitude. Note that the zooplankton peak lags behind the phytoplankton production at the polar latitudes (fig. 13.9a) and middle latitudes (fig. 13.9b). The zooplankton reproduce when sufficient phytoplankton are present to support an increase in herbivores. As grazing by herbivores increases, the phytoplankton biomass decreases. The herbivore population then decreases in its turn as the phytoplankton population declines. The zooplankton population in the tropics (fig. 13.9c) remains nearly constant, with a lower but stable phytoplankton biomass.

There may be many or few rungs on the nutritional ladder or links in the food chain that lead to the top carnivore, a predator on which no other marine organism preys (for example, sharks and killer whales). Food chains are rarely simple and linear; they are more likely to show complex branching or interrelationships among organisms, in which case it is more appropriate to call the interconnecting pattern a **food web.** See the herring food web diagram in figure 13.10 and the overall ocean food web in figure 13.11.

Figure 13.10
The food web of the herring at various stages in the herring's life.

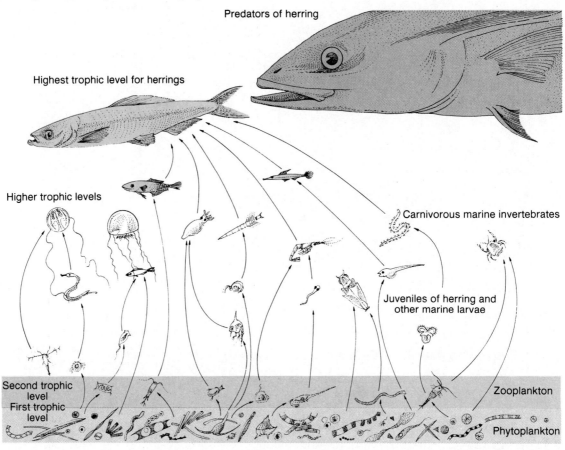

Predators of herring

Highest trophic level for herrings

Higher trophic levels

Carnivorous marine invertebrates

Juveniles of herring and other marine larvae

Second trophic level
First trophic level

Zooplankton

Phytoplankton

Trophic Pyramids

Food chains and food webs represent the pathways followed by nutrients and food energy as they move through the succession of plants, grazing herbivores, and carnivorous predators. These relationships are often demonstrated in the form of a pyramid, as seen in figure 13.12. In this **trophic pyramid,** each **trophic level** represents a link in the food chain. Trophic levels are numbered from the bottom to the top, so that the primary producers are the first trophic level, the herbivo-rous zooplankton are the second trophic level, and carnivores form the upper levels, as needed, on up to the top carnivores. Figure 13.12 includes the sun, the energy source directly necessary to the primary producers and indirectly necessary to all other layers.

In general, as one moves upward from the first trophic level, the size of the organisms increases and the numbers of organisms decrease. The larger numbers of small organisms at the lower trophic levels collectively have a much larger biomass than the smaller numbers of large organisms at the upper levels. Table 13.4 relates the abundance of organisms to their size.

Figure 13.11

*The ocean food web. Plant, animal, and bacterial populations are
dependent on the flow of energy and the recycling of nutrients
through the food web. The initial energy source is the sun, which
fuels the primary production in the surface layers. Herbivores
graze the phytoplankton and benthic algae and are in turn
consumed by the carnivores. Animals at lower depths depend on
organic matter from above. Upwelling recycles nutrients to the
surface, where they are used in photosynthesis.*

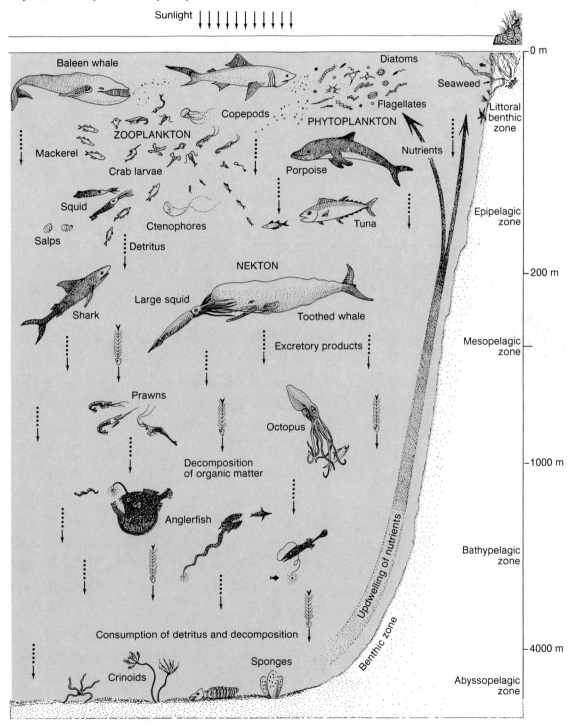

Figure 13.12

A trophic pyramid. Trophic levels are numbered from base to top. The first trophic level requires nutrients and energy. Nutrients are recycled at each level; energy is lost as heat at each level.

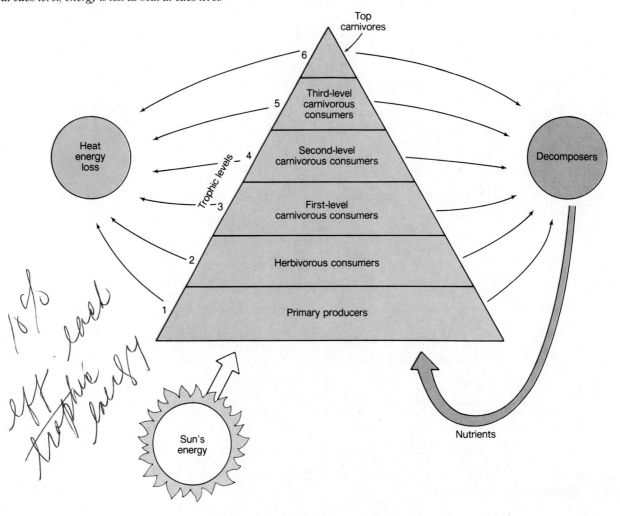

TABLE 13.4

Relative Abundance and Size of Marine Organisms

Organic form	Size range	Relative abundance
fish	10 cm–100 cm	0.01
zooplankton	1 mm–10 cm	1.0
phytoplankton	0.001 mm–0.02 mm	10.0

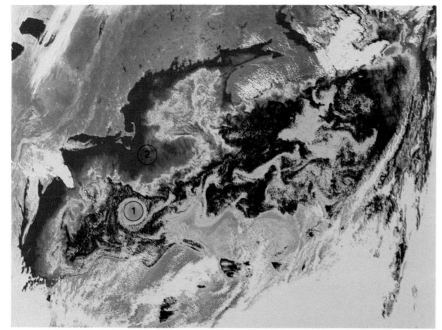

(a)

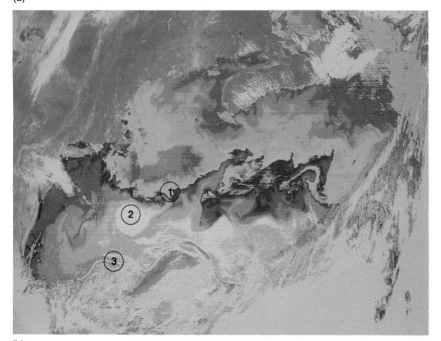

(b)

Colorplate 12

Satellite ocean color observations using specific wavelengths of radiation provide images of phytoplankton abundance (a) and sea surface temperatures (b). (a) In the top image, high phytoplankton pigment concentrations are shown in dark brown, while moderate levels are shown in red through yellow to green. Low levels are in blue. Land areas are depicted in light brown and clouds obscuring the earth's surface are white. High concentrations of phytoplankton are shown in the nutrient-rich coastal waters. The Gulf Stream and Sargasso Sea to the south are nutrient poor and the phytoplankton concentrations are at a minimum. A recently formed warm-core ring from the Gulf Stream (1) has low chlorophyll Sargasso Sea water in its center. Nantucket Shoals and Georges Bank (2) are rich fishing grounds. (b) In the bottom image, the warmer water, about 25° C or 77° F, is red through orange and yellow to green. The coldest water, about 6° C or 43° F, is in blue. The sharp mixing boundary between warm Gulf Stream water and cold Labrador Current water is seen just south of Georges Bank (1). The warm-core ring, noted for its low chlorophyll concentration in (a), is seen in (b) as displaced water of elevated temperature (2). The western edge of the Gulf Stream is shown swinging away from the warm-core ring at (3).

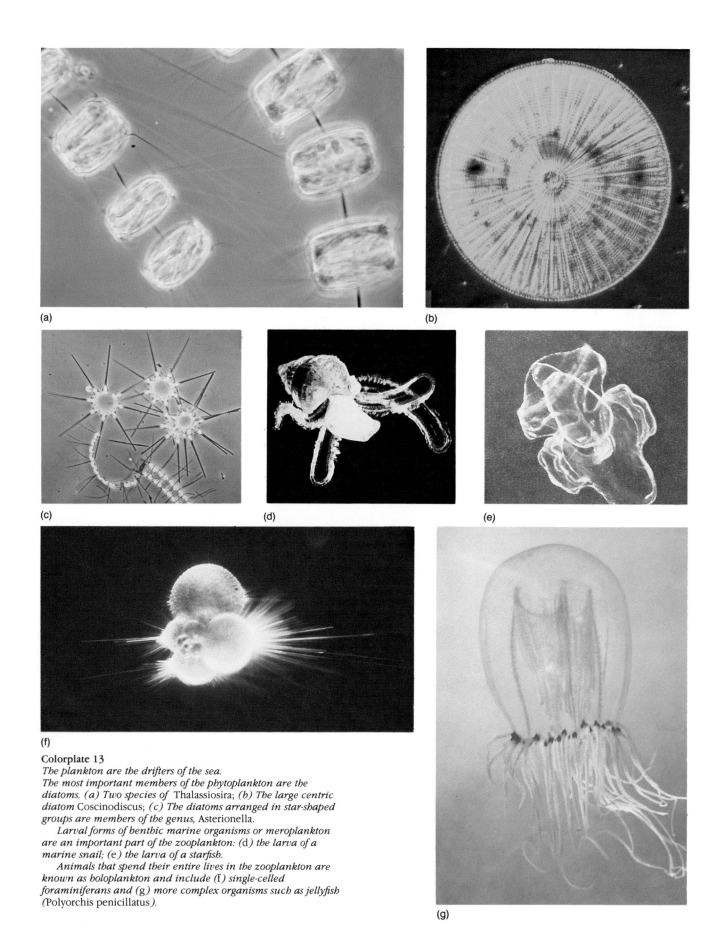

Colorplate 13

The plankton are the drifters of the sea.
The most important members of the phytoplankton are the
diatoms. (a) Two species of Thalassiosira; *(b) The large centric*
diatom Coscinodiscus; *(c) The diatoms arranged in star-shaped*
groups are members of the genus, Asterionella.

Larval forms of benthic marine organisms or meroplankton
are an important part of the zooplankton: (d) the larva of a
marine snail; (e) the larva of a starfish.

Animals that spend their entire lives in the zooplankton are
known as holoplankton and include (f) single-celled
foraminiferans and (g) more complex organisms such as jellyfish
*(*Polyorchis penicillatus*).*

It is estimated that the overall efficiency of energy transfer up each layer of an open-ocean trophic pyramid is about 10%. Thus, in order to add 1 kg of weight, a person must eat 10 kg of salmon. To attain that weight, the salmon had to consume 100 kg of small fish, and the fish needed to consume 1000 kg of carnivorous zooplankton, which in turn required 10,000 kg of herbivorous zooplankton, needing 100,000 kg of phytoplankton to supply the eventual 1 kg gain at the top of the pyramid. The 90% energy loss at each level goes to the metabolic needs of the organisms at that level. These needs include energy required for moving, breathing, feeding, and reproduction, as well as heat loss. In other words, an organism that consumes 100 units from the level directly below will use 90 units for its own metabolic needs and will convert only 10 units to body tissue available to predation from the level above. Therefore, feeding at high trophic levels is less energy efficient than feeding at low trophic levels.

Other Systems

In some areas of the ocean, total productivity is not tied to the primary productivity of the phytoplankton. The shallow waters above a tropic coral reef are clear, indicating a lack of phytoplankton. The reef, however, is a rich and varied community living in an independent, complex association that includes algae, herbivores, and carnivores. (See also the discussion of coral reefs in chapter 16.)

Shallow coastal areas in which the sea bottom is exposed to sunlight support masses of attached seaweed and bottom-living single-celled plants. These function as primary producers in addition to the phytoplankton in the water column. The rapid growth and relatively large size of many seaweeds provide a large amount of organic material to the animal population. Animals graze the living plants or feed from the fragments left after the battering of winter storms and the populations' natural mortality.

The exceedingly high primary production of estuaries has various sources. In deep fjordlike estuary systems, the majority of the primary production comes from the phytoplankton in the water column. In these estuaries, although the concentration of organic carbon per cubic meter of water is highest near the surface, the water below the photic layer may contain a greater total amount of organic carbon because of the active vertical mixing, which displaces plant cells downward, and the large volume of water at depth.

In the broad, shallow estuaries, active production of organic matter by primary producers occurs at all depths, on the estuary floor as well as in the water column. Organic material is also produced in the bordering marshlands, some of which produce two crops of plant growth per year. All of these sources contribute organic matter as food to the animal population in a rather small volume of water. These shallow estuaries enjoy an extraordinary rate of primary productivity, supporting the food chains and webs within the estuary and also contributing organic matter to the coastal waters. Refer back to table 13.3 to compare the gross primary production of deep and shallow estuaries.

In recently discovered deep-water communities that surround hot-water vents on the ocean floor, the primary production is not of plant material based on solar energy; rather, the primary producers are bacteria that take their energy from the chemistry of the hot water flowing from cracks in the sea floor. These richly productive communities and their energy source are discussed with coral reefs and shallow-water communities in chapter 16. Other communities based on oil and gas seeping onto the surface of the sea floor were discussed in the box in chapter 12.

13.6 Practical Considerations: Human Concerns

When human activities supply additional nutrients, the primary production may exceed the ability of the local herbivores to consume it. A continuous and excessively high rate of primary production eventually results in an unnaturally high rate of decomposition, as unconsumed plant matter accumulates at depth, especially in an estuary. The decaying material consumes dissolved oxygen at a rapid rate, and a mat of organic debris may form over the sediments. The waters and the sediments become anoxic, which adversely affects the organisms at the higher levels in the food webs and chains. The section on Chesapeake Bay in chapter 11 illustrates this problem.

Custom, culture, economics, and availability influence the harvesting of the oceans by humans. Many commercial harvests are made from both the higher trophic levels (for example, salmon, tuna, halibut, and swordfish), and the lower levels (for example, herring, shellfish, and anchovy fisheries). Harvesting high in the

Figure 13.13

Trophic level efficiency varies among (a) upwelling areas, (b) coastal regions, and (c) the open ocean. The number of trophic levels and the level at which humans harvest differs with location.

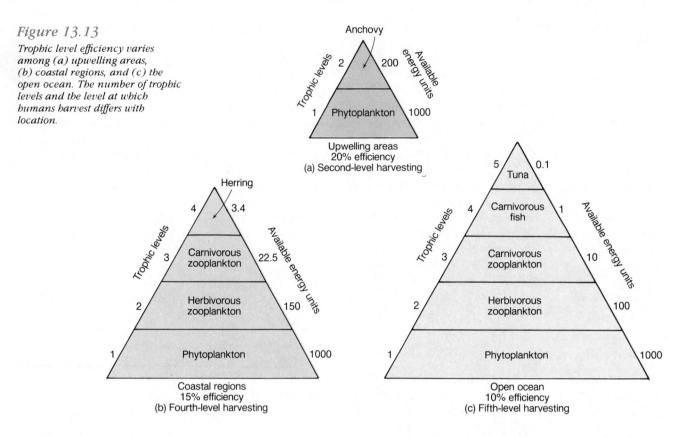

TABLE 13.5

Oceanic Food Production

Area	Plant production (metric tons of carbon/year)	Efficiency of energy transfer per trophic level	Trophic level harvested	Fish production (metric tons/year)
open ocean	16.3×10^9	10%	5	1.6×10^6
coastal regions	3.6×10^9	15%	4	120×10^6
upwelling areas	0.1×10^9	20%	2.5	120×10^6

Source: After J. H. Ryther, 1969.

trophic pyramid is energy inefficient, while overharvesting any level endangers levels both above and below it.

The highest yields of food resources for humans occur when the harvest of marine species is conducted at the lowest possible trophic level. Table 13.5 shows how plant production in the three basic areas of the oceans relates to the production of fish. The third column gives the average efficiency of energy conversion between the trophic levels in each area, and the fourth column shows the trophic level at which humans usually harvest their food. The well-mixed, nutrient-rich waters of the coasts and estuaries support short, efficient

food chains, which are also shown in pyramid form in figure 13.13a and b. Therefore, fish production in these areas is greater and the plant production required to sustain these high levels is less. Compare the efficiency of these food chains with the open-ocean food chain in figure 13.13c and the open-ocean fish production in table 13.5.

We need to learn how to harvest and use the lower trophic levels without depleting them. An understanding of the trophic levels and their relationship to each other is necessary, so that harvesting is controlled to ensure sufficient stock for reproduction and the maintenance of the trophic pyramid.

Summary

Plants use photosynthesis to produce organic compounds from carbon dioxide and water in the presence of sunlight and chlorophyll. Oxygen is formed as a by-product. Respiration breaks down organic compounds with the addition of oxygen to yield energy, water, and carbon dioxide.

Gross primary production is the total amount of organic material produced by photosynthesis per volume per unit of time; net primary production is the gain in organic material per volume per unit of time after the organic material used in plant respiration is deducted. The amount of organic carbon produced by primary production in a volume over a time period is measured as the rate of change in biomass. The biomass available at a location at a specific time is the standing crop.

Primary production is controlled by the interaction of sunlight, nutrients, and the stability of the surface water in an area. At polar latitudes, the availability of light controls phytoplankton growth; nutrients are not limiting. In the tropics, the sunlight is available year-round, and the stability of the water column helps hold the phytoplankton at the surface. However, the nutrient supply is poor and it limits production. At temperate latitudes, light, nutrients, and water column stability vary with the seasons. Nutrients cycle through the land and sea, the plants and animals, and are returned to the water by death and bacterial decomposition.

Where sunlight is not limiting, coastal waters are always more productive than the open ocean. Good mixing, the nutrients in the land runoff, and a water column made stable by the addition of fresh water combine to make shallow coastal regions very productive.

Upwelling areas are three times more productive than coastal water, and six times more productive than the open ocean, but there is ten times more open ocean than coastal water and one hundred times more coastal water than upwelling area. Primary production is concentrated in coastal areas and at upwellings.

Primary production may be estimated by (1) counting the increase in plant cells; (2) determining the change in concentration of chlorophyll in a sample; (3) measuring the rates of nutrient supply and nutrient removal and using the known ratio of nutrients to calculate carbon production; (4) measuring the rate of incorporation of carbon-14 into the plant cells, or measuring the rate that oxygen is produced by phytoplankton growing in light and dark bottles.

The phytoplankton, or primary producers, are preyed upon by herbivorous zooplankton. These are the primary consumers that are preyed upon by the secondary consumers, or carnivorous zooplankton. The relationship between the phytoplankton and the herbivorous zooplankton varies with latitude. The term food web is more appropriate than food chain to describe the interconnecting patterns. Trophic pyramids present the relationship between producers and consumers in terms of the transfer of biomass and energy. Open-ocean transfer efficiency is approximately 10% between trophic levels. Efficiency is higher in coastal and upwelling areas. Nutrients are recycled; energy is not.

Tropic coral reefs, very shallow coastal waters and estuaries, and deep-ocean vents are productive areas in which productivity is not keyed to the phytoplankton.

If human activities supply additional nutrients to seawater, primary production increases, followed by decomposition and anoxia. Harvested food value is increased if the lower trophic levels are used, but care is required to conserve the stock for future harvests and for the requirements of other organisms.

Key Terms

food chain	gross primary production	decomposer	carnivore
phytoplankton	respiration	dark bottle	zooplankton
nutrient	net primary production	light bottle	food web
chlorophyll	biomass	compensation depth	trophic pyramid
photosynthesis	standing crop	herbivore	trophic level

Study Questions

1. Why is the efficiency of energy transfer between trophic levels only about 10% in the open sea? Compare this efficiency of energy transfer with that found in upwelling regions.

2. Explain the general relationship between the abundance and size of organisms shown in table 13.4.

3. Distinguish between the terms in each pair:
   a. Standing crop, biomass.
   b. Photosynthesis, respiration.
   c. Producer, consumer.
   d. Food web, food chain.
   e. Net productivity, gross productivity.

4. Compare the productivity of polar, temperate, and tropic ocean regions. What factor or factors generally limit the productivity of each area?

5. The rate of primary production in the open ocean is less than the rate of primary production along the coasts, but the total primary production of the open ocean exceeds that of the coasts. Explain this apparent contradiction.

6. Which areas of the oceans are most productive? How does the productivity of these areas compare with the productivity of the land?

7. Why are estuaries less important than cultivated land as producers of human foods, although the primary production rate of estuaries approximately equals the most intensively cultivated land?

8. How do the surrounding land areas contribute to an estuary's productivity?

9. Draw a general diagram to explain the movement of (a) a gas and (b) a nutrient through the ocean environment and its plant and animal populations.

Study Problems

1. A sample of water from a depth of 5 m showed 8 mg of dissolved oxygen per liter. Part of this water was placed in a light bottle and part of it was placed in a dark bottle, and the bottles were returned to the 5-m depth. After six hours the oxygen content of the light bottle was found to be 8.9 mg of oxygen per liter; the dark bottle showed 7.4 mg of oxygen per liter. Calculate (a) the respiration rate, (b) net primary production, and (c) gross primary production. How many grams of new carbon were produced per liter in six hours?

2. If the nitrogen available as nitrate is removed from the water of an inlet at the rate of 3.6 mg of nitrogen per liter of water every eight hours, what is the rate of new carbon production by the phytoplankton?

The Plankton
Drifters of the Open Ocean

14

*I*n a sudden awakening, incredible in its swiftness, the simplest plants of the sea begin to multiply. Their increase is of astronomical proportions. The spring sea belongs at first to the diatoms and to all the other microscopic plant life of the plankton. In the fierce intensity of their growth they cover vast areas of ocean with a living blanket of their cells. Mile after mile of water may appear red or brown or green, the whole surface taking on the color of the infinitesimal grains of pigment contained in each of the plant cells.

The plants have undisputed sway in the sea for only a short time. Almost at once their own burst of multiplication is matched by a similar increase in the small animals of the plankton. It is the spawning time of the copepod and the glassworm, the pelagic shrimp and the winged snail. Hungry swarms of these little beasts of the plankton roam through the waters, feeding on the abundant plants and themselves falling prey to larger creatures.

Rachel Carson, from The Sea Around Us

*T*he word plankton comes from the Greek term *planktos,* meaning to wander, and the plant and animal plankton are the wanderers and drifters of the sea. They exist in vast swarms, limited in their mobility, moving with the currents. The diversity of planktonic organisms is so great that it is not possible to discuss all of their life forms here. Instead, representative animals, plants, and groups of animals and plants have been chosen for discussion. Because it is the plant life that is able to use the sun's energy to form the basis of life for all the animals, we begin by considering the ocean's floating plant life, the primary producers of chapter 13, and then go on to describe the animals that drift with the plant plankton, grazing upon it and upon each other.

14.1 The Kinds of Plankton

Although many plankton have the ability to move to some degree, particularly toward and away from the sea surface, they make no purposeful motion against the ocean's currents and are carried from place to place suspended in the seawater. Some plankton are quite large; jellyfish may be the size of a large washtub, trailing 15 m (50 feet) of tentacles. But the phytoplankton and many zooplankton are generally too small for our unassisted vision and must be observed under a microscope.

Bacteria and very small phytoplankton cells are called **ultraplankton.** The ultraplankton are less than 0.005 mm in diameter and can only be studied when large volumes of water are processed to accumulate large enough quantities of these organisms for observation. Slightly larger phytoplankton, called **nannoplankton,** have a size range between 0.005 and 0.07 mm. Zooplankton and phytoplankton between 0.07 and 1 mm are called **microplankton** or **net plankton,** since they can be captured in tow nets made of very fine mesh nylon.

The microscopic phytoplankton are the "grasses of the sea." Just as a land without grass and herbs could not support the insects, small rodents, and birds that serve as food for the larger, meat-eating carnivores, a sea without phytoplankton could not support the zooplankton and the other larger animals. As the British biological oceanographer Sir Alister Hardy has said, "All flesh is grass."

Phytoplankton

The phytoplankton are mainly unicellular (or single-celled) plants known as algae. They are **autotrophic** (or self-feeding) by the process of photosynthesis (refer back to chapter 13). Each cell is an independent individual, and even in the species in which the cells attach together in **filaments** (or long chains) or other aggregations, there is no division of labor between the cells. Each cell functions independently. There is only one large planktonic seaweed, and that is the *Sargassum,* which is found floating in the area of the North Atlantic known as the Sargasso Sea. *Sargassum* reproduces vegetatively by fragmentation to form large mats, which provide shelter and food for a wide variety of organisms, including fish and crabs. The specialized organisms found living in the *Sargassum* mats occur nowhere else.

Groups of organisms belonging to the phytoplankton include the **diatoms, dinoflagellates,** and **coccolithophores.** Other organisms, such as silicoflagellates, cryptomonads, chrysomonads, green algae, and cyanobacteria or blue-green algae, are also represented. Figure 14.1 shows a generalized phytoplankton sample. We shall consider each group separately, with emphasis being placed on the diatoms and the dinoflagellates, as they are the most abundant and most important members of the marine phytoplankton.

Diatoms are single-celled plants found in areas of cold, nutrient-rich water. They are sometimes called golden algae, because of a certain characteristic yellow-brown pigment, **fucoxanthin,** which masks their chlorophyll. Those that have **radial symmetry** are round and shaped like pillboxes and are called **centric diatoms.** Those that show **bilateral symmetry** are elon-

Figure 14.1

Phytoplankton. Diatoms are in chains; dinoflagellates are irregular single cells. Magnification × 110.

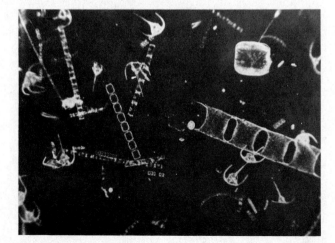

gate and are called **pennate diatoms.** Centric diatoms float better than pennate diatoms; therefore pennate diatoms are often found on the shallow sea floor or attached to floating objects, while centric diatoms are more truly planktonic. Some common diatoms from temperate waters are shown in figure 14.2.

Around the outside of each diatom is a **frustule** (or cell wall) of pectin, a jellylike carbohydrate, impregnated with silica. The frustule is hard, rigid, trans-

parent, and delicately marked with pores that connect the living portion of the cell inside to its outside environment (see fig. 14.3). The two halves of a centric diatom's frustule fit together like a pillbox, and when the cell has grown sufficiently large the cell inside divides, the two halves of the pillbox separate, a new inner half to each pillbox is formed, and two new daughter diatoms are produced. By this process, one of the new cells will remain the same size as the parent, while the other

Figure 14.2

Centric and pennate diatoms. Diatoms exist as single cells or in chains.

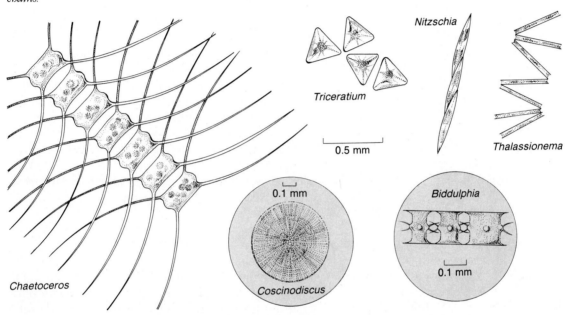

Figure 14.3

Stereoscan micrographs of diatom frustules.

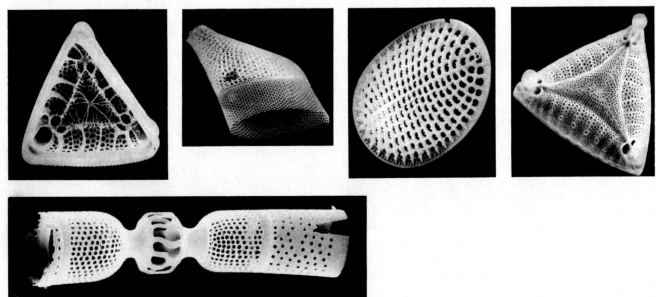

Figure 14.4
The division of a parent centric diatom into two daughter diatoms. The two halves of the pillbox-like cell separate, the cell contents divide, and a new inner half is formed for each pillbox. One daughter cell remains the same size as the parent; the other is smaller.

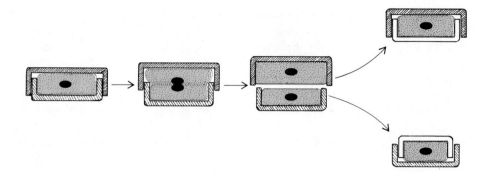

is always smaller, as its larger pillbox half is formed by the smaller bottom half of the parent pillbox. This process is shown in figure 14.4. When a cell reaches a size level of about 25% of the original parent's size, it stops dividing and begins a sexual cycle. In this cycle it produces a naked **auxospore,** which increases in size, forms a new frustule, and begins to divide again. Diatoms divide very rapidly, every twelve to twenty-four hours under conditions of plentiful sunlight and nutrients. This rapid division may increase the population to such a degree that the water becomes discolored by the presence of millions and millions of cells. This rapid population increase is called a **bloom.**

Frustules made with silica are more dense than seawater, but the diatom must stay afloat in the sunlit surface waters to survive. The internal cell material has a low density and increases its buoyancy by the production of oil as a storage product. Fish that feed on large quantities of diatoms may have a distinctly oily taste. The diatom's small size also helps it to stay afloat, because small particles that are only slightly more dense than the liquid in which they are suspended sink very slowly. A small, as opposed to a large, spherical particle has a large surface area in comparison to its volume or mass. This large surface area : volume ratio helps keep the cell afloat, and some diatoms even possess spines, wings, or other projections that increase their surface area still more. This also provides the diatoms with a large area for exposure to sunlight and to water containing the gases and nutrients necessary for photosynthesis and growth.

In the oceans, diatoms are most important as the first level of food production, upon which life in the sea is based. Diatoms that are not consumed by herbivores eventually die naturally and sink to the ocean floor. In shallow areas of the oceans, the cells are deposited with some organic matter still locked inside. Deposits made long ago in this manner have formed petroleum, or oil. Diatom frustules sinking to the greater depths of the oceans build up siliceous sediments under areas of abundant diatom populations. (Refer back to the section on biogenous sediments in chapter 2.) Sometimes geologic processes lift these silica-rich sediments above

the sea, where they are mined as **diatomaceous earth,** which is used in industrial filtration systems, in the filtering of wine as well as swimming pools, and as an abrasive in toothpaste and silver polish.

The dinoflagellates differ from the diatoms in several respects. Dinoflagellates usually have two **flagella,** or whiplike appendages, that beat within grooves in the cell wall. One groove encircles the cell like a belt, and the other lies at right angles to it. The beating of these flagella makes the cells motile and causes them to spin like tops as they move through the water. They also tend to migrate vertically in response to sunlight, but this ability to move is limited, and they are still at the mercy of the waves and currents. Representative dinoflagellates are shown in figure 14.5.

Dinoflagellates are red to green in color and can exist at lower light levels than diatoms, because they can both photosynthesize like a plant and ingest organic material like an animal. They have both autotrophic and **heterotrophic** abilities. Heterotrophic organisms feed on other organisms or on organic substances. Their external walls do not contain silica, but may be armored with plates of cellulose, giving them the appearance of spinning armored helmets. Other dinoflagellates have a smooth, flexible outer surface showing no such plate structures. Some dinoflagellates are called fire algae, because they glow with bioluminescence at night. (Refer back to chapter 12 for a discussion of this phenomenon.)

Dinoflagellates are found over most of the oceans but do not contribute to the bottom sediments, because both the cell walls and the soft parts decay completely. Although they do make up a substantial portion of the phytoplankton, dinoflagellates are not as important as the diatoms as a primary ocean food source. The cells reproduce by a division process similar to that found in diatoms, but without the reduction in size. They can, under favorable conditions, multiply even more rapidly than diatoms to form blooms.

Coccolithophores and silicoflagellates are relatives of the diatoms. The coccolithophores are single-celled plants with **coccoliths,** or outer calcareous plates, that are deposited as sediment when the cells die (see

Figure 14.5
Dinoflagellates. Noctiluca, Ptychodiscus, *and* Gonyaulax *produce red tides.* Noctiluca *is a bioluminescent, nontoxic dinoflagellate.* Ptychodiscus *and* Gonyaulax *produce toxic red tides and paralytic shellfish poisoning.*

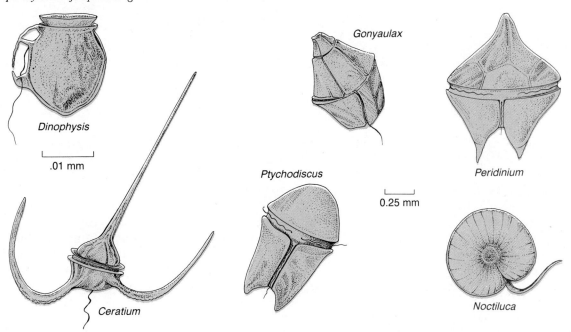

fig. 2.17b). These coccoliths are soluble under the conditions of low temperature and high pressure found in the deep oceans. Therefore, coccolith deposits are limited to shallow regions of less than 4000 m (13,200 ft). Coccolithophores are also limited by the surface water temperatures to the warm tropic regions, and so coccolith deposits are to be found in shallow areas at the lower latitudes. Like the dinoflagellates, they possess two flagellae, and like both diatoms and dinoflagellates, they may reproduce by simple division, although some species do have a form of sexual reproduction. Silicoflagellates are small autotrophic cells with flagella and an internal hollow skeleton of silica. They are not believed to occur in large numbers.

Zooplankton

The animal members of the plankton, or the zooplankton, are either grazers on phytoplankton (herbivores), feeders on other members of the zooplankton (carnivores), or feeders on both plants and animals (omnivores). Many of the zooplankton have some ability to swim and can even dart rapidly over short distances in pursuit of prey or in attempts to escape from predators. They may move verticaly in the water column, but they are still at the mercy of the waves and currents and so are considered planktonic (or drifting) organisms. A general zooplankton sample is shown in figure 14.6.

Figure 14.6
Zooplankton sample. Magnification × 5.

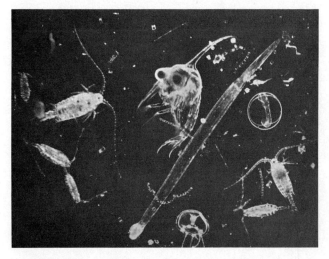

Representatives of nearly every animal phylum are found in the zooplankton. The life histories of organisms that make up the zooplankton are varied. They show a variety of strategies for survival in a world where reproduction rates are high and life spans are short. These animals may produce three to five generations a year in warm waters, where food supplies are abundant and temperatures accelerate the life processes. At high latitudes, where the season for phytoplankton growth is brief, the zooplankton may produce only a single generation in a year. The voracious appetites and rapid growth rates of the carnivorous zooplankton are responsible for liberating nutrients to be recycled by the phytoplankton.

Zooplankton exist in patches of high population density in between areas that are much less heavily populated. The high population patches attract predators. The more sparse populations between the denser patches preserve the stock, as fewer predators feed there. Turbulence and eddies disperse individuals from the densely populated patches to the intervening sparser areas. Convergence zones and boundaries between water types concentrate zooplankton populations, which attract predators.

Plankton accumulate at the density boundaries caused by the layering of the surface waters, and the variation of light with depth and the day-night cycle play additional roles. Members of the zooplankton migrate toward the sea surface at night and return to depth during the day, either in an attempt to maintain their light level or in response to the movement of their food resource. This vertical movement can be as much as 500 m (1650 ft) or may be less than 10 m (33 ft), as in the Arctic, where day-night variations in illumination are small.

Accumulations of organisms in a thin band extending horizontally along a pycnocline or at a preferred light intensity or food resource level are capable of partially reflecting sound waves from depth sounders. The zooplankton layer is seen on a bathymetric recording as a false bottom or a deep scattering layer, the DSL (Refer to the discussion in chapter 4 on sound in the oceans.) Echo-sounding studies are used to record the vertical migration of this layer of plankton and predators and to measure the vertical and horizontal extent of the layer.

Among the most common and widespread zooplankton types worldwide are the small **crustaceans** (or shrimplike animals): the **copepods** and the **euphausiids** (fig. 14.7). They are basically herbivorous and consume more than half their body weight daily. Co-

Figure 14.7
Crustacean members of the zooplankton and an arrowworm.
Calanus *and* Oithona *are copepods.* Euphausia superba *is known as krill.*

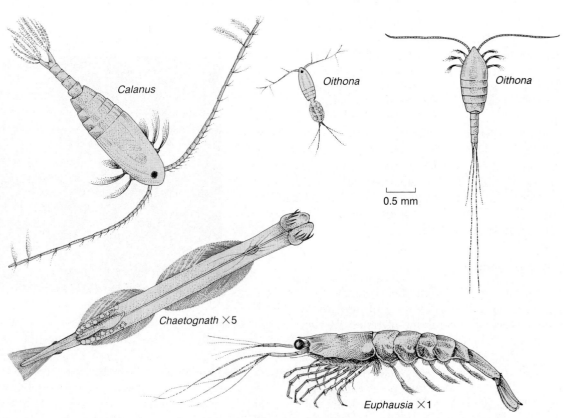

pepods are smaller than euphausiids; euphausiids move more slowly and live longer than the copepods. The euphausiids, because of their size, also eat some of the smaller zooplankton along with the phytoplankton that make up the bulk of their food. These two forms reproduce much more slowly than the diatoms, doubling their populations only three to four times a year. They may make up more than 60% of the zooplankton population in any of the world's oceans and serve as a food source for small fish. In the Arctic and Antarctic, the euphausiids are the **krill,** which occur in such quantities that they provide the main food for the **baleen whales.** Baleen (or whalebone) whales have no teeth; instead they have a netlike strainer of baleen suspended from the roofs of their mouths. After the whales gulp the water and plankton, they expel the water through the baleen, leaving the tiny krill behind. Whales of this type include the blue, right, gray, humpback, sei, and finback whales.

The Antarctic krill, *Euphausia superba,* are present in enormous quantities. Estimates of total biomass have varied from 5 million to 6 billion metric tons, and are probably in excess of 100 million metric tons. Because of their great abundance, the krill in the Southern Ocean have been considered a potential international fishery. Japan, the Soviet Union, West Germany, Poland, and some Latin American nations have participated in fishing for krill. In 1981–82, Japan and the Soviet Union, using large stern trawlers with nets 80 m (264 ft) wide and catching 8–12 tons of krill in a single haul, harvested more than 150,000 metric tons. But it has been found that the krill deteriorate rapidly and must be processed between hauls, limiting a boat's harvest to a maximum of about 50 tons per day. West Germany, Poland, and the Latin American countries have found the fishery only marginally economical. Some vessels have given it up; even the Japanese have reined in their effort. Only the Soviet Union is presently engaged in large-scale harvesting of Antarctic krill.

Marketing the krill for human consumption has not been very successful. Fresh krill are almost flavorless, and when dried the flavor becomes strong and somewhat unpleasant. The shell has been found to contain extremely high amounts of fluoride and must be removed for human consumption. In Russia, the catch is fed mainly to livestock and poultry. The Japanese use their krill as feed on their fish farms. Because of such difficulties, the krill catch for 1982–83 dropped to 80,000 metric tons, and the 1984–85 catch was reduced to 23,000 metric tons.

New peeling machines designed by the Poles may help to find a market for the tail meat. U.S. food authorities have given their approval for its use as an ad-

ditive, and it could be used in seafood salads. When considering the possibilities of a krill fishery, it is important to remember that the krill are not only the food of whales, but also form a basic link in the food web for the seals, penguins and other birds, squid, and fish of the Antarctic region; see the box in this chapter.

Arrowworms, or **chaetognaths** (see fig. 14.7), are abundant in ocean waters from the surface to the great depths. These macroscopic (2–3 cm) (1 in.), nearly transparent, voracious carnivores feed on other members of the zooplankton. Several species of arrowworms are found in the sea, and in some cases a particular species is found only in a certain water mass. The association between organism and water mass is so complete that the species can be used to identify the origin of the water sample in which it is found. (Refer back to the discussion of water masses in chapter 6.)

Simple, microscopic, single-celled, amoebalike protozoans are also found in the zooplankton. These organisms include the **foraminiferans** and the **radiolarians,** as shown in figure 14.8. Foraminiferans, such as the common *Globigerina,* are encased in a compartmented calcareous covering, or shell, while the radiolarians are surrounded by a silica **test,** or shell. The radiolarian tests are ornately sculptured and covered with delicate spines. Openings in the test allow a continuity between internal protoplasm and an external layer of protoplasm. Pseudopodia (false feet), many with skeletal elements, radiate out from the cell. Radiolarians feed on diatoms and small protozoa caught in these pseudopodia. Both foraminiferans and radiolarians are found in the warmer regions of the oceans. After death, their shells and tests accumulate on the ocean floor, contributing to the sediments. Calcareous foraminiferan tests are found in shallow water sediments, while the siliceous radiolarian tests, which are resistant to the dissolving action of the seawater, predominate at greater depths, commonly below 4000 m (13,200 ft) (see chapter 2, fig. 2.17). **Tintinnids** (see fig. 14.8) are tiny protozoans with moving hairlike structures, or **cilia.** These organisms are often called bell animals and are found in coastal waters and in the open ocean.

The **pteropods** (fig. 14.9) are mollusks that are related to the snails and slugs. They may or may not have a small calcareous shell, depending on the species, but all have a foot that is modified into a transparent and gracefully undulating wing. Their hard, calcareous remains contribute to the bottom sediments in shallow tropic regions. Both herbivores and carnivores occur in this group.

Figure 14.8

*Selected members of the radiolaria (*Acanthonia, Acanthometron,
and Aulacantha*), a foraminifera, (*Globigerina*), and a tintinnid.*

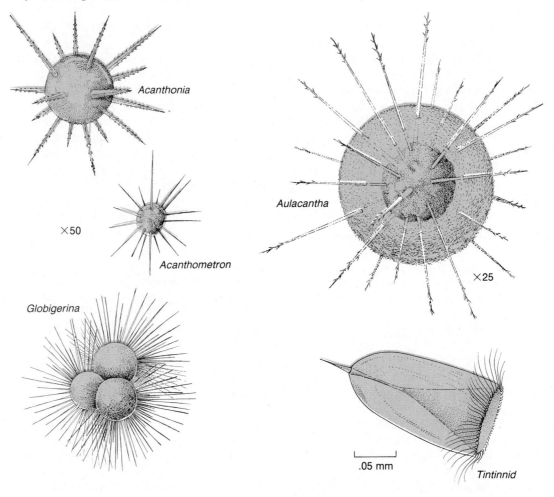

Figure 14.9

The pteropods are planktonic mollusks.

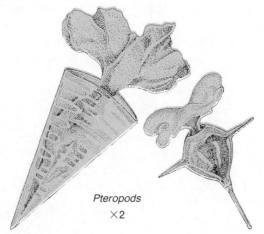

Transparent, delicate, and luminescent, the **ctenophores,** or comb jellies (fig. 14.10), are found floating in the surface waters. Some possess trailing tentacles; all are propelled slowly by eight rows of beating cilia. The small, round forms are familiarly called sea gooseberries or sea walnuts; by contrast, the tropical, narrow, flattened Venus' girdle may grow to 30 cm (12 in.) or more in length. A group of Venus' girdles drifting at the surface and catching the sunlight with their beating cilia is a spectacular sight from the deck of a ship. All ctenophores are carnivores, feeding on other zooplankton.

Another transparent member of the zooplankton is the tunicate, which is related to the more advanced vertebrate animals (animals with backbones) through its tadpolelike larval form. **Salps** (see fig. 14.10), which are pelagic tunicates, are cylindric and transparent and are commonly found in dense patches scattered over many square miles of sea surface.

Both ctenophores and pelagic tunicates, although jellylike and transparent, are not to be confused with jellyfish (fig. 14.11). True jellyfish come from another

Figure 14.10

The comb jellies (ctenophores) Pleurobrachia *and* Beroe.
Pleurobrachia *is often called a sea gooseberry.* Salpa *and* Doliolum
are tunicates.

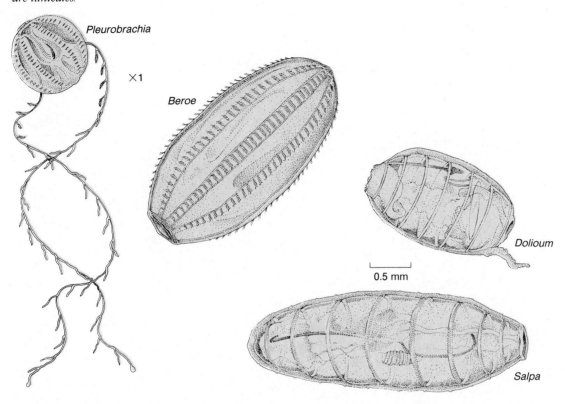

Pleurobrachia

×1

Beroe

Dolioum

0.5 mm

Salpa

and unrelated group of animals, the **Coelenterata** or
Cnidaria. Some jellyfish, such as the common *Aurelia*
and the colorful *Cyanea* with its trailing stinging tentacles, spend their entire lives as drifters. Others, such
as *Gonionemus,* a small jellyfish of the Atlantic and Pacific oceans, and *Aequorea,* found in many temperate
waters, are members of the plankton for only a portion
of their lives, as they eventually settle and change to a
bottom-dwelling, attached form similar to a sea anemone
or coral polyp. Another group of unusual jellyfish are
the **colonial** forms, including the Portuguese man-of-
war, *Physalia,* and the small by-the-wind-sailor, *Velella.*
Both are collections of individual but specialized animals, some of which have the task of gathering food,
reproducing, or protecting the colony with stinging
cells, while some form a float.

All of the zooplankton discussed up to this point,
with the exception of certain jellyfish, spend their entire
lives in the plankton. Such animals are called **holoplankton.** However, there is an important portion of
the zooplankton that spends only part of its life as
plankton; this group is the **meroplankton.** The meroplankton includes the egg, larval, and juvenile stages of
many organisms that spend most of their lives as either
free swimmers (such as fish) or bottom dwellers (such
as crabs and starfish). For a few weeks, the **larvae** (or
young forms) of oysters, clams, barnacles, crabs, worms,

snails, starfish, and many other organisms are a part of
the zooplankton. The larvae of these animals are carried
to new locations by the currents, where they find areas
to settle and food sources. In this way, repopulation of
areas in which a species may have died out occurs, and
overcrowding in the home area is reduced. Sea animals
produce larvae in incredible numbers, and so the meroplankton is an important food source for other members
of the zooplankton and other animals. The parent animals may produce millions of spawn, but only one male
and one female need survive to adulthood in order to
guarantee survival of the stock.

Larvae often look very unlike the adult forms into
which they will develop (see fig. 14.12). Early scientists
who found and described these larvae gave each a name,
thinking they had discovered a new type of animal. We
keep some of these names today, referring, for example,
to the trochophore larva of worms, the veliger larva of
sea snails, the zoea larva of the crab, and the nauplius
larva of the barnacle.

Other members of the meroplankton include fish
eggs, larvae, and juvenile fish. The young fish feed on
other larvae, until they grow large enough to hunt for
other foods. Some large seaweeds release **spores,** or reproductive cells, which drift in the plankton until they
are consumed or settle out to attach to the sea bottom.

Figure 14.11
Jellyfish belong to the Coelenterata or Cnidaria. Velella, *the by-the-wind-sailor, and* Physalia, *the Portuguese man-of-war, are colonial forms.*

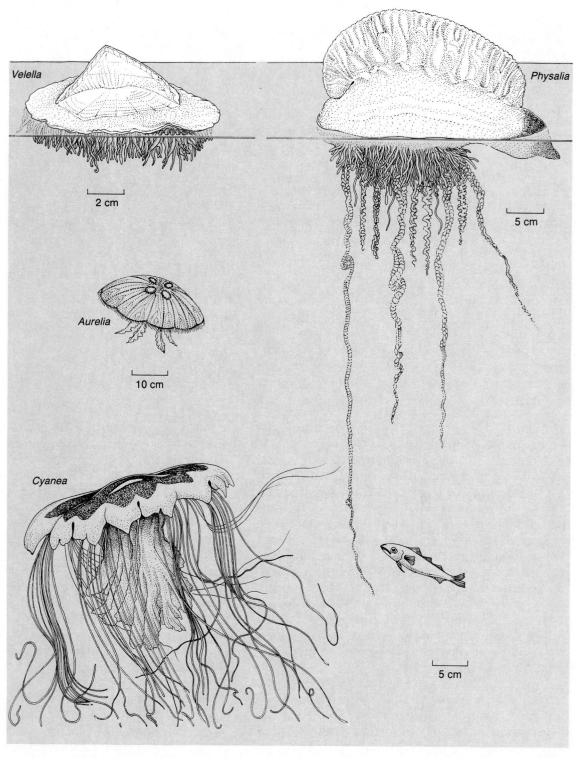

Figure 14.12
Members of the meroplankton. All are larval forms of nonplanktonic adults.

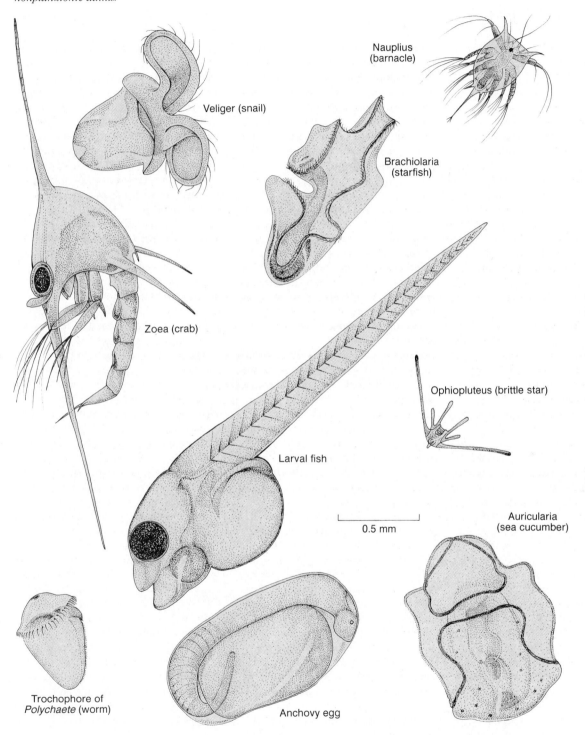

Veliger (snail)

Nauplius
(barnacle)

Brachiolaria
(starfish)

Zoea (crab)

Larval fish

Ophiopluteus (brittle star)

0.5 mm

Auricularia
(sea cucumber)

Trochophore of
Polychaete (worm)

Anchovy egg

A Krill-Based Ecosystem

Almost all the life of Antarctica and the surrounding Southern Ocean depends on the sea. The food web of the Southern Ocean is based on the phytoplankton, mainly diatoms, that harness the energy of the sun. Estimates of primary production vary; in summer, in ice-free areas, the average lies between 20 and 100 g of carbon per square meter per year (gC/m²/year). Among the herbivorous zooplankton one species, *Euphausia superba,* or krill, probably amounts to half the total zooplankton biomass. Krill are the key organisms of the Southern Ocean **ecosystem,** a unit that includes the area's community of organisms and the environment with which it reacts. Although krill are circumpolar in distribution, their concentration is not uniform. There are certain areas of high concentration, the greatest being in the Weddell Sea, between 0° and 60° W. Some swarms have been estimated at more than 2 million tons.

In summer, the ice pack melts back, the primary production increases, and the krill move toward the surface to graze the phytoplankton over large areas. In winter, primary production is practically gone from the surface waters due to reduced light, ice-pack cover, and increased turbulence; at this time, the krill appear to descend into deeper water and feed on the phytoplankton detritus. The annual production of krill biomass in the Southern Ocean has been estimated at 750 million–1300 million tons.

Squid are an important part of the Antarctic food web. There may be more than twenty species, some depending on krill as their most important food. The total annual consumption of squid by whales, birds, and seals is calculated at about 35 million tons. Some species of Antarctic fish stay in the Southern Ocean year-round, feeding on krill; other species migrate into Antarctic waters each summer to feed on krill. It has been estimated that all Antarctic fish combined may consume as much as 100 million tons of krill each year.

There are few species of birds in Antarctica, but populations are usually very large. Birds of Antarctica feed principally on crustacea (mainly krill and copepods), squid, fish, and carion. Krill amounts to 78% of all the food they eat. Recent estimates indicate about 115 million tons of krill are eaten annually by birds, either directly or indirectly. In winter, the birds either switch to a diet of squid or fish or they migrate northward. Most penguin species feed on krill supplemented, in some cases, by fish. The two largest penguins, the emperors and the kings, take fish and squid only.

Seven species of seals are found in the Southern Ocean; four species, the crabeater, leopard, Ross, and Weddell, are almost totally confined to the ice zones. Weddell seals eat mainly fish and squid; the crabeater's diet is about 94% krill; krill makes up 37% of the leopard seal's diet. The crabeater seal, now numbering about thirty million, is the most abundant seal in the world. The fur seals at South Georgia feed almost exclusively on krill. Stocks of Antarctic seals at present total about 33 million with a biomass of about 7 million tons. These seal populations annually consume over 130 million tons of krill (two or three times the current consumption by whales) and at least 10 million tons of squid.

Baleen, or plankton-feeding, whales of the Southern Ocean include the blue, fin, sei, minke, humpback, and southern right whales. Before whaling exploitation, they were probably about four times more abundant and had a biomass five times greater than at present. The major food of the baleen whales is krill. In 1904, their biomass was about 45 million tons and they consumed an estimated 190 million tons of krill. Competition for food probably limited their size and their numbers. By 1973, their numbers had declined to a biomass of about 9 million tons and they ate about 43 million tons of krill. The reduction in the whale population means that some 150 million tons of krill formerly eaten by whales have become available to the remaining whales, other predators, and human harvesters.

In the past thirty years, there has been an increase in the pregnancy rate of fin and sei whales and an apparent, but controversial, decrease in their age at maturity due, presumably, to the greater availability of food. The minke population may be double what it was before whaling. The crabeater seal populations also have experienced a decrease in the age of sexual maturity (4 years in the 1950s to 2.5 years in the early 1960s), also thought to be the result of increased growth rates based on food abundance. The 14–17% increase in the population of the South Georgia fur seals is unusually high, probably related to the abundance of krill.

The three most abundant penguins, the chinstrap, Adelie, and macaroni, have shown increases in population. Large increases in king penguins are thought to be due to their feeding on krill-eating squid. We do not know the response of the fish and squid stocks to the increase in available krill, but it is likely to be similar. The total quantity of krill taken by all predators in the Southern Ocean may be about 500 million tons per year. Populations and numbers may change, but directly or indirectly the Antarctic ecosystem depends on krill, which in turn depends on the primary production of the Southern Ocean waters.

Population figures in this section are from Laws, Richard M., *The Ecology of the Southern Ocean,* American Scientist, 1985.

Bacteria

Another important group of organisms represented in the plankton are the bacteria. They play an important role in the decay and breakdown of organic matter, returning it to the sea as basic chemicals and compounds to be used again by new generations of plants and animals (refer back to chapter 13). In areas where organic debris accumulates at the surface, large populations of bacteria are found. Bacteria exist on every available surface: the sea floor, decaying material, the surface of organisms, pieces of floating wood, and other matter. Planktonic bacteria absorb dissolved organic matter, transforming it into particulate material, which is ingested by planktonic larvae and a variety of single-celled protozoans. A film of bacteria is found on minute particles of floating organic material that are not yet completely broken down. The small size of these particles, with their attached bacterial population, makes them an ideal food for many small zooplankton.

14.2 Classification Summary of the Plankton

The types of plankton can be categorized in the kingdoms, phyla, and classes listed below.

I. *Kingdom Monera:* cells, simple and unspecialized; single cells, some in groups or chains.
   A. *Bacteria:* single cells, in chains or groups; autotrophic and heterotrophic, aerobic and anaerobic; important as food source, in decomposition.
   B. *Cyanobacteria:* blue-green algae; autotrophic single cells, in chains or groups, produce some red blooms in sea; phytoplankton.
II. *Kingdom Protista:* convenience grouping of microscopic and mostly single-celled organisms; autotrophs (algae) and heterotrophs (protozoa).
   A. *Phylum Chrysophyta:* golden-brown algae; yellow to golden autotrophic single cells, in groups or chains; contribute to deep-sea sediments; phytoplankton.
      1. *Class Bacillariophyceae:* diatoms.
      2. *Class Chrysophyceae:* coccolithophores and silicoflagellates.
   B. *Phylum Pyrrophyta:* fire algae; single cells with flagella; produce most red tides; bioluminescence common; usually considered phytoplankton.
      1. *Class Dinophyceae:* dinoflagellates.
   C. *Phylum Protozoa:* microscopic heterotrophs; classified by method of locomotion; zooplankton.
      1. *Class Sarcodina:* radiolarians and foraminiferans.
      2. *Class Ciliophora:* ciliates.
III. *Kingdom Plantae:* plants; primarily nonmotile, multicellular, photosynthetic autotrophs.
   A. *Phylum Phaeophyta:* brown algae; *Sargassum* maintains a planktonic habit in the Sargasso Sea.
IV. *Kingdom Animalia:* Animals; multicellular heterotrophs with specialized cells, tissues, and organ systems; zooplankton. For temporary members of the zooplankton (or meroplankton), see V below.
   A. *Phylum Coelenterata* or *Cnidaria:* radially symmetrical with tentacles and stinging cells.
      1. *Class Hydrozoa:* jellyfish as one stage in the life cycle, including such colonial forms as Portuguese man-of-war.
      2. *Class Scyphozoa:* jellyfish.
   B. *Phylum Ctenophore:* comb jellies; translucent; move with cilia; often bioluminescent.
   C. *Phylum Chaetognatha:* arrowworms; free-swimming, carnivorous worms.
   D. *Phylum Mollusca:* mollusks; the snail-like pteropod is planktonic.
   E. *Phylum Arthropoda:* animals with paired, jointed appendages and hard outer skeletons.
      1. *Class Crustacea:* copepods and euphausiids.
   F. *Phylum Chordata:* animals, including vertebrates, with dorsal nerve cord and gill slits at some stage in development.
      1. *Subphylum Urochordata:* saclike adults with "tadpole" larvae; salps.
V. *Meroplankton:* larval forms from the phyla Annelida (segmented worms), Mollusca (shellfish and snails), Arthropoda (crabs and barnacles), Echinodermata (starfish and sea urchins), and Chordata (fish). See also the classification summaries of the nekton (chapter 15) and the benthos (in chapter 16).

14.3 Sampling the Plankton

The biological oceanographer needs to know what species of plants and animals make up the plankton in a given geographic area, the abundance of these organisms, and where in the water column they are located. Traditionally, sampling is done by means of a conic net of fine mesh material, once silk and now synthetic, which is towed through the sea behind the vessel or dropped straight down over the side of a nonmoving vessel and pulled back up like a bucket (see fig. 14.13). After the net is brought back on board, it is rinsed carefully, and the "catch" is collected in a labelled jar. If the net has been hauled vertically through the water in one place, the volume of water that has passed through the net can be calculated from the area of the net's mouth and the distance through which the net was pulled. If the net was towed behind the vessel, the volume can be calculated from the time of the tow and the speed of the tow, which will tell the distance towed. This information can then be used with the area of the net opening to find the volume of water that passed through the net. However, such a procedure is not accurate for quantitative sampling. To measure the water volume directly and accurately, a flow meter is placed in the mouth of the net. The catch is then known for the total volume of water sample, but there is no indication of how the species are distributed within that volume.

Figure 14.13

A plankton net. The twin nets are used to take duplicate samples. Note the flow meter in each net.

It is also possible to raise and lower the net as it is being towed horizontally. This movement allows sampling to be averaged over both distance and depth. If the sample is to be taken at a specific depth, the net may be lowered closed and only opened when the desired depth is reached. After the towing operation, the net is again closed before it is brought up through the shallower water.

The speed at which the net is towed is important. The tow must be made rapidly enough to catch the organisms but slowly enough to let the water pass through the net. If the tow is too fast, the water will be pushed away from the mouth of the net, and less water than expected will be sampled. The mesh of the net also plays a role, for if a very fine mesh is used and many zooplankton and phytoplankton are caught, they may rapidly clog the net, and water will not pass through it. Fine nets are used for sampling phytoplankton, while coarser mesh nets are reserved for the larger zooplankton.

Plankton may also be sampled by using a water bottle or a submersible pump to collect a water sample at a known depth. The water is then filtered on shipboard to remove the plankton. Whether the sample is taken by net, bottle, or pump, the number and kinds of plankton must be determined. Because of the numbers of organisms involved, in most cases it is impossible to inspect or count the total catch. Instead, the catch is precisely subdivided, and a subsample is inspected directly under the microscope. If the amount of plankton is of greater interest than the kinds of organisms, an electronic particle counter may be used to find the total count, or a subsample may be dried and weighed. The abundance of phytoplankton can also be determined by dissolving out the chlorophyll pigment from the sample and measuring the pigment's concentration.

14.4 Practical Considerations: Marine Toxins

Red Tides

Certain species of dinoflagellates produce the so-called **red tides.** These blooms may or may not be poisonous to fish and other organisms, and may or may not produce symptoms of Paralytic Shellfish Poisoning (PSP) or Neurotoxic Shellfish Poisoning (NSP) in humans eating clams, mussels, and oysters that have ingested the toxic dinoflagellates. In North American waters, several different dinoflagellates produce red tides; *Gonyaulax* and *Ptychodiscus* are toxic; *Noctiluca* is not (see fig. 14.5). Species of *Gonyaulax* are found in temperate latitudes and are generally nonpoisonous to the shellfish themselves. However, the toxins produced by these dinoflagellates are concentrated in the tissues of the shellfish as they feed on the bloom, and it is the toxins that produce PSP or NSP in the humans eating the affected shellfish. *Ptychodiscus* and a species of *Gonyaulax* found in the warmer waters of the Gulf of Mexico kill fish. *Gonyaulax* also kills shrimp and crab; *Ptychodiscus* does not.

No one knows precisely what triggers the sudden bloom of these organisms. It often happens after heavy rains have produced a land runoff of nutrient-rich water and usually occurs in the warmer months of spring and summer. High salinities are thought by some to trigger the blooms in the Gulf waters. Dinoflagellates produce a cyst or thick-walled resting cell that settles to the bottom and mixes with the sediments. Along the New England coast, it has recently been shown that temperature and light play the major roles in activating cysts in the shallow waters of bays and harbors. In deeper coastal waters, the cysts appear to have a natural biological clock or annual cycle. It is thought that sudden disturbances of the bottom by either natural or artificial means (such as slumping of sediments or dredging) could stimulate the cysts into activity and rapid reproduction. Another suggestion is that the cysts may be brought to the surface by upwelling water.

The toxins that produce PSP are powerful nerve poisons that can cause paralysis and death if the breathing centers are affected. Some species of red-tide dinoflagellates produce several toxins. For example, one species of *Ptychodiscus* can produce at least five different toxins. The saxitoxins that are found in butter clams (*Saxidomus*) but are produced by *Gonyaulax* are fifty times more lethal than strychnine. The toxin is not affected by heat, so cooking the shellfish does not neutralize the poison. It is well to remember that even

TABLE 14.1

Worldwide Poisoning in Humans (1976–1983)

Toxin	Lethal dose (mg)	Number of cases
PSP (saxitoxin) (27 deaths reported)	1–2	>400
Ciguatoxin (1% mortality estimated)	<1	>10,000 yearly

Source: Seafood Toxins (1984) American Chemical Society Symposium Series 262

after the visible signs of red water due to a dinoflagellate bloom have disappeared, the shellfish can retain the toxin in their tissues for long periods, and so the beaches are kept closed to shellfish harvesting. Scientists speculating why the toxin is produced believe it is a defense against predators.

In 1935 and again in 1986, major outbreaks of red tide caused by the dinoflagellate *Ptychodiscus brevis* struck the Texas Gulf Coast beaches. In the fall of 1986, the red tide spread 300 miles (484 km) along the Texas coast and killed more than 22 million fish in a period extending over more than two months. During the 1986 outbreak, harvesting of shellfish was banned along three-quarters of the Texas coast south of Galveston.

Toxins are released into the water and then transmitted into the air via sea spray, producing an aerosol effect. When winds blow the particles ashore, people react individually, but irritated eyes and running noses are common; many begin to cough, sneeze, and wheeze. Persons eating the affected shellfish contract NSP with symptoms similar to basic food poisoning. Unlike PSP, few cases have been reported and no known deaths have occurred.

During the red tide, it is reported that business at coastal oyster- and seafood-processing plants dropped as much as 98%. The state of Texas lost at least $1.4 million in oyster production, with a total economic loss of more than $3.7 million. Businesses based on tourism and recreation also suffered severely.

Just as not all red tides are toxic, not all red water is caused by dinoflagellates. The Red Sea received its name because of dense blooms of a nontoxic blue-green alga with large amounts of red pigment. The Gulf of California has been called the Vermillion Sea for the same reason. The Indian Ocean has red tides due to the presence of a toxic blue-green alga, not a dinoflagellate.

Ciguatera Poisoning

It is estimated that between ten thousand and fifty thousand people a year eating fish in the tropical regions of the world are affected by **ciguatera** poisoning. More than four hundred species of fish have been found to be affected and the presence of ciguatoxin in the fish

dramatically affects the development and growth of inshore fisheries in these areas. The situation is complex, because not all of the fish of the same species caught at the same time and in the same place are toxic.

There is no way to prepare an affected fish to make it safe to eat. Symptoms of ciguatera poisoning are extremely variable and may include headache, nausea, vomiting, abdominal cramps, possible irregular pulse beat, reduced blood pressure, and, in severe cases, convulsions, muscular paralysis, hallucinations, and death. Symptoms can occur in various combinations and no proven antitoxin is known.

In the U.S., the number of cases may be over 2000 per year. These are clustered in Florida, Hawaii, the Virgin Islands, and Puerto Rico. Over 80% of the resident adults in the U.S. and British Virgin Islands report having been poisoned at least once. Ciguatera poisoning is thought to be underreported in Hawaii and more serious than generally acknowledged. An outbreak involving twelve persons in Maryland and a single case in Boston involved consumption of grouper shipped from Florida to local restaurants.

It is believed that the source of the toxin is one or more dinoflagellates. A benthic dinoflagellate found in French Polynesia has been implicated in many cases; it has also been identified in Hawaii. This dinoflagellate has been found to flourish in areas of human or natural disturbance or destruction of coral reefs. In the Caribbean, it is considered likely that more than one organism produces a combination of toxins contributing to ciguatera poisoning. This could account for the variety of symptoms and the sporadic and unpredictable outbreaks of the poisoning.

Japanese researchers in the Gilbert Islands found an evolution of toxicity over the years. Initially only a few species are toxic. At the peak of the outbreak almost all reef fish become toxic, and in the final stages only large eels and certain snappers and groupers remain toxic. This cycle appears to take at least eight years and, in this case, points to a food-chain cycle in which the herbivorous fish become toxic first, followed by the carnivores.

Ciguatera is an international problem. It may prevent planned development of Egyptian Red Sea fisheries. Sri Lanka reports hundreds of cases each year. In New Guinea, it is believed that thousands are poisoned each year, but that most cases go unreported as they are attributed to magic. Some islands in the western Pacific have been abandoned because of local ciguatera problems. A bottom fishery in Samoa is required to discard all red snappers (as much as 50% of the catch). It hampers the fledgling Puerto Rico fishing industry, and the loss to the Floridean/Caribbean/Hawaiian seafood industry is estimated at $10 million annually. Other losses include export markets, the loss of use of the banned fish, the cost of treatment and the time lost from employment by victims, and the costs associated with monitoring and implementing fishing and marketing regulations.

Summary

The plankton are the drifting organisms. The microscopic plankton are divided into groups by size. Phytoplankton are autotrophic single cells or filaments. *Sargassum* is the only large planktonic seaweed. The diatoms are found in cold, upwelled water; they are yellow-brown, with a hard, transparent frustule, and they store oil, which increases their buoyancy. Centric forms are round; pennate forms are elongate. Diatoms reproduce rapidly by cell division and make up the first trophic level of the open sea.

Dinoflagellates are single cells with both autotrophic and heterotrophic capabilities. Their cell walls are smooth or are heavily armored with cellulose plates. They, too, reproduce by cell division. These organisms are responsible for much of the bioluminescence in the oceans. Coccolithophores and silicoflagellates are very small autotrophic members of the phytoplankton.

Some herbivorous zooplankton reproduce several times a year, while others reproduce only once, depending on the phytoplankton food supply. Carnivorous zooplankton are important in the recycling of nutrients to the phytoplankton. Heavy concentrations of zooplankton are found at convergence zones and along density boundaries. Zooplankton migrate toward the sea surface at night and away from it during the day, forming the deep scattering layer.

Zooplankton members that spend their entire lives in the plankton are called holoplankton. The copepods and euphausiids are the most abundant members of the holoplankton. Euphausiids are also known as krill; they form a basic food of the baleen whales. Krill is the zooplankton base for all the Antarctic ecosystems; recently it has been harvested for human consumption with mixed success.

Other small members of the holoplankton are the carnivorous arrowworms, the calcareous-shelled foraminiferans, the delicate, silica-shelled radiolarians, the ciliated tintinnids, and the swimming snails, or pteropods. Large zooplankton include the comb jellies, salps, and jellyfish; all are nearly transparent, but each belongs to a different group of animals.

The meroplankton are the juvenile (or larval) stages of nonplanktonic adults. This group comprises fish eggs, very young fish, and the larvae of barnacles, snails, crabs, starfish, and many other nonplanktonic animals. The spores of seaweeds and the marine bacteria are also planktonic.

Plankton sampling is done with a plankton net or with a water bottle. The kinds of organisms in a sample are determined microscopically; the numbers of organisms are counted, or samples are dried and weighed.

Heavy blooms of dinoflagellates and some other kinds of phytoplankton produce red tides. Some red tides are toxic; others are not. The toxin is concentrated in shellfish and produces paralytic shellfish poisoning in humans and sometimes in other animals. Red tides are apparently triggered by the right combination of environmental factors and the disturbance of dormant dinoflagellates. Ciguatoxin is concentrated in fish. Dinoflagellates are also believed to be the source of this toxin, which also affects humans. Ciguatera hampers fishery development around the world.

Key Terms

ultraplankton
nannoplankton
microplankton/net
  plankton
autotrophic
filament
diatom
dinoflagellate
coccolithophore
fucoxanthin
radial symmetry

centric diatoms
bilateral symmetry
pennate diatoms
frustule
auxospore
bloom
diatomaceous earth
flagella
heterotrophic
coccolith
crustacean

copepod
euphausiid
krill
baleen whale
chaetognath
foraminiferan
radiolarian
test
tintinnid
pteropod
ctenophore

salp
coelenterata/cnidaria
colonial forms
holoplankton
meroplankton
larva
spore
red tide
ciguatera
cilia

Study Questions

1. Why does a pycnocline located above the compensation depth promote a phytoplankton bloom?

2. Why are meroplankton produced in such large numbers?

3. When the discoloration has left the water after a PSP red tide, the shellfish may not be safe to eat. Explain why.

4. Patches with abundant populations of zooplankton are frequently found separated by patches with sparse populations. How does this help to assure survival from predators?

5. If the krill of the Southern Ocean were heavily harvested for human consumption, explain the possible effects on the rest of the organisms in that area.

6. Describe four ways to subdivide the plankton.

7. Why are there more planktonic centric diatoms and more benthic pennate diatoms?

8. Discuss what happens to a diatom population if no auxospores form.

9. Why is a planktonic stage important to a nonplanktonic adult?

10. Discuss how plankton may maintain themselves in a given region of the ocean even though there are currents flowing through that region.

11. When you are sampling plankton with a plankton net, how can you determine the quantity of the plankton in a volume of water? Assume that you know (1) the cross-sectional area of the net and (2) the length of time you towed the net and the speed at which you towed the net, or you know (1) the cross-sectional area of the net and (2) the distance you towed the net.

The Nekton
Free Swimmers of the Sea

15

*W*e need another and a wiser and perhaps a more mystical concept of animals. Remote from universal nature, and living by complicated artifice, man in civilization surveys the creature through the glass of his knowledge and sees thereby a feather magnified and the whole image in distortion. We patronize them for their incompleteness, for their tragic fate of having taken form so far below ourselves. And therein we err, and greatly err. For the animal shall not be measured by man. In a world older and more complete than ours they move finished and complete, gifted with extensions of the senses we have lost or never attained, living by voices we shall never hear. They are not brethren, they are not underlings; they are other nations, caught with ourselves in the net of life and time, fellow prisoners of the splendour and travail of the earth.

Henry Beston, from The Outermost House

*T*he nekton are the free swimmers of the oceans. These animals move through the water independent of the motion of currents and waves. The nekton include only a few different kinds of animals, and most of them are fishes. Other members are the marine mammals and birds, the ocean-living reptiles, and that freely moving relative of clams and oysters, the squid. Members of the nekton can move toward their food and away from predators. Many occupy the top trophic levels of the marine food webs, either as herbivores or carnivores. Sizes range from the smallest fish of the tropic reefs to the largest animal ever to have existed on this earth, the blue whale. This chapter examines representative swimmers in coastal waters and in the open sea; it contains additional information on their harvests, including quantities taken, methods used, and current population status.

15.1 The Mammals

Marine **mammals** are warm-blooded air breathers. They may spend all of their lives at sea, or they may return to land to mate and give birth. In either case, the young are born live and are nursed by their mothers. Included in this group are large and small whales (including porpoises and dolphins), seals, sea lions, walruses, sea otters, sea cows, and the polar bear. The polar bear is included in the sea mammal group because under natural conditions, and without the pressure of people, it spends most of its life roaming the sea ice and feeding from the sea, returning to land to bear its young or to migrate to another feeding area.

Whales and Whaling

Whales belong to the mammal group called **cetaceans.** Some cetaceans are toothed, pursuing and catching their prey with their teeth and jaws (for example, the killer whale, the sperm whale, and the porpoises); others have mouths fitted with strainers of whalebone (or **baleen**), through which they filter the seawater and remove the krill. Figure 15.1 compares the mouths of toothed and baleen whales. The blue, finback, right, sei, gray, and humpback whales are baleen whales. The blue, finback, and right whales swim open-mouthed and engulf water and plankton. The tongue acts to push the water through the baleen, and the krill are trapped. The sei whale swims with its mouth partly open and uses its tongue to remove the organisms trapped in the baleen. The gray whale is unique, feeding mainly on small bottom crustaceans and worms. The humpbacks circle an area rich

Figure 15.1

(a) The Pacific white-sided dolphin is a toothed whale. (b) The fin whale is a baleen whale that strains its food from the seawater.

(a)

(b)

in krill and expel air to form a circular screen, or a net of bubbles. This causes the krill to bunch together toward the center of the net, and the whales then pass through the center, scooping up the krill.

Some whales migrate seasonally over thousands of miles; other whales stay in cold water and migrate over relatively short distances. The California gray whale and the humpback whale are both known for their long migratory journeys. In the summer, the California gray whale is found in the shallow waters of the Bering Sea and the adjacent Arctic Ocean. All summer long they feed, building up layers of fat and blubber. In October, when the northern seas begin to freeze, the animals begin to move south. In December, the first gray whales arrive off the west coast of Baja California, where they spend the winter in the warm, calm waters of sheltered lagoons. The gray whales calve and mate in these lagoons but find little food. By the time they leave for their

Figure 15.2
*Migration paths and seasonal distribution of whales. (a) The
California gray whale, (b) the humpback whale, (c) the bowhead
and beluga whales, and (d) the narwhal.*

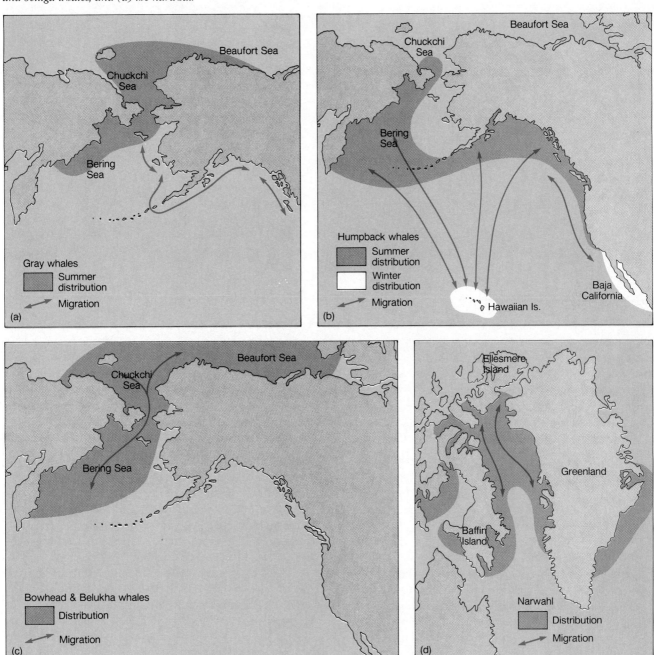

northward migration in February and March, they have
lost 20–30% of their body weight. The animals move
singly or in twos or threes, sometimes in groups of ten
to twelve up the west coast of the United States, Canada,
and Alaska. Moving at about 5 knots day and night, the
whales make their annual 18,000-km (11,000-mi) migratory journey to link areas that provide abundant food
with areas that insure reproductive success. (See fig.
15.2a.)

The humpback whale also has well-defined migration patterns. Humpbacks are found in three geo-
graphically and reproductively isolated populations: in
the North Pacific, North Atlantic, and Southern Ocean.
The North Pacific humpback spends the summer feeding
in the Gulf of Alaska, along the northern islands of Japan,
and in the Bering Sea; in winter, these North Pacific
humpbacks migrate to the Mariana Islands in the west

TABLE 15.1

Principal Characteristics of the Great Whales

	Distribution	Breeding grounds	Average weight (tons)	Greatest length (m)	Food
Toothed whales					
sperm	worldwide; breeding herds in tropic and temperate regions	oceanic	35	18	squid, fish
Baleen whales					
blue	worldwide; large north-south migrations	oceanic	84	30	krill
finback	worldwide; large north-south migrations	oceanic	50	25	krill and other plankton, fish
humpback	worldwide; large north-south migrations along coasts	coastal	33	15	krill, fish
right	worldwide; cool temperate	coastal	(50)	17	copepods and other plankton
sei	worldwide; large north-south migrations	oceanic	17	15	copepods and other plankton, fish
gray	North Pacific; large north-south migrations along coasts	coastal	20	12	benthic invertebrates
bowhead	Arctic; close to edge of ice	unknown	(50)	18	krill
Bryde's	worldwide; tropic and warm temperate regions	oceanic	17	15	krill
minke	worldwide; north-south migrations	oceanic	10	9	krill

Source: After K. R. Allen, *Conservation and Management of Whales,* Washington Sea Grant Program, 1980. Key: ( ) = estimate.

Pacific, the Hawaiian Islands in the central Pacific, and along the west coast of Baja California in the eastern Pacific. At these warmer latitudes, calves are born and mating takes place. (See fig. 15.2b.)

The bowhead, the beluga, and the single-tusked narwahl are whales that remain in cold water but still migrate over short distances each year. The largest population of bowhead whales is found in the Bering, Chukchi, and Beaufort seas. This whale spends nearly all of its life near the edge of the arctic ice pack. Bowheads, singly or in pairs, often accompanied by belugas, migrate north from the Bering Sea to feed in the Beaufort Sea and Chukchi Sea as the ice recedes in the spring and return south to the Bering Sea in groups of up to fifty as the ice begins to extend in the winter. Their mating and reproductive cycles are not well known, but they probably mate during the spring migration and calve sometime during April and May (See fig. 15.2c.)

The narwhal is the most northerly whale and is found only in arctic waters. They are found most commonly on both sides of Greenland. In summer, they move north along the coasts of Ellesmere and Baffin islands. In the autumn, they return south to the waters along the Greenland coast. (See fig. 15.2d.)

The whales that most of us picture in our minds are the "great whales": the blue, sperm, humpback, finback, sei, and right whales (see fig. 15.3 and table 15.1). These whales have been the focus of the whaling industry. The earliest known European whaling was done by the Norse between 800 and 1000 A.D. The Basque people of France and Spain hunted whales in the Bay of Biscay shortly after this period. Recent archeological explorations show that, in the 1500s, Basque whalers set up whaling stations along the Labrador coast to process the blubber of bowhead and right whales into oil for transport back across the Atlantic. In Red Bay, Labrador, the operation reached its peak in the 1560s and

Figure 15.3
Relative sizes of baleen and toothed whales.

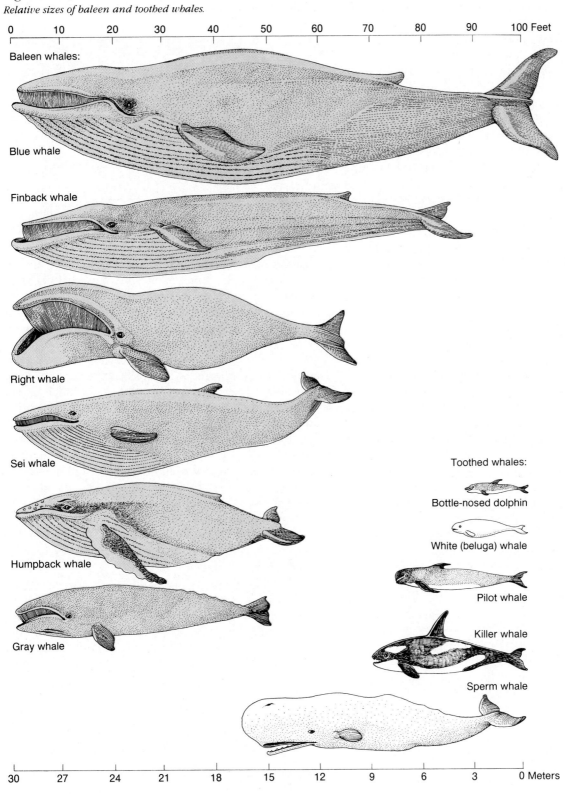

1570s, when one thousand people gathered seasonally to hunt whales and produce 500,000 gallons of whale oil each year. By 1600, whaling had become a major commercial activity among the Dutch and the British, and at about the same time the Japanese independently began harvesting whales. In the 1700s and 1800s, whales were actively pursued far from shore, in sailing vessels and small boats. The whaler's weapon was the hand-held harpoon. During this period the whales were hunted for their oil and baleen, mainly by whalers from northern Europe and the northeast United States. The whales were cut up and processed on land or on board the ships at sea. Long voyages, intense effort, and dangerous combat between whalers and whales characterized these whale hunts.

In 1868, Svend Føyn, a Norwegian, invented the harpoon gun with its explosive harpoon, and changed the character of whaling. Ships were motorized, and in 1925, harvesting was increased further by the addition of great factory ships, to which the small, high-speed whale-hunting vessels brought the dead whales for processing. This system freed the fleets, now centered in the Antarctic, from dependence on shore stations. These methods greatly increased the efficiency of the hunt and rapidly depleted the stocks of these great creatures.

In the 1930s, the annual blue whale harvest was between 4 and 6% of the estimated total original population. Many of the captured females carried young, and harvesting at this rate reduced the population of blue whales to less than 4% of its original numbers, threatening the species with extinction. In 1946, representatives from Australia, Argentina, Britain, Canada, Denmark, France, Iceland, Japan, Mexico, New Zealand, Norway, Panama, South Africa, the Soviet Union, and the United States met in Washington, D.C., to establish the International Whaling Commission (IWC). The regulations drawn up prohibited the killing of the remaining gray, bowhead, and right whales and of cows with calves. Opening and closing dates for whaling were set, and minimum sizes were set for each species harvested. However, the IWC has no police powers, and although each factory ship carried an observer, they could only report offenses and recommend disciplinary action. The government of the country registering the ship involved was responsible for any action. With IWC and its regulations in effect, 31,072 whales were killed in 1951, more than 50,000 in 1960, and in 1962 over 66,000 animals were killed. Figure 15.4 gives the annual

Figure 15.4

Total annual catches of baleen and sperm whales in all oceans from 1910 to 1977.

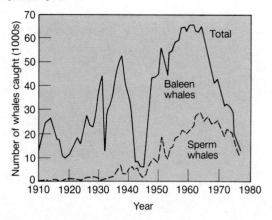

harvest of whales between 1910 and 1977. To understand what these numbers mean to the whale populations, it should be remembered that it took New Bedford, Massachusetts whalers over 150 years to kill 30,000 whales. The current population status of the great whales is given in table 15.2

As the 1970s drew to a close, the era of commercial whaling appeared to be closing as well. In 1979, the IWC placed a moratorium on all whaling in the Indian Ocean and outlawed the use of factory ships as floating bases from which to send out hunter vessels, but whaling continued from land bases in Antarctica. In the spring of 1982, the IWC voted a ten-year moratorium on commercial harvesting of whales, except dolphin and porpoise, in order to study the whale populations and assess their ability to recover. The moratorium began in 1985–86 and continues to be in effect.

In 1982, Japan, Norway, Iceland, Portugal (in the Azores), Spain, Korea, and the Soviet Union filed objections to the moratorium and continued whaling. Recent talks between Japan and the United States have led to an agreement to end all Japanese commercial whaling activities no later than 1988. The last Japanese factory ship began her final expedition to the Southern Ocean in October of 1986. In early 1987, the Soviet Union announced the end of its commercial whaling efforts.

Although commercial whaling is apparently ending, conservation groups are concerned about new plans for scientific whaling. The moratorium calls for assessment of whale stocks to be completed by 1990, but there is no large-scale IWC sampling program in place. Conservation groups charge that research whaling to assess stocks is a return to commercial whaling under

TABLE 15.2
Estimated Population Status of Great Whales

Species	Original population	Current population
sperm	1,377,000	982,200
blue	166,000–226,400	11,700
finback	449,700–452,700	105,200–121,900
humpback	119,400	9599–10,000
right	120,000	3100
sei	108,100–109,400	33,800–53,400
gray	15,000–20,000	18,000
bowhead	54,680	3617–4125
Bryde's	92,000	85,700
minke	320,000	315,800

Sources: After K. R. Allen, 1980 and National Marine Fisheries Service, National Oceanographic and Atmospheric Administration, 1986.

another name. Japan has announced plans for research whaling. Iceland has been catching fin and sei whales and proposes to take minke whales, which are on the IWC protected list. Korea is said to be planning to sample Pacific minke whales, and Norway is to shift from commercial to scientific whaling in 1988.

Under the IWC regulations, aboriginal/subsistence whaling is permitted. IWC regulations attempt to balance the needs of aboriginal peoples in Alaska, Denmark (Greenland), and the Soviet Union who depend on limited whaling for subsistence, cultural, and nutritional needs with the conservation needs of the whales.

The intertwined history of whales and people is not yet ended. In the words of Herman Melville from the pages of *Moby Dick,*

> The moot point is whether the Leviathan can long endure so wide a chase and so remorseless a havoc; whether he must not at last be exterminated from the waters, and the last whale, like the last man, smoke his last pipe, and then himself evaporate in the final puff.

Dolphins and Porpoises

Dolphins and porpoises are small, toothed whales. They are the clowns of the sea, gentle, friendly, and easily trained, and they figure in many folktales, stories, movies, and TV programs. In the open ocean, they are observed traveling at high speeds and in large schools; they occasionally leap clear of the water, in apparent fun

and high spirits, and may even alter their course to keep a vessel company for hours, swimming easily just in front of the bow. Porpoises have been observed swimming at speeds in excess of 30 knots, a feat that interests scientists and researchers. Their abilities to communicate and their intelligence, as demonstrated by their learning and recall abilities, are under study. They have been trained to help divers and to act as messengers between those working at depth and their surface vessels. The U.S. Navy continues to investigate their potential as underwater assistants and has recently used them in the Persian Gulf.

These smaller cetaceans are found in both tropic and temperate waters. Although marine, they will go up rivers and channels into shallow brackish waters. They have been seen moving across the very shallow lakes and canals of the Mississippi delta region in water barely deep enough to support their high speed swimming.

Dolphin and porpoise populations are being affected by the techniques used in the tuna fishery. These air-breathing mammals drown when caught in the great nets used to catch the tuna. This situation is investigated further in the discussion of the tuna fishery at the end of this chapter.

Echolocation and Communication

When submerged marine mammals descend to a world in which little light is available, instead of using sight many of these animals are known to use sound to picture their environment as well as communicate with each other. Perhaps the best known communication between marine mammals is the up to thirty-minute "song" of the humpback whales, which is thought to be an announcement of presence and territory, although some scientists believe it is a secondary sexual characteristic of males in the breeding season. Female gray whales stay in contact with their calves by a series of grunts, and Weddell seals are known to communicate by audible squeaks.

Toothed whales, particularly porpoises and the sperm whale, a few baleen whales, including the gray, blue, and minke, and the Weddell seal, California sea lion, and possibly the walrus are known to make the sharp sound required to produce the reflected echoes that allow certain marine mammals to orient themselves and locate objects. This method of using sound to picture the environment is known as **echolocation.** Although these animals can produce a range of sounds, the most useful sound for echolocation appears to be clicks of short duration released in single pulses or trains of pulses. The abilities of the bottle-nosed porpoise are best known. It produces clicks in frequencies audible to the human ear and higher, each lasting less than a millisecond and repeated up to eight hundred times per second. When each click hits its target, part of the sound is reflected back; the animal continually evaluates the time and direction of return to learn the speed, distance, and direction of the reflecting target. Low-frequency clicks are used to scan the general surroundings, and higher frequencies are used for distinguishing specific objects.

Many of these marine mammals have no vocal cords; how do they produce these sounds? Porpoise and dolphins appear to move air in their nasal passages which vibrate certain structures to produce the clicks; the whistles and squeaks are made by forcing air out of nasal sacs. The bulbous, fatty, rounded forehead of the porpoise acts as a lens to concentrate the clicks into a beam and direct them forward. Sperm whales produce shorter, more powerful, long-range pulses at lower frequencies. These sounds travel more slowly but carry several kilometers. Each pulse is compound, lasting about twenty-four milliseconds and made up of up to nine separate clicks. The sperm whale's massive forehead is filled with oil and may be used to focus the sound pulses.

All marine mammals have good hearing. Humans hear in the range of 16–20,000 vibrations or cycles per second; the bottle-nosed porpoise responds above 150,000 vibrations per second. How do they pick up the faint incoming echoes of their own clicks and screen out the louder outgoing clicks and other sea noises? Sounds enter through the lower jaw and travel through the skull by bone conduction. Within the lower jaw, fat and oil bodies vibrate, and the sound is channeled through the oil directly to the middle ear. Areas on each side of the forehead are also very sensitive to incoming sound. The hearing centers in the brains of marine mammals are extremely well-developed, presumably to analyze and interpret returning sound messages. Their vision centers are less developed, and they are believed to have no sense of smell.

Box Figure 15.1

The pattern of clicks produced by a porpoise for echolocation is shown in black. The reflected echoes by which the porpoise determines the speed, distance, and direction of the target are shown in gray.

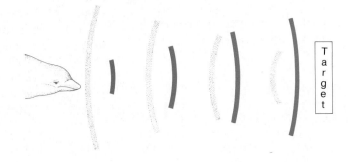

Figure 15.5
Pinnipeds, or feather-footed mammals, and the Arctic walrus.

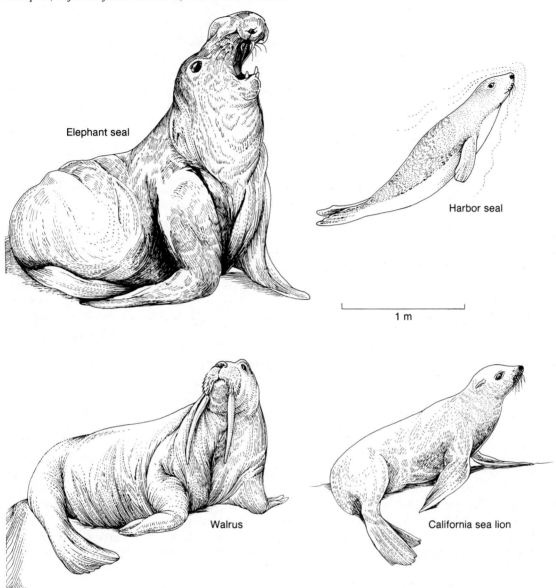

Elephant seal

Harbor seal

1 m

Walrus

California sea lion

Seals, Sea Lions, and Walruses

The seals and sea lions belong to the group called the **pinnipeds,** or "feather-footed" animals, so named for their four characteristic swimming flippers. Representative pinnipeds are shown in figure 15.5. These animals are marine mammals that still retain their ties to land, spending considerable time ashore on rocky beaches, on ice flows, or in caves. They are found from the tropics to the polar seas, ranging from the nearly extinct monk seal of the western Hawaiian Islands and Mediterranean area to the fur seals of the Arctic. The common harbor seal, the harp seal of the northwest Atlantic, the leopard seal of the Antarctic, and the 2-ton male elephant seal, with its great pendulous snout, are all true seals, or seals without external ears and with torpedo-shaped bodies that require them to use a wriggling, wormlike motion to move on land. The northern fur seal and the sea lion (the seal of the circus and amusement park) are eared seals with longer necks and supple forelimbs tipped with broad flippers, which are used for walking and hold the animal's body in a partially erect position on land. Seals may undertake long sea migrations, congregating in spring and summer at specific locations for breeding. For example, the northern fur seal ranges the North Pacific, Bering, and Okhotsk seas, coming ashore to breed in the Pribilof Islands. The habits and natural histories of each of the pinnipeds differ from those of the others, and much is still left to learn.

The walrus is placed in a separate subgroup. It has no external ears and is able to rotate its hind flippers so that it may walk on a hard surface. Its heavy canine teeth (or tusks) are unique and are found in both males and females. These tusks help the walrus to haul itself out of the water, onto the ice, and are probably used to glide over the bottom, like sled runners, while it forages for clams with its heavy muscular whisker pads. It has been said that the tusks are used for digging clams, but the tusks show wear on the front surfaces, not the ends and back surfaces as they would if used for the clam-digging function.

Some seals and sea lions are currently enjoying a period of relative peace, compared to the sealing days of the nineteenth and early twentieth centuries. The Guadalupe fur seal of southern California was hunted to the brink of extinction, until, in 1892, only seven animals were thought to survive; its remarkable recovery to a current population of 1500 is due to protection by both the United States and Mexican governments and to luck. Between 1870 and 1880, hunters for furs and oil reduced the northern elephant seal population to one hundred and cut the walrus population in half. Recently, an increasing number of walrus have been killed for their ivory. In 1986, between ten thousand and twelve thousand walrus were killed by Alaskan natives and the Soviets. There is a growing concern that this kill rate may be more than the slowly reproducing walrus population can sustain.

The northern fur seal's 1870 population of 2 million to 2.5 million was reduced to 300,000 by 1914. The present population of northern fur seals is approximately 1.1 million and declining. We do not have a complete understanding of their decline but entanglement with nets, lines, plastic strapping rings, and other debris kills an estimated thirty thousand fur seals each year. In the past, northern fur seal populations have been managed and harvested under international agreement between the U.S., Canada, Japan, and the Soviet Union. At present, fur seals are covered by the Marine Mammal Protection Act, which is discussed at the end of this section. A subsistence harvest of 3500 seals was allowed for the Pribilof islanders in 1985. There is currently conflict between animal protection groups and the Aleut people of the Pribilof Islands regarding the commercial harvesting of the fur seals.*

*Population figures in this and subsequent sections are from W. Nigel Bonner, *Seals and Man,* 1981, and the National Marine Fisheries Service of the National Oceanographic and Atmospheric Administration, 1986.

The Canadian harp seal population has dropped from an estimated initial stock of 3 million to 4 million to its present level of 1 million to 1.5 million over the last 150 years. The annual commercial harvest of white-fur harp seal pups in eastern Canada, Greenland, and Norway fell from 350,000 in 1963 to 30,000 in 1984. This harvest had become the focus of a public outcry against sealing that resulted in no commercial hunting of harp seal pups in 1985 and 1986 because of a lack of buyers. Conservationists, however, fear the hunt may begin again if the European Economic Community can be persuaded to lower their ban on the import of harp seal pup pelts.

Sea Otters

Sea otters (fig. 15.6) are related to river otters but are larger and live always in salt water. This animal lives in coastal areas, taking shellfish and other food from the bottom in relatively shallow water. It differs from both seals and whales in having no insulating layer of blubber beneath the skin and so must depend entirely on its dense fur for warmth. This soft, thick fur became the sea otters' death warrant in the eighteenth and nineteenth centuries, when they were hunted nearly to extinction. At one time, prime pelts sold for more than $1000 each. The sea otter was brought under protection in 1911, and many scientists doubted that the species could survive. The Alaskan population in the mid-1970s had increased to between 100,000 and 120,000. In recent years overcrowded populations have been thinned. In California in 1977, 1800 to 2000 animals were counted, but the population there is still officially designated as "threatened" due to increasing human activity, such as offshore oil development, and to the sea otter's appetite for the commercially valuable abalone and clams. The species, although now reestablished, will need continued watching and management in this area.

Sea Cows

Sea cows, manatees, and **dugongs** (see fig. 15.6) are members of the **Sirenia** and are thought to be the possible source of the mythic mermaid. All are herbivores and have been used to help control the water hyacinth plant that threatens to choke the slow-moving bayous of tropic regions. Manatees are found in the brackish coastal bays and waterways of the warm southern Atlantic coasts and in the Caribbean. Dugongs are found in Southeast Asia, Africa, and Australia.

In former times the Steller sea cow existed in the shallow waters off the Commander Islands in the Bering Sea. The last Steller sea cow was killed for its meat in

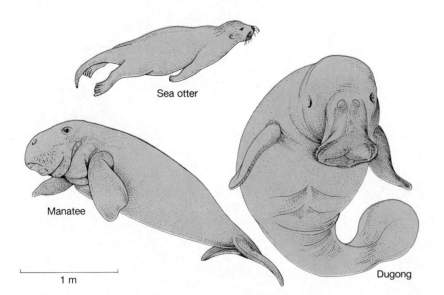

1 m

Figure 15.6
The sea otter of the North Pacific, the tropic manatee of the Caribbean, and the dugong of Southeast Asia are all marine mammals.

about 1768. These sea cows were slow-moving, docile, totally unafraid, and present in only limited numbers. These characteristics, coupled with a low reproduction rate, made them unable to withstand the human hunting pressure, and so this largest of the sirenians became extinct. At present, the growth of human populations and their need for protein is putting increasing hunting pressure on the dugong in the southern Pacific. Manatees in the coastal waters of the Caribbean and south Atlantic are frequently injured and killed by collisions with the propellers of large and small vessels; they are protected along the Florida coast.

Marine Mammal Protection Act

In 1972, the Congress of the United States established the Marine Mammal Protection Act, Public Law 92–522. Implementation of the act included a ban on the taking or importing of any marine mammals or marine mammal product. "Taking" is defined by the act as harvesting, hunting, capturing, or killing any marine mammal or attempting to do so. The act covers all United States territorial waters and fishery zones. It is also unlawful "for any person subject to the jurisdiction of the United States or any vessel or any convoy once subject to the jurisdiction of the United States to take any marine mammals on the high seas" except as provided under preexisting international treaty.

The act effectively removed the animals and their products from commercial trade in the United States. Only under strict permit procedures and with the approval of the Marine Mammal Commission can a few individual marine mammals be caught for scientific research and public display.

Alaskan natives are exempted from the act for purposes of subsistence hunting and of creating and selling authentic native articles of handicraft and clothing. If it is determined that a species or stock being hunted under this exemption is being depleted, further regulations may be established to conserve the animals.

In some cases, the Marine Mammal Protection Act has been very successful in its mission of protecting marine mammal populations; some would argue that it has been too successful. Interactions between people and marine mammals competing for the same resource and/or habitat can be difficult problems to solve. Before this act, marine mammal populations and distribution were controlled by harassment. Since passage of the act, the numbers of some species have increased and the animals have become bolder. This is particularly true of the harbor seals and sea lions. Harbor seal populations have increased 7–10% per year along the U.S. west coast. The seals feed on fish that are harvested commercially and increasingly rob fishing nets directly. Under the provisions of the act, those who fish can only attempt to discourage the animals, and this is not always successful. In Oregon, an acoustic device called a "seal-chaser" was initially successful in keeping the seals away from the nets, but in a few weeks the seals became accustomed to the sound and used it to find the nets. Along the central California coast, the California sea lion population increases each year. Sea lions numbering between 80,000 and 125,000, consume 100,000–250,000 tons of Pacific hake (whiting) annually along the west coast of the United States.

Figure 15.7
Marine reptiles.

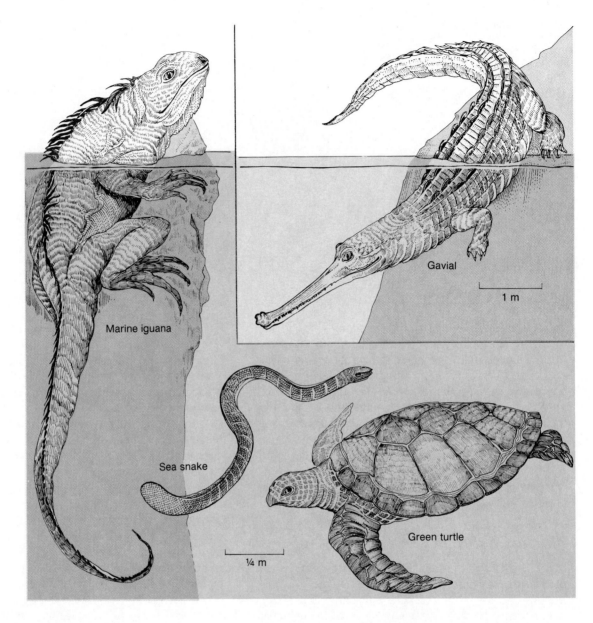

15.2 The Reptiles

Although many land reptiles visit the shore to feed, mainly on crabs and shellfish, few reptiles are found in the sea. Examples of marine reptiles are shown in figure 15.7. The only modern marine lizard is the big, gregarious marine iguana of the Galápagos Islands. It lives along the shore and dives into the water at low tide to feed on the algae. It has evolved a flattened tail for swimming and has strong legs with large claws for climbing back up on the cliffs. It regulates its buoyancy by expelling air, allowing it to remain underwater. The large monitor lizards of the Indian Ocean islands, known as the Komodo dragons, are capable of swimming but do so only under duress.

Alligators may enter shallow shore water, and crocodiles are known to go to sea. The estuarine crocodile of Asia is found in the coastal waters of India, Sri Lanka (Ceylon), Malaya, and Australia. The Indian gavial and false gavial are slender-nosed, fish-eating crocodilians found in near-shore waters.

Sea Snakes

There are about fifty different kinds of sea snakes found in the warm waters of the Pacific and Indian oceans. There are no sea snakes in the Atlantic Ocean. Sea snakes are extremely poisonous with small mouths, flattened tails for swimming, and nostrils on the upper surface of the snout that can be closed when the snakes are submerged. The sea snakes' skin is nearly impervious to salt but it is permeable to gases. Nitrogen gas passes through the skin as well as carbon dioxide and oxygen. The snake is able to lose nitrogen gas to the water and this allows the snake to dive as deep as 100 m (300 ft), stay submerged as long as two hours, and surface rapidly with no decompression problems. See chapter 12 for a discussion of the way in which marine mammals handle diving and decompression. Sea snakes eat fish, and most reproduce at sea by giving birth to live young.

Sea Turtles

Sea turtles live in the ocean but nest on land. The four large sea turtles are the green, hawksbill, leatherback, and loggerhead turtles. Green sea turtles are herbivores; they weigh 300 pounds or more. The hawksbill is a tropic species found on reefs and feeding primarily on sponges. The loggerhead, weighing between 150 and 400 pounds, is usually found around wrecks and reefs, where it feeds on crabs, mollusks, and sponges. The leatherback is the largest sea turtle, weighing up to 1400 pounds. This giant feeds only on jellyfish, following them over large areas of the ocean. These animals are all migratory and make long sea journeys of many hundreds of miles between their nesting sites and their foraging areas. The green turtles of the Brazil coast migrate 2250 km (1400 mi) to Ascension Island in the South Atlantic to lay their eggs. The female turtle lays a hundred or more eggs in a scooped-out depression in the warm sand, covers them, and returns to the sea. While the eggs incubate, they are an easy prey for humans and their associates, dogs and rats. Other wild carnivores take their share, and in more and more cases no eggs survive to replenish the population. If the young turtles hatch, they must make their way across the sand while hungry birds attack and must then survive the waiting predatory fish when they do find the water.

The greatest threat to the survival of all sea turtles is the human. Turtle eggs and turtle meat are prized by humans throughout the Pacific. Turtle nests are raided by poachers, and, until recently, 5000 Caribbean sea turtles were used each year by a New York soup company to produce 600,000 liters of turtle soup. Hawksbill turtles are hunted for "tortoiseshell," which is actually turtle shell. It is used to form ornaments of many kinds: combs, boxes, and jewelry. Turtles are also hunted for their skins for leather, and their fat and oils for cosmetics.

The Kemp's ridley turtle is a small sea turtle found in the Gulf of Mexico and the South Atlantic. It is the rarest and most endangered sea turtle. No one knew where Kemp's ridley turtles nested until 1947, when their only natural nesting site was found at Rancho Nuevo on Mexico's eastern coast. That year, forty thousand females came ashore to lay their eggs. In spite of protection by the Mexican government, in 1985 there were less than five hundred egg-laying females. The Kemp's ridley is under seige; their eggs are stolen to make aphrodisiac cocktails; they are killed for food and for their hides; they are suffocated in shrimp trawls; and they are preyed on by raccoons, skunks, and seabirds as the young try to reach the sea. Scientists are trying to start a new colony on Padre Island, Texas, by collecting the eggs, hatching the baby turtles, and then imprinting them with a chemical signature to assure their return to Padre Island. It is believed that the Kemp's ridley turtles reach sexual maturity in eight to ten years. The program has been releasing turtles only since 1978, and so the extent of its success is not yet known.

Worldwide, the slaughter of these sea creatures is enormous. Although placing species on the United States endangered species list prevents the legal importing of turtle products into this country, according to the U.S. Fish and Wildlife Service there is a vast international illegal trade in turtle products which rivals the illegal ivory trade in dollar value. The problem of turtles drowning in shrimp nets is believed to play a significant role in the depletion of sea turtle populations. Because U.S. fishermen accidently catch an estimated 48,000 sea turtles each year in the Gulf of Mexico and South Atlantic, the shrimp fishermen are required to use Turtle Excluding Devices (TEDs) that will allow the turtles but not the shrimp to escape the nets. Fishermen resisted the initial outlay for the equipment that they say reduce shrimp yields and continue to lobby against enforcement of the regulation. National Marine Fisheries Service studies of vessels outfitted with TEDs show no loss in shrimp yields and a reduction in the incidental catch of sea turtles of 97%.

15.3 The Squid

Squid (fig. 15.8) are abundant in deeper water, travel in large schools, and, like many of the mesopelagic fish, migrate toward the surface at night. It is not unusual in the open ocean to see schools of squid rapidly swimming near the surface at night and often shooting out of the water at the wave crests. There are giant squid

Figure 15.8
The squid is a swimming mollusk considered to be a member of the nekton.

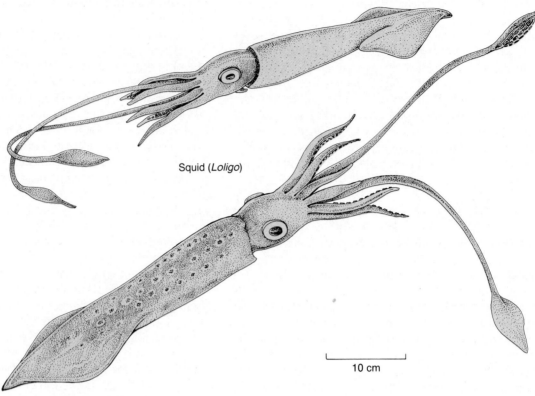

Squid (*Loligo*)

10 cm

(*Architeuthus*) more than 15 m (49 ft) long in the oceans. They are known to be the food of the sperm whale, and long before a single giant squid was ever seen, bits and pieces were found in the stomachs of sperm whales, so that their existence was anticipated. Scientists know very little about these huge creatures but suspect that there may be very large numbers of them in the mesopelagic zone.

15.4 The Fish

The fish dominate the nekton. Fish are found at all depths and in all the oceans, but their distribution patterns are determined directly or indirectly by their dependency on the oceans' primary producers. Fish are concentrated in the upwelling areas, the shallow coastal areas, and the estuaries. The surface waters support much greater populations per unit of water volume than the deeper zones.

Fish come in a wide variety of shapes related to their environment and behavior. Some are streamlined, designed to move rapidly through the water (tuna and mackerel). Others are flattened for life on the sea floor (sole and halibut), while still others are elongate for living in soft sediments and under rocks (some eels).

Fins provide the push or thrust for locomotion and also occur in a variety of shapes and sizes. Fins are used to change direction, turn, balance, and brake. Flying fish use their fins to glide above the sea surface; mudskippers and sculpins walk on their fins.

Schooling is common among certain types of fish (herring, mackerel, menhaden). Schools may consist of a few fish in a small area, or they may cover several square kilometers; for example, herring in the North Sea have been seen in schools 15 km long and 5 km wide (9 × 3 mi). Usually the fish are all of the same species and similar in size. Fish schools have no definite leaders, and the fish change position continually. Schooling fish keep their relationship to one another constant as the school moves or changes direction. Most schooling fish have wide angle eyes and the ability to sense changes in water displacement that allow them to keep their place with respect to their neighbors. Schooling probably developed as a means of protection; each fish has less chance of being eaten in the school than alone. The school may also keep reproductive members of the population together.

Ocean fish are divided into two groups: (1) fish with skeletons of cartilage and (2) fish with skeletons of bone. Cartilagenous fish, sharks and their relatives, the skates, rays, and ratfish, are considered more primitive than bony fish. The fish that are used for food are mainly those with skeletons of bone.

Sharks and Rays

The shark is an ancient fish; it predates the mammals, first appearing in the earth's oceans 450 million years ago. Sharks are set apart from other fish by their cartilagenous skeleton and also by their tooth-like scales. Shark scales have a covering of dentine similar to vertebrate teeth. Because of its abrasive texture, sharkskin has been used as a sandpaper and polishing material. The shark's teeth, which are modified scales, are rapidly replaced if they are lost, and they occur in as many as seven overlapping rows in the jaw. Sharks are actively aware of their environment through good eyesight, excellent senses of smell, hearing, mechanical reception, and electrical sense. A shark's vision is more important than previously thought; sharks see well under dim-light conditions. They have the ability to sense chemicals in their environment through smell, taste, a general chemical sense, and unique pit-organs distributed over their bodies. These pit-organs contain clusters of sensory cells resembling taste buds. The shark's sense of smell is acute. It has a pair of nasal sacs located in front of its mouth; water flows into the sacs as the shark swims forward and water passes over a series of thin folds with many receptor cells. Sharks are most sensitive to chemicals associated with their feeding; they are able to detect such chemicals in amounts as dilute as one part per billion. Receptors along the shark's sides are sensitive to touch, vibration, currents, sound, and pressure. The movement of water from currents or from an injured or distressed fish are sensed by the shark's lines of cells which communicate with the watery environment by a series of tubes; water displacement stimulates nerve impulses along these systems. The shark is able to hear and uses its hearing in the location of its prey. Pores in the shark's skin, especially around the head and mouth, are sensitive to small electric fields. Fish and other small marine organisms produce electric fields around themselves; therefore the shark is able to use its electroreception sense to locate prey and recognize food. As the shark swims through the earth's magnetic field, an electric field is produced that varies with direction, giving the shark its own compass. We still do not understand completely how all these senses function or how they affect the behavior of sharks; we do know that the shark is extremely well-tuned to its environment.

There are some three hundred species of sharks widely spread through the oceans and found in rivers more than a hundred miles from the sea, as well as one species landlocked in fresh water, the Lake Nicaragua shark. Some of these sharks are shown in figure 15.9. The whale shark is the world's largest fish, reaching lengths of more than 15 m (50 ft). This graceful and passive animal feeds on plankton and is harmless to other fish and mammals. The docile basking shark, 5 to 12 m (15–40 ft) long, is another plankton feeder. It is found commonly off the California coast and in the North Atlantic, where it has been harvested for its oil-rich liver. A third species of large (4m (14 ft) long and weighing approximately 680 kg (1500 lbs)) plankton-feeding shark, nicknamed megamouth, is known from two specimens, one caught in 1976 and one in 1984.

Many sharks are swift and active predators, able to kill large prey. A shark attacks quickly and efficiently, using its rows of serrated teeth to remove massive amounts of tissue or whole limbs and body portions. They also play an important role as scavengers, and, like wolves and the large cats on land, eliminate the diseased and aged animals. These sharks can and do attack humans, although the reasons for these attacks and the periodic frenzied feeding observed in groups of sharks is not understood. A human swimming inefficiently at the surface may look like a struggling, ailing animal and be attacked; a diver swimming completely submerged may appear as a more natural part of the environment and be ignored.

Studies of shark attacks on humans in South African waters indicate that most shark attacks occur in shallow waters and in rivers (where there are more swimmers), in murky water, and during the darker parts of the day, such as at dawn and at dusk. Shark attacks have increased along the rocky northern California coast during the last fifteen years. Fourteen attacks upon humans on surfboards have occurred since 1972 along this coast. In the summer of 1982, beaches were closed north of San Francisco. In 1984, four shark attacks occurred in fifteen days, but in 1985 no humans were attacked but surfboards were bitten. One shark attack on a swimmer was reported in 1986. This increase in shark attacks is considered by many to be related to the increase in the sharks' prey, the seal and sea lion populations protected by the Marine Mammal Protection Act. Although surfing is more popular along the beaches of the southern California coast, the shark attacks occur along the rocky coasts where seals and sea lions congregate. It has also been observed that modern surfboard designs appear very similar in shape to a seal when viewed from below.

Researchers have spent considerable time trying to perfect a shark repellent. During World War II, aircraft and ship crew members were provided with a so-called shark repellent made from copper acetate and a soluble ammonium compound, which had a positive psychologic effect on the men but had little effect on the sharks. Current research in shark repellents includes experiments with underwater sound as well as with chemicals. Recent studies indicate that an excre-

Figure 15.9
Representative members of the cartilaginous fish. The leathery egg case, or mermaid's purse, is that of a skate.

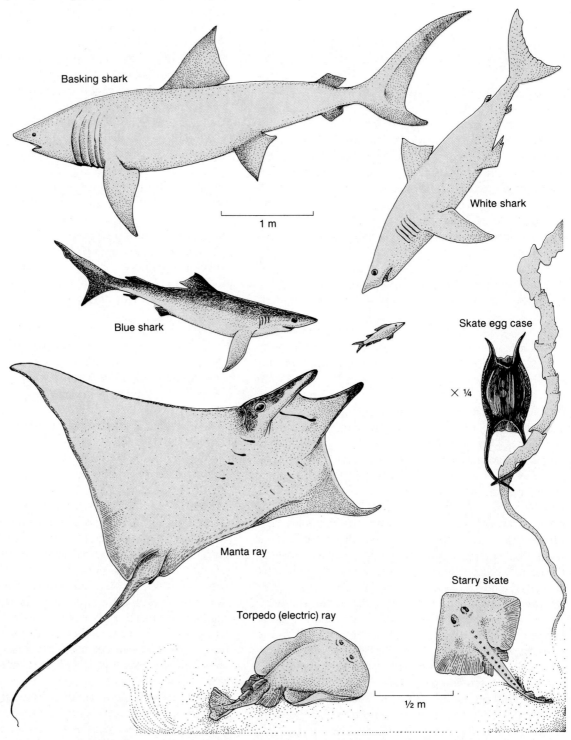

Basking shark

White shark

1 m

Blue shark

Skate egg case

× ¼

Manta ray

Starry skate

Torpedo (electric) ray

½ m

tion found on the surface of a certain flat fish may act as a shark deterrent.

Although not usually considered a major food fish, the shark is consumed in many countries. The flesh is very good when treated properly to remove its high urea content, and shark fins are considered a delicacy in Asia.

Skates and rays, also shown in figure 15.9, are flattened, shark-type fish that have adopted a bottom-dwelling life style. They move by undulating their large side fins, which gives them the appearance of flying through the water. Their tails are usually thin and whip-like and, in the case of stingrays, carry a poisonous barb

Figure 15.10
Commercially harvested members of the bony fish.

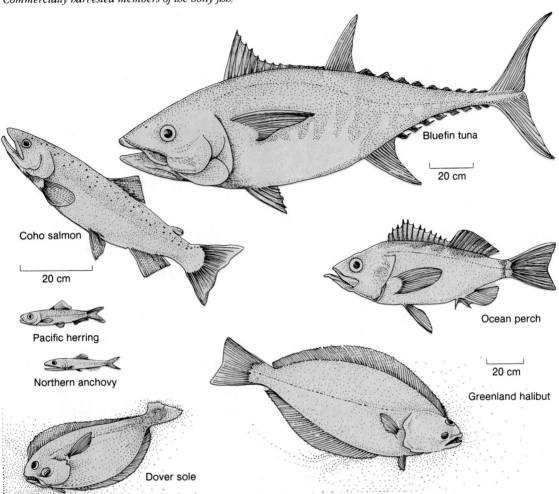

at the base. The large manta rays are plankton eaters, but most rays and skates are carnivorous, eating fish but preferring crustaceans, mollusks, and other bottom dwellers. Some of the skates and a few of the rays have electric organs that can produce a shock. These organs are located along the sides of the tail in skates and on the wings of the rays. Their purpose appears to be mainly defensive. Like the sharks, most rays bear their young live. Skates enclose the fertilized eggs in a leathery capsule, which is ejected into the sea and from which the young emerge in a few months. When found along the shore, these egg cases are called sea purses or mermaid's purses (see fig. 15.9).

Commercial Species of Bony Fish

Most food fish are found in the ocean's surface layers to a depth of approximately 200 m (660 ft), or in the epipelagic zone. Most of these fish are streamlined, active, predatory, and capable of high-speed, long-distance travel.

Among the most important species fished commercially are the enormously abundant, small, herring-type fish, such as sardine, anchovy, menhaden, and herring. These fish feed directly on plankton and are found in large schools in areas of high primary productivity. Other valuable fish that are harvested commercially include mackerel, pompano, swordfish, and tuna. These fish are caught at sea, out of sight of land.

Fish that live on or near the bottom, known as **demersal fish** or bottom fish, do not swim as rapidly as those that live in the water column above. The flounder, halibut, turbot, and sole are commercially valuable bottom fish. Perch and snapper tend to congregate along the sea floor in the shallower, nearshore areas. They hide among the rocks and live in the cracks of underwater reefs. Because of their habitat, they are called rockfish. Representative fish are shown in figure 15.10. In a later section of this chapter, we will discuss the commercial exploitation and economic significance of some of these food fish.

Figure 15.11
Fish from the deep sea.

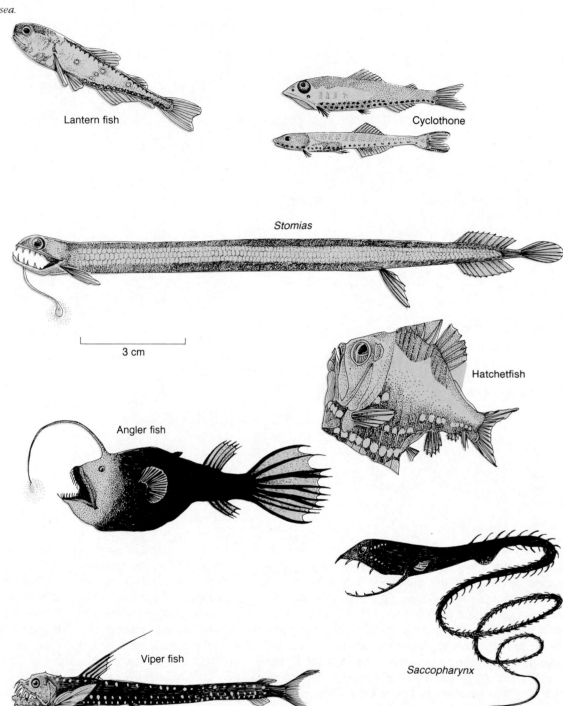

Lantern fish

Cyclothone

Stomias

3 cm

Hatchetfish

Angler fish

Viper fish

Saccopharynx

Deep-Sea Species of Bony Fish

The fish of the deep sea are not well known. The zones below the epipelagic zone are difficult and expensive to sample. No fish from these depths have been exploited commercially. A variety of deep-sea types is shown in figure 15.11.

In the dim, transitional mesopelagic layer, between 200 and 1000 m (660–3300 ft) below the surface, the waters support vast schools of small luminous fish. One genus, *Cyclothone,* is believed to be the most common fish in the sea. There are many species of *Cyclothone,* each living at a relatively fixed depth. The deeper-living species are black, while the shallower-living species are silvery, to blend with the dim light.

Another mesopelagic fish is the lantern fish, with a worldwide distribution. Some two hundred species are distinguished by the pattern of light organs along their sides. These fish are a major item in the diet of tuna, squid, and porpoises. *Stomias,* a fish with a huge mouth, long, pointed teeth, and light organs along its sides, is also found at this depth, as are the large-eyed hatchet fish. These fish prey on the great clouds of euphausiids and copepods found at this depth and support a large array of predators across the world's oceans.

Other fish move freely through these deeper waters. Sharks, sablefish, and grenadiers have all been observed at these depths. Often, the species found at great depths at low latitudes are the same as the cold-water species found at shallow depths at high latitudes. The sablefish is caught commercially in the waters off Alaska, British Columbia, and Washington State; it is also found in large numbers off southern California, between 800 and 1500 m (2640–4950 ft).

Until recently, it was assumed that although fish did exist at depths below 200 m (660 ft), there were very few of them. Researchers are now finding that life is more abundant at these depths than was previously believed. Submersibles, which allow direct observation and better underwater photography techniques, have helped to change our ideas. Underwater cameras, with attached bait, show populations of scavenging fish, including sharks, congregating at a baited target over a period of a few hours. Probably these species follow scentlike clues carried by the deep water currents. Particularly abundant deep water populations have been found in areas of the Pacific where the surface populations are sparse. John Isaacs of the Scripps Institution of Oceanography observed more than forty large fish attracted to the bait at 6000 m (19,800 ft) under the least productive waters of the Pacific gyre, while at only 4000 m under the most productive waters of the Antarctic the bait was visited by only a few small fish. A partial explanation may be that in the least populated, shallower areas food descends to the bottom, while in productive, highly populated regions it is consumed on the way down, leaving little to feed the deeper dwellers.

In the perpetually dark bathypelagic zone are predators with highly specialized equipment for catching their prey. These fish are equipped with light-producing organs used as lures, large teeth that in some species fold backward toward the gullet so the prey cannot escape, and gaping mouths with jaws that unhinge to allow the catching and eating of fish larger than themselves. Among the most famous of these predators is *Macropharynx longicaudatus* and the related *Saccopharynx,* which have funnellike throats and tapering bodies ending in whiplike tails. When the stomach is empty the fish appears slender, but it expands to accept anything the great mouth can swallow. The female angler fish, *Ceratias halboelli,* has a dorsal fin modified into a fishing rod, which slides back and forth along a canal in the back of the fish. An illuminated lure dangles from the tip of the rod. Other fish are attracted to the lighted bait, which the angler fish moves forward along her back until the bait is just above her jaws.

Although fierce and monstrous in appearance, most of these fish are small, between 2 and 30 cm (1–12 in) in length. They breathe slowly, and the tissues of their small bodies have a high water and low protein content. The fish go for long periods between feedings and use their food for energy production rather than for increased tissue production.

15.5 Classification Summary of the Nekton

The nekton, all members of the kingdom animalia, can be classified in the phyla and classes listed below.

I. Kingdom Animalia: animals; multicellular heterotrophs with specialized cells, tissues, and organ systems.
  A. *Phylum Mollusca:* mollusks.
    1. *Class Cephalopoda:* squid; the octopus is a member of the benthos.
  B. *Phylum Chordata:* animals with a dorsal nerve cord and gill slits at some stage in development.
    1. *Subphylum Vertebrata:* animals with a backbone of bone or cartilage.
      a. *Class Agnatha:* jawless fish; lampreys and hagfish.
      b. *Class Chondrichthyes:* jawed fish with cartilagenous skeletons; sharks and rays.
      c. *Class Osteichthyes:* bony fish; all other fish, including commercial species such as salmon, tuna, herring, and anchovy.
      d. *Class Reptilia:* air breathers with dry skin and scales; young develop in self-contained eggs; turtles, sea snakes, iguanas, and crocodiles.
      e. *Class Mammalia:* warm-blooded; body covering of hair; produce milk; young born live.
        (1). *Order Cetacea:* whales.
          (a). *Suborder Mysticeti:* baleen whales, including blue, gray, right, sei, finback, and humpback whales.
          (b). *Suborder Odonticeti:* toothed whales, including the killer and sperm whale, the porpoise, and the dolphin.
        (2). *Order Carnivora:* sea otters, seals, sea lions, and walruses.
        (3). *Order Sirenia:* dugongs and manatees.

15.6 Practical Considerations: Commercial Fisheries

In 1950, the total world marine fish catch was approximately 21 million metric tons. During the next twenty years, as human populations exploded, the fishing effort by all nations intensified, and the technology and gear used to hunt and catch the fish improved dramatically. In 1955, the catch was 28 million metric tons, and by 1960 it was nearly twice what it had been ten years previously, at 40 million metric tons. The catch increased again in 1965 to 47 million metric tons, and by 1970 it was 70 million metric tons. Since then, it has oscillated between 62 and 70 million metric tons, climbing to 74.8 million metric tons in 1985. At the same time, there has been a shift in the use made of the catch. In 1950, 90% of the world fish catch was consumed as food by humans, and 10% was used to make fish-meal products to feed the poultry and livestock of the more-developed countries. By 1980, approximately 60% of the world catch was going to feed people, and 40% was being used as feed for domestic animals. The use of fish meal in this way is only about 20% efficient; that is to say, for every ton of fish meal fed, only 0.2 ton of additional animal protein is produced.

At the present time, total world fish catches continue to increase due to fishing effort, opening of new fisheries, such as the North Pacific bottom fishery and the South African pilchard fishery, and an increased consumer demand for fish and fish products. For example, Alaskan pollock, a bottom fish, is processed to remove the fats and oils which give the fish its flavor, and a highly refined fish protein called **surimi** is produced. Surimi is processed again and flavored to form artificial crab, shrimp, and scallops. Surimi, a major fish product in Japan, is the fastest growing product in the U.S. seafood market. Meanwhile, many traditional and valuable fisheries have declined. Four of these fisheries are to be discussed in the following sections. Each case illustrates the difficulties faced by those fishing, fishery managers, and fishing nations, as well as by fish consumers, as the catches decline and the costs rise.

Anchovies

The anchovy fishery concentrated in the upwelling zone off the coast of Peru has produced the world's greatest fish catches for any single species. The anchovy is a small, fast-growing fish that feeds directly on the phytoplankton in the upwelling. Anchovies travel in dense schools, which makes it possible to net and harvest them in large quantities. The anchovies in this fishery are not caught for human consumption. The fish are processed into fish meal and are exported as feed for domestic animals.

The fish meal fishery began in 1950 with a harvest of 7000 metric tons. By 1962, the harvest had risen to 6.5 million metric tons, and as the world demand for fish meal rose the anchovy fishing intensified. In 1970, the peak year, about 12.3 million metric tons were taken. As the fishing increased, the average size of the fish taken decreased, and it took more and more fish (smaller and younger) to make one ton of catch. During the period between 1950 and 1972, the Peruvian coast was visited three times by El Niño. (Refer back to chapter 7.) When the general wind pattern reverses, the winds become westerly. This change diminishes the upwelling and moves warmer, nutrient-poor surface water eastward into the areas of normally cold and productive upwelled water. The decrease in nutrients and the increase in temperature result in the destruction of the plant and animal populations. The decomposing organic material strips oxygen from the water and further reduces the populations. El Niño devastates the fishing and kills large numbers of seabirds that feed on the schools of anchovies.

The heavy fishing of 1970–1971 and the severe effect of El Niño in 1972 resulted in a radical decrease in the fishery to a harvest of only 2 million metric tons in 1973. After 1973, government quotas were placed on the anchovy catch. Many experts thought the fishery would never recover, even with government controls. It did increase slowly in 1974 and 1975, but El Niño struck again in 1976 and catches dropped again. In 1981 and 1982, catches increased to between 1 and 2 million metric tons. Then, in the winter of 1982–83, El Niño appeared once more, and once more the fishery declined. The 1983 through 1985 catches were kept below 150 thousand metric tons. (See fig. 15.12.)

Figure 15.12
The anchovy catch of Peru and Ecuador by calendar year given in millions of metric tons. Years of El Niño are 1957, 1965, 1972, 1976, and 1982–83.

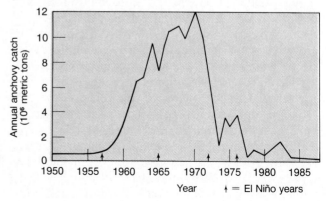

Tuna

The tuna fishery is an expensive, sophisticated business that involves many nations, including the United States. The United States tuna industry accounts for about 50% of the world tuna catch and operates out of southern California. A modern tuna boat costs 5 million dollars or more and takes advantage of the tuna's schooling habit to catch the fish in huge seine nets 1100 m (3600 ft) long and 180 m (595 ft) deep. A single set of a net brings in over 150 tons of tuna. Although the modern open-ocean netting of tuna is an efficient method of harvesting, it has a significantly undesirable side effect. Tuna are accompanied in their movement across the oceans by porpoises. The porpoises swim just above the schools of tuna and are used by the tuna boats to locate the schools. The net is set in a circle around the school of tuna and pulled up into a bag. If porpoises are caught in the net they can drown. The total cost of the tuna harvest in porpoises is unknown, but since the passage of the 1972 Marine Mammal Protection Act, conscious and legally enforced efforts have been made to reduce the porpoise kill. It is estimated that the largest kill occurred in 1970, with a loss of more than 500,000 porpoises. The world quota of incidental porpoise kill for 1980 was set at no more than 31,000, with United States boats being allowed less than 15,000.

It has been mandatory that all United States tuna boats carry an observer from the Marine Mammal Protection Agency of NOAA during at least part of their fishing season. These observers check fishing methods and techniques used to free the porpoises. Design changes for gear have included fine mesh panels around the edge of the net, in which the porpoises do not become entangled, and the fishing method of backing down on the net and allowing the front end to sink is used to let the porpoises escape. Small boats and crew members in scuba gear also work inside the gathering net to guide the porpoises to freedom. The observers count the porpoise kills, and if the number killed rises above the set quota, the entire United States fleet can be recalled, as happened in 1977.

Salmon

Another fishery in which the United States has an important role and investment is the salmon industry of the Pacific Northwest and Alaska (see fig. 15.13). These fish spawn in fresh water and remain there as juveniles for a year or less, depending on the species. Then they begin their long journey to the sea, where they will live and grow for one to four years. At the end of this time, the fully mature adults return to the same rivers, lakes, and streams in which they.were born to spawn in their turn. As they return, the salmon are fished both com-

mercially and for sport. The demand for salmon is high, and the fish bring a high price per pound. But like many other heavily exploited fish stocks, the salmon are becoming scarcer, and the fishing is now tightly controlled. The fishing season is shortened, and the catches decrease, causing profits for those who fish to fall; and management requires more and more regulations in attempts to ensure a population for the next year. An additional difficulty in maintaining wild salmon runs is the degradation of their freshwater environment. The salmon depend on high-quality, pollution-free fresh water and clean, gravel-bottomed, shady, cool streams for successful natural spawning. People have routinely dammed rivers for power usage and for flood control, cutting off the salmon from their home streams. People have harvested trees up to the stream edge, removing shade and allowing mud and silt eroded from the exposed land to cover the streambed. People have used streams to carry away their wastes, degrading the water quality and damaging both the juvenile fish and the returning runs. In many areas, these activities are being corrected, but the naturally spawning fish are now few in number, and people must pay both the cleanup costs and the costs of hatchery programs to supplement the dwindling natural stocks.

Redfish

The red drum, or redfish, fishery of the Gulf of Mexico is an example of a local, small-scale fishery that suddenly expanded due to national demand and without adequate safeguards for the fish stock. Stocks of redfish occur in both the inshore and offshore waters of the Gulf of Mexico, with the larger fish, believed to be the spawning stock, gathering in large schools offshore. In

Figure 15.13
A salmon purse seiner hauling in its net.

1983, 7.6 million pounds were caught in the recreational fishery and 3 million in the commercial fishery. Only 7% of the commercially caught fish came from offshore waters. In 1984, recreational fishing took 6.5 million pounds; commercial fishing took 4.3 million pounds, 25% from offshore waters. By 1985, the commercial catch had grown to 6.5 million pounds, 55% from offshore waters, and in the first half of 1986, 6 million pounds were fished commercially, nearly all from offshore waters.

The rapid growth of this commercial fishery is attributed to the sudden popularity of the Cajun dish, blackened redfish. Commercial fishing responded to the increased demand for redfish by developing a purse-seiner fleet assisted by spotter planes and able to fish the offshore schools. Each set of the purse seine nets takes 50,000–150,000 pounds of redfish. In 1986, to forestall collapse of the fishery by overfishing, federal agencies placed a one-million pound emergency quota on redfish. The states of Texas, Louisiana, Alabama, Mississippi, and Florida have now passed laws to prevent the landing of fish caught by purse seiners. Whether the rage for Cajun cooking will dissipate and the demand for redfish will drop is unknown. However, while the demand is high, careful management of the redfish stocks is required to assure sustained yield of this resource.

Policies

In 1977, the United States expanded the coastal areas over which it controls its fisheries to a zone that extends 200 miles out to sea. Within this zone, other countries fish only under the terms of agreements negotiated with the United States. The species fished, the catch limits, and the areas for fishing are all negotiated. Enforcement is by the Coast Guard, and fishery observers sail on foreign ships to count and record the catch. Despite such efforts, and even with increased management and regulation of fish stocks, the general trend of ocean fishing is down and the costs are up. In the United States, the fishing boat and gear, the crew's wages, and the fuel needed are all increasing the cost of commercial fishing.

In the United States, those who fish are usually independent operators who sell their fish directly to the processor. The United States consumer prefers fish fillets and fish steaks, not low-cost minced fish products; therefore, the fishing is for high-cost fish such as salmon, swordfish, and halibut. The filleting process uses only a part of the fish's flesh, between 20% and 50% of the live body weight. By contrast, fish caught by the large, subsidized fleets of other countries are processed cheaply at sea, and much more of the fish is used. In some cases, it is cheaper to import fish products to the United States than to catch and process the fish ourselves. Some United States boats fish in what are called joint ventures with other countries, selling their catch to be processed in foreign factory ships. Still, the cost of fishing and, therefore, of fish increases, and the fish catches decrease.

Fisheries scientists around the world are beginning to realize that large numbers of marine animals die each year only because they are caught incidentally while fishing for other species. **Incidental catch,** or by-catch, or what are often called "trash fish," represent a tremendous waste of marine resources. During a single year, more than 25,000 porpoise have been killed as incidental catch in the tuna fishery. The Japanese high-seas gillnet salmon fishery in the North Pacific yearly is reported to claim 10,000–15,000 Dall porpoise and 500,000–700,000 seabirds trying to feed on the netted fish. It is estimated that shrimp trawlers catch more than 45,000 sea turtles and that more than 12,000 of them do not survive. The world's shrimp fishery is estimated to have an annual catch of 1.1 million–1.5 million tons; the associated by-catch of fin-fish is 5 million–21 million tons, of which 3–5 million tons are discarded. The discard rate on by-catch varies from place to place; if incidental catch does not bring a high enough price, and if processors are not available, these "trash fish" will be returned to the sea, usually dead. In the Gulf of Mexico, 1 pound of shrimp results in 15 pounds of by-catch, nearly all discarded. In Southeast Asia, the need for protein is so great that most of the by-catch is marketed locally.

Fish Farming

A possible alternative way to increase the fish harvest from the sea is by farming it. An old Chinese proverb states, "If you give a person a fish, he will have food for one day, but if you teach him to raise fish, he will have food for a lifetime." Farming the water, known as **mariculture** or **aquaculture,** began in China some four thousand years ago. The Chinese were culturing common carp in 1000 B.C., and a book was written in 500 B.C. giving directions on fish farming, including methods for building the pond, selecting the stock, and harvesting it. In China, Southeast Asia, and Japan, fish farming in fresh and salt water has continued to the present as a practical and productive method of raising large quantities of fish. Carp, milkfish, *Tilapia,* and catfish are raised. The methods used are labor intensive, and most fish farms are small, family-run operations. In **monoculture,** fish are raised as a single species. In **polyculture,** they are raised in ponds they share with other species. Surface feeders and bottom feeders are raised together, making use of the total volume of the pond; this type of system may even include ducks.

Israel's requirements for protein to feed a large population concentrated in a small area have resulted in the development of acres of fish ponds, which produce annually 130 million pounds of carp, *Tilapia,* and mullet in polyculture, with plans for further development. Fish farming in the United States, however, produces only 2% of our fishery products, but that amount includes 50% of the catfish and nearly all the trout. In order to be successful in the United States, fish farming must be profitable, which requires a proven market and a large-scale operation, along with the development of a technology to keep costs down. Another factor to be considered is that a profitable operation requires careful choice of the species. To be economically productive, a species must reproduce in captivity, have juvenile forms that survive well under controlled conditions, gain weight rapidly, eat a cheap and available food, and fetch a high market price.

Salmon farming is on the rise worldwide. Norway and Scotland have been particularly successful. Farmed or pen-reared salmon production in Norway has increased from 170 tons in 1973 to 23,000 tons in 1985. Enthusiasts predict international production will reach 110,000 to 200,000 tons by 1990. The Canadian government has adopted a national policy of encouraging salmon farms, and the farms are proliferating rapidly along the British Columbia coast, north of Vancouver. Maine, Washington, California, and Oregon allow salmon farming, but the permit procedure is long and complex. Although pen-rearing of salmon has been tried and is successful in the protected waters of Puget Sound in Washington State, (fig. 15.14), it is an expensive operation that raises problems in the sound's multiple-use waterways. Salmon pens are not particularly attractive to those with homes along the shore, the pens get in the way of recreational boating and water sports, and the fish produce waste products into the water.

In Texas, the craze for redfish discussed in the preceding section has opened the way to redfish mariculture. At present, there is only one commercial redfish farm in the state and several development projects are underway at the universities. Hatcheries are beginning to produce, but initial startup costs are estimated to be between $300,000 and $400,000, which may limit individual ventures and allow only corporate investment.

Hatchery rearing and release of seagoing fish (salmon and steelhead trout) is called **ocean ranching** or **sea ranching.** Federal and state hatcheries provide millions of fish, which are caught in both the commercial and recreational fisheries along the coasts of Oregon, Washington, and Alaska. Because pen culture of salmon is expensive, the release from private hatcheries of salmon that will return to the hatcheries to be netted has been proposed. This private sea ranching is being tried in Oregon, where the Weyerhauser Company's salmon-rearing facility is associated with their paper plant. The warm water from the plant is mixed with river water to accelerate the salmon's early growth. These juveniles are then transported to holding areas adjacent to the sea and imprinted with a chemical label in the water (fig. 15.15). The fish are released to the ocean. Upon maturing, the salmon return to the holding areas, seeking them by the chemical label. Here, they

Figure 15.15
The ocean ranching release and recapture facility at Coos Bay, Oregon, was originally constructed by Ore-Aqua Farms and is now operated by Anadromous, Inc.

Figure 15.14
The Domsea Farms salmon-rearing pens in Puget Sound, Washington.

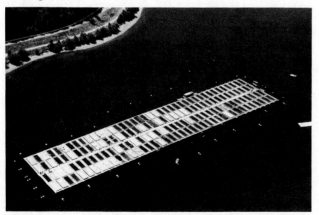

TABLE 15.3
Survival Rates of Salmonid Spawners

	Eggs (1 male 1 female)	Eggs to fry stage (survivors)	Fry to adult (survivors)	Caught in fishing	Left to spawn (survivors)
natural spawners	2000 100%	200 10%	6 0.3%	4 0.2%	2 0.1%
hatchery spawners	2000 100%	1600 80%	48 2.4%	32 1.6%	16 0.8%

are harvested for the market. Table 15.3 compares natural survival with hatchery survival and returns for spawning.

A population of natural spawners must produce at least 2000 eggs per pair to assure the survival of a spawning pair to sustain the natural run. Hatcheries increase the survival at the egg to fry stage eightfold. Thus they have the ability to assure continuance and even build a run of fish by spawning more than one pair of returnees. The greater return of mature fish to the

hatchery because of the increased juvenile survival is what makes ocean ranching appear attractive as an enterprise.

The idea of private ocean ranching has not proved popular with those who fish commercially, because they fear the intrusion and control of the large companies and a possible drop in salmon prices. Most economists and fishery biologists believe that an ocean-ranching program in the private sector would add fish to the harvest without any great effect on market prices.

Summary

The nekton swim freely and independently. The marine mammals include whales, seals, sea lions, walruses, sea otters, and sea cows. The whalers hunted baleen whales and the sperm whales, which are toothed whales. As hunting techniques changed, harvesting became more intensive, and some species of whales have been threatened with extinction. The International Whaling Commission regulates whaling on a voluntary basis; a ten-year moratorium on commercial whaling has begun, but conservationists fear the so-called scientific whaling. Dolphins and porpoises are small, toothed whales.

The fur seals and sea otters were hunted heavily during the nineteenth and early twentieth centuries. Their populations are now protected or hunted by quota under international treaty. The Steller sea cow is extinct; manatees and dugongs are found in the warm waters of the Indian and Atlantic oceans. The Marine Mammal Protection Act was passed in 1972 to protect all marine mammals and to prohibit commercial trade in marine mammal products.

Sea snakes, the marine iguana, seagoing crocodiles, and sea turtles are the reptile members of the nekton. Sea turtle populations are under great pressure by hunters and poachers. The squid also belong to the nekton.

Fish are found at all depths in all oceans. Sharks, rays, and skates are primitive fish with cartilagenous skeletons. All other fish have bony skeletons, including the commercially

fished species and the highly specialized types of the deeper ocean.

World fish catches have increased dramatically between 1950 and 1980, due to the increased fishing effort and improved technology. Although stock abundance has declined, new fisheries have developed, and the fishing effort has continued high to keep the catch stable. At the same time, more of the fish is used as domestic animal feed and less is used for human consumption.

The problems of the traditional commercial fisheries include declining fish catches, increased regulation, and increased costs. The Peruvian anchovy fishery had produced the world's greatest fish catches until it was hit by the effects of overfishing and El Niño. The United States tuna fishery has had to cope with the killing of porpoises as a side effect of fishing technology, and the Pacific Northwest and Alaskan salmon fisheries are experiencing difficulty due to the degradation of the salmon's freshwater environment and the loss of natural salmon runs. The Gulf Coast redfish fishery has been threatened with collapse due to its recent explosive growth.

Aquaculture has proved to be practical and productive in Asia and Israel. In the United States, it is a small industry that requires large-scale operations, proven markets, and cost-cutting technology. Ocean ranching is the release of hatchery-raised seagoing fish to increase fishing stocks.

(a)

(b)

(c)

(d)

Colorplate 14
*(a) Marine mammals. The harbor seal (*Phoca vitulina*) is a
friendly, curious animal that coexists well with humans.*

*(b) This large male Northern Fur Seal (*Callorhinus ursinus*)
stands vigilant watch over his harem of females in the Pribilof
Islands.*

*(c) Sea otter (*Enhydra lutris*) and her pup dine on crab.*

*(d) The walrus (*Odobenus rosmarus*) feeds from the bottom
and relaxes on the arctic ice floes.*

*(e) Pacific white-sided dolphins (*Lagenorhynchus obliguidens*)
play in the bow wave of a research vessel.*

(e)

Colorplate 15

The benthos are a large, varied group of animals living on or in the sea floor, especially the organisms of the tide pools and the intertidal areas of the rocky coasts.

The anemone, Tealia crassicornis, *is a sessile carnivore (a). The graceful and beautiful nudibranchs or sea slugs include (b) the white* Dirona albolineata, *(c) the orange-flecked* Triopha carpenteri, *and (d)* Hermissenda crassicornis *with its orange-and-white striped tips.*

*Starfish come in a remarkable diversity of shapes and sizes: (e) the bright orange bloodstar (*Herrica leviuscula*), the slender armed* Evasterias troschelli, *the young, multiarmed* Solaster dawsoni, Mediaster aequalis *with its wide disk and broad arms, the leather star (*Dermasterias imbricata*), and the purple, rough-skinned* Pisaster ochraceus. *All are carnivores and use their tube feet to hold and open the shellfish on which they feed (f).*

*(g) The purple sea urchin (*Strongylocentrotus purpuratus*) and the green urchin (*S. droebachiensis*) (a) are closely related to the sea stars but they are herbivores, clipping off the algae with their especially constructed mouth parts (h).*

*The pink sea scallop (*Chlamys hastata hericia*) lies open, as it filters organic particles from the sea water (i).*

*The clown fish (*Amphiprion*) enjoys a commensal relationship with a sea anemone (j).*

(j)

Key Terms

mammal	manatee	surimi	polyculture
cetaceans	dugong	incidental catch	monoculture
baleen	sea cow	mariculture	ocean ranching/sea
echolocation	sirenia	aquaculture	ranching
pinniped	demersal fish		

Study Questions

1. The Marine Mammal Protection Act of 1972 has allowed the recovery of certain populations to such an extent that the marine mammals are now competing for food resources of economic value to humans. Under these circumstances should any adjustments be made to this act? Explain why.

2. What sensory abilities help a shark to find its prey?

3. In the last twenty-five years the world's marine fish catch has grown from 40 million metric tons to 75 million metric tons. During this period the catch from many large fisheries has declined. How do you explain this contradictory situation?

4. Fish farming is growing slowly in the United States in comparison to other countries. What reasons can you give for this situation?

5. In what ways could fish utilization be increased without increasing the fish catch from the oceans?

6. Why are bottom fish more likely to show the first effects of a degraded nearshore environment, rather than fish that inhabit the water column?

7. How will the ability to predict El Niño help manage the Peruvian anchovy fishery?

8. Why are more toothed whales found at low latitudes and why do more baleen whales reside in temperate and polar latitudes? Consider food requirements only.

9. Why is it difficult to increase a pinniped population in the northern temperate latitudes after hunting has ceased? Do you think the situation would be the same in the southern temperate latitudes?

10. Sea snakes are thought to have their origin in the land snakes of India and Southeast Asia. Why are there no sea snakes in the Atlantic Ocean? What might be the effect of a sea level canal through the isthmus of Panama?

11. If equal numbers of natural-run fish and hatchery fish are spawned, the returning fish will be in the ratio of 6 natural run:48 hatchery fish (see table 15.3). If a quota of 36 of the returning fish is assigned to a commercial fishery, why is there a risk of completely destroying the natural run? Discuss the impact of hatchery-enhanced fish runs on natural fish runs as hatchery production is increased and fish catch quotas are based on the total number of mixed stock returns.

The Benthos
Dwellers of the Sea Floor

16

*T*he exposed rocks had looked rich with life under the lowering tide, but they were more than that: they were ferocious with life. There was an exuberant fierceness in the littoral here, a vital competition for existence. Everything seemed speeded-up; starfish and urchins were more strongly attached than in other places, and many of the univalves were so tightly fixed that the shells broke before the animals would let go their hold. Perhaps the force of the great surf which beats on this shore has much to do with the tenacity of the animals here. It is noteworthy that the animals, rather than deserting such beaten shores for the safe cove and protected pools, simply increase their toughness and fight back at the sea with a kind of joyful survival. This ferocious survival quotient excites us and makes us feel good, and from the crawling, fighting, resisting qualities of the animals, it almost seems that they are excited too.

John Steinbeck
from The Log from the Sea of Cortez

*T*he animals and plants that live on the sea floor or in the sediments are members of the benthos. Because the plants require light, they are found only along shallow coastal areas and in the intertidal zone. Benthic animals are found at all depths. The benthos form a remarkably rich and diverse group of organisms, which includes the luxuriant, colorful life of the coral reef and the bizarre, newly discovered organisms surrounding deep-sea vents. Benthic organisms are also important food resources (for example, oysters, clams, crabs, and lobster). In this chapter, we first present an overview of the benthos by group and habitat, with special focus on some of the world's more intriguing benthic communities, and then we consider the harvesting of the benthos, its problems, and its potential. This chapter is intended to serve only as an introduction to this topic, for no single chapter can do justice to the profusion of organisms present in this group. If you wish to establish a greater appreciation and understanding of the benthos, we suggest additional readings in books and magazines devoted to marine biology and biological oceanography (see the selected references at the end of this book).

Seaweeds are members of a large group called **algae.** The unicellular planktonic algae were considered in chapter 14. In this chapter, we consider the large benthic algae, conspicuous members of coastal intertidal and subtidal communities. Biologists have different opinions on including the algae in the plant kingdom. Because they photosynthesize and contain chlorophyll pigments, some consider them plants; others consider them plantlike, but prefer to place them in other categories, because of their body form, methods of reproduction, accessory pigments, and storage products, which generally differ from those of land plants. For our purposes, we shall consider the large benthic algae (or seaweeds) as primitive members of the plant kingdom. These algae have simple tissues; they do not produce flowers or seeds; and their pigments and storage compounds vary from group to group.

General Characteristics of Benthic Algae

Although seaweeds are often found floating at the water's edge or thrown up on shore after a storm, these seaweeds are not floating (or planktonic) algae. They are true benthic forms that grow attached to rocks, shells, or even beer cans and soda bottles. The benthic algae have a basal organ called a **holdfast,** which anchors the plant firmly to the rocky bottom; the holdfast is not a root, in that it does not absorb water and nutrients. Because the seaweeds grow attached, they are found in areas of rocky substrate and not in areas of mud or sand, where the holdfast has nothing to which it can attach. Above the holdfast is a stemlike portion known as the **stipe.** The stipe may be so short that it is barely identifiable, or it may be up to 35 m (115 ft) in length. These long stipes belong to certain brown seaweeds commonly known as **kelps,** which are found in the temperate and colder waters of the world. The stipe in most algae does not have tissues that transport water and nutrients or organic compounds, in contrast with the stem of a land plant. However, some larger algae (the kelps) form a transport tissue that is similar to that of land plants. This tissue has been shown to transport photosynthetic products. The stipe acts as a flexible connection between the holdfast and the **blades,** which are the leaflike portions of the seaweeds. Although a blade serves the same purpose as a leaf (it is the plant's photosynthetic organ), it does not possess the specialized conducting tissues seen as veins in leaves. These tissues are not needed by the algae, as their blades are thin and bathed on all sides by water. The algal cells take up water directly and do not require a supply from the ground conducted up a stem and through veins to leaves. The blades may be flat, ruffled, feathery, or even encrusted with calcium carbonate. Because seaweed is surrounded by water, with its dissolved carbon dioxide and nutrients, and because it exposes a large blade area to both the water and the sun, it is an efficient producer. The general characteristics of a benthic alga are shown in figure 16.1.

Since the benthic algae are primary producers, dependent on sunlight, they are confined to the shallow, sunlit areas of the oceans. Sometimes during a storm, the plants are dislodged, taking with them the rocks to which the holdfasts are attached. The plants, carrying their rock anchors, often drift for some time before they sink back to the bottom. If they sink too deep for sufficient light, they die, but if their blades remain in the photic zone, they will continue to grow. This process is also responsible for spreading seaweeds into new areas.

In the sea, the quality of light as well as the quantity of light changes with depth (see chapter 4). The red end of the visible light spectrum is quickly absorbed at the surface, and it is the blue-green light that penetrates to the greatest depths. On land and at the sea surface, the full spectrum of visible light is available. Green algae

Figure 16.1

Benthic algae are attached to the sea floor by a holdfast. A stipe connects the holdfast to the blade. Laminaria *is a kelp and a member of the brown algae.*

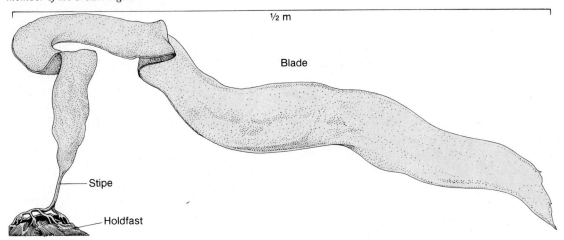

are found in shallow water, and their chlorophylls absorb both long and short wavelengths. Algae found at moderate depths have a brown pigment, which is more efficient at trapping the shorter wavelengths of solar light. At maximum growing depths, the algae are red, for the red pigment can best absorb the remaining blue-green light. Both the brown algae and the red algae possess the green chlorophyll, but the red and brown pigments mask its color, so that the seaweeds appear red or brown. Therefore, the characteristic pattern for seaweed growing on a rocky shore is first the green algae, then the brown algae, and last the red algae as depth increases. Few large red algae are seen at low tide, for they are primarily sublittoral species.

Seaweeds provide food and shelter for many animals. They act in the sea much as the forest and shrubs do on land. Fish and other animals, such as sea urchins, limpets, and some snails, feed directly on the algae; other animals feed on shreds and pieces as they settle to the bottom. Some organisms use large seaweeds as a place of attachment; some of the smaller algae grow on the large kelps. Algae that produce calcareous outer coverings are important in the building of tropical coral reefs, which are discussed later in this chapter.

Kinds of Seaweeds

The pigment colors are used as a basis for classifying algae. Green algae are moderate in size and may be finely branched or may occur in thin, flat sheets. They are mostly freshwater organisms, but a small number do occur in the sea, including *Ulva,* the sea lettuce, and *Codium,* known as dead man's fingers. Green algae are like the land plants; they have the same chlorophyll pigments as land plants and store starch as a food reserve.

All brown algae are marine and range from simple microscopic filaments to the kelps, which are the largest

of all the algae. Kelps have more structure than most algae but are much simpler than flowering land plants. The kelps have strong stipes and effective holdfasts that allow them to colonize rocky points in fast currents or heavy surf, a habitat favored by the sea palm, *Postelsia.* Other kelps grow with holdfasts well below the depth of wave action and float their blades at the surface attached to gas-filled floats; for example, the bull kelp, *Nereocystis,* is especially abundant along the Alaskan, British Columbian, and Washington coasts, and the great kelp, *Macrocystis,* is found along the California coast. Other species are found off Chile, New Zealand, northern Europe, and Japan. Their brown color comes from fucoxanthin pigment, which masks the green chlorophyll. Storage products include the carbohydrates laminarin and mannitol, but not starch.

The red algae are almost exclusively marine; they are the most abundant and widespread of the large algae. Their body forms are varied and often beautiful, flat, ruffled, lacy, or intricately branched. Their life histories are specialized and complex, and they are considered the most advanced of the algae. All contain the pigments phycoerythrin and phycocyanin. Their storage product is floridean starch.

Although the algae are commonly classified by color, it is a mistake to depend only on color for classification, because the visible color can be misleading. While some red algae appear brown, green, or violet, some brown algae appear black or greenish. Some representative types of algae are shown in figure 16.2.

We considered diatoms as part of the plankton (see chapter 14), yet there are benthic diatoms as well. These diatoms are usually of the pennate type and grow on rocks, muds, and docks, where they produce a slippery brown coating that is treacherous for the walker.

Figure 16.2
Representative benthic algae. Ulva *and* Codium *are green algae.*
Postelsia, Nereocystis, *and* Macrocystis *are kelps. The kelps and*
Fucus *are brown algae. The red algae are* Corallina, Porphyra,
and Polyneura. Corallina *has a hard, calcareous covering.*

Other Plants

Algae do not produce seeds or flowers, but there are a few flowering plants with true roots, stems, and leaves that have made a home in the sea. Eelgrass, with its strap-shaped leaves, is found growing on mud and sand in the quiet waters of bays and estuaries along the Pacific and Atlantic coasts, and turtle grass is common along the Gulf Coast. Surf grass flourishes in more turbulent areas exposed to waves and surge action. In the tropics, mangrove trees grow in swampy intertidal areas, while temperate-area salt marshes are dominated by marsh grasses able to tolerate the brackish water. The sea grasses are rich sources of food and shelter for animals and act as attachment surfaces for some algae. The intertwining roots of the mangrove trees provide shelter and also trap sediments and organic material, helping eventually to fill in the swamps and move the shore seaward.

16.2 The Animals

Benthic marine animals, unlike benthic plants, are found at all depths and are associated with all substrates. There are fifty times more benthic than pelagic animal species in the sea, in that there are more than 150,000 benthic species, as opposed to about 3000 pelagic species. About 80% of the benthic animals belong to the **epifauna,** which are the animals that live on or attached to rocky areas or firm sediments on the bottom. Animals that live buried in the substrate belong to the **infauna** and are associated with soft sediments such as mud and sand. There are only about 30,000 species of infauna, as compared to 120,000 different species of epifauna. Some animals of the sea floor are **sessile,** or attached to the sea floor, as adults (for example, barnacles, sea anemones, and oysters), while others are motile all their lives (for example, crabs, starfish, and snails). Although most benthic forms produce motile larval stages that spend a few weeks of their lives as members of the plankton (refer to the discussion of the meroplankton in chapter 14), the planktonic larval stage is of particular importance to the sessile species. This mobility of the juvenile stages allows the species to relieve overcrowding and to colonize new areas. Sessile adults must wait for their food to come to them, either under its own power or carried in by waves, tides, and currents. Motile organisms are able to pursue their prey, scavenge over the bottom, or graze quietly on the seaweed-covered rocks.

The distribution of benthic animals is not governed by any single controlling factor such as light or pressure. In fact, the increasing pressure that occurs with depth does not have much effect on animals that are more than 90% water and have no internal gas storage chambers or lungs. Animal distribution is controlled by a complex interaction of factors, creating living conditions that are extremely variable. The substrate may be solid rock, movable cobbles, shifting sand, or soft mud. Temperature is nearly constant in deep water but changes abruptly in shallow areas covered and uncovered by the daily tidal cycle; salinity, pH, exposure to air, oxygen content of the water, and water turbulence all vary as well in the intertidal zone. Benthic animals exist at all depths and are as diverse as the conditions under which they live. Their life-styles are related to their varied habitats, and the following sections discuss both habitat and life-style.

Animals of the Rocky Shore

The rocky coast is a region of rich and complex plant and animal communities living in an area of environmental extremes. It is the meeting place between the more variable land conditions and the more stable sea conditions. As the water moves in and out on its daily tidal cycle, rapidly changing combinations of temperature, salinity, moisture, pH, dissolved oxygen, and food supply must be endured. At the top of the littoral zone, organisms must cope with long periods of exposure, heat, cold, rain, snow, and predation by land animals and seabirds as well as with the waves and turbulence of the returning water. At the littoral zone's lowest reaches, the plants and animals are rarely exposed but have their own problems of competition for space and predation by other organisms. Figure 16.3 shows the exposure endured by marine life at different levels in the littoral zone.

The distribution of the plants and animals is governed by their ability to cope with the stresses that accompany exposure, turbulence, and loss of water. Along the rocky coasts of such areas as North America, Australia, and South Africa, biologists have noted the pattern formed as the plants and animals sort themselves

Figure 16.3

The time of exposure to air for intertidal benthic organisms is determined by their location above and below mean sea level and by the tidal range. (MLLW = mean lower low water range; MLW = mean low water; MHW = mean high water; MHHW = mean higher high water.)

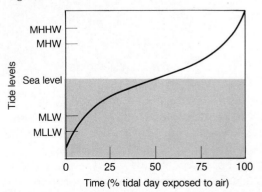

Figure 16.4

A typical distribution of benthic plants and animals on a rocky shore at temperate latitudes. Vertical zonation is the result of the relationships of the organisms to their intertidal environment.

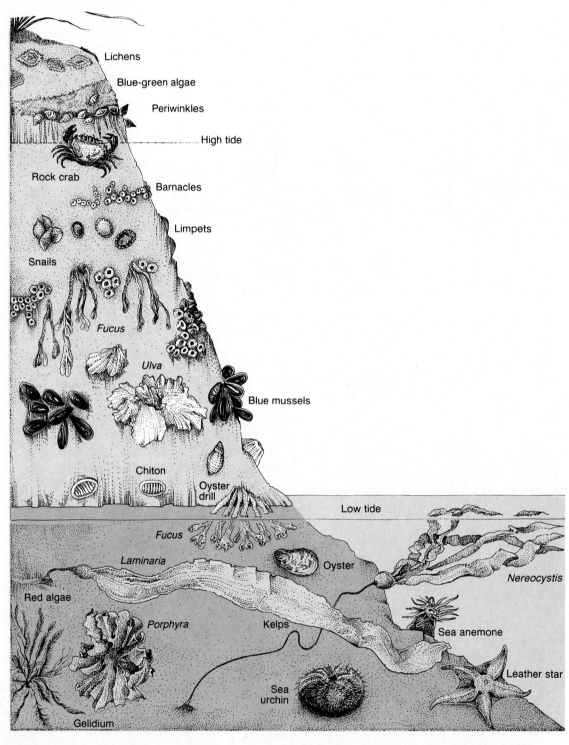

out over the intertidal zone. This grouping is termed zonation, **vertical zonation,** or **intertidal zonation** and is shown in figure 16.4. The distribution of the seaweeds, with the green algae in shallow water, the brown algae in the intertidal zone, and the red algae in the subtidal area, is an example of such zonation.

The following is a general discussion of some of the more conspicuous life forms found in these rocky intertidal zones. It is not a guide to the marine life of rocky shores, as substantial differences occur from region to region. If you wish to explore the rocky shores in your area, obtain a manual written specifically for your location.

Figure 16.5

Organisms of the supralittoral zone. The limpets and the snail,
Littorina, *are herbivores. The barnacles feed on particulate matter
in the water.* Ligia *is a scavenger.*

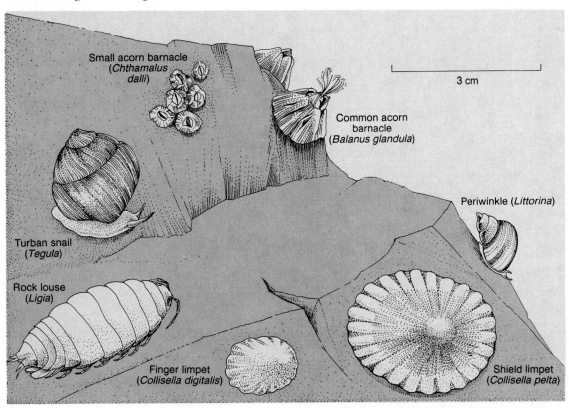

In the supralittoral (or splash) zone, which is above the high water level and is covered with water only during extreme storms and high water, the animals and plants occupy an area that is as nearly land as it is ocean bed. At the top of this area, patches of dark lichens and blue-green algae appear as crusts on rocks and stones and are often nearly indistinguishable from the rock itself. Scattered vegetation, including tufts of green algae, provides grazing for the small snails and limpets of the supralittoral zone. The snail *Littorina,* the periwinkle, is so well adapted to an environment that is more dry than wet that, unlike its marine relatives, it is an air breather, and some species will drown if caught underwater. *Littorina* is an herbivore and so are the limpets. Snails withdraw into their shells to prevent moisture loss and seal themselves off from the air with a horny disk called an operculum. Limpets are able to use their single muscular foot to press themselves tightly to the rocks to prevent drying out.

Just below the zone of snails and limpets are found small acorn barnacles, which must filter food from the seawater and are able to survive even though they are covered with water only during the few days of high water or spring tides each month. Barnacles are crustaceans, related to crabs and lobsters but cemented firmly in place. They have been described as animals that lie on their backs and spend their lives kicking food into their mouths with their feet. In some areas, the rocky splash zone is the home of another crustacean, the large (3–4 cm (1–2 in)) isopod *Ligia.* Organisms of the supralittoral zone are shown in figure 16.5.

The width of the supralittoral zone varies. The slope of the rocks, variations in light and shade, exposure to waves and spray, tidal range, and the frequency of cool days and damp fogs all contribute to establishing its limits.

Conspicuous members of the upper littoral zone are illustrated in figure 16.6. These organisms include several other species of barnacles, limpets, snails, and two other **mollusks:** the bivalved (or two-shelled) mussels and the chitons, which may appear to resemble limpets but on closer inspection will be seen to have shells of eight separate plates. Chitons, like limpets, are grazers that scrape algae from the hard surfaces. Mussels are filter feeders. Food removed from the water is trapped in a heavy mucus and moved to the mouth by liplike palps. The animals are well anchored to the rocks by a variety of mechanisms: a muscular foot in chitons and limpets, strong cement in barnacles, and special threads in mussels. Their profiles tend to be rounded, so as to present little resistance to the breaking waves. Tightly closed shells protect many of these organisms from drying out during periods of low tide. Species of brown algae, typically rockweed (*Fucus*), are in this zone. The algae of the area have strong holdfasts and flexible stipes.

Figure 16.6
*Representative organisms from the midlittoral zone. The mussels
and barnacles filter their food from the water. The chitons graze
on the algae covering the rocks.* Thais, *a snail, is a carnivore.
Small shore crabs and hermit crabs are scavengers.* Balanus
cariosus *is a larger and heavier barnacle than the barnacles of the
supralittoral zone.* Nereis *is often found in the mussel beds. The
anemones huddle together to keep moist when exposed.*

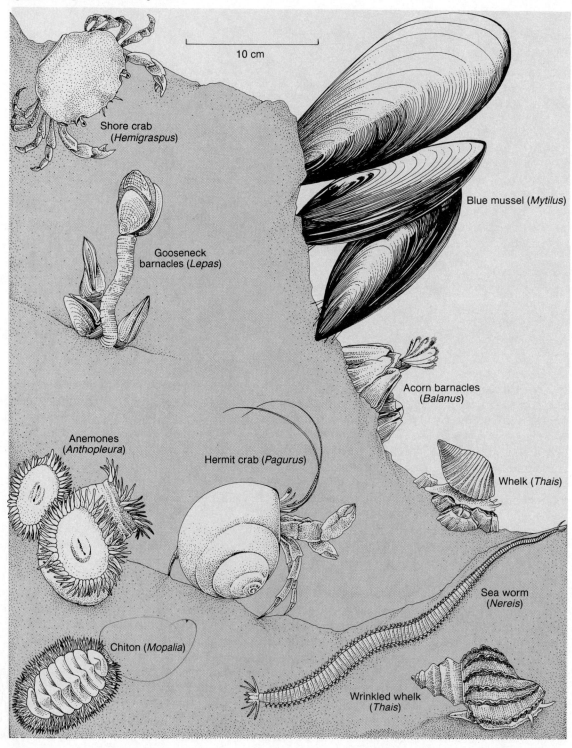

Gooseneck barnacles are found attached to rocks where wave action is strong. They have evolved an interesting feeding style, facing shoreward and feeding by taking particulate matter from the runback of the surf rather than facing the sea, as might be expected.

The mussel beds provide shelter for less conspicuous animals, such as the segmented sea worm, *Nereis,* and small crustaceans. Shore crabs of varied colors and patterns are found in the moist shelter among the rocks. The crabs are active predators as well as important scavengers. Small sea anemones, huddled together in large groups to conserve moisture, are also found in the higher regions of the midlittoral zone.

The area is crowded, the competition for space appears extreme. The free-swimming juveniles (or larval forms) of these animals settle and compete for space, each with its own special set of requirements. New space in an inhabited area becomes available as the whelks, which are carnivorous snails, prey on some species of thinner-shelled barnacles, and the sea stars move up with the tide to feed on the mussels. This predation restricts the thinner-shelled barnacles to the upper levels of the littoral zone, which are too dry for the predator snails. Other species of barnacles inhabit the lower midlittoral zone in association with the predator snails, which cannot pierce the heavier plates of the mature individuals. Seasonal die-offs of algae and the battering action of strong seas and floating logs also act to clear space for newcomers.

A selection of organisms from the lower littoral zone is found in figure 16.7. The larger anemones are common inhabitants of the midlittoral and lower littoral zones. These delicate-looking, flower-like animals attach firmly to the rocks and spread their tentacles, loaded with poisonous darts called nematocysts. The darts are fired when small fish, shrimp, or worms brush the tentacles. The prey is paralyzed, the tentacles grasp, and the prey is pushed down into the anemone's central mouth. A finger inserted into the tentacles of most anemones will feel only a stickiness and a tingling sensation, but some large tropic species can produce a painful injury. Some snails and sea slugs prey on the anemones. They are unaffected by the nematocysts, and certain sea slugs store the anemone's nematocysts in their tissues and use them for their own defense.

Starfish of many colors and sizes make their home in the lower littoral zone. These slow-moving but voracious carnivores prey on shellfish, sea urchins, and limpets. Their mouths are on their undersides, at the center of their central disks, surrounded by strong arms that are equipped with hundreds of tiny suction cups, or tube feet. The tube feet are operated by a water-vascular system, a kind of hydraulic system that attaches the animal very firmly to a hard surface. In feeding, the tube feet attach to the shell of the intended prey and, by a combination of holding and pulling, open the shell sufficiently to insert the starfish's stomach, which can be extruded through its mouth; enzymes are released, and digestion begins. Shellfish sense the approach of a starfish by substances it liberates into the water, and some execute violent escape maneuvers. Scallops swim jerkily; clams and cockles jump away; even the slow-moving sea urchins and limpets move as rapidly as possible. Starfish are so effective as predators that oyster farmers must take care to exclude them from their oyster beds.

The filter-feeding sponges, some flat and some vase-shaped, encrust the rocks. On a minus tide, delicate, free-living flatworms and long **nemerteans,** or ribbon worms, armed with poison-injecting mouth parts, are found keeping moist under the mats of algae. Snails and crabs inhabit this zone; the scallop, another filter-feeding bivalve, and red algae are found as well, along with beds of kelp, eelgrass, and surf grass. Calcareous red algae encrust some rocks and are seen as tufts on others. Occasionally **brachiopods,** or lampshells, are found in the lower littoral zone. They resemble clams but are completely unrelated to them. The shell encloses a coiled ridge of tentacles used in feeding.

Beautiful, graceful, and colorful sea slugs, or **nudibranchs,** are active predators, feeding on sponges, anemones, and the spawn of other organisms. Although soft-bodied, they have few if any enemies, apparently due to the poisonous, acid secretions they produce. Herbivores are also present in the lower intertidal region. Species of chitons and limpets as well as the sea urchin graze on the algae covering the rocks. Sea cucumbers are found wedged in cracks and crevices; some types are identifiable by their brightly colored tentacles, which act as mops to remove food particles from the water and are then thrust into the animal's mouth. Tube worms secrete the leathery or calcareous tubes in which they live and extend only their graceful, feathered tentacles to strain their food from the water.

The octopus is an interesting but shy member of the benthos. Octopuses are seen occasionally from shore on very low tides. These eight-armed carnivorous animals are soft-bodied mollusks. They feed on crabs and shellfish and live in caves or dens identifiable by the piles of waste shells outside. They are known for their ability to flash color changes and move gracefully and swiftly over the bottom and through the water. The world's largest octopus is found in the coastal waters of the eastern North Pacific. It commonly measures 2 to 3 m (10–16.5 ft) in diameter and weighs 20 kg (45 lb) but specimens in excess of 7 m (23 ft) and 45 kg (100 lb) have been observed. They are not aggressive, although they are curious and intelligent; the terrible and dangerous devilfish of adventure stories is only fiction.

Figure 16.7

Lower littoral zone organisms. A variety of related organisms inhabit the area. The starfish feed on the oysters; the related sea urchins are herbivores; and the sea cucumbers feed on detritus suspended in the water. Among the mollusks are the oysters, scallops, snails, abalone, nudibranchs, and octopuses. The oysters and scallops are filter feeders. Calliostoma *and the abalone are grazers; the nudibranchs, the octopus, and the triton snail are predators. Note the size of the anemones in this zone.*

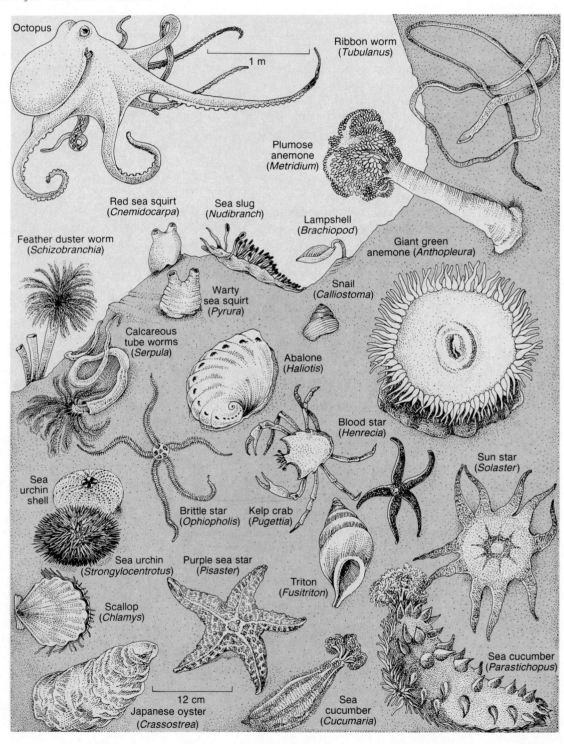

Symbiotic Relationships

Competition for food and space and the predator and prey relationship are common in the intertidal region, but other, cooperative relationships are also present. The beautiful green anemone of the Pacific coast, *Antho-pleura xanthogrammica,* is green because a small, single-celled alga grows in the animal's cells. This mutually beneficial relationship or **mutualism,** in which the alga receives protection as well as carbon dioxide and nitrogenous compounds from the anemone, and the anemone acquires organic compounds and some oxygen manufactured by the alga, is a type of **symbiosis.** A symbiotic relationship is one in which the two dissimilar organisms live in a close, intimate relationship that benefits both.

Another interesting symbiotic relationship is found in tropical waters between the sea anemone and the clown fish. The clown fish acquires protection by nestling among the anemone's stinging tentacles while acting to lure other fish within the anemone's grasp. Shellfish of all types play host to various worms and small crustaceans. Rather than being mutually beneficial, these relationships are often beneficial to one partner and of no harm to the other, a relationship known as **commensalism,** for example, shrimp fish acquire protection by hovering, head down, among the spines of the sea urchin. The small fish *Nomeus* takes shelter within the tentacles of the Portuguese man-of-war.

A relationship in which one partner is harmed by the other is **parasitism,** and parasitic flatworms, roundworms, and bacteria similar to those that infect land organisms are found in the sea. These relationships are not confined to benthic organisms. Cleaner fish feed by removing the parasites and damaged tissues from other fish. Sea mammals are also often heavily parasitized.

Tide Pools

The zonation of the benthic forms varies with local conditions. Zones are generally narrow where the beach is steep or the tidal range is small, while in areas where the beach is flat and the range of tides is large the zones are wide. The orientation of the shore to sunlight and shade, wave action, and beach topography often results in the apparent displacement of benthic zones. A tide pool formed from water left by the receding tide in a rock depression or basin provides a habitat for animals common to the lower littoral zone. However, some isolated tide pools provide a very specialized habitat due to evaporation and the resulting increase in salinity of the remaining water, as well as to the accompanying rise in temperature. On a summer day, the water in a tide pool may feel quite warm to the touch. On the other hand, the tide pool is also a catch basin for rainwater, and so the salinity of the water may be much less than that of the sea, and its temperature may become very cold in the wintertime. Isolated tide pools of this type often support blooms of microscopic algae that give the water the appearance of pea soup, and in various parts of the world, a tiny, bright-red copepod, *Tigripus,* is also found.

The size and depth of a tide pool are also important. The deeper the tide pool and the greater the volume of water, the more stable its environment during its isolation by the receding tide. The larger the tide pool, the more slowly it will change temperature, salinity, pH, and carbon dioxide-oxygen balance. Animals such as starfish, sea urchins, and sea cucumbers cannot survive in pools such as that described in the preceding paragraph. These animals do not tolerate significant changes in their chemical and physical environment; they require large, deep pools. A few fish, such as the small sculpins, can be found in tide pools. These fish are patterned and colored to match the rocks and the algae within the pool. They spend much of their time resting on the bottom, swimming in short spurts from one resting place to another. Each tide pool is unique; it is a specialized environment populated with organisms that are able to survive under the conditions established in that particular pool.

The bottom of the littoral (or intertidal) zone merges into the beginning of the sublittoral (or subtidal) zone extending across the continental shelf. If the shallow areas of the subtidal zone are rocky, many of the same lower littoral zone organisms will be found. When soft sediments begin to collect in protected areas or deeper water, the population types change, and animals of the rocky bottom are replaced by those commonly found on mud and sand substrates.

Animals of the Soft Substrates

Mud, sand, and gravel shores are usually less stable habitats than the rocky areas discussed in the previous section. A pebble or gravel beach does not support much plant or animal life, for the pebbles roll and shift under the waves, scrubbing off any attached growth. Open-coast beaches frequently lose much of their sand under winter storm conditions and regain it with the gentler waves of summer, making these beaches difficult areas for benthic organisms to colonize.

When currents deposit the sands and muds in quiet coves and bays, the energy for movement of the substrate is greatly reduced, and the habitat is mechanically more stable. In these areas, it is the size and shape of the sediment particles and the organic content of the sediment that determine the quality of the environment. The size and shape of the particles, and therefore the size of the spaces between the particles, regulate the flow of water between particles and also the availability of oxygen dissolved in the water. Beach sand is fairly coarse and thus porous to water, both gaining and losing it quickly, while fine particles of mud hold more water and replace the water more slowly. Therefore, sand beaches exchange water, dissolved wastes, and organic particles more quickly than mud flats, which act as traps for organic material. The finer the mud particles, the tighter they pack together and the slower the exchange of water; oxygen is not resupplied quickly nor are wastes removed rapidly. The fragrance of mud flats is due to the decomposition of the organic material, and the odor of hydrogen sulfide may be quite strong. Digging into the mud will generally show a black layer 1 or 2 cm (.5–1 in) below the surface. Above this layer the water between the sediment particles contains dissolved oxygen; below the black layer, organisms (mainly bacteria) function without oxygen, producing hydrogen sulfide, the rotten egg smell. Lack of oxygen may restrict the depth to which infauna species can be found, but some animals, such as clams, are commonly found below the black layer. They use their siphons to obtain food and oxygen from the water above the sediments.

Waves along gravel and sandy shores produce an unstable environment. Few plants can attach, and therefore few grazing animals are found. In locations protected from wave and current energy, eelgrass and surf grass help stabilize the small particle sediments and provide shelter, substrate, and food, creating a special community of plants and animals. Most sand and mud animals are **detritus** feeders, living on bits and pieces of organic material washed in from the sea or off the land. A familiar sand creature, the sand dollar, feeds on detritus particles found between the sand grains. Most detritus is formed from plant material that is degraded by bacteria and fungi. Some animals, such as clams, cockles, and certain polychaete worms, are filter feeders, feeding on the detritus and microscopic organisms suspended in the water. Others are deposit feeders that engulf the sediment and process it in their gut to extract organic matter in a manner similar to that of earthworms. Deposit feeders are usually found in muds or muddy sands with a high organic content; for example, burrowing sea cucumbers and the lugworm, *Arenicola,* which produces the coiled castings seen outside its burrow. Small crustaceans, crabs, and some worm species are scavengers, preying on any available plant or animal material, while still other worms and snails are carnivores. The moon snail is a clam eater that drills a hole in the shell of its prey and then sucks out the flesh.

Bacteria not only play the major role in the decomposition of plant and animal material; they also serve as a major protein source. It is estimated that 25% to 50% of the organic material the bacteria decompose is converted into bacterial cell material, which is consumed by microscopic protozoans, which in turn serve as food for tiny worms, clams, and crustaceans. Areas of mud that are high in organic detritus produce large quantities of bacteria.

The intertidal area of a soft-sediment beach shows some zonation of benthic organisms, but it is not nearly as clear-cut as the zonation along a rocky cliff. In temperate latitudes, small crustaceans called sand hoppers are found at the high intertidal region; they are replaced by ghost crabs in the tropics. Lugworms, mole crabs, and ghost shrimp occupy the midbeach area, while clams, cockles, polychaete worms, and sand dollars are found in the lower intertidal region. The subtidal zone is home to sea cucumbers, sea pens, more crabs and clams, and some species of worms, snails, and sea slugs. The distribution of life in soft sediments is shown in figure 16.8. A selection of animals from this region is found in figure 16.9.

Figure 16.8
Zonation on a soft-sediment beach is less conspicuous than that found on a rocky beach. Animals living at the higher tide levels burrow to stay moist.

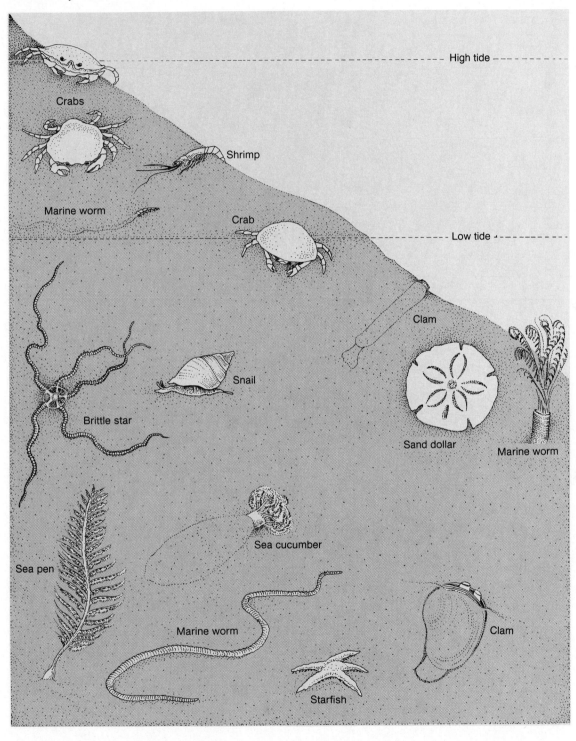

Figure 16.9

Organisms of the soft sediments. Infauna types include Upogebia, Arenicola, *the clam, the cockle, and the burrowing sea cucumber. The sand dollar feeds on detritus; the moon snail drills its way into shellfish; the sea pens feed from the water above the soft bottom. The sea mouse, like* Nereis, *is a polychaete worm.*

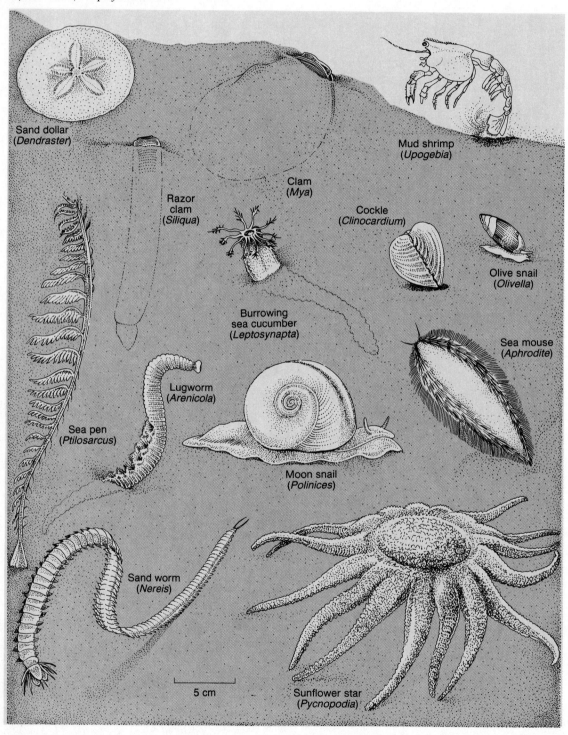

Sand dollar (*Dendraster*)

Mud shrimp (*Upogebia*)

Razor clam (*Siliqua*)

Clam (*Mya*)

Cockle (*Clinocardium*)

Olive snail (*Olivella*)

Burrowing sea cucumber (*Leptosynapta*)

Sea mouse (*Aphrodite*)

Lugworm (*Arenicola*)

Sea pen (*Ptilosarcus*)

Moon snail (*Polinices*)

Sand worm (*Nereis*)

5 cm

Sunflower star (*Pycnopodia*)

The Deep Sea Floor

The deep sea floor includes the flat abyssal plains, the trenches, and the rocky bases of seamounts and mid-ocean ridges. The seafloor sediments are more uniform and their particle size is smaller than those of the shallower regions close to land sources. The environment of the bathyal, abyssal, and hadal zones is uniformly cold and dark.

Many of the animals of the deep sea floor are very like their benthic relatives in the shallow waters of the continental shelf. But in general, it appears that while the population of an organism decreases with depth, the number of species within that population increases. That is to say, diversity of species, which is determined from the ratio of number of species to number of organisms in a sample, increases with depth. Samples show that few species of burrowing copepods are found to be duplicated between samples. This diversity of species between samples is also evident in the protozoan foraminiferans. Thirty species of planktonic foraminiferans are known, whereas one thousand species of benthic foraminiferans have been described.

The stable conditions of the sea floor appear to have favored deposit-feeding infaunal animals of many species. Many members of the deep sea infauna are very small. They are known as the **meiobenthos** and measure 2 mm or less; dominant members include nematode worms, burrowing crustaceans, and segmented worms. At 7000 m (23,000 ft), tusk shells lie buried in the ooze with tentacles at the sediment surface to feed on the foraminiferans. Acorn worms are found frequently in samples taken at 4000 m (13,000 ft). Hagfish burrow into the sediment at the 2000 m (6600 ft) depth. Detritus-eating worms and bivalve mollusks have been found on the sea floor in all the oceans.

Among the epifauna, protozoans are abundant and are widely distributed. Glass sponges attach to the scattered rocks on oceanic ridges and seamounts (see fig. 16.10); so do sea squirts and sea anemones. Their stalks lift them above the soft sediments into the water, where they feed by straining out organic matter. Stalked barnacles attach to the stalks of glass sponges and sea squirts as well as to shells and boulders. Tube worms are common, ranging in size from a few millimeters to 20 cm (8 in), and sea spiders with four pairs of very long legs that span up to 60 cm (27 in) are found at depths to 7000 m (23,000 ft). Snails are found to the greatest of depths; those in the deepest trenches frequently have no eyes or eye-stalks.

The beard worms, or **poganophora,** are found in more productive areas at depths to 10,000 m (33,000 ft). They secrete a close-fitting tube and stand erect, with only their lower portion buried in the sediment. They have no mouth, no gut, and no anus and absorb their

Figure 16.10

(a) Benthic marine life 500 meters (1650 feet) under the surface on the Explorer Sea Mount in the northeast Pacific. Forms shown include vase-shaped glass sponges, corals, a hydroid, and small crabs. (b) This underwater photo, taken at 2300 meters (7500 feet) on the Juan de Fuca Ridge, is typical of this region of the ridge, showing little or no marine life present. (c) When the underwater camera approaches a geothermal vent on the same ridge, there is an increase in benthic marine forms. The marine life shown has not been identified, but researchers believe the reflecting forms to be mussels and the tube worms to be poganophorans.

(a)

(b)

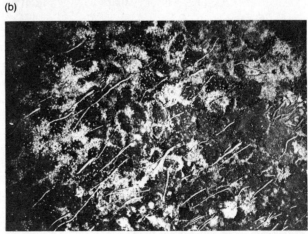

(c)

Figure 16.11

Unprotected wood, such as driftwood, is subject to destruction by marine borers. This beach log has been riddled by shipworm.

food through their skin. The poganophora associated with the vent communities are discussed in a later section of this chapter.

Horny corals, or sea fans, which resemble plants more than animals, grow at depths of 5000 to 6000 m (16–20,000 ft); so do solitary stone corals, which grow larger at these depths than do the coral organisms in the surface waters. Sea lilies, or crinoids, which are related to starfish, are also found at this depth, as are brittle stars and sea cucumbers. Sea cucumbers are found in areas of sediments that are rich in organic substances; they are a dominant and widely spread organism of the deep sea floor.

Bottom photographs and cores show that the deep sediments are continually disturbed and reworked by sea cucumbers and worms in the process of extracting organic matter. This reworking of the sediment destroys layering and results in the uniform sediments that cover vast areas of the ocean floor. This process is called **bioturbation.**

Fouling and Boring Organisms

Organisms that settle and grow on pilings, docks, and boat hulls are said to foul these surfaces. Fouling organisms include barnacles, anemones, tube worms, sea squirts, and algae. When these organisms grow on a vessel's hull, they slow its movement through the water and add to the costs in shipping time and haul-out for cleaning. Fouling organisms also make it difficult to operate equipment in the ocean environment. Much research goes into experiments with paints and metal alloys to discover methods of discouraging and controlling fouling organisms.

Other organisms naturally drill or bore their way into the substrate. Sponges bore into scallop and clam shells; snails bore into oysters; some clams bore into rock. Organisms that bore into wood are costly, for they destroy harbor and port structures as well as wooden boat hulls. Two organisms are responsible for most of the wood boring: the shipworm (*Teredo*) is a mollusk, and the gribble is a small crustacean. *Teredo* is a worm-like mollusk with one end covered by a bivalved shell. These mollusks secrete enzymes that break down and at least partially digest the wood fibers, which are scraped away by the shell's rocking and turning movement. This motion allows the animal to bore deep and extremely destructive holes all through a piece of wood. The gribble gnaws more superficial and smaller burrows, but it, too, is very destructive. Figure 16.11 shows the destructive effects of these organisms.

16.3 Classification Summary of the Benthos

Benthic organisms include members of the following kingdoms and phyla.

I. **Kingdom Monera:** includes the bacteria, an important food source in mud and sand environments; also blue-green algae, especially abundant on coral reefs.
II. **Kingdom Protista:** convenience grouping of microscopic, usually unicellular plants and animals.
  A. *Phylum Chrysophyta:* golden-brown algae; mainly planktonic, but benthic diatoms are abundant.
  B. *Phylum Protozoa:* infauna include many members among sand and mud particles, including amoeboid and foraminiferan species.

III. **Kingdom Plantae:** plants; primarily nonmotile, multicellular, photosynthetic autotrophs.
   A. *Phylum Chlorophyta:* green algae; seaweeds such as *Ulva* and *Codium.*
   B. *Phylum Phaeophyta:* brown algae; coastal seaweeds, including the large kelps *Nereocystis, Macrocystis,* and *Postelsia. Sargassum* maintains a planktonic habit in the Sargasso Sea.
   C. *Phylum Rhodophyta:* red algae; littoral and sublittoral seaweeds of varied form.
   D. *Phylum Tracheophyta:* plants with vascular tissue; true roots, stems, and leaves present.
      1. *Class Angiospermae:* flowering plants; seeds enclosed in fruit; eelgrass, surfgrass, and mangrove trees.

IV. **Kingdom Animalia:** animals; multicellular heterotrophs with specialized cells, tissues, and organ systems.
   A. *Phylum Porifera:* sponges; simple, nonmotile filter feeders.
   B. *Phylum Coelenterata or Cnidaria:* radially symmetric, with tentacles and stinging cells.
      1. *Class Anthozoa:* sea anemones, corals, and sea fans.
   C. *Phylum Platyhelminthes:* flatworms; free-living and parasitic forms.
   D. *Phylum Nemertina or Nemertea:* ribbon worms.
   E. *Phylum Aschelminthes:* round worms, or nematodes.
   F. *Phylum Brachiopoda:* lampshells.
   G. *Phylum Annelida:* segmented worms.
      1. *Class Polychaete:* free-living marine carnivores, including tube worms, *Nereis,* and clam worms.
   H. *Phylum Pogonophora:* deep-sea tube-dwelling worms; giant members found around hot vents on the sea floor.
   I. *Phylum Mollusca:* mollusks.
      1. *Class Polyplacophora:* chitons.
      2. *Class Scaphopoda:* tusk shells.
      3. *Class Gastropoda:* single-shelled mollusks; snails, limpets, and sea slugs.
      4. *Class Pelecypoda or Bivalvia:* bivalved mollusks; clams, mussels, scallops, and oysters.
      5. *Class Cephalopoda:* octopus. The squid is a member of the nekton.
   J. *Phylum Arthropoda:* animals with paired, jointed appendages and outer skeletons.
      1. *Class Crustacea:* crabs, shrimp, and barnacles.
   K. *Phylum Echinodermata:* all marine, radially symmetric, spiny-skinned animals with water-vascular system, including starfish, sea urchins, sand dollars, and sea cucumbers.
   L. *Phylum Hemichordata:* acorn worms.
   M. *Phylum Chordata:* animals, including vertebrates, with dorsal nerve cord and gill slits at some stage in development.
      1. *Subphylum Urochordata:* filter-feeding, saclike adults with "tadpole" larvae; sea squirt.

16.4 The Tropical Coral Reefs

Coral reefs are of interest as the most luxuriant and complex of all benthic communities (see fig. 16.12). Corals are animals; the corals of the curio shop and jeweler are the calcareous skeletons of these animals. Individual coral animals are called **polyps,** and large numbers of these polyps grow together in colonies of delicately branched forms or rounded masses. The coral polyp is very similar to the sea anemone. Like the anemone the polyp is sessile and has tentacles and stinging cells; unlike the anemone, it extracts calcium carbonate from the water and forms a calcareous skeletal cup, in which it sits and to which it adds as it grows.

Although various corals are found in shallow and deep water and in temperate and tropic climates, reef-building corals have specialized requirements restricting their range. These reef-building corals require warm, clear, shallow, clean water and a firm substrate to which they can attach. The water temperature must not go below 18° C; optimum is 23° to 25° C. This temperature requirement restricts their growth to tropical waters between 30°N and 30°S and away from cold-water currents. Waters at depths greater than about 50 m (165 ft) are also too cold for reef-building corals. Most reefs are found in the South Pacific and Indian oceans. The largest coral reef in the world stretches more than 2000 km (1220 mi), from New Guinea along the east coast of Australia. It is known as the Great Barrier Reef.

Clear, shallow water is required by the reef-building coral, because within the tissues of the polyps are masses of single-celled dinoflagellate algae called **zooxanthellae.** The zooxanthellae require light for photosynthesis and must stay in the photic zone. This is a symbiotic relationship, similar to that described for the green anemone. The coral provides the algal cells with a protected environment, carbon dioxide, and nitrate and phosphate nutrients. The algal cells return oxygen and remove waste. Experiments have shown that the zooxanthellae supply the corals with substantial amounts of the organic products of photosynthesis. In addition, the algae enhance the ability of the coral to extract the calcium carbonate from the sea that is required to produce its skeleton. The degree of interdependence probably varies from species to species. The polyps feed actively at night, extending their tentacles to feed on zooplankton. During the day, their tentacles are contracted to expose the outer layer of cells containing zooxanthellae to the sunlight.

The corals require a firm base to which they can cement their skeletons. The classic reef types of the tropic sea—fringing reefs and barrier reefs—are attached to existing islands or landmasses. Atolls are attached to submerged seamounts. (For a review of reef formation around a seamount, see chapter 2.) Delicate, finely branched corals inhabit the more protected areas

Figure 16.12
A coral reef is a complex, interdependent, but self-contained community. Members include corals, clams, sponges, sea urchins, anemones, tube worms, algae, and fish.

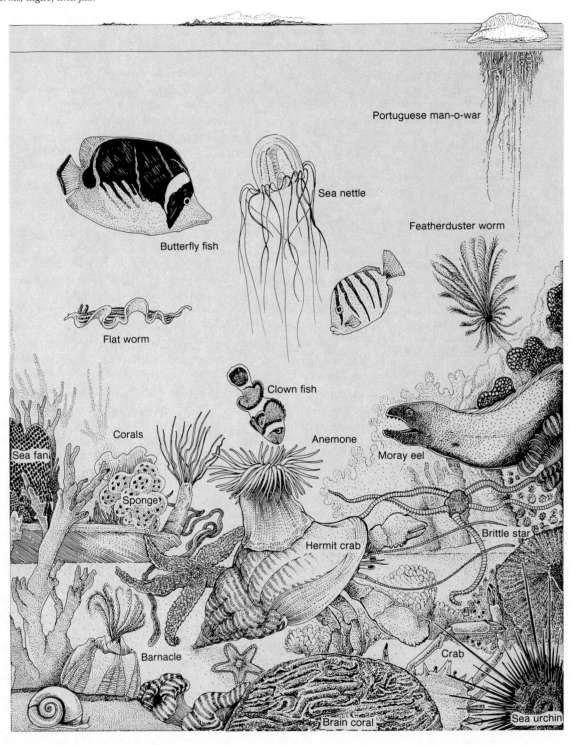

Portuguese man-o-war

Sea nettle

Featherduster worm

Butterfly fish

Flat worm

Clown fish

Corals

Anemone

Moray eel

Sea fan

Sponge

Hermit crab

Brittle star

Barnacle

Crab

Brain coral

Sea urchin

of the reef, while the more massive rounded (or brain) corals are found in regions of heavy surf. The same coral may be found in different shapes and sizes, depending on the depth and the wave action of an area. Corals are slow-growing organisms. Some species grow less than 1 cm (.39 in) in a year; others add between 1 and 5 cm (.39–2 in) each year.

Coral reefs show a vertical zonation and a profile that is largely the product of wave action and water depth. On the sheltered (or lagoon) side of the reef, the shallow **reef flat** is covered with a large variety of corals and other organisms. Fine coral particles broken off from the reef top produce sand, which fills the sheltered lagoon floor. On the reef's windward side, the reef's highest point, or **reef crest,** is exposed at low tide and is pounded by the breaking waves of the surf zone. Below the low tide line to a depth of 10 to 20 m (35–65 ft) on the seaward side is a zone of steep, rugged buttresses, which alternate with grooves in the reef face. Masses of large corals grow here, and many large fish frequent the area. The buttresses dissipate the wave energy, and the grooves drain off fine sands and debris, which would smother the coral colonies. At depths of 20 to 30 m (65–100 ft), there is little wave energy, and the light intensity is only about 25% of its surface value, although it is adequate to support reef algae and corals. The corals are less massive at this depth, and more delicately branched forms are found here. Between 30 and 40 m (100–130 ft), the slope is gentle and the level of light is very reduced; sediments accumulate at this depth, and the coral growth becomes patchy. Below 50 m (165 ft) the slope drops off sharply into the deep water. The reef exists as a balance between the growth of the organisms on the reef surface, as they build on top of old, dead, calcareous remains, and the wearing away of the reef by mechanical and biological forces.

Coral reefs are complex assemblages of many different types of plants and animals. Competition for space and food is intense. Algae, sponges, and corals are constantly growing over and killing each other. Some species are active only at night: some fishes, snails, shrimp, the octopus, fireworm, and moray eel. During the day, other species depend on color and vision to make their way. It has been estimated that as many as 3000 animal species may live together on a single reef. The giant clam, *Tridacna,* is among the most conspicuous. These clams measure up to a meter in length and weigh over 150 kg (330 lb). They also possess zooxanthellae in large numbers in the colorful tissues that line the edges of the shell. Crabs, moray eels, colorful reef fish, poisonous stonefish, long-spined sea urchins, sea horses, shrimp, lobsters, sponges, and many more organisms are all found living here together. The total reef community can be considered to be an autotrophic association based on the sun's energy as transformed by its photosynthetic members. On some reefs, the zooxanthellae have been shown to produce several times more organic material per unit of space than the phytoplankton. This is probably due to the rapid recycling of nutrients between the corals and the zooxanthellae.

The reefs are not formed exclusively from the calcium carbonate skeletons of the coral. Encrusting algae that produce an outer calcareous covering contribute; so do the minute shells of foraminifera, the shells of bivalves, the calcareous tubes of polychaete worms, and the spines and plates of sea urchins. All are compressed and cemented together to form new places for more organisms to live. At the same time, some sponges, worms, and clams bore into the reef; some fish graze on the coral and the algae, and the sea cucumbers feed on the broken fragments, reducing them to sandy sediments. Periodically, a dramatic population increase of the sea star *Acanthaster,* or crown of thorns, which feeds on the coral polyps occurs. No one knows why the *Acanthaster* population increases so rapidly. Some evidence points to an increase in rainy weather (low salinity) and runoff (increased nutrients) increasing the survival of crown of thorns larvae. Others are concerned that humans are in some way responsible, for example harvesting the large conchs that prey on the starfish. However, the reefs are resilient and are able to regenerate in a few years. The opening up of areas on the reef by the starfish may even allow slower-growing species to expand.

The greatest threat to the reefs at the present time is human. Coral reefs are mined for building materials, despoiled by shell collectors, and poisoned by the wastes of advancing industrial and recreational complexes. If pollutants upset the reef's chemical balance, the results can be disastrous. Too little oxygen and the animals die; too many nitrates and the algae may overgrow and smother the corals. If the grazers decline, the predators decline, upsetting the balance once again. To ensure the continued existence of these beautiful and productive areas requires both an increased understanding of the complex nature of reef communities and the development of policies designed to protect them from human interference.

16.5 *High Energy Environments*

Recent work has shown that intertidal communities that are constantly battered by the waves are more productive than the world's lush, green rain forests. On the average, waves deliver 0.335 watts of energy per square centimeter of coastline, about fifteen times more energy than comes from the sun. Even during calm periods, wave energy is 100% greater than solar energy. The algae have little woody tissues; instead, the kelps of the rocky intertidal have 2.5 times more photosynthetic area per square meter of growing surface and are from two to ten times more productive than rain forest vegetation. The mussels, which are consumers, have been found to match or exceed the rain forest productivity when growing in areas of high wave action.

The method of harnessing the energy is indirect. Wave action reduces predators, such as starfish and sea urchins, allowing more mussels and more kelp to live in a unit area. The moving water brings a constant supply of nutrients to the algae and keeps their blades in motion, so they are never in the shade for long. Lastly, the waves can dislodge mussels, allowing more kelp to move in. The waves allow the primary producers to become highly concentrated, and so the consumers can grow and expand their population as well.

16.6 Deep-Ocean Chemosynthetic Communities

Before 1977, deep-sea benthic communities were thought to be made up of small numbers of deposit-feeding animals living on and in the soft sediments. These animals grow slowly in the cold water and depend for food on the slow descent of decayed organic material from the surface layers. Since then, the communities of animals found along the Galapagos Rift and East Pacific Rise between 2500–2600 m (8000–8500 ft), on the Juan de Fuca Ridge at 1570 m (5000 ft), off the west coast of Florida at 3266 m (11,000 ft), and in the central Gulf of Mexico at 700–800 m (2300–2600 ft) have been shown to be completely different (see fig. 16.13). Animals in these areas include filter-feeding clams and mussels, in addition to anemones, worms, barnacles, limpets, crabs, and fish. The clams are very large and show the fastest growth rate of any known deep-sea animal, up to 4 cm (2 in) per year. The tube worms are startling in size, up to 3 m (10 ft) long. These worms belong to the phylum Pogonophora, which is a small and, until now, relatively obscure group related to segmented worms.

Dense clouds of bacteria are the base of the food pyramid for these communities. These bacteria are able to utilize dissolved chemicals in the seawater to obtain the energy they require to live and grow in a process known as **chemosynthesis.** They have no link to the sun through photosynthesis, either directly or indirectly. The bacteria are able to do in the dark what plants do in the sunlight, that is, fix carbon from carbon dioxide into organic molecules such as sugars, which can be used in the metabolism of other organisms. Along the Galapagos Rift and the East Pacific Rise, the bacteria are able to oxidize the sulfides, particularly hydrogen sulfide, that issue from hydrothermal vents. Refer back to

Figure 16.13
Tube worms surround a vent area on the Galapagos Rift. This photograph was made from the submersible Alvin. *The tube worms are approximately 2 meters in length.*

chapter 3 for a discussion of the hydrothermal vent environment. Off Florida and Oregon, communities based on chemosynthesis have been found that are associated with cold seepage areas.

Small animals, perhaps small crustacea, may graze the bacteria directly. Soft-bodied organisms may absorb dissolved organic molecules released by dead bacteria. Other organisms, including the large tube worms, the giant white clam, and the vent mussels have been shown to harbor the bacteria in their tissues in a symbiotic relationship similar to that between dinoflagellates and corals.

The tube worms, which may reach more than a meter in length and several centimeters in diameter, have no mouth and no digestive system; the soft tissue filling their internal body cavity is filled with bacteria. The mussels have only a rudimentary gut, and the clams and mussels have large numbers of bacteria in their gills. How dependent the tube worms, mussels, and clams are on the bacteria is not known; the animals may also absorb organic or particulate material from the water, but what type of particulates or in what quantity is not known. Both the clams and the tube worms have red flesh and red blood. The color is due to the oxygen-binding molecule, hemoglobin. The oxygen is needed for the oxidation of the hydrogen sulfide and to maintain body tissues and high growth rates.

Finding the remains of clam shells at some inactive vent sites, and knowing that it takes the shells approximately fifteen years to dissolve, researchers have estimated that active vents last about twenty years. This has been confirmed by radiometric dating. Therefore, if the colonies are to survive, they must colonize new vents as they develop. The clams have been shown to have high growth rates, becoming mature in four to six years. Their large body size is thought to allow them to produce many larvae as well as harbor large quantities of bacteria. It is estimated that larvae drifting for periods of several weeks to several months could be transported over hundreds of kilometers by the abyssal currents. Some suggest that the larvae might rise to the surface and be dispersed more rapidly by the surface currents before sinking back to the sea floor.

These self-contained communities are among the most profuse and productive in the world. The presence of a plentiful food supply allows their rapid growth, despite the surrounding low temperatures and their remoteness from the photosynthetic layer at the sea's surface. The deep sea floor will never again be considered sparsely populated or inhospitable to life.

16.7 Sampling the Benthos

In the littoral (or intertidal) zone, sampling is done directly during a low tide. Beaches all over the world are sampled by researchers armed with buckets and shovels. To understand the relationships between the organisms and their environment, one needs to know where the plants and animals are found on the beach with respect to the tide levels, or the vertical zonation pattern. In order to establish this pattern, it is necessary to determine the slope of the beach and to mark locations relative to mean sea level; this determination is done by surveying a line directly up the beach from low tide level to a point above high tide. This is known as a **beach transect line.** Trenches paralleling the transect line are dug on sandy beaches to reveal the infauna populations below a measured surface area. On a rocky shore, surface counts of individuals within an area of specific size are made along the transect line. Once the transect line has been established, it is possible to return to the same place, season after season or year after year, to study seasonal changes in the populations or to determine the number of juveniles added yearly. Another type of study compares a natural area to an area stripped of all benthic organisms. The rate of repopulation and the sequence in which the plants and animals return give information on the relationships that exist among species as well as the recovery rates in the event of a catastrophe due to human error or to natural occurrences.

To develop management techniques for the harvesting of benthic organisms such as shellfish, comparative studies are made between natural areas and harvested areas. Surveys determine the average size and age of the shellfish and the density of the populations under both conditions. From such data, researchers determine whether the harvesting is reducing the total population too quickly, whether it is removing too many individuals of spawning age or too many juveniles, and for how long an area should be open to repeated harvests.

Figure 16.14
A biological dredge used to collect epibenthic organisms from rocky substrate.

The classic method of obtaining information on subtidal species is by using a bottom dredge towed by a slowly moving ship. A dredge is a metal frame to which a heavy net bag is attached; it is dragged over the sea floor, scraping up the organisms in its path (see fig. 16.14). This is a strictly qualitative sampling method, as the number and kinds of organisms cannot be accurately related to the actual area sampled. It is possible to attach a measuring wheel to the dredge frame, so that when the dredge is on the bottom, the wheel measures the distance over which it is dragged. Knowing this distance and the width of the dredge, it is possible to compute the approximate size of the area sampled, assuming the dredge did not bounce or skip over a rough bottom.

Soft bottoms can be sampled with a bottom grab or a box corer (refer back to chapter 2). The grab can be designed to penetrate to a specific depth and to take a bite with a specific surface area. The sediment collected is washed through a series of mesh screens of varying sizes, and the organisms are collected and counted. Many samples need to be taken in order to determine the community structure of any single area of the sea floor.

Divers wearing scuba gear can sample and photograph bottom populations directly in depths to approximately 35 m (105 ft). A diver can place a frame of a specific area on the bottom, identify the species within it, and count individuals of each type. If the bottom is not disturbed, it is possible to return to the same plot and check it again. Other methods of observing the bottom without disturbing it include underwater photography and television cameras that are operated remotely from a ship or from a submersible.

Sampling the benthos of the bathyal, abyssal, and hadal zones is difficult and expensive, for it requires large research vessels, with or without submersibles, and extended periods of time at sea. The areas to be sampled are large and remote. Our knowledge of them is still incomplete; for example, the recent discoveries of the vent communities have produced totally unexpected results and have forced a reconsideration of the abundance and composition of deep-sea life.

16.8 *Practical Considerations: Harvesting the Benthos*

The Animals

Benthic animals are a valuable part of the seafood harvest. The crustaceans (crabs, shrimp, prawns, and lobsters) and the mollusks (mainly the bivalved clams, mussels, and oysters) are the most important of these animals as food resources. World aquaculture estimates for 1980 include 3.4 million metric tons of shellfish, and 61,000 metric tons of crustaceans. Because of the demand for shellfish and crustaceans, the catches are much more important in dollar value than the weight of the catches would suggest. In the United States, oysters are harvested in the Gulf of Mexico, southern New England, Chesapeake Bay, and Puget Sound; lobsters are fished in New England; crabs and clams are caught along all our coasts; and shrimp are caught in the Gulf area and off other coasts as well. In many cases, these fisheries make important contributions to the local economy.

Many of the problems of the finfish fisheries are repeated in the benthic fisheries. For example, in the period between 1975 and 1980, the growth of the king crab fishery of the Bering Sea was phenomenal. More and more boats entered the fishery, and in 1979 and 1980, in excess of 45,000 metric tons of crab meat were harvested (see fig. 16.15). In the early 1980s, the harvest dropped to less than 16,000 metric tons. The rapid decrease in catch is apparently due to the classic ills of

Figure 16.15
Alaska king crab being unloaded at dockside.

overfishing and to an insufficient knowledge of the crab's natural history. King crabs migrate across the floor of the Bering Sea, but how many crabs are migrating in any direction and at what stage in their life is at present a frustrating mystery to both those fishing and those engaged in benthic research.

Attempts to increase the harvests of crustaceans and mollusks have focused on aquaculture, or mariculture. Oysters, mussels, and clams are raised on aquaculture farms around the world. In Asia and Europe, raft culture of mollusks is popular. The larvae attach to ropes trailing below the rafts. This attachment keeps the shellfish in the water column, with abundant food and fewer predators. In Spain, raft culture produced over 125,000 metric tons of mussels in 1984. Raft culture in Japan was responsible for the successful harvest of 250,000 metric tons of oysters in 1984. The Japanese culture scallops in hanging net cages, as well as on the bottom, resulting in a scallop industry of 200,000 metric tons a year. Raft culture methods are beginning to be used in the United States (see fig. 16.16), but mussel and oyster farmers suffer from some of the same difficulties encountered by fish farmers, discussed in chapter 15. In Japan, commercial shrimp culture has proved very successful, although the production costs are high. To produce 1 kg (2.2 lbs) of shrimp during the spring-fall growing season requires 10 to 12 kg (22–26.41 lbs) of diatoms, which must be raised to feed the young shrimp. However, the demand for the less than fully grown shrimp exceeds the supply. Shrimp are also being grown in Latin America by U.S. corporations taking advantage of the lower costs in those countries.

Figure 16.16
Experimental raft culture of bay mussels in Puget Sound. Mussels adhere to ropes suspended from floating rafts.

On a model aquaculture farm at Woods Hole Oceanographic Institution, biological oceanographer John Ryther developed a pilot plant in which the sewage effluent from a town of 50,000 contributed its nutrients to produce the algae needed to feed oysters, so as to harvest 800 metric tons of oyster meat. Because the oysters in the model farm produced large quantities of solid waste, marine worms were introduced to feed on the oyster's waste. The worms were later harvested and sold for bait. Another waste product, industrial heat, may also have a use in aquaculture. In Long Island Sound, 30° C water from industrial cooling systems was used experimentally to increase oyster production. The success of these experimental projects has led to the idea of locating an aquaculture project in the 30°N–30°S latitude belt. Here, there is the advantage of warm water and year-round sunlight, which can be used with nutrients produced either from sewage or from an artificial upwelling created by pumping nutrient-rich deep water into the aquaculture pens. Such a project could be conducted in conjunction with ocean thermal energy conversion (OTEC), which is discussed in chapter 6.

Aquaculture projects for benthos in the United States are faced with the same difficulties as those facing fish farms. The high costs, the licensing policies, the need for technology to replace hand labor, and the need for research to improve diet and disease control require considerable attention if we are to increase our seafood harvest in this way. The public must first want or need the products of aquaculture if these obstacles are to be overcome.

The Algae

Seaweeds are gathered from the wild in northern Europe, Japan, China, and Southeast Asia; they are also an important part of the Japanese aquaculture industry. The 1980 world estimate for seaweed harvest from aquaculture is 2.4 million metric tons. Algae are good sources of vitamins and minerals but not of food calories, as most of the cellular material is indigestible. Certain species of green algae, known as sea lettuces, are used in seaweed soups, in salads, and as a flavoring in other dishes. A species of kelp, *Laminaria japonica,* is the **kombu** of Asia. Its blade is used in soups and stews. It is used fresh, dried, pickled, and salted; it is also sweetened and shredded for use in candy and cakes. Another species of *Laminaria* is used in Europe in the same way. Along the northwest coast of the United States and Canada, the stipes of bull kelp are made into pickles. Historically along coastal areas, kelp was used as winter fodder for sheep and cattle and to mulch and fertilize the fields. The red alga *Porphyra* is the **nori** of Japan and the **laver** of the British Isles. It has been cultivated in Japan since 1700 and, in 1984, 400,000 metric tons were harvested. Nori is used in soups and stews and is rolled around portions of rice and fish for flavor. Laver is fried or used in salads.

There are also important industrial uses for some algal products. **Algin** is extracted from brown algae, and **agar** and **carrageenan** are obtained from red algae. Algin derivatives are used as stabilizers in dairy products and candies as well as in paints, inks, and cosmetics. Agar-producing algae are harvested in Japan, Africa, Mexico, and South America. Agar is used as a medium for bacterial culture in laboratories and hospitals, in addition to serving as an ingredient in desserts and in pharmaceutical products. Algae rich in carrageenan are gathered in the wild in New England and northeastern Canada; it is a stabilizer and emulsifier used to prevent separation in ice creams, salad dressings, soups, puddings, cosmetics, and medicines. About 1 million pounds of agar and 10 million pounds of carrageenan are used in the United States each year.

Kelp Bioconversion

Because kelp grows at a great rate and is found over large areas of the world's oceans, some people have proposed that it be cultivated, harvested, and placed in tanks for decomposition, which will produce methane gas for use as fuel and other chemical by-products. The use of a plant crop to harness the sun's energy for conversion to a fuel or energy source is called **bioconversion** (in this case, kelp bioconversion). The energy produced in such a system would be used first to power the harvesting and fuel production process. The remaining fuel either would be made available for other uses at the extraction site or it could be piped to other areas. Suggestions have been made to grow the kelp harvest on offshore raft systems, but no pilot plant of any size has been built, and the dollar and energy costs required are not known.

Drugs from the Benthos

Many benthic organisms contain biologically active compounds that have potential practical use. Extracts of certain sponges yield antibiotic substances; an anticoagulant has been isolated from red algae; and the antibiotic substance acrylic acid has been extracted from other seaweeds. Some corals produce antimicrobial compounds, and the sea anemone, *Anthopleura,* provides a cardiac stimulant. Certain polychaete worms produce a substance that kills some kinds of insects, and extracts of abalone and oyster act as antibacterial agents. A muscle relaxant has been isolated from the snail *Murex.* Even the organic adhesives used by organisms to attach themselves to hard surfaces are under investigation. The strength and properties of these organic substances make them possibilities for closing wounds without stitches. Corals are being investigated to discover why, when alive, they are not encrusted by larvae, algae, and other organisms. Perhaps corals produce substances that will be useful as antifouling compounds for boats, docks, pipes, and underwater equipment. Collecting, extracting, identifying, testing, evaluating, and ultimately synthesizing active substances from the world's benthos is a time-consuming task. However, scientists and pharmaceutical firms are continuing to look at the highly active compounds produced by marine organisms in the hopes that they may serve as models for the eventual development of new drugs.

The benthos is a remarkably diverse grouping of plants and animals. It is easily accessible in the shallow littoral zone, where it has been studied by scientists and students for hundreds of years. Yet we know very little about the benthos over most of the deeper ocean floor. As we are able to see, sample, and understand more about the deep-sea benthos, we may expect to make other discoveries as surprising and as unexpected as the vent communities in the rift areas.

Many of the benthos have additional value to us as food resources, and some may even serve as energy resources. As we consider the uses we make of the oceans at present and the uses for which we need the oceans in the future, we must remember their impact on the benthic environment and its inhabitants. We gain little if we use one resource at the expense of another. The balance within the marine environment is a fragile one, easy to degrade and difficult to reconstruct. The health of the oceans' plants and animals is important to us as a sign of the health of our planet. We are all linked together, and each of us affects the others.

Genetic Manipulation of Fish and Shellfish

Fisheries scientists in North America, Japan, and Northern Europe are using new techniques to manage and enhance the potential of ocean and coastal fisheries and to keep coastal waters and fish farms stocked with species of rapid development and high growth rate. Chromosome manipulation techniques from the research laboratory are being used to produce sterile fish, fish of selected gender, and fish carrying an extra third set of **chromosomes** known as **triploids.**

Triploid salmon and trout are produced by exposing the eggs to temperature, pressure, or chemical shock shortly after fertilization. The shock interferes with the division of the egg nucleus; the egg retains an extra or third set of chromosomes. The third set is retained throughout the fishes' development and inhibits its sexual maturation. Salmon with a normal number of chromosomes grow, mature sexually, spawn, and die, while triploid fish do not mature sexually, continue to eat, and reach a larger size since they are not spending energy on egg production. Triploid coho salmon have been produced in Washington State for experimental net pen culture and release to the sea. The effect of sterility on migration of released triploid coho has yet to be determined. The improved growth and survival of fish after sexual maturity has led to widespread production of farmed triploid trout in the United Kingdom.

This same technique allows hybridization of salmon and trout. While normal chromosome-number hybrids between different species of salmon and trout do not survive, triploid hybrids do. In Idaho, a hybrid between rainbow trout and coho salmon has been shown to resist a common viral disease which causes serious economic loss to trout producers.

The gender of fish can be determined by chromosome manipulation and hormone treatment. Using a technique called gynogenesis, researchers expose fish sperm to ultraviolet light, denaturing its chromosomes. Eggs fertilized with the inactivated-chromosome sperm are briefly chilled, causing the eggs to develop with only the female's chromosomes. The result is all female offspring. Many female food fish grow bigger and live longer than males. Female flounder grow to twice the size of males, and an all-female brood of sturgeon would bring higher profits to caviar producers. Gynogenesis also magnifies genetic effects, making it possible to produce a superior strain in a much shorter period of time than is required for standard hybridization techniques. An alternative to gynogenesis is androgenesis, in which the egg chromosomes are inactivated and the sperm chromosomes are doubled.

It is possible to reverse the sex of female fish produced by gynogenesis by treating them with hormones. This procedure yields fish that are genetically female but functionally male, producing sperm with only female chromosomes. Matings of females with these males will result in all-female fish.

Future research includes plans to extract genes for a specific characteristic from one fish and build it into another. For example, if the genes that cause salmon to return to their river birthplace to spawn could be transferred into species of fish that do not normally have this homing instinct, the new species would also come home after growing wild for several years.

The commercial production of triploid, or "four-seasons," oysters has begun in the United States. Summer mortality and fluctuations in the marketability of oysters have kept the oyster industry seasonal. Normal oysters enter a summer reproductive phase during which their meat is poor in quality and unmarketable. Triploid oysters do not produce sperm and egg but continue to grow steadily, becoming significantly larger than normal oysters and ready for the summer market. The blue mussel is the next mollusk considered ready to be changed to the triploid form.

Fish raised in closed pens and ponds are not considered a threat to natural stock. However, before substantial numbers of sterile fish are released into the natural environment, a number of questions need to be answered: Will the rapid inbreeding possible through gynogenesis produce populations more susceptible to disease or environmental problems? Do the sterile fish grow at rates similar to normal fish? If the fish are not completely sterile, could they migrate upriver and attempt to spawn? If intermixing occurred between the experimental fish and normally spawning fish, could it endanger the spawning population? Although these questions must be answered, chromosome manipulation as a useful tool in fish management, and aquaculture has arrived and is in use.

Summary

Benthic algae are anchored to firm substrates. These algae have a holdfast, a stipe, and photosynthetic blades but no roots, stems, or leaves. Algal growth along a rocky beach ranges from green algae at the surface through brown algae at moderate depths to red algae, which are found primarily below the low tide level. Each group's pigments trap the available sunlight at these depths. Algae are generally classified by their principle pigment. The brown algae include the large kelps. Seaweeds provide food, shelter, and substrate for other organisms in the area. There are also benthic diatoms and a few seed plants, including eelgrass and mangroves.

Benthic animals are subdivided into the epifauna, which live on or attached to the bottom, and the infauna, which live buried in the substrate. Animals that inhabit the rocky littoral region are sorted by the stresses of the area into a series of zones. Organisms that live in the supralittoral (or splash) zone spend long periods of time out of water. The animals of the midlittoral zone experience nearly equal periods of exposure and submergence. These animals have tight shells or live close together to prevent drying out. The area is crowded, and competition for space is great. The lower littoral zone is a less stressful environment. It is home to a wide variety of animals. The organisms of the littoral zone are herbivores and carnivores, and each has its specialized life-style and adaptations for survival.

Symbiotic relationships are intimate, cooperative relationships between two dissimilar organisms. Mutualism, commensalism, and parasitism are types of symbiosis found in the marine environment.

The zonation of the organisms in the benthic region varies with local conditions. Tide pools provide homes for lower littoral zone organisms; they can also become extremely specialized habitats.

Mud, sand, and gravel areas are less stable than rocky areas. The size of the spaces between the substrate particles determines the porosity of the sediments. Some beaches have a higher organic content than others. Few algae can attach to soft sediments, and thus few grazers are found here. Eelgrass and surf grass provide food and shelter for specialized communities. Most organisms that live on soft sediments are detritus feeders or deposit feeders. Zonation patterns are not conspicuous along soft bottoms. Bacteria play an important role in the decomposition of plant material and its reduction to detritus. The bacteria themselves represent a large food resource.

The environment of the deep sea floor is very uniform. The diversity of species increases with depth, but the population density decreases. The microscopic members of the deep-sea infauna are the meiobenthos. Larger burrowers like sea cucumbers continually rework the sediments. The organisms of the epifauna are found at all depths.

Some organisms specialize in attaching to surfaces and others bore into them. Wood-borers are very destructive.

Tropical coral reefs are specialized, self-contained systems. The coral animals require warm, clear, clean, shallow water and a firm substrate. Photosynthetic dinoflagellates called zooxanthellae live in the cells of the corals and the giant clams. The reef exists in a complex but delicate biologic balance, which can be easily upset. Reefs have a typical zonation and structure associated with depth and wave exposure. High energy benthic environments are two to ten times more productive than rain forest vegetation. Self-contained, deep-ocean benthic communities, made up of large, fast-growing animals, depend on chemosynthetic bacteria for the first step in their food chains.

Benthic organisms are sampled by hand in the intertidal zone; deeper samples are obtained with dredges, grabs, and corers. Sampling is both quantitative and qualitative. Divers, cameras, and television can be used to identify and count organisms without disturbing them.

Shellfish are valuable world food resources. Aquaculture can be used to increase the shellfish and shrimp harvests. Experimental aquaculture projects have used nutrients from sewage and heat from industrial resources.

Algae are gathered in many countries and are cultivated in Japan. Some are used directly as food; others are used as stabilizers and emulsifiers in foods and other products. It has been suggested that kelp be cultivated, harvested, and used to produce fuel and other substances. This process is known as kelp bioconversion. Biologically active substances with potentially practical uses have been isolated from benthic organisms.

Key Terms

alga/algae	intertidal zonation	detritus	beach transect line
holdfast	mollusk	meiobenthos	kombu
stipe	nemerteans	poganophora	nori/laver
kelp	brachiopod	bioturbation	algin
blade	nudibranch	polyp	agar
epifauna	mutualism	zooxanthellae	carrageenan
infauna	symbiosis	reef flat	bioconversion
sessile	commensalism	reef crest	chromosome
vertical zonation	parasitism	chemosynthesis	triploid

Study Questions

1. Explain the relationship between hydrogen sulfide gas, bacteria, tube worms, and clams in a hydrothermal vent environment.

2. Distinguish between mutualism, commensalism, and parasitism in marine communities. Give an example of each.

3. In what ways are the benthic algae (seaweeds) adapted for life in the littoral and sublittoral zones? Consider their structure, pigments, and life requirements.

4. In what ways are the benthic algae important in the ocean environment?

5. Discuss the food-gathering strategies of motile and sessile organisms in the littoral and sublittoral zones.

6. Discuss the factors that are responsible for the littoral zonation of marine organisms along a rocky shore.

7. Design an original organism to inhabit the supralittoral, the littoral, or the sublittoral zone. Consider its requirements for food, shelter, and protection from predators, its adaptations to its environment, and its life history.

8. Why are some subtidal forms found in a tide pool high on a rocky beach, while other subtidal forms are not?

9. Why are there few benthic organisms on a beach made up of noncohesive sediments in a wave and surf area?

10. Discuss the importance of bacteria to benthic organisms.

11. Compare a square meter sample of deep-sea benthos with a similar size sample from the rocky intertidal zone.

12. How are coral reefs able to support a rich and varied population, when the water surrounding the reef is clear and devoid of planktonic primary producers?

Appendix

Scientific [or Exponential] Notation

The writing of very large and very small numbers is simplified by using exponents, or powers of 10, to indicate the number of zeroes required to the left or to the right of the decimal point. The numbers that are equal to some of the powers of 10 are as follows:

$$
\begin{aligned}
1,000,000,000. &= 10^9 &&= \text{one billion} \\
1,000,000. &= 10^6 &&= \text{one million} \\
1,000. &= 10^3 &&= \text{one thousand} \\
100. &= 10^2 &&= \text{one hundred} \\
10. &= 10^1 &&= \text{ten} \\
1. &= 10^0 &&= \text{one} \\
0.1 &= 10^{-1} &&= \text{one tenth} \\
0.01 &= 10^{-2} &&= \text{one hundredth} \\
0.001 &= 10^{-3} &&= \text{one thousandth} \\
0.000001 &= 10^{-6} &&= \text{one millionth} \\
0.000000001 &= 10^{-9} &&= \text{one billionth}
\end{aligned}
$$

149,000,000 is rewritten by moving the decimal point eight places to the left and multiplying by the exponential number 10^8, to form 1.49×10^8. In the same way, 605,000 becomes 6.05×10^5.

A very small number such as 0.000032 becomes 3.2×10^{-5} by moving the decimal point five places to the right and multiplying by the exponential number 10^{-5}. Similarly, 0.00000372 becomes 3.72×10^{-6}.

To add or subtract numbers written in exponential notation, convert the numbers to the same power of 10. For example,

$$
\begin{array}{rcccr}
1.49 \times 10^3 & & 1.490 \times 10^3 & & 14.90 \times 10^2 \\
+\ 6.05 \times 10^2 & = & 0.605 \times 10^3 & = & 6.05 \times 10^2 \\
\hline
& & 2.095 \times 10^3 & & 20.95 \times 10^2
\end{array}
$$

$$
\begin{array}{rcccr}
2.36 \times 10^3 & & 23.60 \times 10^2 & & 2.360 \times 10^3 \\
-\ 1.05 \times 10^2 & = & -\ 1.05 \times 10^2 & = & -0.105 \times 10^3 \\
\hline
& & 22.55 \times 10^2 & & 2.255 \times 10^3
\end{array}
$$

To multiply, the exponents are added and the numbers are multiplied:

$$
\begin{array}{r}
4.6 \times 10^3 \\
\times\ 2.2 \times 10^2 \\
\hline
10.12 \times 10^5 = 1.012 \times 10^6
\end{array}
$$

To divide, subtract the exponents and divide the numbers:

$$
\frac{6.0 \times 10^8}{2.5 \times 10^3} = 2.4 \times 10^5
$$

The following prefixes correspond to the powers of 10 and are used in combination with metric units:

Exponential value	Prefix	Symbol
10^{12}	tera	T
10^9	giga	G
10^6	mega	M
10^3	kilo	k
10^2	hecto	h
10^1	deka	da
10^{-1}	deci	d
10^{-2}	centi	c
10^{-3}	milli	m
10^{-6}	micro	μ
10^{-9}	nano	n
10^{-12}	pico	p
10^{-15}	femto	f
10^{-18}	atto	a

SI Units

The *Système international d'unités,* or International System of Units, is a simplified system of metric units (known as SI units) adopted by international convention for scientific use.

Basic SI Units

Quantity	Unit	Symbol
length	meter	m
mass	kilogram	kg
time	second	s
temperature	Kelvin	°K

Derived SI Units

Quantity	Unit	Symbol	Expression in SI base units
area	meter squared	m²	m²
volume	meter cubed	m³	m³
density	kilogram per cubic meter	kg/m³	kg/m³
speed	meter per second	m/s	m/s
acceleration	meter per second per second	m/s²	m/s²
force	newton	N	(kg)(m)/s²
pressure	pascal	Pa	N/m²
energy	joule	J	(N)(m)
power	watt	W	J/s;(N)(m)/s

Length: The Basic SI Unit Is the Meter

Unit	Metric equivalents	English equivalents	Other
meter (m)	100 centimeters 1000 millimeters	39.37 inches 3.281 feet	0.546 fathom
kilometer (km)	1000 meters	0.621 land mile	0.540 nautical mile
centimeter (cm)	10 millimeters 0.01 meter	0.394 inch	
millimeter (mm)	0.1 centimeter 0.001 meter	0.0394 inch	
land mile (mi)	1609 meters	5280 feet	0.869 nautical mile
nautical mile (nm)	1852 meters	1.151 land miles 6076 feet	
fathom (fm)	1.8288 meters	6 feet	

Area: Derived from Length

Unit	Metric equivalents	English equivalents	Other
square meter (m²)	10,000 square centimeters	10.76 square feet	
square kilometer (km²)	1,000,000 square meters	0.386 square land mile	0.292 square nautical mile
square centimeter (cm²)	100 square millimeters	0.151 square inch	

Volume: Derived from Length

Unit	Metric equivalents	English equivalents	Other
cubic meter (m³)	1,000,000 cubic centimeters 1000 liters	35.32 cubic feet 264 U.S. gallons	
cubic kilometer (km³)	1,000,000,000 cubic meters	0.239 cubic land mile	0.157 cubic nautical mile
liter (L)	1000 cubic centimeters	1.06 quarts, 0.264 U.S. gallon	
millimeter (mL)	1.0 cubic centimeter		

Mass: The Basic SI Unit Is the Kilogram

Unit	Metric equivalents	English equivalents
kilogram (kg)	1000 grams	2.205 pounds
gram (g)		0.053 ounce
metric ton, or tonne (t)	1000 kilograms 1,000,000 grams	2205 pounds
U.S. ton	907 kilograms	2000 pounds

Time: The Basic SI Unit Is the Second

Unit	Metric and English equivalents
minute	60 seconds
hour	60 minutes; 3600 seconds
day	86,400 seconds ⎫ 24 hours ⎬ mean solar day
year	31,556,880 seconds ⎫ 8765.8 hours ⎬ mean solar year 365.25 solar days ⎭

Temperature: The Basic SI Unit Is the Kelvin

Reference points	Kelvin (°K)	Celsius (°C)	Farenheit (°F)
absolute zero	0	−273.2	−459.7
seawater freezes	271.2	−2.0	28.4
fresh water freezes	273.2	0.0	32.0
human body	310.2	37.0	98.6
fresh water boils	373.2	100.0	212.0
conversions	$°K = °C + 273.2$	$°C = \dfrac{(°F - 32)}{1.8}$	$°F = (1.8 \times °C) + 32$

Speed (Velocity): The Derived SI Unit Is the Meter per Second

Unit	Metric equivalents	English equivalents	Other
meter per second (m/s)	100 centimeters per second 3.60 kilometers per hour	3.281 feet per second 2.237 land miles per hour	1.944 knots
kilometer per hour (km/hr)	0.277 meter per second	0.909 foot per second	0.55 knot
knot (kt)	0.51 meter per second	1.151 land miles per hour	1 nautical mile per hour

Acceleration: The Derived SI Unit Is the Meter per Second per Second

Unit	Metric equivalents	English equivalents
meter per second per second (m/s²)	12.960 kilometers per hour per hour 1000 centimeters per second per second	3.281 feet per second per second 8053 miles per hour per hour

Force: The Derived SI Unit Is the Newton

Unit	Metric equivalents	English equivalents
newton (N)	100,000 dynes	0.2248 pound
dyne (dyn)	0.00001 newton	0.000002248 pound

Pressure: The Derived SI Unit Is the Pascal

Unit	Metric equivalents	English equivalents	Other
pascal (Pa)	1 newton per square meter 10 dynes per square centimeter		
bar	100,000 pascals 1000 millibars	14.5 pounds per square inch	0.927 atmosphere 29.54 inches of mercury
standard atmosphere (atm)	1.013 bars 101,300 pascals	14.7 pounds per square inch	29.92 inches of mercury

Energy: The Derived SI Unit Is the Joule

Unit	Metric equivalents	English equivalents
joule (J)	1 newton-meter 0.2389 calorie	0.0009481 British thermal unit
calorie (cal)	4.186 joules	0.003968 British thermal unit

Power: The Derived SI Unit Is the Watt

Unit	Metric equivalents	English equivalents
watt (W)	1 joule per second 0.2389 calorie per second 0.001 kilowatt	0.0569 British thermal unit per minute 0.001341 horsepower

Density: The Derived SI Unit Is the Kilogram per Cubic Meter

Unit	Metric equivalents	English equivalents
kilogram per cubic meter	0.001 grams per cubic centimeter (g/cm³)	0.0624 pounds per cubic foot

Credits

CHAPTER 1

Opener, 1.1: courtesy of NASA; **1.7, 1.13:** courtesy The National Maritime Museum, London.

CHAPTER 2

Opener: courtesy Ken Adkins, School of Fisheries, University of Washington; **2.6:** courtesy Official U.S. Navy Photo by R. F. Diel; **2.13:** courtesy of Alyn and Alison Duxbury; **2.14:** courtesy H. Paul Johnson, School of Oceanography, University of Washington; **2.15:** courtesy Dr. Robert Burns, retired NOAA; **2.16:** courtesy Dr. Vigil E. Barnes; **2.17 both:** courtesy Ken Adkins, School of Fisheries, University of Washington; **2.20:** courtesy of Joe Creager and M. G. Gross, "Varied Marine Sediments in a Stagnant Fjord" *Science* 141 Sept. 1963 pp. 918–19; **2.23, 2.24:** courtesy Ken Adkins, School of Fisheries, University of Washington; **2.25A:** courtesy of Alyn and Alison Duxbury; **2.26:** courtesy Exxon Corporation; **2.27:** courtesy Deep Sea Ventures Inc., Glouster Point, VA.

CHAPTER 3

Opener: courtesy of U.S. Geological Survey; **3.11:** courtesy Ocean Drilling Program, Texas A&M University; **3.18B, 3.21:** courtesy of U.S. Geological Survey; **3.26:** © courtesy Woods Hole Oceanographic Institution; **3.27:** © Rod Catanch/Woods Hole Oceanographic Institution; **3.28:** © courtesy Woods Hole Oceanographic Institution.

CHAPTER 4

Opener: © Kent Reno/Jeroboam; **4.9:** courtesy of Kathy Newell, School of Oceanography, University of Washington; **4.10:** courtesy of Ken Adkins, School of Oceanography, University of Washington; **4.12:** courtesy of E.P.C. Laboratories Inc., Danvers, MA; **4.15 all:** courtesy of Arnold M. Hanson;

4.17A: courtesy of Ed. Joshberger, U.S. Geological Survey, **B:** courtesy of Arnold M. Hanson; **4.18:** © Kent Reno/Jeroboam.

CHAPTER 5

Opener: courtesy of Alyn and Alison Duxbury; **5.4:** courtesy Ken Adkins; **5.9:** courtesy of U.S. Geological Survey.

CHAPTER 6

Opener: courtesy William Von Arx, *Introduction to Physical Oceanography,* © 1962. Addison-Wesley, Reading, MA Fig. 316. Reprinted with permission; **6.17:** courtesy School of Oceanography, University of Washington; **6.19:** courtesy of J. R. Postel, School of Oceanography, University of Washington; **6.21, 6.22 both:** courtesy of Naval Oceanographic Manuals; **6.25A:** courtesy of Lockheed Missle and Space Company.

CHAPTER 7

Opener: courtesy Willard Bascom, Long Beach, CA; **7.19 all:** courtesy Aanderaa Instruments, Woburn, Mass.

CHAPTER 8

Opener: © Bob Barbour; **8.1:** courtesy William Von Arx, *Introduction to Physical Oceanography,* © 1962. Addison-Wesley, Reading, MA. Fig. 316. Reprinted with permission; **8.2:** courtesy D. M. Owen; **8.7:** courtesy Leonard Wolhüter, by permission Exxon Corporation; **8.14, 8.17 both:** courtesy of Alyn and Alison Duxbury; **8.18:** © Bob Barbour; **8.19:** © courtesy Willard Bascom, Long Beach, CA.

CHAPTER 9

Opener: © courtesy Dr. Francis P. Shepard, Prof. Emeritus, University of California, Scripps Institution of Oceanography; **9.15 both, 9.16:** courtesy Tourism New Brunswick, Canada Photo; **9.18:** courtesy Nova Scotia Power Corporation.

CHAPTER 10

Opener, 10.1: From K. Reedway Allen, *Conservation and Management of Whales,* Washington Sea Grant Program, 1980; **10.2:** courtesy Austin Post/U.S. Geological Survey; **10.3:** courtesy NASA; **10.4:** courtesy of Alyn and Alison Duxbury; **10.5:** courtesy Sherwood Maynard, University of Hawaii; **10.7A:** courtesy of Alyn and Alison Duxbury; **B:** From K. Reedway Allen, *Conservation and Management of Whales,* Washington Sea Grant Program, 1980; **10.8:** © Alex S. Maclean of Landsides; **10.9:** From K. Reedway Allen, *Conservation and Management of Whales,* Washington Sea Grant Program, 1980; **10.11:** courtesy U.S. Army Corps of Engineers; **10.13, 10.14:** courtesy of Alyn and Alison Duxbury; **10.15:** courtesy of U.S. Army Corps of Engineers; **10.16:** From K. Reedway Allen, *Conservation and*

Management of Whales, Washington Sea Grant Program, 1980; **10.17 both:** courtesy of Dr. Francis P. Shepard, Prof. Emeritus, University of California, Scripps Institution of Oceanography; **10.22:** courtesy U.S. Army Corps of Engineers; **10.23:** courtesy Steve Malone, Santa Barbara News Press; **10.24:** courtesy U.S. Army Corps of Engineers.

CHAPTER 11

Opener: courtesy Sherwood Maynard, University of Hawaii; **11.8:** courtesy of J. Galt, NOAA; **11.9:** courtesy Clean Sound Corporation, Seattle, WA.; **11.10:** courtesy Port of Seattle; **11.11:** courtesy Alex S. Maclean of Landsides.

CHAPTER 12

Opener: © The Field Museum of Natural History, Chicago; **p. 271:** Excerpt from "The Dry Salvages" in FOUR QUARTETS, copyright 1943 by T. S. Eliot; renewed 1971 by Esme Valerie Eliot. Reprinted by permission of Harcourt Brace Jovanovich, Inc. **12.6:** © James Sumich; **12.7:** © The Field Museum of Natural History, Chicago.

CHAPTER 13

Opener: From A. Hardy, *The Open Sea,* Houghton-Mifflin, Boston 1956.

CHAPTER 14

Opener: From A. Hardy, *The Open Sea,* Houghton-Mifflin, Boston 1956; **14.3 all:** courtesy Enge's Equipment Company, Morton Grove, IL.; **14.6:** From A. Hardy, *The Open Sea,* Houghton-Mifflin, Boston 1956; **14.13:** courtesy Washington Sea Grant Program.

CHAPTER 15

Opener, 15.1 both: courtesy Moclips Cetotogical Society; Chuck Flaherty; **15.13:** courtesy Washington Sea Grant Program; **15.14:** courtesy C. Gunnar Safsten National Marine Fisheries Services, NOAA; **15.15:** courtesy Anadronous Inc.

CHAPTER 16

Opener, 16.10 all: courtesy John R. Delaney, School of Oceanography, University of Washington; **16.11:** courtesy of Alyn and Alison Duxbury; **16.13:** © Woods Hole Oceanographic Institution; **16.14:** courtesy School of Oceanography, University of Washington; **16.15:** courtesy Dan C. Schneringer, Schneringer and Associates, Kirkland, Washington; **16.16:** courtesy Dr. Kenneth Chen, School of Fisheries, University of Washington.

COLORPLATES

Plates 1A,E: © courtesy Jet Propulsion Laboratory, **B:** © courtesy of NOAA, **C,F:** © courtesy Inter Network, Inc., **D:** © courtesy United States Navy; **plate 2A:** © W. Haxby Lamont-Doherty Geological

Glossary

absorption taking in of a substance by chemical or molecular means; change of sound or light energy into some other form, usually heat, in passing through a medium or striking a surface.

abyssal pertaining to the great depths of the ocean below approximately 4000 m.

abyssal hill low, rounded submarine hill less than 1000 m high.

abyssal plain flat ocean-basin floor extending seaward from the base of the continental slope and continental rise.

abyssopelagic oceanic zone from 4000 m to the deepest depths.

accretion natural or artificial deposition of sediment along a beach, resulting in the buildup of new land.

adsorption attraction of ions to a solid surface.

advection horizontal or vertical transport of seawater, as by a current.

agar substance produced by red algae; the gelatinlike product of these algae.

algae marine and freshwater plants (including most seaweeds) that are single-celled, colonial, or multicelled, with chlorophyll but no true roots, stems, or leaves and with no flowers or seeds.

algin complex organic substance found in or obtained from brown algae.

alluvial plain flat deposit of terrestrial sediment eroded by water from higher elevations.

amphidromic point point from which cotidal lines radiate on a chart; the nodal, or low-amplitude, point for a rotary tide.

amplitude for a wave, the vertical distance from sea level to crest or from sea level to trough, or one-half the wave height.

anaerobic living or functioning in the absence of oxygen.

andesite type of volcanic rock that is intermediate in composition between basalt and granite, associated with subduction zones.

anion negatively charged ion.

anoxic deficient in oxygen.

Antarctic Circle *see* Arctic and Antarctic Circles.

antinode portion of a standing wave with maximum vertical motion.

aphotic zone that part of the ocean in which light is insufficient to carry on photosynthesis.

aquaculture cultivation of aquatic organisms under controlled conditions. *See also* Mariculture.

Arctic and Antarctic Circles latitudes 66 ½°N and 66 ½°S, respectively, marking the boundaries of light and darkness during the summer and winter solstices.

asthenosphere upper, deformable portion of the earth's mantle, the layer below the lithosphere; probably partially molten; may be site of convection cells.

atmospheric pressure pressure exerted by the atmosphere as a consequence of gravitational force exerted upon the column of air lying directly above any point on earth.

atoll ring-shaped coral reef that encloses a lagoon in which there is no exposed preexisting land and which is surrounded by the open sea.

attenuation decrease in the energy of a wave or beam of particles occurring as the distance from the source increases, caused by absorption, scattering, and divergence from a point source.

autotrophic pertaining to organisms able to manufacture their own food from inorganic substances. *See also* Chemosynthesis; Photosynthesis.

autumnal equinox *See* Equinoxes.

auxospore naked cell of a diatom, which grows to full size and forms a new siliceous covering.

backshore beach zone lying between the foreshore and the coast, acted upon by waves only during severe storms and exceptionally high water.

baleen whalebone; horny material growing down from the upper jaw of plankton-feeding whales; forms a strainer, or filtering organ, consisting of numerous plates with fringed edges.

bar offshore ridge or mound of sand, gravel, or other loose material, which is submerged, at least at high tide; located especially at the mouth of a river or estuary, or lying a short distance from and parallel to the beach.

barrier island deposit of sand, parallel to shore and raised above sea level; may support vegetation and animal life.

barrier reef coral reef that parallels land but is some distance offshore, with water between reef and land.

basalt fine-grained, dark igneous rock, rich in iron, magnesium, and calcium; characteristic of oceanic crust.

basin large depression of the sea floor having about equal dimensions of length and width.

bathyal pertaining to ocean depths between approximately 1000 m and 4000 m.

bathymetry study and mapping of seafloor elevations and the variations of water depth; the topography of the sea floor.

bathypelagic oceanic zone from 1000 m to 4000 m.

bathythermograph (BT)   instrument used to determine water temperature as a function of depth (pressure).

beach   zone of unconsolidated material between the mean low water line and the line of permanent vegetation, which is also the effective limit of storm waves; sometimes includes the material moving in offshore, onshore, and longshore transport.

beach face   section of the foreshore normally exposed to the action of waves.

Beaufort scale   scale of wind forces by range of velocity; scale of sea state created by winds of these velocities.

benthic   of the sea floor, or pertaining to organisms living on or in the sea floor.

benthos   organisms living on or in the ocean bottom.

berm   nearly horizontal portion of a beach (backshore) with an abrupt face; formed from the deposition of material by wave action at high tide.

berm crest   ridge marking the seaward limit of a berm.

bilaterally symmetric   having right and left halves that are approximate mirror images of each other.

bioconversion   conversion of a plant crop storing the sun's energy to a fuel or energy source.

biogenous sediment   sediment having more than 30% material derived from organisms.

bioluminescence   production of light by living organisms as a result of a chemical reaction either within certain cells or organs or outside the cells in some form of excretion.

biomass   amount of living matter, expressed in weight units, per unit of water surface or volume.

bioturbation   reworking of sediments by organisms that burrow and ingest them.

blade   flat, photosynthetic, "leafy" portion of an alga or seaweed.

bloom   high concentration of phytoplankton in an area, caused by increased reproduction; often produces discoloration of the water. *See also* Red tide.

bore   *see* Tidal bore.

breaker   sea surface-water wave that has become too steep to be stable and collapses.

breakwater   structure protecting a shore area, harbor, anchorage, or basin from waves; a type of jetty.

buffer   substance able to neutralize acids and bases, therefore able to maintain a stable pH.

bulkhead   structure separating land and water areas, primarily designed to resist earth sliding and slumping or reduce wave erosion at the base of a cliff.

buoy   floating object anchored to the bottom or attached to another object; used as a navigational aid or surface marker.

buoyancy   ability of an object to float due to the support of the fluid the body is in or on.

calcareous   containing or composed of calcium carbonate.

calorie   amount of heat required to raise the temperature of 1 gram of water 1° C.

calving   breaking away of a mass of ice from its parent glacier, iceberg, or sea-ice formation.

capillary waves   waves with wavelengths less than 1.5 cm in which the primary restoring force is surface tension.

carnivore   flesh-eating organism.

carrageenan   substance produced by certain algae that acts as a thickening agent.

cation   positively charged ion.

cat's-paw   patch of ripples on the water's surface, related to a discrete gust of wind.

centrifugal force   outward-directed force acting on a body moving along a curved path or rotating about an axis.

chaetognaths   free-swimming, carnivorous, pelagic, wormlike, planktonic animals; arrow worms.

chemosynthesis   formation of organic compounds with energy derived from inorganic substances such as ammonia, sulfur, and hydrogen.

chloride   atom of chlorine in solution, forming an ion with a negative charge.

chlorinity ($Cl^0/_{00}$)   measure of the chloride content of seawater in grams per kilogram.

chlorophyll   group of green pigments that are active in photosynthesis.

chromosome   one of the bodies in a cell that carry the genes in a linear order.

chronometer   portable clock of great accuracy used in determining longitude at sea.

ciguatera   toxin found in fish of tropical regions.

cilia   microscopic, hairlike processes of living cells, which beat in coordinated fashion and produce movement.

coast   strip of land of indefinite width that extends from the shore inland to the first major change in terrain that is unaffected by marine processes.

coastal circulation cell (drift sector, littoral cell)   longshore transport cell pattern of sediment moving from a source to a place of deposition.

coccolithophore   microscopic, planktonic alga surrounded by a cell wall with embedded calcareous plates (coccoliths).

cohesion   molecular force between particles within a substance that acts to unite them.

colonial organism   organism consisting of semi-independent parts that do not exist as separate units; groups of organisms with specialized functions that form a coordinated unit.

commensalism   an intimate association between different organisms in which one is benefited and the other is neither harmed nor benefited.

compensation depth   depth at which there is a balance between the oxygen produced by algae through photosynthesis and that consumed by them through respiration; net oxygen production is zero.

condensation   process by which a vapor becomes a liquid or a solid.

conduction   transfer of heat energy through matter by internal molecular motion; also heat transfer by turbulence in fluids.

conservative constituent   component or property of seawater whose value changes only as a result of mixing, diffusion, and advection and not as a result of biological chemical processes; for example, salinity.

consumer animal that feeds on plants (primary consumer) or on other animals (secondary consumer).

continental crust crust forming the continental land blocks; mainly granite and its derivatives.

continental margin zone separating the continents from the deep-sea bottom, usually subdivided into shelf, slope, and rise.

continental rise gentle slope formed by the deposition of sediments at the base of a continental slope.

continental shelf zone bordering a continent, extending from the line of permanent immersion to the depth at which there is a marked or rather steep descent to the great depths.

continental shelf break zone along which there is a marked increase of slope at the outer margin of a continental shelf.

continental slope relatively steep downward slope from the continental shelf break to depth.

contour line on a chart or graph connecting the points of equal value for elevation, temperature, salinity, and so on.

convection transmission of heat by the movement of a heated gas or liquid; vertical circulation resulting from changes in density of a fluid.

convergence situation in which fluids of different origins come together, usually resulting in the sinking, or downwelling, of surface water and the rising of air.

copepod small, shrimplike members of the zooplankton.

coral colonial animal that secretes a hard, outer, calcareous skeleton; the skeletons of coral animals form in part the framework for warm-water reefs.

corange lines in a rotary tide, lines of equal tidal range about the amphidromic point.

core vertical, cylindrical sample of bottom sediments, from which the nature of the bottom can be determined; also the central zone of the earth, thought to be liquid or molten on the outside and solid on the inside.

corer device that plunges a hollow tube into bottom sediments to extract a vertical sample.

Coriolis effect apparent force acting on a body in motion, due to the rotation of the earth, causing deflection to the right in the northern hemisphere and to the left in the southern hemisphere; the force is proportional to the speed and latitude of the moving body.

cosmogenous sediment sediment particles with an origin in outer space; for example, meteor fragments and cosmic dust.

cotidal lines lines on a chart marking the location of the tide crest at stated time intervals.

covalent bond chemical bond formed by the sharing of one or more pairs of electrons.

cratons large pieces of the earth's crust that form the centers of continents.

crest *see* Wave crest.

crust outer shell of the solid earth; the lower limit is usually considered to be the Mohorovičić discontinuity.

crustacean member of a class of primarily aquatic organisms with paired jointed appendages and a hard outer skeleton, including lobsters, crabs, shrimp, and copepods.

ctenophores transparent, planktonic animal, spherical or cylindrical in shape with rows of cilia; comb jellies.

current horizontal movement of water.

current meter instrument for measuring the speed and direction of a current.

deadweight ton (DWT) capacity of a vessel in tons of cargo, fuel, stores, and so on; determined by the weight of the water displaced.

declinational tide *see* Diurnal tide.

decomposer heterotrophic; microorganisms (usually bacteria and fungi) that break down nonliving organic matter and release nutrients, which are then available for reuse by autotrophs.

deep exceptionally deep area of the ocean floor, usually below 6000 m.

deep scattering layer (DSL) layer of organisms that move away from the surface during the day and toward the surface at night; the layer scatters or returns vertically directed sound pulses.

deep-sea reversing thermometer (DSRT) mercury-in-glass thermometer that records seawater temperature upon being inverted and retains its reading until returned to its upright position.

deep water wave wave in water, whose depth is greater than one-half its wavelength.

delta area of unconsolidated sediment deposit, usually triangular in outline, formed at the mouth of a river.

demersal fish fish living near and on the bottom.

density property of a substance defined as mass per unit volume and usually expressed in grams per cubic centimeter or kilograms per cubic meter.

depth recorder *see* Echo sounder.

desalination process of obtaining fresh water from seawater.

detritus any loose material, especially decomposed, broken, and dead organic materials.

dew point temperature to which air must be cooled for its water vapor to condense.

diatom microscopic alga with an external skeleton of silica.

diatomaceous ooze sediment made up of more than 30% skeletal remains of diatoms.

diffraction process that transmits energy laterally along a wave crest.

diffusion movement of a substance from a region of higher concentration to a region of lower concentration (movement along a concentration gradient); may be due to molecular motion or turbulence.

dinoflagellate one of a class of planktonic organisms possessing characteristics of both plants and animals.

dispersion (sorting) sorting of waves as they move out from a storm center; occurs because long waves travel faster in deep water than short waves.

diurnal inequality   difference in height between the two high waters or two low waters of each tidal day; the difference in speed between the two flood currents or two ebb currents of each tidal day.

diurnal tide (declinational tide)   tide with one high water and one low water each tidal day.

divergence   horizontal flow of fluids away from a common center, associated with upwelling in water and descending motions in air.

doldrums   nautical term for the belt of light, variable winds near the equator.

downwelling   sinking of water toward the bottom, usually the result of a surface convergence or an increase in density of water at the sea surface.

dredge   cylindrical or boxlike sampling device made of metal, net, or both, which is dragged across the bottom to obtain biological or geological samples.

drift bottle   bottle released into the sea for use in studying currents; contains a card identifying date and place of release and requesting the finder to return it with date and place of recovery.

drift sector   *see* Coastal circulation cell.

dugong   *see* Sea cow.

dune   wind-formed hill or ridge of sand.

dynamic equilibrium   state in which the sum of all changes is balanced and there is no net change.

earth sphere depth   uniform depth of the earth below the present mean sea level, if the solid earth surface were smoothed off evenly, 2440 m.

ebb current   movement of a tidal current away from shore or down a tidal stream as the tide level decreases.

ebb tide   falling tide; the period of the tide between high water and the next low water.

echolocation   use of sound waves by some marine animals to locate and identify underwater objects.

echo sounder (depth recorder)   instrument used to measure the depth of water by measuring the time interval between the release of a sound pulse and the return of its echo from the bottom. *See also* Precision depth recorder (PDR).

ecosystem   the organisms in a community and the nonliving environment with which they interact.

eddy   circular movement of water.

Ekman spiral   in a theoretical ocean of infinite depth, unlimited extent, and uniform viscosity, with a steady wind blowing over the surface, the surface water moves 45° to the right of the wind in the northern hemisphere. At greater depths the water moves farther to the right with decreased speed, until at some depth (approximately 100 m) the water moves opposite to the wind direction. Net water transport is 90° to the right of the wind in the northern hemisphere. Movement is to the left in the southern hemisphere.

electrodialysis   separation process in which electrodes of opposite charge are placed on each side of a membrane to accelerate the diffusion of particles across the membrane.

electromagnetic radiation   waves of energy formed by simultaneous electric and magnetic oscillations; the electromagnetic spectrum is the continuum of all electromagnetic radiation from low-energy radiowaves to high-energy gamma rays, including visible light.

El Niño   wind-driven reversal of the Pacific equatorial currents resulting in the movement of warm water toward the coasts of the Americas, so called because it generally develops just after Christmas.

entrainment   mixing of salt water into fresh water overlying salt water, as in an estuary.

epifauna   animals living attached to the sea bottom or moving freely over it.

epipelagic   upper portion of the oceanic pelagic zone, extending from the surface to about 200 m.

episodic wave   abnormally high wave unrelated to local storm conditions.

equator   0° latitude, determined by a plane that is perpendicular to the earth's axis and is everywhere equidistant from the north and south poles.

equilibrium tide   theoretical tide formed by the tide-producing forces of the moon and sun on a nonrotating, water-covered earth.

equinoxes   times of the year when the sun stands directly above the equator, so that day and night are of equal length around the world. The vernal equinox occurs about March 21, and the autumnal equinox occurs about September 22.

escarpment   nearly continuous line of cliffs or steep slopes caused by erosion or faulting.

estuary   semi-isolated portion of the ocean, which is diluted by freshwater drainage from land.

euphausiid   planktonic, shrimplike crustacean. *See also* Krill.

evaporation   process by which a liquid becomes a vapor.

evaporite   deposit formed from minerals left behind by evaporating water, especially salt.

fast ice   sea ice that is anchored to shore or the sea floor in shallow water.

fathom   a unit of length equal to 1.8 m or 6 ft; used to measure water depth.

fault   break or fracture in the earth's crust, in which one side has been displaced relative to the other.

fetch   continuous area of water over which the wind blows in essentially a constant direction.

filament   chain of living cells.

fjord   narrow, deep, steep-walled inlet of the ocean formed by the submergence of a mountainous coast or by the entrance of the ocean into a deeply excavated glacial trough after the melting of the glacier.

flagellum   long, whiplike extension from a living cell's surface that by its motion moves the cell.

floe   discrete patch of sea ice moved by the currents or by the wind.

flood current   movement of a tidal current toward the shore or up a tidal stream as the tide level increases.

flood tide   rising tide; the period of the tide between low water and the next high water.

flushing time   length of time required for an estuary to exchange its water with the open ocean.

fog   visible assemblage of tiny droplets of water formed by condensation of water vapor in the air; a cloud with its base at the surface of the earth.

food chain   sequence of organisms in which each is food for the next member in the sequence. *See also* Food web.

food web   complex of interacting food chains; all the feeding relations of a community taken together; includes production, consumption, decomposition, and the flow of energy.

foraminifera   minute, one-celled animals that usually secrete calcareous shells.

foraminiferal ooze   sediment made up of 30% or more skeletal remains of foraminifera.

forced wave   wave generated by a continuously acting force and caused to move at a speed faster than it freely travels.

foreshore   portion of the shore that includes the low-tide terrace and the beach face.

fouling   attachment or growth of marine organisms on underwater objects, usually objects that are made or introduced by humans.

fracture zone   large, linear zone of irregular bathymetry of the sea floor, characterized by asymmetrical ridges and troughs; commonly associated with fault zones.

free wave   wave that continues to move at its natural speed after its generation by a force.

friction   resistance of a surface to the motion of a body moving (e.g., sliding or rolling) along that surface.

fringing reef   reef attached directly to the shore of an island or a continent and not separated from it by a lagoon.

frustule   siliceous external shell of a diatom.

generating force   disturbing force that creates a wave, such as wind or a landslide.

geostrophic flow   horizontal flow of water occurring when there is a balance between gravitational forces and the Coriolis effect.

glacier   mass of land ice, formed by the recrystallization of compacted old snow, flowing slowly from an accumulation area to an area of ice loss by melting, sublimation, or calving.

graben   a portion of the earth's crust that has moved downward and is bounded by steep faults.

grab sampler   instrument used to remove a piece of the ocean floor for study.

granite   crystalline, coarse-grained, igneous rock composed mainly of quartz and feldspar.

gravitational force   mutual force of attraction between particles of matter (bodies).

gravity wave   water wave form in which gravity acts as the restoring force; waves with wavelengths greater than 2 cm.

groin   protective structure for the shore, usually built perpendicular to the shoreline; used to trap littoral drift or retard erosion of the shore; a type of jetty.

group speed   speed at which a group of waves travels (in deep water, group speed equals one-half the speed of an individual wave); the speed at which the wave energy is propagated.

guyot   submerged, flat-topped seamount.

gyre   circular movement of water, larger than an eddy; usually applied to a larger system.

habitat   place where a plant or animal species naturally lives and grows.

hadal   pertaining to the greatest depths of the ocean.

half-life   time required for half of an initial quantity of a radioactive isotope to decay.

halocline   water layer with a large change in salinity with depth.

harmonic analysis   process of separating astronomical tide-causing effects from the tide record, in order to predict the tides at any location.

heat budget   accounting for the total amount of the sun's heat received on earth during one year as being exactly equal to the total amount lost from the earth due to radiation and reflection.

heat capacity   quantity of heat needed to raise the temperature of a unit mass (1 g) of a substance by 1° C.

height, of wave   *see* Wave height.

herbivore   animal that feeds only on plants.

heterotrophic   pertaining to organisms requiring preformed organic compounds for food; unable to manufacture food from inorganic compounds.

higher high water   higher of the two high waters of any tidal day in a region of mixed tides.

higher low water   higher of the two low waters of any tidal day in a region of mixed tides.

high water   maximum height reached by a rising tide.

holdfast   organ of a benthic alga that attaches the alga to the sea floor.

holoplankton   organisms living their entire life cycle in the floating (planktonic) state.

hook   spit turned landward at its outer end.

horse latitudes   regions of calms and variable winds that coincide with latitudes at approximately 30° to 35°N and S.

hot spot   surface expression of a persistent rising jet of molten mantle material.

hurricane   severe, cyclonic, tropic storm at sea, with winds of 120 km/hr (73 mph) or more; generally applied to Atlantic Ocean storms. *See also* Typhoon.

hydrocast   process of obtaining water samples at predetermined depths.

hydrogen bond   in water, the weak attraction between the positively polar hydrogen of one water molecule and the negatively polar oxygen of another water molecule.

hydrogenous sediment   sediment formed from substances dissolved in seawater.

hydrologic cycle   movement of water among the land, oceans, and atmosphere due to changes of state, vertical and horizontal transport, evaporation, and precipitation.

hydrothermal vent   seafloor outlet for high-temperature groundwater and associated minerals; a hot spring.

hydrowire   measured cable to which oceanographic instruments are attached to be lowered into the sea.

hypsographic curve   graph of land elevation and ocean depth versus area.

iceberg   mass of land ice that has broken away from a glacier and floats in the sea.

igneous rock   rock formed by congealing rapidly or slowly from molten magma.

infauna   animals that live buried in the sediment.

internal wave   wave created below the sea surface at the boundary between two density layers.

intertidal   *see* Littoral.

intertidal volume   in an embayment, the volume of water gained or lost due to the rise and fall of the tide.

ion   positively or negatively charged atom or group of atoms.

island arc   chain of volcanic islands formed when plates converge at a subduction zone.

isobath   contour of constant depth.

isohaline   having a uniform salt content.

isopycnal   having a uniform density.

isostasy   mechanism by which areas of the earth's crust rise or subside until their masses are in balance, "floating" on the mantle. It is theorized that the continents and mountains are supported by low-density crustal "roots."

isothermal   having a uniform temperature.

isotope   atoms of the same element having different numbers of neutrons.

jellyfish   semitransparent, bell-shaped pelagic organism, often with long tentacles bearing stinging cells.

jetty   structure located to influence currents or protect the entrance to a harbor or river from waves (U.S. terminology). *See also* Breakwater; Groin.

kelp   any of several large, brown algae, including the largest known algae.

kinetic energy   energy produced by the motion of an object.

knot   a unit of speed equal to 0.51 m/sec or 1 nautical mile per hour.

krill   term used by whalers for the small, shrimplike crustaceans found in huge masses in polar waters and eaten by baleen whales.

lag deposits   large particles left on a beach after the small particles are washed away.

lagoon   shallow body of water, which usually has a shallow, restricted outlet to the sea.

larva   immature juvenile form of an animal.

latent heat of fusion   amount of heat required to change the state of 1g of water from ice to liquid.

latent heat of vaporization   amount of heat required to change the state of 1g of water from liquid to gas.

lava   magma, or molten rock, that has reached the earth's surface; the same material solidified after cooling.

lee   shelter; the part or side sheltered from wind or waves.

lithogenous sediment   sediment composed of rock particles eroded mainly from the continents by water, wind, and waves.

lithosphere   outer, rigid portion of the earth; includes the continental and oceanic crust and the upper part of the mantle.

littoral (intertidal)   area of the shore between mean high water and mean low water; the intertidal zone.

littoral cell   *see* Coastal circulation cell.

littoral drift   *see* Longshore transport.

longshore current   current produced in the surf zone by waves breaking at an angle with the shore; runs roughly parallel to the shoreline.

longshore transport (littoral drift)   movement of sediment by the longshore current.

loran   navigational system in which position is determined by measuring the difference in the time of reception of synchronized radio signals; derived from the phrase "long-range navigation."

lower high water   lower of the two high waters of any tidal day in a region of mixed tides.

lower low water   lower of the two low waters of any tidal day in a region of mixed tides.

low tide terrace   flat section of the foreshore seaward of the sloping beach face.

low water   minimum height reached by a falling tide.

lunar month   time required for the moon to pass from one new moon to another new moon (approximately twenty-nine days).

magma   molten rock material that forms igneous rocks upon cooling. Magma that reaches the earth's surface is referred to as lava.

magnetic pole   either of the two points on the earth's surface where the magnetic field is vertical.

manatee   *see* Sea cow.

manganese nodules   rounded, layered lumps found on the deep-ocean floor that contain as much as 20% manganese and smaller amounts of iron, nickel, and copper; a hydrogenous sediment.

mantle   main bulk of the earth between the crust and the core; increasing pressure and temperature with depth divide the mantle into concentric layers.

mariculture   cultivation of marine organisms under controlled conditions. *See also* Aquaculture.

maximum sustained yield   maximum number or amount of a species that can be harvested each year without steady depletion of the stock; the remaining stock is able to replace the harvested members by natural reproduction.

meander   turn or winding curve of a current that may detach from the main stream as an eddy.

mean sea level   average height of the sea surface, based on observations of all stages of the tide over a nineteen-year period in the United States.

mean solar day   *see* Solar day.

mechanical bathythermograph (BT)   *see* Bathythermograph (BT).

meiobenthos   very small animals living buried in the sediments of the sea floor.

meridian   circle of longitude passing through the poles and any given point on the earth's surface.

meroplankton   floating developmental stages (eggs and larvae) of organisms that as adults belong to the nekton and benthos.

mesopelagic   oceanic zone from 200 m to 1000 m.

messenger   weight, usually hinged and with a latch so that it can be fastened around a wire, used to activate oceanographic instruments after they have been lowered to the desired depth.

microplankton   net plankton, composed of individuals below 1 mm in size but large enough to be retained by a small mesh net.

mitigation   coastal management concept requiring developers to replace developed areas with equivalent natural areas or to reengineer other areas to resemble areas prior to development.

mixed tide   type of tide in which large inequalities between the two high waters and the two low waters occur in a tidal day.

Mohorovičić discontinuity (Moho)   boundary between crust and mantle, marked by a rapid increase in seismic wave speed.

mollusks   marine animals, usually with shells; includes mussels, oysters, clams, snails, and slugs.

monoculture   cultivation of only one species of organism in an aquaculture system.

monsoon   name for seasonal winds; first applied to the winds over the Arabian Sea, which blow for six months from the northeast and for six months from the southwest; now extended to similar winds in other parts of the world. In India, the term is popularly applied to the southwest monsoon and also to the rains that it brings.

moon tide   portion of the tide generated solely by the moon's tide-raising force, as distinguished from that of the sun.

moraine   glacial deposit of rock, gravel, and other sediment left at the margin of an ice sheet.

mutualism   an intimate association between different organisms in which both organisms benefit.

nannoplankton   plankton smaller than 10 μ, which will pass through an ordinary plankton net but can be removed from the water by centrifuging water samples.

nautical mile   unit of length equal to 1852 m or 1.15 land miles.

neap tides   tides occurring near the times of the first and last quarters of the moon, when the range of the tide is least.

nekton   pelagic animals that are active swimmers; for example, adult squid, fish, and marine mammals.

nephelometer   device that measures the scattering of light by particulate material in the sea.

neritic   shallow water marine environment extending from low water to the edge of the continental shelf. *See also* Pelagic.

net plankton   *see* Microplankton.

node   point of least or zero vertical motion in a standing wave.

nonconservative constituent   component or property of seawater whose value changes as a result of biological or chemical processes as well as by mixing, advection, and diffusion; for example, nutrients and oxygen in seawater.

nudibranchs   soft bodied, gastropod mollusks; sea slugs.

nutrient   in the ocean, any one of a number of inorganic or organic compounds or ions used primarily in the nutrition of primary producers; nitrogen and phosphorus compounds are examples.

oceanic   pertaining to the ocean water seaward of the continental shelf; the "open ocean." *See also* Pelagic.

oceanic crust   crust below the deep ocean sediments; mainly basalt.

ocean ranching   raising of salmon to a juvenile stage, release of the juveniles to sea, and harvest of adults on their return.

offshore   direction seaward of the shore.

offshore current   any current flowing away from the shore.

offshore transport   movement of sediment or water away from the shore.

onshore   direction toward the shore.

onshore current   any current flowing toward the shore.

onshore transport   movement of sediment or water toward the shore.

ooze   fine-grained deep ocean sediment composed of at least 30% sand or silt-sized calcareous or siliceous remains of small marine organisms, the remainder being clay-sized material.

orbit   in water waves, the path followed by the water particles affected by the wave motion; also, the path of a body subjected to the gravitational force of another body, such as the earth's orbit around the sun.

orographic effect   precipitation patterns caused by the flow of air over and around mountains.

osmosis   tendency of water to diffuse through a semipermeable membrane to make the concentration of water on one side of the membrane equal to that on the other side.

osmotic pressure   pressure that builds up in a confined fluid because of osmosis.

overturn   sinking of more dense water and its replacement by less dense water from below.

oxygen minimum   zone in which respiration and decay reduce dissolved oxygen to a minimum, usually between 800 and 1000 m.

parallel   circle on the surface of the earth parallel to the plane of the equator and connecting all points of equal latitude; a line of latitude.

parasitism   an intimate association between different organisms in which one is benefited and the other is harmed.

pelagic   primary division of the sea, which includes the whole mass of water subdivided into neritic and oceanic zones; also pertaining to the open sea.

period   *see* Tidal period; Wave period.

pH   measure of the concentration of hydrogen ions in a solution; the concentration of hydrogen ions determines the acidity of the solution;
$pH = -\log_{10}(H^+)$; where H^+ is the concentration of hydrogen ions in gram atoms per liter.

photic zone   layer of a body of water that receives ample sunlight for photosynthesis; usually less than 100 m.

photo cell   a device that converts light energy to electrical energy; used to determine solar radiation below the sea surface.

photophore   luminous organ found on fish.

photosynthesis manufacture by plants of organic substances and release of oxygen from carbon dioxide and water in the presence of sunlight and the green pigment chlorophyll.

physiographic portrayal of the earth's features by perspective drawing.

phytoplankton microscopic plant forms of plankton.

pinniped member of the marine mammal group, characterized by four swimming flippers; for example, seals and sea lions.

plankton passively drifting or weakly swimming organisms.

plate tectonics theory and study of the earth's lithospheric plates, their formation, movement, interaction, and destruction; the attempts to explain the earth's crustal changes in terms of plate movements.

polar easterlies winds blowing from the poles toward approximately 60°N and S; winds are northeasterly in the northern hemisphere and southeasterly in the southern hemisphere.

polychaetes marine segmented worms, some in tubes, some free-swimming.

polyculture cultivation of more than one species of organism in an aquaculture system.

polyp sessile stage in the life history of certain members of the phylum coelenterata (Cnidaria); sea anemones and corals.

potential energy energy that an object has because of its position or condition.

precipitation falling products of condensation in the atmosphere, such as rain, snow, or hail; also the falling out of a substance from solution.

precision depth recorder (PDR) instrument used to obtain a continuous pictorial record of the ocean bottom by timing the returning echoes of sound pulses. *See also* Echo sounder.

primary coast coastline shaped primarily by land forces rather than sea forces.

primary production (primary productivity) amount of organic material synthesized by organisms from inorganic substances in unit time in a unit volume of water or in a column of water of unit area cross section and extending from the surface to the bottom.

prime meridian meridian of 0° longitude, used as the origin for measurements of longitude; internationally accepted as the meridian of the Royal Naval Observatory, Greenwich, England.

productivity amount of organic material synthesized by organisms in unit time in a unit volume of water. *See also* Primary production (primary productivity).

progressive tide tide wave that moves, or progresses, in a nearly constant direction.

progressive wave wave that moves, or progresses, in a certain direction.

protozoa minute, mostly one-celled animals.

pseudopodium flowing, temporary extension of the protoplasm of a cell, used in locomotion or feeding.

pteropod pelagic snail whose foot is modified for swimming.

pteropod ooze sediment made up of more than 30% shells of pteropods.

P-waves primary, or faster, waves moving away from a seismic event; can penetrate solid rock; consist of energy transmitted by alternate compressions and dilations of the material. *See also* S-waves.

pycnocline water layer with a large change in density with depth.

radar system of determining and displaying the distance of an object by measuring the time interval between transmission of a radio signal and reception of the echo return; derived from the phrase "Radio detecting and ranging."

radially symmetric having similar parts regularly arranged around a central axis.

radiation energy transmitted as rays or waves without the need of a substance to conduct the energy.

radiolarians single-celled protozoans with siliceous skeletons.

radiolarian ooze sediment made up of more than 30% skeletal remains of radiolarians.

radiometric dating determining ages of geological samples by measuring the relative abundance of radioactive isotopes and comparing isotope systems.

rafting transport of sediment, rocks, silt, and other land matter out to sea by ice, logs, and the like, with the deposition of the rafted material when the carrying agent disintegrates.

range *see* Tidal range.

red clay red to brown fine-grained lithogenous deposit, of predominantly clay size, which is derived from land, transported by winds and currents, and deposited far from land and at great depth. Also known as brown mud and brown clay.

red tide red coloration, usually of coastal waters, caused by large quantities of microscopic organisms (generally dinoflagellates); some red tides result in mass fish kills, others contaminate shellfish, and still others produce no toxic effects.

reef offshore hazard to navigation made up of consolidated rock, with a depth of 20 m or less.

reef crest highest portion of a coral reef on the exposed seaward edge of the reef.

reef flat portion of a coral reef landward of the reef crest and seaward of the lagoon.

reflection rebounding of light, heat, sound, waves, and so on, after striking a surface.

refraction change in direction or bending of a wave.

relict sediments sediments deposited by processes no longer active.

residence time mean time that a substance remains in a given area before replacement, calculated by dividing the amount of a substance by its rate of addition or subtraction.

respiration metabolic process by which food or food-storage molecules yield the energy on which all living cells depend.

restoring force force that returns a disturbed water surface to the equilibrium level, such as surface tension and gravity.

ridge   long, narrow elevation of the sea floor, with steep sides and irregular topography.

rift valley   trough formed by faulting along a zone in which plates move apart and new crust is created, such as along the crest of a ridge system.

rip current   strong surface current flowing seaward from shore; the return movement of water piled up on the shore by incoming waves and wind.

rise   long, broad elevation that rises gently and generally smoothly from the sea floor.

rotary current   tidal current that continually changes its direction of flow through all points of the compass during a tidal period.

rotary tide   tide that is the result of a standing wave moving around the central node of a basin.

salinity   measure of the quantity of dissolved salts in seawater. It is formally defined as the total amount of dissolved solids in seawater in parts per thousand ($^o/_{oo}$) by weight when all the carbonate has been converted to oxide, all the bromide and iodide have been converted to chloride, and all organic matter is completely oxidized.

salinity bridge (salinometer)   instrument for determining the salinity of water by measuring the electrical conductivity of a water sample of a known temperature.

salt budget   balance between the rates of salt supply in and removal from a body of water.

salt wedge   intrusion of salt water along the bottom; in an estuary, the wedge moves upstream on high tide and seaward on low tide.

sand spit   *see* Spit.

satellite   body that revolves around a planet; a moon; a device launched from earth into orbit around a planet of the sun.

scarp   elongated and comparatively steep slope separating flat or gently sloping areas on the sea floor or on a beach.

scattering   random redirection of light or sound energy by reflection from an uneven sea bottom or sea surface, from water molecules, or from particles suspended in the water.

sea   same as the ocean; subdivision of the ocean; surface waves generated or sustained by the wind within their fetch, as opposed to swell.

sea cow (dugong, manatee)   large herbivorous marine mammal of tropic waters; includes the manatee and dugong.

seafloor spreading   movement of crustal plates away from the midocean ridges; process that creates new crustal material at the midocean ridges.

sea level   height of the sea surface above or below some reference level. *See also* Mean sea level.

seamount   isolated volcanic peak that rises at least 1000 m from the sea floor.

sea smoke   type of fog caused by dry, cold air moving over warm water.

sea stack   isolated mass of rock rising from the sea near a headland from which it has been separated by erosion.

sea state   numerical or written description of the roughness of the ocean surface relative to wave height.

Secchi disk   white disk used to measure the transparency of the water by observing the depth at which the disk disappears from view.

secondary coast   coastline shaped primarily by marine forces or marine organisms.

sediment   particulate organic and inorganic matter that accumulates in loose, unconsolidated form.

seiche   standing wave oscillation of an enclosed or semienclosed body of water that continues, pendulum fashion, after the generating force ceases.

seismic   pertaining to or caused by earthquakes or earth movements.

seismic sea wave   *see* Tsunami.

seismic tomography   using seismic data to produce computerized, detailed, three-dimensional maps of the boundaries between the earth's layers.

semidiurnal tide   tide with two high waters and two low waters each tidal day.

semipermeable membrane   membrane that allows some substances to pass through it but restricts or prevents the passage of other substances.

sessile   permanently fixed or sedentary; not free-moving.

set   direction in which the current flows.

shallow water wave   wave in water whose depth is less than one-twentieth the average wavelength.

shingles   flat, water-worn pebbles or cobbles found in beds along a beach.

shoal   elevation of the sea bottom comprising any material except rock or coral (in which case it is a reef) and which may endanger surface navigation.

shore   strip of ground bordering any body of water, which is alternately exposed and covered by tides and waves.

sidereal day   time period determined by one rotation of the earth relative to a far distant star, about four minutes shorter than the mean solar day.

sigma-t   abbreviated value of the density of seawater neglecting pressure and at a given temperature and salinity; $\sigma_t = (\text{density} - 1) \times 1000$.

siliceous   containing silica.

sill   shallow area that separates two basins from one another or a coastal bay from the adjacent ocean.

slack water   state of a tidal current when its velocity is near zero; occurs when the tidal current changes direction.

slick   area of smooth surface water.

sofar channel   natural sound channel in the oceans, in which sound can be transmitted for very long distances; the depth of minimum sound velocity; derived from the phrase "sound fixing and ranging."

solar constant   rate at which solar radiation is received on a unit surface that is perpendicular to the direction of incident radiation just outside the earth's atmosphere at the earth's mean distance from the sun; equal to 2 cal/cm²/min.

solar day   time period determined by one rotation of the earth relative to the sun; the mean solar day is twenty-four hours.

solstices   times of the year when the sun stands directly above 23 ½°N or S latitude. The winter solstice occurs about December 22, and the summer solstice occurs about June 22.

sonar   method or equipment for determining, by underwater sound, the presence, location, or nature of objects in the sea; derived from the phrase "sound navigation and ranging."

sorting   *see* Dispersion.

sounding   measurement of the depth of water beneath a vessel.

sound shadow   area of the ocean into which sound does not penetrate because the density structure of the water refracts the sound waves.

specific gravity   ratio of the density of a substance to the density of 4° C water.

specific heat   ratio of the heat capacity of a substance to the heat capacity of water.

sphere depth   thickness of a material spread uniformly over a smooth sphere having the same area as the earth.

spit (sand spit)   low tongue of land, or a relatively long, narrow shoal extending from the shore.

spoil   dredged material.

spore   minute, unicellular, asexual reproductive structure of an alga.

spreading center   region along which new crustal material is produced.

spring tides   tides occurring near the times of the new and full moon, when the range of the tide is greatest.

standing crop   biomass of a population present at any given time.

standing wave   type of wave in which the surface of the water oscillates vertically between fixed points called nodes, without progression; the points of maximum vertical rise and fall are called antinodes.

steepness, of wave   *see* Wave steepness.

stipe   portion of an alga between the holdfast and the blade.

storm center   area of origin for surface waves generated by the wind; an intense atmospheric low pressure system.

storm tide (storm surge)   along a coast, the exceptionally high water accompanying a storm, owing to wind stress and low atmospheric pressure, made even higher when associated with a high tide and shallow depths.

subduction zone   plane descending away from a trench and defined by its seismic activity, interpreted as the convergence zone between a sinking plate and an overriding plate.

sublimation   transition of a substance from its solid state to its gaseous state without becoming a liquid.

sublittoral (subtidal)   benthic zone from the low tide line to the seaward edge of the continental shelf.

submarine canyon   relatively narrow, V-shaped, deep depression with steep slopes, the bottom of which grades continuously downward across the continental slope.

submersible   a research submarine, designed for manned or remote operation at great depths.

subsidence   sinking of a broad area of the crust without appreciable deformation.

substrate   material making up the base on which an organism lives or to which it is attached.

subtidal   *see* Sublittoral.

summer solstice   *see* Solstices.

sun tide   portion of the tide generated solely by the sun's tide-raising force, as distinguished from that of the moon.

supralittoral   benthic zone above the high tide level that is moistened by waves, spray, and extremely high tides.

surf   wave activity in the area between the shoreline and the outermost limit of the breakers.

surface tension   tendency of a liquid surface to contract owing to bonding forces between molecules.

surimi   refined fish protein used to form artificial crab, shrimp, and scallop meat.

swash zone   beach area where water from a breaking wave rushes.

S-waves   secondary, or slower, transverse seismic waves; cannot penetrate a liquid; consist of elastic vibrations perpendicular to the direction of travel. *See also* P-waves.

swell   long and relatively uniform wind-generated ocean waves that have traveled out of their generating area.

symbiosis   living together in intimate association of two dissimilar organisms.

tectonic   pertaining to processes that cause large-scale deformation and movement of the earth's crust.

tektites   particles with a characteristic round shape, derived from cosmic material.

terranes   fragments of the earth's crust bounded by faults, each fragment with a history distinct from each other fragment.

terrigenous   of the land; sediments composed predominantly of material derived from the land.

thermocline   water layer with a large change in temperature with depth.

thermohaline circulation   vertical circulation caused by changes in density; driven by variations in temperature and salinity.

tidal bore   high tide crest that advances rapidly up an estuary or river as a breaking wave.

tidal current   alternating horizontal movement of water associated with the rise and fall of the tide.

tidal datum   reference level from which ocean depths and tide heights are measured; the zero tide level.

tidal day   time interval between two successive passes of the moon over a meridian, approximately 24 hours and 50 minutes.

tidal period   elapsed time between successive high waters or successive low waters.

tidal range   difference in height between consecutive high and low waters.

tide   periodic rising and falling of the sea surface that results from the gravitational attractions of the moon and sun acting on the rotating earth.

tide wave   long-period gravity wave that has its origin in the tide-producing force and is observed as the rise and fall of the tide.

tombolo   deposit of unconsolidated material that connects an island to another island or to the mainland.

topography   general elevation pattern of the land surface (or the ocean bottom). *See also* Bathymetry.

toxicant   substance dissolved in water that produces a harmful effect on organisms, either by an immediate large dose or by small doses over a period of time.

trade winds   wind systems occupying most of the tropics, which blow from approximately 30°N and S toward the equator; winds are northeasterly in the northern hemisphere and southeasterly in the southern hemisphere.

transform fault   fault with horizontal displacement connecting the ends of an offset in a midocean ridge. Some plates slide past each other along a transform fault.

transverse ridge   ridge running at nearly right angles to the main or principal ridge.

trench   long, deep, and narrow, depression of the sea floor with relatively steep sides, associated with a subduction zone.

triploid   condition in which cells have three sets of chromosomes.

trophic   relating to nutrition; a trophic level is the position of an organism in a food chain or food (trophic) pyramid.

Tropics of Cancer and Capricorn   latitudes 23 ½°N and 23 ½°S, respectively, marking the maximum angular distance of the sun from the equator during the summer and winter solstices.

trough   long depression of the sea floor, having relatively gentle sides; normally wider and shallower than a trench. *See also* Wave trough.

T-S diagram   graph of temperature versus salinity, on which seawater samples taken at various depths are used to describe a water mass.

tsunami (seismic sea wave)   long-period sea wave produced by a submarine earthquake or volcanic eruption. It may travel across the ocean for thousands of miles unnoticed from its point of origin and build up to great heights over shallow water at the shore.

tube worm   any worm or wormlike organism that builds a tube or sheath attached to a submerged substrate.

turbidite   sediment deposited by a turbidity current, showing a pattern of coarse particles at the bottom, grading gradually upward to fine silt.

turbidity   loss of water clarity or transparency owing to the presence of suspended material.

turbidity current   dense, sediment-laden current flowing downward along an underwater slope.

typhoon   severe, cyclonic, tropic storm originating in the western Pacific Ocean, particularly in the vicinity of the South China Sea. *See also* Hurricane.

ultraplankton   plankton that are smaller than nannoplankton; they are difficult to separate from water.

upwelling   rising of water rich in nutrients toward the surface, usually the result of diverging surface currents.

vernal equinox   *see* Equinoxes.

viscosity   property of a fluid to resist flow; internal friction of a fluid.

water bottle   device used to obtain a water sample at depth.

water budget   balance between the rates of water added and lost in an area.

water mass   body of water identified by similar patterns of temperature and salinity from surface to depth.

water type   body of water identified by a specific range of temperature and salinity from a common source.

wave   periodic disturbance that moves through or over the surface of a medium with a speed determined by the properties of the medium.

wave crest   highest part of a wave.

wave height   vertical distance between a wave crest and the adjacent trough.

wavelength   horizontal distance between two successive wave crests or two successive wave troughs.

wave period   time required for two successive wave crests or troughs to pass a fixed point.

wave ray   line indicating the direction waves travel; drawn at right angles to the wave crests.

wave steepness   ratio of wave height to wavelength.

wave train   series of similar waves from the same direction.

wave trough   lowest part of a wave.

westerlies   wind systems blowing from the west between latitudes of approximately 30° and 60°N and 30° and 60°S; they are southwesterly in the northern hemisphere and northwesterly in the southern hemisphere.

wind wave   wave created by the action of the wind on the sea surface.

winter solstice   *see* Solstices.

zenith   point in the sky that is immediately overhead.

zonation   parallel bands of distinctive plant and animal associations found within the littoral zones and distributed to take advantage of optimal conditions for survival.

zooplankton   animal forms of plankton.

zooxanthellae   symbiotic microscopic organisms (dinoflagellates) found in corals and other marine organisms.

Suggested Readings

General References

Atlas of the Oceans. 1977. Rand McNally, New York, 208 pp.

Bowditch, N. 1984 ed. *American Practical Navigator,* vol. 1. U.S. Defense Mapping Agency Hydrographic Center, Washington, D.C. 1414 pp. (Sections on many topics.)

Oceans, bimonthly journal of the Oceanic Society, Fort Mason Center, San Francisco.

Ocean Science, readings from *Scientific American.* 1977. Freeman, San Francisco, 307 pp.

Oceanus, published March, June, September, and December by the Woods Hole Oceanographic Institution, Woods Hole, Mass.

Sea Frontier-Sea Secrets, bimonthly magazine of the International Oceanographic Foundation, Miami.

Sverdrup, H. U., M. W. Johnson, and R. H. Fleming. 1942. *The Oceans.* Prentice-Hall, Englewood Cliffs, N.J. 1087 pp. (Classic text in oceanography.)

Von Andel, T. 1981. *Science at Sea, Tales of an Old Ocean.* Freeman, San Francisco. 186 pp. (Writings on various topics.)

Prologue

Bailey, H. S. 1953. The Voyage of the Challenger. In *Ocean Science,* readings from *Scientific American* 188 (5): 8–12.

Charnock, H. 1973. H.M.S. Challenger and the Development of Marine Science. In *Oceanography, Contemporary Readings in Ocean Sciences,* 2d ed., Pirie, R. G., ed. (1977). Oxford. Univ., New York. pp. 24–33.

Deacon, M. 1971. *Scientists and the Sea 1650–1900, A Study of Marine Science.* Academic Press, New York. 445 pp.

Deacon, M., ed. 1978. *Oceanography, Concepts and History.* Dowden, Hutchinson & Ross, Stroudsburg, Pa. 394 pp. (Source book of milestone papers in facsimile form.)

Deacon, G., and M. Deacon. 1982. *Modern Concepts of Oceanography.* Hutchinson & Ross, Stroudsburg, Pa. 386 pp. (More recent papers in facsimile form.)

Ryan, P. R., ed. Spring 1982. *Oceanus* 25 (1). (Issue devoted to research vessels.)

Schlee, S. 1973. *The Edge of an Unfamiliar World, A History of Oceanography.* Dutton, New York. 398 pp.

Sears, M., and D. Merriman, eds. 1980. *Oceanography: The Past.* Springer-Verlag, New York. 812 pp.

Chapter 1
The Water Planet

Abell, G. O., D. Morrison, and S. C. Wolff. 1987. *Exploration of the Universe,* 5th ed. Saunders, Philadelphia. 768 pp. (General astronomy text.)

Baker, J. D., and W. S. Wilson. 1986. Spaceborne Observations in Support of Earth Science. *Oceanus* 29 (4): 76–85.

Bowditch, N. 1984 ed. *American Practical Navigator,* vol. 1. U.S. Defense Mapping Agency Hydrographic Center, Washington, D.C. 1414 pp. (History of navigation, chart projections, and navigation aids are covered in chapters 1–5 and 41–46.)

Gore, R. 1985. The Planets, Between Fire and Ice. *National Geographic* 167 (1): 4–51.

McClintock, J. 1987. Remote Sensing, Adding to Our Knowledge of Oceans and Earth. *Sea Frontiers* 33 (2): 105–13.

Ryan, P. R., ed. Fall 1981. *Oceanus* 24 (3). (Issue devoted to oceanography from space.)

Stephenson, A. G. 1970. Hydrographic Surveying. In *Oceanography, Contemporary Readings in Ocean Sciences,* 2d ed., Pirie, E. G., ed. (1977). Oxford Univ., New York, pp. 34–38.

Chapter 2
The Sea Floor

Geology of the Sea Floor

Ballard, R. D. 1985. How We Found the Titanic. *National Geographic* 168 (6): 696–719.

Ballard, R. D. 1986. A Long Last Look at Titanic. *National Geographic* 170 (6): 698–727.

Emery, K. O. 1969. The Continental Shelves. In *Ocean Science,* readings from *Scientific American* 221 (3): 32–44 (1977).

Heezen, B. C., and C. D. Hollister. 1971. *The Face of the Deep.* Oxford Univ., New York. 659 pp. (An illustrated natural history approach.)

Kennett, J. 1982. *Marine Geology.* Prentice-Hall, Englewood Cliffs, N.J. 813 pp.

Menard, H. W. 1969. The Deep-Sea Floor. In *Ocean Science,* readings from *Scientific American* 221 (3): 55–64 (1977).

Montgomery, C. 1985. *Environmental Geology.* Wm. C. Brown, Dubuque, Ia. 454 pp. (General geology text.)

Press, F., and R. Siever. 1982. *Earth,* 3d ed. Freeman, San Francisco. 613 pp. (General geology text.)

Ryan, P. R., ed. Winter 1985–86. *Oceanus* 28 (4). (Issue devoted to the discovery of the *Titanic.*)

Sea Technology. 1986. 27 (12) (Issue devoted to submersibles.)

Seibold, E., and W. H. Berger. 1982. *The Sea Floor, An Introduction to Marine Geology.* Springer-Verlag, New York. 288 pp.

Shepard, F. P. 1973. *The Earth Beneath the Sea,* rev. ed. Atheneum, N.Y. 242 pp. (Introductory text.)

Shepard, F. P. 1973. *Submarine Geology,* 3d ed. Harper & Row, New York. 517 pp. (Classic text in geological oceanography.)

Stephenson, A. G. 1970. Hydrographic Surveying. In *Oceanography, Contemporary Readings in Ocean Sciences,* 2d ed., Pirie, R. G., ed. (1977). Oxford Univ., New York. pp. 34–38.

Seabed Resources

Blissenbach, E., and Z. Nawab. 1982. Metalliferous Sediments of the Seabed: The Atlantic II–Deep Deposits of the Red Sea. In *Ocean Yearbook 3,* Borgese, E., and N. Ginsburg, eds. Univ. of Chicago, Chicago. pp. 77–104.

Borgese, E. M. 1983. The Law of the Sea. *Scientific American* 248 (3): 42–49.

Broadus, J. M. 1987. Seabed Materials. *Science,* 235 (4791): 853–60. (Complete update on seabed resources.)

Dubs, M. 1986. Minerals of the Deep Sea: Politics and Economics in Conflict. In *Ocean Yearbook 6,* Borgese, E. M., and N. Ginsburg, eds. Univ. of Chicago, Chicago. pp. 55–83.

Edmond, J. M., and K. Von Damm. 1983. Hot Springs on the Ocean Floor. *Scientific American* 248 (4): 78–93.

Ellers, F. S. 1982. Advanced Offshore Oil Platforms. *Scientific American* 246 (4): 39–49.

Harrison, P. 1983. Offshore Oil and Gas: Development, Transportation, and Coastal Impacts. In *Ocean Yearbook 4,* Borgese, E. M., and N. Ginsburg, eds. Univ. of Chicago, Chicago. pp. 319–46.

Kent, P. 1980. *Minerals from the Marine Environment.* Edward Arnold, London. 88 pp.

Koski, R. A., W. R. Normark, J. L. Morton, and J. R. Delaney. 1982. Metal Sulfide Deposits on the Juan de Fuca Ridge. *Oceanus* 25 (3): 43–48.

Marine Technology Society Journal. 1985. 19 (4). (Issue devoted to marine minerals.)

Mottl, M. J. 1980. Submarine Hydrothermal Ore Deposits. *Oceanus* 23 (2): 18–27.

Ryan, P. R., ed. Fall, 1983. *Oceanus.* 26 (3). (Issue devoted to offshore oil and gas.)

Chapter 3
The Not-So-Rigid Earth

Interior of the Earth
Montgomery, C. 1985. *Environmental Geology.* Wm. C. Brown, Dubuque, Ia. 454 pp. (General geology text.)

Press, F., and R. Siever. 1982. *Earth,* 3d ed. Freeman, San Francisco. 613 pp. (General geology text.)

The Moving Continents
Ballard, R. 1975. Dive into the Great Rift. *National Geographic* 147 (5): 604–15. (Project FAMOUS and photographs from *Alvin.*

Ballard, R. D., and J. F. Grassle. 1979. Return to Oases of the Deep. *National Geographic* 156 (5): 689–705. (Photographs of vent areas and life forms.)

Bambach, R. K., C. R. Scotese, and A. M. Ziegler. 1980. Before Pangea: The Geographies of the Paleozoic World. *American Scientist* 68 (1): 26–38.

Bonatti, E. 1987. The Rifting of Continents. *Scientific American* 256 (3): 97–103.

Bullard, Sir E. 1969. The Origin of the Oceans. In *Ocean Science,* readings from *Scientific American* 221 (3): 45–54 (1977).

Continents Adrift and Continents Aground. Readings from *Scientific American,* 1976. Freeman, San Francisco. 230 pp.

Decker, R., and B. Decker. 1981. The Eruptions of Mount St. Helens. *Scientific American* 244 (3): 68–80.

Gore, R. 1985. Our Restless Planet Earth. *National Geographic* 168 (2): 142–81.

Howell, D. G. 1985. Terranes. *Scientific American* 253 (5): 116–25.

Jones, D. L., A. Cox, P. Coney, and M. Beck. 1982. The Growth of Western North America. *Scientific American* 247 (6): 70–84.

MacDonald, K. C., and B. P. Luyendyk. 1981. The Crest of the East Pacific Rise. *Scientific American* 245 (5): 100–16.

Marvin, U. 1973. *Continental Drift, The Evolution of a Concept.* Smithsonian, Washington, D.C. 239 pp.

Molnar, P. 1986. The Structure of Mountain Ranges. *Scientific American* 255 (1): 70–79.

Mutter, J. C. 1986. Seismic Images of Plate Boundaries. *Scientific American* 254 (2): 66–75.

Rabinowitz, P. S., S. Herrig, and K. Riedel. 1986. Ocean Drilling Program Altering Our Perception of Earth. *Oceanus* 29 (3): 36–41. (Report on the first nine voyages of the *JOIDES Resolution* drill ship.)

Vink, G. E., W. Morgan, and P. Vogt. 1985. The Earth's Hot Spots. *Scientific American* 252 (4): 50–57.

Chapter 4
The Properties of Water

The Seawater
Bowditch, N. 1984 ed. *American Practical Navigator,* vol. 1. U.S. Defense Mapping Agency Hydrographic Center, Washington, D.C. 1414 pp. (Soundings are covered in chapter 28 and sound in chapter 35; ice and weather in chapters 36 and 39.)

Duxbury, A. C. 1971. *The Earth and Its Oceans.* Addison-Wesley, Reading, Mass. 381 pp.

Hewitt, P. G. 1981. *Conceptual Physics,* 4th ed. Little, Brown, Boston. 637 pp. (General physics text with little mathematical emphasis.)

MacLeish, W. H., ed. Spring 1977. *Oceanus* 20 (2). (Issue includes articles on uses of sound in navigation, warfare, animal behavior, and seismics.)

Neumann, G., and W. J. Pierson. 1966. *Principles of Physical Oceanography.* Prentice-Hall, Englewood Cliffs, N.J. 545 pp.

Chapter 5
The Salt Water

The Seawater
Anderson, A. T. 1982. The Ocean Basins and Ocean Water. In *Ocean Yearbook 3,* Borgese, E., and N. Ginsburg, eds. Univ. of Chicago, Chicago. pp. 135–56. (The origin of seawater and the interaction of seawater and ocean crust.)

Broecker, W. S. 1974. *Chemical Oceanography.* Harcourt, Brace, Jovanovich, New York. 214 pp.

MacIntyre, F. 1970. Why the Sea Is Salt. In *Ocean Science,* readings from *Scientific American* 223 (5): 104–15 (1977).

Palais, J. M. 1986–87. Polar Ice Cores. *Oceanus* 29 (4): 55–60.

Riley, J. P., and R. Chester, eds. 1983. *Chemical Oceanography,* Vol. 8. Academic Press, New York. 398 pp.

Chemical Resources
Anderson, A. T. 1982. The Ocean Basins and Ocean Water. In *Ocean Yearbook 3,* Borgese, E., and N. Ginsburg, eds. Univ. of Chicago, Chicago. pp. 135–56.

Blissenbach, E., and Z. Nawab. 1982. Metalliferous Sediments of the Seabed: The Atlantis II–Deep Deposits of the Red Sea. In *Ocean Yearbook 3,* Borgese, E., and N. Ginsburg, eds. Univ. of Chicago, Chicago. pp. 77–104.

Charlier, R. H. 1983. Water, Energy, and Nonliving Ocean Resources. In *Ocean Yearbook 4,* Borgese, E. M., and N. Ginsburg, eds. Univ. of Chicago, Chicago. pp. 75–120. (An overview of mineral resources.)

Hotta, H. 1987. Recovery of Uranium from Seawater. *Oceanus,* 30 (1): 44–46.

Kent, P. 1980. *Minerals from the Marine Environment.* Edward Arnold, London. 88 pp.

Rona, P. 1986. Mineral Deposits from Seafloor Hot Springs. *Scientific American* 254 (1): 84–92.

Desalination
Levine, S. N., ed. 1968. *Selected Papers on Desalination and Ocean Technology.* Dover, New York. 437 pp.

Spiegler, K. S. 1977. *Salt Water Purification.* Plenum, New York. 189 pp.

Chapter 6
Structure of the Oceans

Surface Processes and Ocean Structure
Bryan, K. 1978. The Ocean Heat Balance. *Oceanus* 21 (4): 19–26.

Kitano, Y., and M. Tanaka. 1987. The Atmospheric Carbon Dioxide Problem. *Oceanus,* 30 (1): 78–82.

Moore III, B., and B. Bolin. 1986–87. The Oceans, Carbon Dioxide and Global Climate Change. *Oceanus* 29 (4): 9–15.

Neumann, G., and W. J. Pierson. 1966. *Principles of Physical Oceanography.* Prentice-Hall, Englewood Cliffs, N.J. 545 pp.

Pickard, G. L., and W. J. Emery. 1982. *Descriptive Physical Oceanography, An Introduction.* 4th (SI) ed. Pergamon Press, New York. 249 pp.

Scientific Plan for the World Ocean Circulation Experiment. 1986. World Climate Research Program, WCRP Publication Series 6, WMO/TD No. 122. 83 pp.

Von Arx, W. S. 1962. *An Introduction to Physical Oceanography.* Addison-Wesley, Reading, Mass. 422 pp.

Williams, J., J. J. Higginson, and J. P. Rohrbough. 1968. Oceanic Water Masses and Their Circulation. In *Oceanography, Contemporary Readings in Ocean Sciences*, 2d ed., Pirie, R. G., ed. (1977). Oxford Univ., New York. pp. 24–33.

Energy

Brin, A. 1981. *Energy and the Oceans.* Westbury House, Surrey, England. 133 pp.

Bruce, M. 1986. Ocean Energy: Some Perspectives on Economic Viability. In *Ocean Yearbook 6,* Borgese, E. M., and N. Ginsburg, eds. Univ. of Chicago, Chicago. pp. 58–78.

Charlier, R. H. 1983. Water, Energy, and Nonliving Ocean Resources. In *Ocean Yearbook 4,* Borgese, E. M., and N. Ginsburg, eds. Univ. of Chicago, Chicago. pp. 75–120.

Isaacs, J. D., and W. R. Schmitt. 1980. Ocean energy: Forms and Prospects. *Science* 207 (4428): 265–73.

Lavi, A., ed. 1980. *Ocean Thermal Energy Conversion.* Pergamon Press, New York. 569 pp.

Merriam, M. F. 1978. Wind, Waves and Tides. In *Annual Review of Energy,* Vol. 3, Hollander, J. M., ed. Annual Reviews, Palo Alto, Calif. pp. 29–56. (Overview of energy sources.)

Penny, T. R., and D. Bharathan. 1987. Power from the Sea. *Scientific American* 256 (1): 86–92. (Review ocean thermal energy conversion (OTEC).)

Simeons, C. 1980. *Hydro-Power, The Use of Water As an Alternative Source of Energy.* Pergamon, Elmsford, N.Y. 549 pp.

Chapter 7
Winds and Currents

The Atmosphere in Motion

Bowditch, N. 1984 ed. (American Practical Navigator, Vol. 1. U.S. Defense Mapping Agency Hydrographic Center, Washington, D.C. 1414 pp. (Winds and weather are in chapter 24 and chapters 37–39.)

Simpson, R. H., and H. Riehl. 1981. *The Hurricane and Its Impact.* Louisiana State Univ., Baton Rouge, La. 398 pp.

Strahler, A. N., and A. H. Strahler. 1973. *Environmental Geoscience.* Hamilton, Santa Barbara, Calif. 575 pp. (Part 1 covers energy systems of the atmosphere and hydrosphere.)

Webster, P. J. 1981. Monsoons. *Scientific American* 245 (2): 108–18.

The Water in Motion

Barkley, R. A. 1970. The Kuroshio Current. In *Oceanography, Contemporary Readings in Ocean Sciences,* 2d ed., Pirie, R. G., ed. Oxford Univ., New York. pp. 70–80.

Bowditch, N. 1984 ed. *American Practical Navigator,* Vol. 1. U.S. Defense Mapping Agency Hydrographic Center, Washington, D.C. 1414 pp. (Ocean currents are covered in chapters 31 and 32.)

Canby, T. Y. 1984. El Niño's Ill Wind. *National Geographic* 165 (2): 144–83.

Joyce, T., and P. Wiebe. 1983. Warm-Core Rings of the Gulf Stream. *Oceanus* 26 (2): 34–44.

Ramage, C. S. 1986. El Niño. *Scientific American* 254 (6): 76–83.

Ryan, P. R., ed. Summer 1984. *Oceanus* 27 (2). (Issue devoted to El Niño.)

Tolmazin, D. 1985. *Elements of Dynamic Oceanography.* Allen and Unwin, Boston. 181 pp.

Von Arx, W. S. 1962. *An Introduction to Physical Oceanography.* Addison-Wesley, Reading, Mass. 422 pp.

Wiebe, P. H. 1982. Rings of the Gulf Stream. *Scientific American* 246 (3): 60–70.

Wooster, W., and D. Fluharty, eds. 1985. *El Niño North.* Washington Sea Grant Program, Univ. of Washington, Seattle. 312 pp.

Energy

Brin, A. 1981. *Energy and the Oceans.* Westbury House, Surrey, England. 133 pp.

Bruce, M. 1986. Ocean Energy: Some Perspectives on Economic Viability. In *Ocean Yearbook 6,* Borgese, E. M., and N. Ginsburg, eds. Univ. of Chicago, Chicago. pp. 58–78.

Charlier, R. H. 1983. Water, Energy, and Nonliving Ocean Resources. In *Ocean Yearbook 4,* Borgese, E. M., and N. Ginsburg, eds. Univ. of Chicago, Chicago. pp. 75–120. (Overview of energy sources.)

Merriam, M. F. 1978. Wind, Waves and Tides. In *Annual Review of Energy,* Vol. 3, Hollander, J. M., ed. Annual Reviews, Palo Alto, Calif. pp. 29–56. (Overview of energy sources.)

Simeons, C. 1980. *Hydro-Power, The Use of Water As an Alternative Source of Energy.* Pergamon, Elmsford, N.Y. 549 pp.

Chapter 8
The Waves

Waves and Wave Motion

Bascom, W. 1980. *Waves and Beaches: The Dynamics of the Ocean Survey,* rev. ed. Doubleday, Garden City, N.J. 366 pp.

Bowditch, N. 1984 ed. *American Practical Navigator,* Vol. 1. U.S. Defense Mapping Agency Hydrographic Center, Washington, D.C. 1414 pp. (Waves, breakers, and surf are covered in chapters 33 and 34.)

Duxbury, A. C. 1971. *The Earth and Its Oceans.* Addison-Wesley, Reading, Mass. 381 pp.

Lissau, S. 1975. Ocean Waves. In *Oceanography, Contemporary Readings in Ocean Sciences,* 2d ed., Pirie, R. G., ed. (1977). Oxford Univ., New York. pp. 88–97.

Russell, R. C. H., and D. H. MacMillan. 1954. *Waves and Tides,* 2d ed. Hutchinson's, London. 348 pp.

Simpson, R. H., and H. Riehl. 1981. *The Hurricane and Its Impact.* Louisiana State Univ., Baton Rouge, La. 398 pp.

Von Arx, W. S. 1962. *An Introduction to Physical Oceanography.* Addison-Wesley, Reading, Mass. 422 pp.

Energy

Brin, A. 1981. *Energy and the Oceans.* Westbury House, Surrey, England. 133 pp.

Bruce, M. 1986. Ocean Energy: Some Perspectives on Economic Viability. In *Ocean Yearbook 6,* Borgese, E. M., and N. Ginsburg, eds. Univ. of Chicago, Chicago. pp. 58–78.

Changery, M. J., and R. G. Quayle. 1987. Coastal Wave Energy, *Sea Frontiers* 33 (4): 259–62.

Charlier, R. H. 1983. Water, Energy, and Nonliving Ocean Resources. In *Ocean Yearbook 4,* Borgese, E. M., and N. Ginsburg, eds. Univ. of Chicago, Chicago. pp. 75–120. (Overview of energy sources.)

Isaacs, J. D., and W. R. Schmitt. 1980. Ocean Energy: Forms and Prospects. *Science,* 207 (4428): 265–73.

McCormick, M. E. 1986. Ocean Wave Energy Conversion. *Sea Technology* June: 32–34.

Merriam, M. F. 1978. Wind, Waves and Tides. In *Annual Review of Energy,* vol. 3, Hollander, J. M., ed. Annual Reviews, Palo Alto, Calif. pp. 29–56. (Overview of energy sources.)

Miyazaki, T. 1987. Wave Power Generator Kaimei. *Oceanus* 30 (1): 43–44.

Ross, D. 1979. *Energy from Waves.* Pergamon, Elmsford, N.Y. 121 pp.

Shaw, R. 1982. *Wave Energy: A Design Challenge.* Halsted, N.Y. 202 pp.

Simeons, C. 1980. *Hydro-Power, The Use of Water As an Alternative Source of Energy.* Pergamon, Elmsford, N.Y. 549 pp.

Chapter 9
The Tides

Theory and Dynamics

Bowditch, N. 1984 ed. *American Practical Navigator,* vol. 1. U.S. Defense Mapping Agency Hydrographic Center, Washington, D.C. 1414 pp. (Chapters 12 and 31 are on tide prediction, tides, and tidal currents.)

Darwin, G. H. 1898. *The Tides and Kindred Phenomena in the Solar System.* Freeman, San Francisco (1962). 378 pp.

Duxbury, A. C. 1971. *The Earth and Its Oceans.* Addison-Wesley, Reading, Mass. 381 pp.

Lynch, D. K. 1982. Tidal Bores. *Scientific American* 247 (4): 146–56.

National Oceanic and Atmospheric Administration. *National Ocean Survey—Tide and Tidal Current Predictions.* (Published annually for specific geographic areas.)

Neumann, G., and W. J. Pierson. 1966. *Principles of Physical Oceanography.* Prentice-Hall, Englewood Cliffs, N.J. 545 pp.

Redfield, A. C. 1980. *Introduction to Tides.* Marine Science International, Woods Hole, Mass. 108 pp.

Russell, R. C. H., and D. H. McMillan. 1954. *Waves and Tides,* 2d ed. Hutchinson's, London. 348 pp.

Von Arx, W. S. 1962. *An Introduction to Physical Oceanography.* Addison-Wesley, Reading, Mass. 422 pp.

Energy

Booda, L. 1985. River Rance Tidal Power Plant Nears Twenty Years in Operation. *Sea Technology* September: 22–26.

Brin, A. 1981. *Energy and the Oceans.* Westbury House, Surrey, England. 133 pp.

Bruce, M. 1986. Ocean Energy: Some Perspectives on Economic Viability. In *Ocean Yearbook 6,* Borgese, E. M., and N. Ginsburg, eds. Univ. of Chicago, Chicago. pp. 58–78.

Charlier, R. H. 1983. Water, Energy, and Nonliving Ocean Resources. In *Ocean Yearbook 4,* Borgese, E. M., and N. Ginsburg, eds. Univ. of Chicago, Chicago. pp. 75–120. (Overview of energy sources.)

Greenberg, D. A. 1987. Modeling Tidal Power. *Scientific American* 257 (5): 128–31.

Isaacs, J. D., and W. R. Schmitt. 1980. Ocean Energy: Forms and Prospects. *Science,* 207 (4428): 265–73.

Merriam, M. F. 1978. Wind, Waves and Tides. In *Annual Review of Energy,* vol. 3, Hollander, J. M., ed. Annual Reviews, Palo Alto, Calif. pp. 29–56. (Overview of energy sources.)

Simeons, C. 1980. *Hydro-Power, The Use of Water As an Alternative Source of Energy.* Pergamon, Elmsford, N.Y. 549 pp.

Chapter 10
Coasts, Shores and Beaches

Bascom, W. 1960. Beaches. In *Ocean Science,* readings from *Scientific American* 203 (2): 171–81 (1977).

Bascom, W. 1980. *Waves and Beaches: The Dynamics of the Ocean Survey,* rev. ed. Doubleday, Garden City, N.J. 366 pp.

Bird, E. C. F. 1985. *Coastline Changes: A Global Review.* Wiley & Sons, New York. 219 pp.

Coleman, E., ed. 1986. Is the Sea Rising? In *Aquanotes* 15 (3): 4–5. Center for Wetland Resources, Louisiana State Univ., Baton Rouge, La.

Conservation Foundation. 1976. *Barrier Islands Workshop,* Technical Proceedings. Washington, D.C. 149 pp.

Dolan, R. and H. Lins. 1987. Beaches and Barrier Islands. *Scientific American* 257 (1): 68–77.

Fox, W. T. 1983. *At the Sea's Edge: An Introduction to Coastal Oceanography for the Amateur Naturalist.* Prentice-Hall, Englewood Cliffs, N.J. 317 pp.

Jackson, T. C., ed. 1981. *Coast Alert, Scientists Speak Out.* Friends of the Earth, San Francisco. 181 pp.

Kaufman, W., and O. Pilkey. 1979. *The Beaches Are Moving.* Doubleday, Garden City, N.J. 326 pp.

Lowenstein, F. 1985. Beaches or Bedrooms—The Choice as Sea Level Rises. *Oceanus* 28 (3): 20–29.

MacLeish, W. H., ed. Winter 1980–81. *Oceanus* 23 (4). (Articles on coastal zone management, barrier islands, and conservation.)

Mitchell, J. 1986. Coastal Management since 1980: The U.S. Experience and Its Relevance for Other Countries. In *Ocean Yearbook 6,* Borgese, E. M., and N. Ginsburg, eds. Univ. of Chicago, Chicago. pp. 319–45.

Ringold, P. L., and J. Clark. 1980. *The Coastal Almanac.* Freeman, San Francisco. 172 pp. (Data reference.)

Shepard, F. P. 1973. *The Earth Beneath the Sea,* rev. ed. Atheneum, New York 242 pp. (Introductory text.)

Shepard, F. P. 1973. *Submarine Geology,* 3d ed. Harper & Row, New York. 517 pp. (Classic text in geological oceanography.)

Shepard, F. P. 1977. *Geological Oceanography: Evolution of Coasts, Continental Margins and Deep Sea Floor.* Crane, Russak, New York. 214 pp.

Shepard, F. P. and H. R. Wanless. 1971. *Our Changing Coastlines.* McGraw-Hill, New York. 579 pp.

Simon, A, W. 1978. *The Thin Edge: Coast and Man in Crisis.* Harper & Row, New York. 180 pp.

Trefil, J. 1984. *A Scientist at the Seashore.* Scribner, New York. 208 pp.

Chapter 11
Bays and Estuaries

Estuaries and Circulation

Caspers, H..1967. Estuaries: Analysis of Definitions and Biological Considerations. In *Estuaries,* Lauff, G. H., ed. Publication No. 83, American Association for the Advancement of Science, Washington, D.C. pp. 6–8.

Cloern, J., and F. Nichols. 1985. *Temporal Dynamics of an Estuary,* Dr. W. Junk, Boston. 237 pp.

Emery, K. O. 1967. Estuaries and Lagoons in Relation to Continental Shelves. In *Estuaries,* Lauff, G. H., ed. Publication No. 83, American Association for the Advancement of Science, Washington, D.C. pp. 9–11.

McLusky, D. S. 1981. *The Estuarine Ecosystem.* Blackie, Glasgow and London. 150 pp.

Mitchell, R. 1982. Coastal Zone Management: A Comparative Analysis of National Programs. In *Ocean Yearbook 3,* Borgese, E., and N. Ginsburg, eds. Univ. of Chicago, Chicago. pp. 258–319.

Nichols, F., J. Cloern, S. Luoma, and D. Peterson. 1986. Temporal Dynamics of an Estuary: San Francisco Bay. *Science,* 231 (4738): 567–73.

Pritchard, D. W. 1967. What Is an Estuary: Physical Viewpoint. In *Estuaries,* Lauff, G. H., ed. Publication No. 83, American Association for the Advancement of Science, Washington, D.C. pp. 3–5.

Ringold, P. L., and J. Clark. 1980. *The Coastal Almanac.* Freeman, San Francisco. 172 pp. (Data reference.)

Water Quality and Pollution

Breuel, A., ed. 1981. *Oil Spill Cleanup and Protection Techniques for Shorelines and Marshlands.* Noyes Data Corporation, Park Ridge, N.J. 404 pp.

Bruce, M. 1986. The London Dumping Convention, 1972: First Decade and Future. In *Ocean Yearbook 6,* Borgese, E. M., and N. Ginsburg, eds. Univ. of Chicago, Chicago. pp. 298–318.

Farrington, J. 1985. Oil Pollution: A Decade of Monitoring. *Oceanus* 28 (3): 2–12.

Fingas, M., W. Duval, and G. Stevenson. 1979. *The Basics of Oil Spill Cleanup.* Environmental Protection Service, Environment Canada. 155 pp.

Hain, J. 1986. Low-Level Radioactivity in the Irish Sea. *Oceanus* 29 (3): 16–27.

Ketchum, B. H., J. M. Capuzzo, W. V. Burt, I. W. Duedall, P. K. Park, and D. R. Kester, eds. 1985. *Wastes in the Ocean,* Vol. 6. Wiley & Sons, New York. 534 pp.

Kuiper, J., and W. J. Van Den Brnik, eds. 1987. *Fate and Effects of Oil in Marine Ecosystems.* Martinus Nijhoff, Boston. 338 pp.

MacLeish, W.H., ed. Spring 1981. *Oceanus* 24 (1). (Collected papers on toxicants, heavy metals, sewage, and radioactive waste.)

Meyers, E. P., and E. T. Harding, eds. 1983. *Ocean Dispersal of Municipal Waste Water—Impacts on the Coastal Environment.* Sea Grant College Program, Massachusetts Institute of Technology, Cambridge, Mass. 1115 pp.

Moriarity, F. 1983. *Ecotoxicology: The Study of Pollutants in Ecosystems.* Academic Press. New York. 233 pp.

National Oceanic and Atmsopheric Administration. *National Maritime Marine Pollution Program Plan—Federal Plan for Ocean Pollution Research, Development, and Monitoring, Fiscal Years 1981–1985.* 1981. Interagency Committee on Ocean Pollution Research, Development and Monitoring. 185 pp.

Palmer, H. D., and M. G. Gross. 1979. *Ocean Dumping and Marine Pollution.* Dowden, Hutchinson & Ross, Stroudsburg, Pa. 202 pp.

Ragotzkie, R., ed. 1983. *Man and the Marine Environment.* CRC Press, Boca Raton, Fla. 180 pp.

Scott, S. 1987. Painting with Pesticides, The Controversial Organotin Paints. *Sea Frontiers* 33 (6): 414–21.

Southern California Coastal Water Research Project Authority. 1986. *Southern California Coastal Water Research Project.* Long Beach, Ca. 56 pp.

Wood, L. B. 1982. *The Restoration of the Tidal Thames.* Adam Hilger, Bristol, England. 202 pp.

Chapter 12
Oceans: Environment for Life

Brooks, J., M. Kennicutt II, R. Bidigare, T. Wade, E. Powell, G. Denoux, R. Fay, J. Childress, C. Fisher, I. Rossman, and G. Boland. 1987. Hydrates, Oil Seepage, and Chemosynthetic Ecosystems on the Gulf of Mexico Slope: An Update. *EOS* 68 (18): 498–99.

Childress, J., C. Fisher, J. Brooks, M. Kennicutt II, R. Bridigare, and A. Anderson. 1986. A Methanotophic Marine Molluscan (Bivalvia, Mytilidae) Symbiosis: Mussels Fueled by Gas. *Science* 233 (4770): 1306–8.

Hedgpeth, J., ed. 1957. *Treatise on Marine Ecology and Paleoecology,* vol. 1: Ecology. Geological Society of America, New York. 1296 pp.

Lerman, M. 1986. *Marine Biology, Environment, Diversity and Ecology.* Benjamin-Cummings, Menlo Park, Calif. 535 pp.

Leschine, T. M. 1981. The Panamanian Sea-Level Canal. *Oceanus* 24 (2): 20–30.

Life in the Sea, readings from *Scientific American,* 1982. Freeman, San Francisco. 248 pp. (Articles on marine organisms, the condition in which they live, and their food resources.)

MacLeish, W. H., ed. Fall 1980. *Oceanus* 23 (3). (Issue includes articles on the various senses of organisms of the sea.)

Nealson, K., and C. Arneson. 1985. Marine Bioluminescence: About to See the Light. *Oceanus* 28 (3): 13–18.

Nybakken, J. 1988. *Marine Biology, An Ecological Approach,* 2d ed. Harper & Row, New York. 514 pp.

Russell, F. S., and M. Yonge. 1975. *The Seas* 4th ed. Frederick Warne, London. 283 pp.

Siezen, R. 1986. Cuttlebone: The Bouyant Skeleton. *Sea Frontiers* 32 (2): 115–22.

Sumich, J. L. 1984. *An Introduction to the Biology of Marine Life,* 3d ed. Wm. C. Brown, Dubuque, Ia. 386 pp.

Thorson, G. 1971. *Life in the Sea.* McGraw-Hill, New York. 256 pp.

Ward, P., L. Greenwald, and O. E. Greenwald. 1980. The Buoyancy of the Chambered Nautilus. *Scientific American,* 243 (4): 190–203.

Chapter 13
Production and Life

Cushing, D. H. 1975. *Marine Ecology and Fisheries.* Cambridge Univ., Cambridge, England. 278 pp.

Cushing, D. H., and J. J. Walsh. 1976. *The Ecology of the Seas.* Saunders, Philadelphia. 467 pp.

Isaacs, J. D. 1969. The Nature of Ocean Life. In *Life in the Sea,* readings from *Scientific American,* 1982. Freeman, San Francisco. pp. 4–17.

Lerman, M. 1986. *Marine Biology, Environment, Diversity and Ecology.* Benjamin-Cummings, Menlo Park, Calif. 535 pp.

Parsons, T. R., M. Takahashi, and B. Hargrave. 1984. *Biological Oceanographic Processes,* 3d ed. Pergamon, Elmsford, N.Y. 330 pp.

Pomeroy, L. R. 1974. The Ocean's Food Web, A Changing Paradigm. In *Oceanography, Contemporary Readings in Ocean Sciences,* 2d ed. Pirie, R. G., ed. (1977). Oxford Univ., New York. pp. 105–15.

Raymont, J. E. G. 1980. *Plankton and Productivity in the Oceans,* 2d ed., Vol. 1: Phytoplankton. Pergamon, Elmsford, N.Y. 489 pp.

Rhyther, J. H. 1969. Photosynthesis and Fish Production in the Sea. *Science,* 166 (3901): 72–76.

Steele, J. H., ed. 1970. *Marine Food Chains.* Univ. of California, Berkeley. 552 pp.

Sumich, J. L. 1984. *An Introduction to the Biology of Marine Life,* 3d ed. Wm. C. Brown, Dubuque, Ia. 386 pp.

Chapter 14
Plankton: Drifters of the Open Ocean

General

Lerman, M. 1986. *Marine Biology, Environment, Diversity and Ecology.* Benjamin-Cummings, Menlo Park, Calif. 535 pp.

Life in the Sea, readings from *Scientific American,* 1982. Freeman, San Francisco. 248 pp.

Niesen, T. M. 1982. *The Marine Biology Coloring Book.* Harper & Row, New York. 218 pp.

Russell, F. S., and M. Yonge. 1975. *The Seas,* 4th ed. Frederick Warne, London. 283 pp.

Sumich, J. L. 1984. *An Introduction to the Biology of Marine Life,* 3d ed. Wm. C. Brown, Dubuque, Ia. 386 pp.

Plankton

Barnes, R. D. 1980. *Invertebrate Zoology,* 4th ed. Saunders, Philadelphia. 870 pp.

Beddington, J. R., and R. M. Beddington. May. 1982. The Harvesting of Interacting Species. *Scientific American* 247 (5): 62–69. (Krill: harvesting and its effect on the ecosystem.)

Coleman, B. C., R. N. Doetsch, and R. D. Sjoblad. 1986. Red Tide: A Recurrent Marine Phenomenon. *Sea Frontiers* 32 (3): 184–92.

Doorenbos, N. 1984. Ciguatera Toxins: Where Do We Go from Here? In *Seafood Toxins,* Ragelis, E. P., ed. American Chemical Society Symposium Series 262, Washington, D.C. pp. 69–73.

Hammer, W. 1984. Krill, Untapped Bounty from Sea. *National Geographic* 165 (5): 627–43.

Hardy, A. 1958. *The Open Sea: Its Natural History,* part 1; The World of Plankton. Houghton Mifflin, Boston. 335 pp.

Laws, R. 1985. The Ecology of the Southern Oceans. *American Scientist* 73: 26–40.

Lewis, N. 1984. Ciguatera in the Pacific: Incidence and Implications for Marine Resource Development. In *Seafood Toxins,* Ragelis, E. P., ed. American Chemical Society Symposium Series 262, Washington, D.C. pp. 289–306.

Ragelis, E. 1984. Ciguatera Seafood Poisoning. In *Seafood Toxins,* Ragelis, E. P., ed. American Chemical Society Symposium Series 262, Washington, D.C. pp. 25–36.

Smith, D. L. 1977. *A Guide to Marine Coastal Plankton and Marine Invertebrate Larvae.* Kendall/Hunt, Dubuque, Ia. 161 pp.

Chapter 15
Nekton: Free Swimmers of the Sea

General

Lerman, M. 1986. *Marine Biology, Environment, Diversity and Ecology.* Benjamin-Cummings, Menlo Park, Calif. 535 pp.

Life in the Sea, readings from *Scientific American,* 1982. Freeman, San Francisco. 248 pp.

Niesen, T. M. 1982. *The Marine Biology Coloring Book.* Harper & Row, New York. 218 pp.

Russell, F. S., and M. Yonge. 1975. *The Seas,* 4th ed. Frederick Warne, London. 283 pp.

Sumich, J. L. 1984. *An Introduction to the Biology of Marine Life,* 3d ed. Wm. C. Brown, Dubuque, Ia. 386 pp.

Mammals

Allen, K. R. 1980. *Conservation and Management of Whales.* Washington Sea Grant, Univ. of Washington, Seattle. 107 pp.

Bonner, W. N. 1982. *Seals and Man, A Study of Interactions.* Washington Sea Grant, Univ. of Washington, Seattle. 170 pp.

Burton, R. 1980. *The Life and Death of Whales,* 2d ed. Deutsch, London. 185 pp.

Gentry, R. 1987. Seals and Their Kin. *National Geographic* 171 (4): 475–501.

Haley, D., ed. 1978. *Marine Mammals.* Pacific Search, Seattle. 256 pp.

Hall, A. 1984. Man and Manatee. *National Geographic* 166 (3): 400–418.

Hoyt, E. 1984. The Whales Called Killer. *National Geographic* 166 (2): 220–37.

Nelson, C., and K. Johnson. 1987. Whales and Walruses as Tillers of the Sea Floor. *Scientific American,* 256 (2): 112–17.

Scammon, C. M. 1874. *The Marine Mammals of the Northwestern Coast of North America.* Dover, N.Y. (1968). 324 pp.

Tuck, J. A., R. Grenier, and R. Laxalt. 1985. Sixteenth Century Basque Whaling in America. *National Geographic* 168 (1): 40–57.

Whitehead, H. 1984. The Unknown Giants: Sperm and Blue Whales. *National Geographic* 166 (6): 774–89.

Zopal, W. 1987. Diving Adaptations of the Weddell Seal. *Scientific American* 256 (6): 100–105.

Squid

Roper, C. F. E., and K. J. Boss. 1982. The Giant Squid. *Scientific American* 246 (4): 96–105.

Reptiles

Carr, A. 1984. *So Excellent A Fishe: A Natural History of Sea Turtles.* Scribner, New York. 248 pp.

Minton, S. A., and H. Heatwole. 1978. Snakes and the Sea. *Oceanus* 11 (2): 53–56.

Fish

Eastman, J. T., and A. L. DeVries. 1986. Antarctic Fishes. *Scientific American* 255 (5) 106–14.

Hardy, A. 1959. *The Open Sea: Its Natural History,* Part 2: Fish and Fisheries. Houghton Mifflin, Boston. 322 pp.

Horn, M. H., and R. N. Gibson. 1988. Intertidal Fisheries. *Scientific American* 258 (1): 64–70.

Idyll, C. P. 1976. *Abyss—The Deep Sea and the Creatures That Live In It,* rev. ed. Crowell, New York. 428 pp.

Isaacs, J. D., and R. A. Schwartzlose. 1975. Active Animals of the Deep-Sea Floor. In *Ocean Science,* readings from *Scientific American* 233 (4): 202–9 (1977).

MacLeish, W. H., ed. Winter 1981–1982. *Oceanus* 24 (4). (Issue devoted to sharks: vision, feeding, reproduction, and behavior.)

McCosker, J. E. 1985. White Shark Attack Behavior: Observations of and Speculations About Predator and Prey Strategies, *Memoirs, Southern California Academy of Sciences* 9: 123–25.

Moss, S. A. 1984. *Sharks, An Introduction for the Amateur Naturalist.* Prentice-Hall, Englewood Cliffs, N.J. 246 pp.

Partridge, B. L. 1982. The Structure and Function of Fish Schools. *Scientific American* 246 (6): 114–23.

Wheeler, A. 1975. *Fishes of the World, An Illustrated Dictionary.* Macmillan, New York. 366 pp.

Wood, L. 1986. Megamouth, New Species of Shark. *Sea Frontiers* 32 (3): 192–98.

Fisheries

Bardach, J. 1986. Fish Far Away: Comments on Antarctic Fisheries. In *Ocean Yearbook 6,* Borgese, E. M., and N. Ginsburg, eds. Univ. of Chicago, Chicago. pp. 38–54.

Bauereis, E. I., and J. N. Kraeuter. 1984. Power Plants and Striped Bass: A Partnership, *Oceanus* 27 (1): 40–45.

Borgese, E. M. 1980. *The Story of Aquaculture.* Abrams, New York. 236 pp.

Brown, L. R. 1985. Maintaining World Fisheries. In *State of the World 1985,* Brown, L. R., ed. Norton, New York. pp. 73–96.

Food and Agricultural Organization of the United Nations. 1985. *Yearbook of Fisheries Statistics, Catches, and Landings.* Vol. 60. Rome, Italy.

Iverson, E. S., and J. Z. Iverson. 1987. Salmon Farming Success in Norway. *Sea Frontiers* 33 (5): 354–61.

Lee, C. M. 1984. Surimi Gel and the U.S. Seafood Industry. *Oceanus* 27 (1): 35–39.

Limburg, P. R. 1980. *Farming the Waters.* Beaufort, New York. 223 pp.

Ling, S. 1977. *Aquaculture in Southeast Asia, A Historical Overview.* Washington Sea Grant, Univ. of Washington, Seattle. 108 pp.

Michael, R. G., ed. 1987. *Managed Aquatic Ecosystems.* Elsevier, New York. 166 pp.

Sainsbury, J. C. 1975. *Commercial Fishing Methods.* Fishing News (Books), Surrey, England. 119 pp.

Chapter 16
The Benthos: Dwellers of the Sea Floor

General

Lerman, M. 1986. *Marine Biology, Environment, Diversity and Ecology.* Benjamin-Cummings, Menlo Park, Calif. 535 pp.

Life in the Sea, readings from *Scientific American,* 1982. Freeman, San Francisco. 248 pp.

Niesen, T. M. 1982. *The Marine Biology Coloring Book.* Harper & Row, New York. 218 pp.

Russell, F. S., and M. Yonge. 1975. *The Seas,* 4th ed. Frederick Warne, London. 283 pp.

Sumich, J. L. 1984. *An Introduction to the Biology of Marine Life,* 3d ed. Wm. C. Brown, Dubuque, Ia. 386 pp.

The Plants and Animals

Barnes, R. D. 1980. *Invertebrate Zoology,* 4th ed. Saunders, Philadelphia. 870 pp.

Barth, R., and R. Broshears. 1982. *The Invertebrate World,* Saunders, Philadelphia. 646 pp.

Carefoot, T. 1977. *Pacific Seashores, A Guide to Intertidal Ecology.* Univ. of Washington, Seattle. 208 pp.

Childress, J., H. Felbeck, and G. Somero. 1987. Symbiosis in the Deep Sea. *Scientific American* 256 (5): 114–20.

Hardy, A. 1959. *The Open Sea: Its Natural History,* Part 2: Fish and Fisheries. Houghton Mifflin, Boston. 322 pp.

Idyll, C. P. 1976. *Abyss—The Deep Sea and the Creatures That Live In It,* rev. ed. Crowell, New York. 428 pp.

Lobban, C. S., and M. J. Winne, eds. 1982. *The Biology of Seaweeds.* Univ. of California, Berkeley, Calif. 786 pp.

MacGinitie, G. E., and N. MacGinitie. 1968. *Natural History of Marine Animals,* 2d ed. McGraw-Hill, New York. 523 pp.

Marshall, N. B. 1979. *Developments in Deep-Sea Biology.* Blandford, Poole, Dorset, England. 566 pp.

Ricketts, E. F., J. Calvin, and J. Hedgpeth. 1985 ed. *Between Pacific Tides.* Stanford Univ., Stanford, Calif. 678 pp.

Ryan, P. R., ed. Fall, 1984. *Oceanus* 27 (3). (Issue devoted to hot springs and cold seeps.)

Ryan, P. R., ed. Summer, 1986. *Oceanus* 29 (2). (Issue devoted to coral reefs.)

Harvesting the Benthos

Borgese, E. M. 1980. *The Story of Aquaculture.* Abrams, New York. 236 pp.

Chapman, V. J., and D. C. Chapman. 1980. *Seaweeds and Their Uses,* 3d. ed. Chapman and Hall, New York. 334 pp.

Faulkner, D. J. 1979. The Search for Drugs from the Sea. *Oceanus* 22 (2): 44–50.

Isaacs, J. D., and W. R. Schmitt. 1980. Ocean Energy: Forms and Prospects. *Science* 207 (4428): 265–73.

Limburg, P. R. 1980. *Farming the Waters.* Beaufort, New York. 223 pp.

McPeak, R. H., and D. A. Glantz. 1984. Harvesting California's Kelp Forests. *Oceanus* 27 (1): 19–26.

Michael, R. G., ed. 1987. *Managed Aquatic Ecosystems.* Elsevier, New York. 166 pp.

Miura, A. 1980. Seaweed Cultivation: Present Practices and Potentials. In *Ocean Yearbook 2,* Borgese, E., and N. Ginsburg, eds. Univ. of Chicago, Chicago. pp. 57–68.

Ryther, J. H., and J. C. Goldman, C. E. Gifford, J. E. Huguenin, A. S. Wing, J. P. Clarner, L. D. Williams, and B. E. La Pointe. 1975. Physical Models of Integrated Waste Recycling—Marine Polyculture Systems. *Aquaculture* 5: 163–77.

Thorgaard, G. H., and S. K. Allen Jr. 1987. Chromosome Manipulation and Markers in Fishery Management. In *Population Genetics and Fishery Management,* Ryman, N., and F. Utter, eds. Washington Sea Grant Program, Univ. of Washington, Seattle. pp. 319–31.

Index